2nd International Conference on Social Development and Media Communication (SDMC 2022)

AA004098

SHS Web of Conferences Volume 155 (2023)

Sanya, China
25 - 27 November 2022

Editors:

Jamilah Hj Ahmad
James Guo

ISBN: 978-1-7138-6636-7

Printed from e-media with permission by:

Curran Associates, Inc.
57 Morehouse Lane
Red Hook, NY 12571

Some format issues inherent in the e-media version may also appear in this print version.

For additional information, please contact EDP Sciences – Web of Conferences at the address below.

EDP Sciences – Web of Conferences
17, Avenue du Hoggar
Parc d'Activité de Courtabœuf
BP 112
F-91944 Les Ulis Cedex A
France

Phone: +33 (0) 1 69 18 75 75
Fax: +33 (0) 1 69 28 84 91

contact-edps@webofconferences.org

Additional copies of this publication are available from:

Curran Associates, Inc.
57 Morehouse Lane
Red Hook, NY 12571 USA
Phone: 845-758-0400
Fax: 845-758-2633
Email: curran@proceedings.com
Web: www.proceedings.com

TABLE OF CONTENTS

Preface

This issue of SHS Web of Conferences contains the proceedings of the 2022 2nd International Conference on Social Development and Media Communication (SDMC 2022). The conference took place in Sanya, China during November 25th-27th, 2022 (virtual conference), with the attendance of about100 scholars, experts, researchers and scientists involved in related areas.

The Conference aims at creating an active and stimulating forum where current research results and technical advances can be exchanged and the development of social development and media communication can be promoted. For this reason, scholars, experts, scientists and researchers in social development and media communication have come from all over the world to present their activities and innovative approaches to related issues.

To make this conference on schedule, we have a three-part plenary conference including a keynote speech part delivered by three speakers from different countries. Among them, A. Prof. Anthony Pak-Hin KONG from The University of Hong Kong, China performed a keynote speech on Training Speech-and-language Pathologists during the Time of COVID: Some Insights from Practice in USA and Hong Kong. In his presentation, he aimed to summarize and discuss some challenges and opportunities of COVID-19 for higher education in the discipline of speech-and-language pathology (SLP) by highlighting how different academic SLP programs in the US and Hong Kong have introduced and implemented alternative solutions to address the learning and education needs of SLP students beyond the conventional classrooms/clinics amid the COVID-19 pandemic. We are very optimistic that all participants will acquire new ideas and knowledge from this conference to enhance the capability toward social development and media communication.

We have received tremendous paper submissions from all over China and abroad. Through a rigorous peer-review process, all submissions were performed double blind review to check their quality of content, level of innovation, significance, originality and legibility. Based on the expertise review comments, some excellent papers were accepted and the corresponding authors were invited to submit their papers for final publication. The accepted papers included topics on: Social Development Theory Research, New Era Theory and Thought, Cross-cultural Communication, Media Communication and News Spread, Systems and Technologies in E-learning, etc.

We would like to thank the Conference General Chair and the Technical Program Committees for the procedures that lead to the organization of this event, to our colleagues that contributed with its realization, to the SHS Web of Conferences members for their patience while waiting for the final versions of the papers and to every author who has contributed with an article in these proceedings. We hope you will benefit from all the manuscripts of the proceedings and learn new and novel ideas.

The Committee of SDMC 2022

Committee member

Conference General Chair
Prof. Augustin F. C. Holl, School of Sociology and Anthropology, Xiamen University, China

Technical Program Committee Chairs
Prof. Shen Chaoqun, Collage of Art Design and Media, Sanda University, China
Prof. SHAKILA YACOB, Faculty of Arts & Social Sciences, University of Malaya, Malaysia
A. Prof. NG YUE HOONG, IRENE, Department Of Social Work, National University of Singapore, Singapore

Local Organizing Chairs
Prof. Junhua Du, Chongqing University, China
A. Prof. Adeel Luqman, College of Management, Shenzhen University, China
A. Prof. Nadeem Akhtar, School of Urban Culture, South China Normal University, China

Publication Chairs
Prof. Jamilah Hj Ahmad, School of Communication, Universiti Sains Malaysia, Malaysia
Prof. Junhua Du, Chongqing University, China
Prof. Wenming Cao, Shenzhen University, China
A. Prof. James Guo, School of Journalism and Communication, Chongqing University, China

Technical Program Committees
Prof. Intan Hashimah Mohd Hashim, Universiti Sains Malaysia, Malaysia
Prof. KHONG, YUEN FOONG, National University of Singapore, Singapore
Prof. Jacob Dahl Rendtorff, Department of Society and Business, Roskilde University, Denmark
Prof. Carlotta Viti, Research Center for History and Culture, Beijing Normal University (Zhuhai Campus) & UIC, China
A. Prof. Azman Azwan Azmawati, School of Communication, Universiti Sains Malaysia, Malaysia
A. Prof. Andrea Lorenzo Baldini, School of Arts of Nanjing University, China
A. Prof. LAI CHEE SERN, University Tun Hussein Onn Malaysia, Malaysia
A. Prof. Peter Thomas ZABIELSKIS, Department of Sociology, University of Macau, China
A. Prof. Deepanjali Mishra, School of Humanities, KIIT University, Bhubaneswar

Statement of Peer review

In submitting conference proceedings to *Web of Conferences*, the editors of the proceedings certify to the Publisher that

1. They adhere to its **Policy on Publishing Integrity** in order to safeguard good scientific practice in publishing.
2. All articles have been subjected to peer review administered by the proceedings editors.
3. Reviews have been conducted by expert referees, who have been requested to provide unbiased and constructive comments aimed, whenever possible, at improving the work.
4. Proceedings editors have taken all reasonable steps to ensure the quality of the materials they publish and their decision to accept or reject a paper for publication has been based only on the merits of the work and the relevance to the journal.

Title, date and place of the conference

2022 2nd International Conference on Social Development and Media Communication (SDMC 2022)
November 25th-27th, 2022
Sanya, China

Proceedings editor(s):

Prof. Jamilah Hj Ahmad, Universiti Sains Malaysia, Malaysia
A. Prof. James Guo, Chongqing University, China

Date and editor's signature

Jamilah Hj Ahmad James Guo

Analysis on the status quo of construction industry and workers' professional literacy

Junxia Sun, Yan Zhang, Qin Tao, Ying Wang, Nana Liu*

Chongqing College of Architecture and Technology, Chongqing, China.

Abstract. Under the background of the new building industrialization in China, the traditional cast-in-place construction mode begins to change to assembly construction mode. Such subversive changes make all regions face various assembly construction quality problems caused by traditional workers' lack of assembly construction knowledge, low awareness and unskilled operation. Therefore, improving workers' professional quality is very important to improve the quality of assembly construction. Through the analysis of a large number of domestic and foreign literature, this paper found that when studying prefabricated buildings from different dimensions such as quality, cost, schedule and safety, Will involve the "operation proficiency", "education level", "professional skills", "vocational training", "qualification", "working years", "professional experience", "lack of knowledge", "reputation", "professional pride", "mobile operation", "work attitude" and other aspects of the workers, These factors focus on safety awareness, operation norms, vocational training, work experience, professional knowledge, cultural level, quality awareness, technical skill level. All these factors will affect the quality of assembly construction, and the quality of workers is one of the reasons for poor construction performance, frequent safety accidents and low construction quality.

1 Instructions

In 2017, The General Office of the State Council issued "The opinions on Promoting the Sustainable and Healthy Development of the Construction Industry", pointed out the problems such as backward construction technology and low quality of workers. In the first half of 2021, the added value of the construction industry increased by 8.6 percent year-on-year, accounting for 6.26 percent of the country's GDP. By the end of June 2021, the number of construction enterprises with construction activities had reached 41.734 million, up 1.25 percent year-on-year. At present, China's urbanization development is facing a comprehensive transformation period of innovation-driven development. Construction industry recruitment demand increased 10.7% from 2020, among which skilled talent recruitment demand increased 10.9% from the previous year. In particular, the demand for assembly building design, production, construction and management personnel, BIM talents and intelligent building talents had increased significantly, and the market prospect for high-skilled applied talents of construction engineering specialty was broad. On the basis of the industrialization of construction, in November 2017, the Ministry of Housing and Urban-Rural Development of China issued a draft of the opinions on workers on "Training and Guiding Construction Workers in the New Era", pointing out that the skills and quality of construction workers should be effectively improved.

*Corresponding author: ldnnana@sina.com

2 Quality development status of China's construction industry

At present, China's comprehensive national strength is in the rising, the national economic construction and rapid is development, from construction nearly five years development statistical data analysis (Figure 1), the added value of construction industry in China in nearly five years still maintain rise but growth began to slow, this means that the scale of construction by the development of rapid expansion, is becoming a thing of the past, High quality and high efficiency will become the focus of future development. Large-scale and backward extensive operations and on-site wet operations have been gradually abandoned [1]. To keep up with the pace of industrialization and intelligent development, the construction industry should improve labor productivity, comprehensive economic benefits, overall personnel and technical management level, preliminary design and in-depth design level, energy consumption of engineering processing and production process, environmental protection capacity of construction site and other aspects [2]. This suggests that our country construction has entered a stage of high quality development, the future development direction of change in order to promote our country construction industrialization level of development, the developed countries construction labor productivity, to dare to chase propulsion with prefabricated construction as the technical features of

https://doi.org/10.1051/shsconf/202315501001

new-type industrialization, achieve the traditional labor-intensive enterprises to intensive industry chain of industrialization.

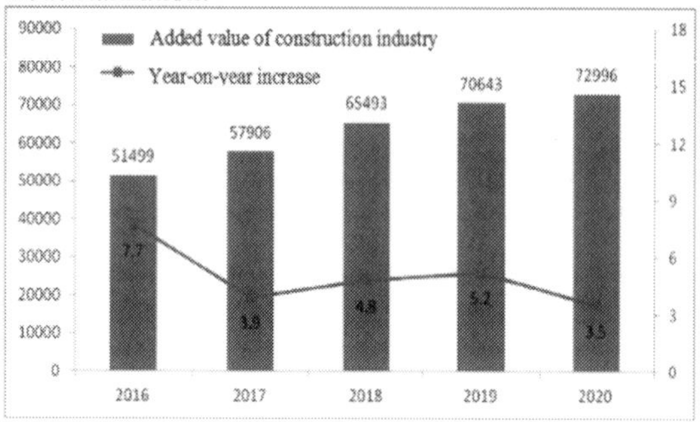

Figure1 Added value and growth rate of construction industry from 2016 to 2020
(Data source: 2020 National Economic and Social Development Bulletin)

Since 2016, prefabricated buildings had been thoroughly implemented in China (Figure 2). The newly started construction area of prefabricated buildings in China increased from 114 million square meters in 2016 to 630 million square meters in 2020, with an annual increase of nearly 50%. Its share of new floor space jumped from 4.9% in 2016 to 20.5% in 2020. However, due to the relatively short time of the introduction and used of prefabricated buildings, the construction industry had insufficient grasp of its technology, leading to the current many construction industry to implement prefabricated buildings with traditional buildings, which led to prefabricated buildings in the design, production, construction and other links there were many quality problems. For example, in the process of component production, the connection of embedded pipelines within the components is prone to quality deviation, and the interface between embedded pipelines is not properly handled, resulting in incomplete sealing [3]. In the construction process, the lack of full grouting is the most prominent problem [4]. The dislocation of the connected sleeve is also one of the more important invisible defects. In addition, because the construction site practitioners lack of management of finished products or its attention is not enough to produce improper quality problems such as product protection.

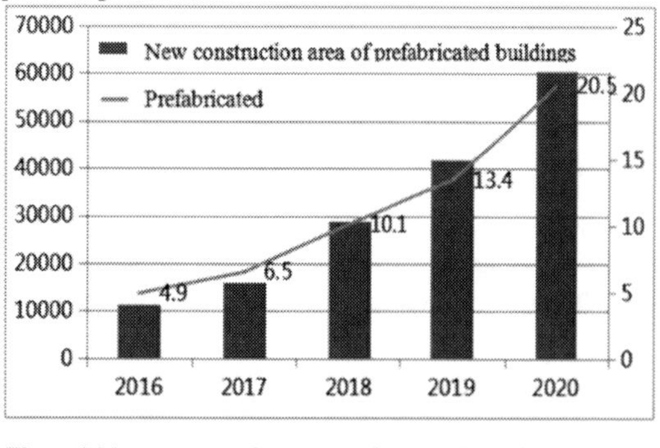

Figure2 New construction area and proportion of prefabricated buildings in China from 2016 to 2020
(Data source: Research Institute of Forward-looking Industry, Ministry of Housing and Urban-Rural Development)

3 Status quo of Chinese workers' professional literacy

Based on the industrialization development of new building, the low quality of workers' vocational skills was the main factor that restricts the sustainable and healthy development of construction industry. There was a serious shortage of front-line operators for existing workers who were able to engage in prefabricated buildings, and the low technical skills and comprehensive professional quality of key components such as lifting and joint construction had become the "choking" problem of prefabricated buildings [5]. In the field of workers, there was a serious shortage of high-quality innovative technical talents with craftsman spirit who could combine technology, economy and management.

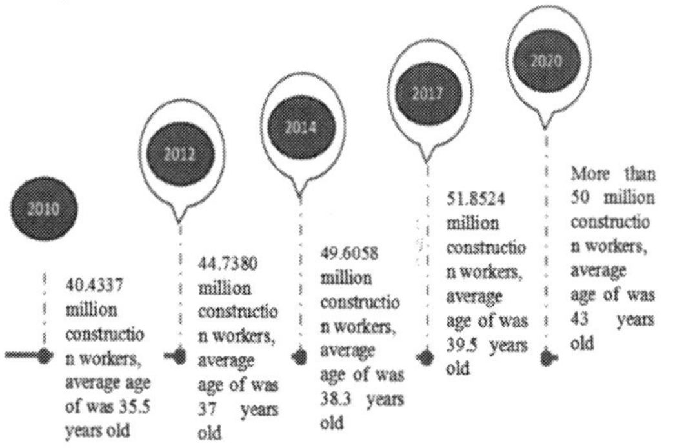

Figure3 Number and age of construction employees in 2010-2020
(Data source: China Construction Industry Association)

Most projects currently in the process of building a severe lack of their own professional team work assignments, much depends on the local contractor or mobile operation construction group, multiple levels of its personnel basically is the age of 40 to 60 years old age range of migrant workers (Figure3), in 2020, Front-line workers of building construction work the crowd with an average age of 43 years old. The age groups of migrant workers due to the economic pressure to engage in this industry. first of all, due to the effect of construction industry in vulgar operation mode, the one-sided pursuit of progress, and requirements for engineering quality is not high, influence the workers only pursue quantity rather than the pursuit of quality, secondly is influenced by the social education, living environment quality, relative knowledge level is not high; Quality awareness and safety awareness is not high; Without systematic professional learning, most of them can only achieve basic operating skills; As I have been doing repetitive work for a long time, and as I get older, the phenomenon of solidified thinking becomes more and more serious, which leads to the inability to communicate with new technologies, new knowledge and new methods, and the difficulty to accept new working modes. Therefore, my overall professional accomplishment is low. All these problems indicate that workers, the basic human level, must be reformed in terms of improving professional literacy and build a group of high-quality and skilled workers, so as to adapt to the high-quality development of the construction industry transformation.

4 Analysis of the influence of Workers' professional quality on prefabricated building quality

The development of prefabricated building is imperative. The working capacity of workers in the traditional extensive operating environment cannot match it. Prefabricated building components need workers to do fine work in production, production, transportation and other links, otherwise they may not be able to proceed normally in the construction and installation stage. Quality problems caused by workers' professional knowledge, work attitude, quality awareness, safety awareness, non-standard operation and other professional qualities are not high and many factors are more common and significant.

Through literature review and the use of work logs, quality inspection, stand-by and communication with technical personnel in the daily management of engineering projects, this study found that the construction quality of prefabricated buildings in the construction process is mostly related to the low professional literacy of workers.

Domestic and foreign literature on the impact of workers' professional literacy on the quality of prefab construction is relatively scattered, and direct research on the impact of workers on the quality of prefab construction is relatively scarce. This study is mainly based on the retrieval of CNKI database. From 148 literatures, it is found that domestic researches on workers and construction quality are mainly reflected in the analysis of construction workers' safety consciousness and construction project quality management. There is no separate discussion and research on the influencing factors of workers' professional literacy on the quality of prefabricated construction, but the importance of workers is recognized by many scholars. Many scholars have mentioned the construction quality problems caused by the low quality of workers. Many researchers have pointed out that the main improper products observed on construction sites are caused by human error. Of the 200 problematic products examined at 50 sites across the UK, 45% were due to negligence; About half are due to lack of knowledge or experience. In the analysis and induction of domestic research results, this study tries to correlate workers' occupational literacy with prefabricated construction quality, and extracts the influencing factors of workers' occupational literacy on prefabricated construction quality.

Through research data at home and abroad, it is found that when prefabricated buildings are studied from different dimensions such as quality, cost, schedule and safety, Will involve the "operation proficiency", "education level", "professional skills", "vocational training", "qualification", "working years", "professional experience", "lack of knowledge", "reputation", "professional pride", "mobile operation", "work attitude" and other aspects of the workers, These factors focus on safety awareness, operation norms, vocational training, work experience, professional knowledge, cultural level, quality awareness, cost awareness and technical skill level. All these factors will affect the quality of assembly construction, and the quality of workers is one of the reasons for poor construction performance, frequent safety accidents and low construction quality.

5 Conclusion

Starting from the construction quality and based on the high quality development of prefabricated buildings, this study finds out the factors influencing workers' professional literacy, and discovers the key factors that can improve prefabricated construction quality. The model analyzes the relationship between workers' professional literacy and prefabricated construction quality, and the conclusion of empirical theoretical analysis. In the process of human resource management of workers in construction production enterprises and construction enterprises, suggestions are put forward to strengthen workers' professional quality and cultivate working teams with craftsman spirit. It provides reference for improving the quality of assembly construction fundamentally from the manpower of workers, and provides basis for workers to improve their professional quality.

References

1. Sun Yu-Fang, Wu Xia, HE Meng-lin, CONG Xu-hui (2021). Research on quality Management of Prefabricated Building Based on BIM+ Internet of Things Technology [J]. Building Economy, 42(05):58-61.
2. Liang Xianchao (2020). Prefabricated building Engineering Quality Management System and Strategy under EPC Mode [J]. Building economics, 41(11):73-78.]
3. Su Yangyue, Zhao Jinkai, Xu Youquan, SI Hongyun (2016). Construction economy, 37(11):43-48.
4. Sun Qianjin (2021). Construction Technology and Quality Management of Prefabricated Buildings [J]. China Housing Facilities (05):8-9.
5. Fan jun, Yang sixin(2017). Thoughts and suggestions on realizing prefabricated building [J]. Building technology, 48(02):118-122.
6. Tao Yu, Qingpeng Man, Yaowu Wang, Geoffrey Qiping Shen, Jingke Hong, Jiasheng Zhang, Jia Zhong (2019). Evaluating different stakeholder impacts on the occurrence of quality defects in offsite construction projects: A Bayesian network-based model[J]. Journal of Cleaner Production, 241: 104-112.
7. Liu Guangchen, Wen Zhendi, Shen Jing, An Lu, Liang Yue (2019). Research on the influencing factors of prefabricated concrete building quality based on factor analysis [J]. Building economy, 40(08):97-101.
8. Xiaolong Gan, Ruidong Chang, Jian Zuo, Tao Wen (2018), George Zillante. Barriers to the transition towards off-site construction in China: An

 https://doi.org/10.1051/shsconf/202315501001

Interpretive structural modeling approach[J]. Journal of Cleaner Production,197: 8-18.

9. Yadi Duan, Duan Yadi, Li Guanlei (2020). Analysis on the Quality Problems and Preventive Measures of Prefabricated Building Construction[J]. Journal of Physics: Conference Series,1648(3):
10. Aynur Kazaz, M. Talat Birgonul (2005). Determination of Quality Level in Mass Housing Projects in Turkey[J]. Journal of Construction Engineering and Management, 131(2): 195-202.
11. Gul Polat (2010). Precast concrete systems in developing vs. industrialized countries[J]. Journal of Civil Engineering and Management, 16(1): 85-94.
12. Bon-Gang Hwang, Ming Shan, Kit-Ying Looi (2018). Key constraints and mitigation strategies for prefabricated prefinished volumetric construction[J]. Journal of Cleaner Production, 183: 183-193.
13. Guobin Wu, Ru Yang, Ling Li, Xing Bi, Bingsheng Liu, Shaoyan Li, Shixiang Zhou (2019). Factors influencing the application of prefabricated construction in China: From perspectives of technology promotion and cleaner production[J]. Journal of Cleaner Production,219: 753-762.

A Study on the Comparison and Enhancement of Health Literacy of College Students in Guangdong Province in 2020 and 2022

Yintong Guo, Dan Jiang, Jian Dong*, Mengli Zhuang, Jieping Lin, Xiaodan Wang, Juan Feng, and Zhuangwei Li

Guangdong Food and Drug Vocational College, 321 Longdong Bei Road, Tianhe District, Guangzhou, Guangdong, China

Abstract. In order to compare the health literacy level of college students in Guangdong province in 2020 and 2022, so as to provide a scientific basis for targeted health literacy intervention and policy formulation for college students in Guangdong province, surveys were respectively conducted in 2020 and 2022. The data collation and analysis were performed using SPSS 19.0 statistical software. The χ^2 test was used to compare different health literacy, and logistic regression was performed to analyse the factors influencing health literacy. The results show that the general health literacy level of college students in Guangdong province in 2022 is 46.5%, 6.3% higher than 40.2% in 2020, which difference is statistically significant. The three dimensions and six aspects of health literacy all have improved. The results of both years showed that health skills, basic medical literacy and health information literacy were at a low level. According to logistic regression analysis, the health literacy level of senior students is higher than that of junior students, and those who have taken health related courses have higher health literacy level. The most desirable type of health knowledge is prevention and treatment of infectious diseases, and the new media access is becoming more popular among students to gain health knowledge. In conclusion, Guangdong college students' health literacy is relatively high, but still needs to be improved, especially in health skills, basic medical care and health information literacy. The government, colleges and universities should work together to improve college students' health literacy.

1 Introduction

The concept of health literacy was first proposed by Simonds SK at the Bangkok International Health Conference in 1974[1]. It refers to the ability of individuals to acquire, understand and process basic health information and services, and to use these information and services to make decision-making for improving and maintaining their own health [2].

Since 2008, Chinese government has attached great importance to residents' health literacy, and a series of documents and policies have been issued to bring the improvement of residents' health literacy into the goal of building a healthy China.

Especially, in 2017, The Ministry of Education issued *The Guidelines for Health Education in Universities and Colleges*, calls for improving the health literacy of college students. In 2021, The Ministry of Education and other five departments joint release the *Opinions on Comprehensively Strengthening and Improving School Hygiene and Health Education in the New Era*, proposed that schools adhere to the education concept of health first, and ensure that health education runs through all areas of education.

The students in our country account for 17.79% (61.23 million) of the youth population, and college students in Guangdong province is 2.3 million, which is a huge population. College students are an important force to promote the development of our country and society. The level of their health literacy is not only related to their own, but also affects the future of the country.

2 Objects and methods

2.1 Object

A total of 7370 college students from 15 colleges and universities in Guangdong Province were selected by stratified cluster random sampling for the study. Inclusion criteria: students in grades 1-3 in college, students in grades 1-4 in undergraduate, who were willing to cooperate with the investigation and informed consent; exclusion criteria: unwilling to participate in the investigation.

2.2 Methods

2.2.1 Survey tools

The *2019 National Residents' Health Literacy Monitoring Questionnaire* was modified and applied for the survey. The questionnaire contains mainly two parts, one part is basic information of students objects, the

* Corresponding author: 315287054@qq.com

other part is the main body, namely the health literacy. The main body includes three dimensions: basic health knowledge and concept, healthy lifestyle and behaviour, and health skills, and six aspects, namely scientific health concept, infectious disease prevention and control literacy, chronic disease prevention and control literacy, safety and first aid literacy, basic medical literacy, and health information literacy. The field survey used mobile terminal devices (mobile phones, iPads, etc.) to conduct face-to-face survey in the form of online questionnaires.

2.2.2 Scoring criteria

The main body of questionnaire includes 10 judgment questions, 26 single choice questions, 16 multiple choice questions and 4 situational questions, total 56 questions. One point will be given for each correct answer to the judgment question and single choice question, and two points to the multiple-choice question (no score will be given for more or less choices). A total of 73 points will be given. People whose score is equal to or higher than 80% of the total score ($\geqslant$ 58 points) are considered to have health literacy.

2.2.3 Statistical analysis

SPSS 19.0 statistical software was used for general descriptive analysis, χ^2 test and logistic regression analysis.

2.2.4 Quality control

The investigators received normative training before the survey began, and 20% of the questionnaires were randomly selected for logic check after the survey. Double-blind entry of survey data was performed using Epidata 3.1.

3 Objects and methods

3.1 General data

In 2020, 3610 questionnaires were distributed, out of which 3215 questionnaires were effective, with an effective rate of 89.06%. In 2022, a total of 3760 questionnaires were distributed, out of which 3411 questionnaires were effective, with an effective rate of 90.72%. Specific data is shown in Table 1.

3.2 The health literacy level of college students in Guangdong province

The general health literacy level of college students in Guangdong province in 2022 is 46.5%, 6.3% higher than 40.2% in 2020, which difference is statistically significant($P<0.05$). From the three dimensions of health literacy, in 2020, the literacy level of health knowledge and concept, healthy lifestyle and behaviour and health skills were respectively 49.7%, 44.1% and 39.2%, while in 2022, they are 54.6%, 49.8% and 41.3%, all of which differences are statistically significant($P<0.05$). The results of the two-year survey showed that the level of basic knowledge and concept literacy are the highest, followed by healthy lifestyle and behavior and health skills. From six aspects of health literacy, in 2020, the top 3 aspects were scientific health concept (55.7%), infectious disease prevention and control literacy (49.8%) and chronic disease prevention and control literacy (44.2%). While in 2022, the top 3 aspects are infectious disease prevention and control literacy (58.6%), scientific health concept (58.1%), and safety and first aid literacy (51.4%). The differences on infectious disease prevention and control, and literacy safety and first aid literacy in 2020 and 2022 are statistically significant($P<0.05$), as shown in Table 2.

3.3 Influencing factors of health literacy

First, taking the health literacy level of college students as the dependent variable, and referring to the factors with statistical significance (gender, grade, home address, whether taking health related courses and self-rated health status) as the independent variable, a multi-factor logistic regression analysis for data in 2020 and 2022 was conducted. The results of 2020 show that gender, grade, home address and whether taking health related courses are the influencing factors of health literacy. While in 2022, the influencing factors are grade and whether taking health related courses, as shown in Table 3. In summary, grade and whether taking health related courses are the influencing factors of health literacy for college students in Guangdong province.

3.4 Health knowledge needs of college students in Guangdong province

In 2020, the top 3 wanted types of health knowledge are prevention and treatment of infectious diseases, reasonable nutrition and a balanced diet, and TCM health care. The top 3 desirable accesses to gain health knowledge are: books, health lecture, and Wechat (a popular social networking APP in China). In 2022, the top 3 wanted types of health knowledge remain the same, while the top 3 desirable accesses turn out to be: books, Tiktok (a popular short video APP in China) and health lecture, as shown in Table 4.

Table 1. General data of the subjects

Demographic characteristics	2020		2022		Total	
	Numbers	Percentage	Numbers	Percentage	Numbers	Percentage
Gender						
Men	1539	47.9%	1662	48.7%	3201	48.3%
Women	1676	52.1%	1749	51.3%	3425	51.7%
Grade						
Freshman	1015	31.6%	986	28.9%	2001	30.2%
Sophomore	939	29.2%	1263	37.0%	2202	33.2%
Junior	862	26.8%	913	26.8%	1775	26.8%
Senior	399	12.4%	249	7.3%	648	9.8%
Major						
Natural science	1865	58.0%	1937	56.8%	3802	57.4%
Humanities and social science	1350	42.0%	1474	43.2%	2824	42.6%
Health related course						
Taken	1531	47.6%	1496	43.9%	3027	45.7%
Not taken	1684	52.4%	1915	56.1%	3599	54.3%
Home address						
Countryside	1759	54.7%	1682	49.3%	3441	51.9%
City and town	1456	45.3%	1729	50.7%	3185	48.1%
Monthly living expenses(RMB)						
<1500	1236	38.4%	1293	37.9%	2529	38.2%
>1500	1979	61.6%	2118	62.1%	4097	61.8%
Father's educational level						
illiteracy	69	2.1%	75	2.2%	144	2.2%
Primary and secondary education	2458	76.5%	2587	75.8%	5045	76.1%
Higher education	688	21.4%	749	22.0%	1437	21.7%
Mother's educational level						
illiteracy	86	2.7%	93	2.7%	179	2.7%
Primary and secondary education	2687	83.6%	2572	75.4%	5259	79.4%
Higher education	442	13.7%	746	21.9%	1188	17.9%
Self-rated health status						
Good	1997	62.1%	1863	54.6%	3860	58.3%
Poor	1218	37.9%	1548	45.4%	2766	41.7%

Table 2. Health literacy level of college students in Guangdong Province

Demographic characteristics	Health literacy level (%)		Three-year cumulative increase	χ^2	*P*
	2020	2022			
Three dimensions					
basic knowledge and concept	49.7	54.6	4.9	5.843	<0.05
healthy lifestyle and behavior	44.1	49.8	5.7	4.972	<0.05
basic skills	39.2	41.3	2.1	1.263	>0.05
Six aspects					
scientific health concept	55.7	58.1	2.4	2.762	>0.05
infectious disease prevention and control literacy	49.8	58.6	8.8	8.571	<0.05
chronic disease prevention and control literacy	44.2	47.3	3.1	3.015	>0.05
safety and first aid literacy	43.6	51.4	7.8	7.903	<0.05
basic medical literacy	39.2	42.6	3.4	3.364	>0.05
health information literacy	36.1	38.2	2.1	2.315	>0.05
Total health literacy	40.2	46.5	6.3	6.361	<0.05

Table 3. The multi-factor Logistic regression analysis of influencing factors of health literacy of college students

Variables	2020			2022		
	P	*OR*	95%*CI*	*P*	*OR*	95%*CI*
Gender						

Men		1		-	-	-
Women	<0.05	1.435	1.102-1.859	-	-	-
Grades						
Lower grades		1			1	
Senior grades	<0.05	1.301	0.573-2.959	<0.05	1.412	1.142-1.732
Home address						
Countryside		1			1	
City and town	<0.05	1.339	1.1456-1.565	>0.05	1.307	0.986~1.674
Health related course						
Taken		1			1	
Not taken	<0.05	0.421	0.251-0.811	<0.05	0.447	0.379-0.527
Self-rated health status						
Good		1			1	
Poor	>0.05	0.638	0.045~1.005	>0.05	0.836	0.336-2.066

Table 4. health knowledge needs of college students in Guangdong Province

Items	2020		2022	
	Numbers	Percentage	Numbers	Percentage
Type of health knowledge desired				
Prevention and treatment of infectious diseases	1296	40.32%	1560	45.74%
Reasonable nutrition and a balanced diet	1241	38.61%	1338	39.22%
Traditional Chinese medicine health care	1198	37.26%	1308	38.34%
First aid	1139	35.43%	1159	33.97%
Desirable access to gain health knowledge				
Books	1933	60.12%	2171	63.66%
Tiktok	1621	50.42%	1847	54.14%
Health lecture	1648	51.26%	1783	52.26%
Wechat	1652	51.37%	1714	50.25%
Bilibili Website	1596	49.65%	1658	48.62%
Health promotion in public places(posters, advertisements, etc.)	1327	41.28%	1505	44.11%

4 Conclusions

The results of surveys show that the overall level of health literacy of college students in Guangdong province rises from 40.2% of 2020 to 46.5% of 2022, significantly higher than the national residents' health literacy monitoring level of 25.4% [3] and the national residents' health literacy level target 30.0% in 2030 [4], and also higher than the level of 28.6% of the residents in the central urban area of Guangzhou, provincial capital of Guangdong province [5]. This proves that the work of college students' health literacy in Guangdong province has achieved good results. However, taking data in 2022 for instance, compared with colleges and universities in other provinces and cities in China, it is lower than the survey results of 68.9% in Yunnan province [6], but higher than 41.3% in Henan province [7]. The overall health literacy of college students in Guangdong province is relatively high, but there is still a certain gap compared with students in some other provinces and cities in China.

From Table 2, the level of basic knowledge and concept literacy are the highest, followed by healthy lifestyle and behavior and health skills, which comply with the theoretical model of Knowledge, Attitude and Practice (KAP model). Students may possess the knowledge and concept, but find it hard to turn them into daily life behavior.

Also from Table 2, the No.1 aspect of health literacy of both years is scientific health concept, indicating colleges and universities have achieved good results in health concept education. The No.2 aspect of both years is infectious disease prevention and control literacy, which may due to COVID-19 pandemic. The No.3 aspect in 2022 changes from chronic disease prevention and control literacy to safety and emergency literacy. This may also be related to the COVID-19 pandemic, in that people stay in their home more often due to the quarantine, and there are many accidents happened, such as students drowning in wild swimming near their homes. The television news reported many accidents, and colleges and universities intensified education on personal safety and first aid, contributing to the level rise of safety and emergency literacy. However, the level of basic medical literacy and health information literacy are always the lowest, suggesting that colleges and universities should appropriately increase the content of basic medical knowledge to guide students to correctly access to medical and health information when making health education plans.

4.1 Discussions on the influencing factors of health literacy

Students from high grades are better than lower grades,

which is consistent with a survey in Shandong province [8], indicating maturity and long term study can contribute to the health literacy enhancement. Students who have taken health-related courses have higher health literacy than those not, and it is same with survey held in Chengdu[9] and Dezhou city[10]. This demonstrates the importance of health-related courses in improving health literacy. The 2020 and 2022 regression model were not consistent in gender, home address and self-rated health status, which may due to the difference of sample and survey period. But more attention can still be paid to male students and students from countryside.

4.2 Health literacy enhancement strategy

The level of health literacy is directly related to health outcomes [11], and improving the health literacy of the people is one of the most fundamental, economical and effective measures to improve the health quality of the population [12]. College students are the future of a country and society, so it is of great importance to improve their health literacy.

First of all, the government should hold the responsibility to enhance health literacy. A series of policies like *The Healthy China 2030 Plan* and so on have been promulgated as mentioned at the beginning of this essay, however, at provincial level, there is no specific policies or regulations to students' overall health enhancement, let alone the health literacy. Provincial regulation can be formulated, together with the funding added, can colleges and universities have rules to follow and money to spend on the health literacy enhancement.

Colleges and universities are the main places for college students' health education, which have the advantages of high organizational efficiency, high plasticity of educational objects, relative low cost of health education and greater benefit of social communication, as well as academic authority [7]. Colleges and universities should first reform the existing health education mode. Besides the access of traditional health lecture, books, health promotion in public places (posters, advertisements, etc.), health education can be added as a compulsory course; and official accounts can be set in TikTok, Wechat, and Bilibili website, on which health education can be carried out. In the accounts' initial stage, knowledge of infectious diseases and TCM should be emphasized because they are students' favorite. Later, students' weak knowledge of chronic diseases and so on should be added, thus to improve their overall health literacy. On the other hand, an evaluation mechanism can be set up to promote the change of health knowledge into health behaviour. Health knowledge contest, healthy class rating can all be effective methods. In this way, knowledge and practice, online and offline education can be bind effectively to enhance the health level of college students.

Acknowledgments

This research is a 2020 scientific research project "Research on the improvement of college s tudents' health literacy level in Guangdong province" funded by Guangdong Provincial Department of Education, and "Innovative Talents Training Program for Outstanding Young People in Guangdong Universities" (NO. 2020KQNCX171). Also, it is supported by a project of "Science and Technology Resources Sharing Service Innovation Center" funded by Guangdong Food and Drug Vocational College's "Double-high Plan" project (SG03-02.6), and a project of "Research on community intelligent nursing system for the aged based on Internet of things" funded by Guangdong Food and Drug Vocational College's Pan Deng project(pdjh2021b0801).

References

1. Speros C. (2005) Health Literacy: concept Analysis. Journal of Advanced Nursing, 50(6): 633-640.
2. Sorensen K, Stephan VB, James F, et al.(2012) Health literacy and public health: a systematic review and integration of definitions and models. BMC Public Health, 12(1):80-80.
3. The Central People's Government of the PRC.(2016) The CPC Central Committee and the State Council printed and distributed the outline of the Healthy China 2030 Plan. http://www.gov.cn/zhengce/2016-10/25/content_5124174.htm.
4. The National Health Commission, PRC.(2021) The National Health Commission: national residents' health literacy levels to 25.40% in 2021. http://www.nhc.gov.cn/xcs/s3582/202206/5dc1de46b9a04e52951b21690d74cdb9.shtml.
5. Luo M, Chen J, Sun A, Luo L.(2021) Analysis of health literacy level and influencing factors of residents in central urban area of Guangzhou. Chinese Journal of Social Medicine, 04: 435-440.
6. Sun X, Sun Y, Song J, Xue J, Peng Y, Liu H. (2022) Research on the effect and optimization path of college students' health literacy cultivation. Journal of Kunming University of Science and Technology Science (social science edition), 5(22):114-122.
7. Si J, Wang X, Guo Q. (2022) A study on the influencing factors and promotion paths of health literacy of college students in Henan province. Soft science of health, 36(09):90-96.
8. Lou P, Luo S, Zhang J, and Wang Y. (2019) The relationship between health literacy and medical seeking behavior of college students in Shandong province. School Health in China, 12:1820-1823 + 1827.
9. Wang Y, Hu Q, Liu R, Zhang H; Chen J. (2022) Analysis on the current situation and influencing factors of health literacy of college students in Chengdu-Chongqing Medical College. Journal of Chengdu Medical College, 03:371-376.
10. Dong M, Jia X, Yin L & Teng X. (2022) Research on the influencing factors and promotion strategies of college students' Health Literacy. Journal of the Dezhou University, 02:35-39.

11. Kang L, Liang X, Li H, et al. (2020) Evaluation of intervention effect of Wechat and participatory classroom on health literacy of college students. School Health in China, 41(7):1015-1018.
12. Sarhan M, Yu F, Kiriya J, et al. (2021) Exploring Health Literacy and Its Associated Factors among Palestinian University Students: A Cross-sectional Study. Health Promotion International, 36(3):854-865.

Research on The Offensive Characteristics of La Liga Team Based on Social Network Analysis

Wei Du[1,*], Jingwen Li[2], Feiyu Li[3]

[1] Manchester University Oxford Road, Manchester M13 9PL, UK
[2] Manchester University Oxford Road, Manchester M13 9PL, UK
[3] Manchester University Oxford Road, Manchester M13 9PL, UK

Abstract. To explore the difference of social network parameters between the network of passing before scoring and the network of passing before missing the goal, and to explore the correlation between social network parameters and team performance, this paper establishes the offensive pass network of 20 teams in the La Liga from 2017 to 2018, and 11 social network parameters are calculated. The Pearson correlation test is used to explore the linear correlation between 11 social network parameters and team performance. The results show that the linear correlation between the network parameters of passing before scoring and team performance is stronger than the network parameters of passing before missing the goal. According to the results, we can provide reliable and effective information to the football coaches to help improve the performance of football matches.

1 Introduction

Social network analysis is widely used in the performance analysis of football matches, but only a few scholars focus on the passing network in the process of attack. In football matches, the formation of every efficient attack requires multiple accurate passes, and the more effective the attacks are, the more likely the game is to win. The use of social network analysis methods in foreign countries to explore the performance of football has been very mature, and the relationship between corresponding network parameters such as network density, network centrality of individual players and team performance has been systematically sorted out [1]. There are relatively little researches on offensive patterns and social network analysis of offensive processes. Here are some related studies. Based on the passing position information, Rein et al. proposed spatial control parameters to explore the efficiency of the offensive link [2]. Mclean et al. used social network analysis (SNA) and notational analysis (NA) to study the goal-scoring passing networks (GSPN) of all UEFA Champions League matches in 2016. They found the GSPN network. Density, coherence, connectivity, and continuity are all relatively low. At the same time, it is found that there is a certain statistical relationship between network parameters and game conditions [3]. Kubayi studied the scoring tactics of all teams in the 2018 FIFA World Cup and analyzed their scoring patterns [4].

Although social network analysis has been widely used in the academic analysis of football sports, my country's research in this area is still in its infancy. Some scholars have only explored the feasibility of social network analysis in passing performance analysis. [5]; Some scholars, like those abroad, use the World Cup as the research object to do some offensive strategy classification and offensive technical and tactical analysis of football matches [6]. The most relevant study used a complex network analysis method to explore the impact of the Chinese Football Super League home and away games on sports performance [7]. However, at present, few researchers are using social network analysis to simply analyze the offensive passing network that forms shots in football matches.

*Corresponding author: 1280538273@qq.com

2 Research objects and methods

2.1 Research object

This article takes as the research object all the passing incidents of 20 teams in the Spanish First Division Football League for a total of 380 games throughout the season and analyzes the offensive passing networks of all 20 teams in La Liga to explore different social situations. The relationship between network parameters and team record.

2.2 Research methods

2.2.1 Offensive effect evaluation standard

This study uses the conversion of the ball as a sign of the beginning of an offence, and the completion of a shot represents the end of an offence. All consecutive passes of a team during this period are defined as the pass attack network before the shot. It should be noted that in this process, whether it is through the pass organization, the process of preparing for the attack or the process of executing the offence through the pass, it is a part of the football offence [8]. Therefore, the definition of an offensive pass before shooting is reasonable.

Besides, this study adopts the shooting result as the evaluation standard of an offensive effect. Furthermore, whether a shot is hit or not can most intuitively and scientifically reflect the pros and cons of an offensive effect. Therefore, this study uses whether the shot hits or not as an evaluation criterion for an offence.

2.2.2 The method of establishing the passing network

Considering the the data sample size is too large, it is not suitable to use social network analysis software such as Ucinet or Gephithat manually enter data. Social network establishment and calculation of social network parameters. This research is based on the networkX of the Python platform, which processes a large amount of data in batches. First, perform data cleaning and preprocessing on the relational data in JSON format from figureshare, extract and identify the model objects analyzed in this research, and then divide the model objects according to 20 teams and 2 kinds of offensive effects (shooting goals and Factors such as shooting but not scoring) are classified into 40 subsets, respectively imported into networkX, and 40 directed graphs are generated (example shown in Figure 1), networkX can visualize graphs and calculate social network parameters, and data analysis The efficiency has been significantly improved.

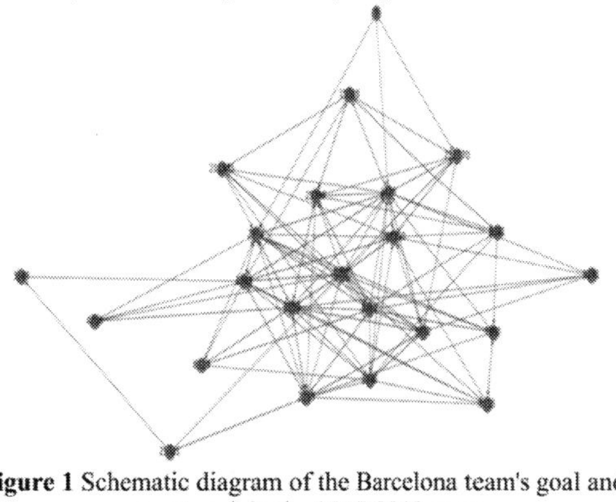

Figure 1 Schematic diagram of the Barcelona team's goal and pass network in the 2017-2018 season

3 Research results

3.1 Statistical analysis results

3.1.1 The statistical difference analysis results of the two types of shooting and passing network parameters

The average value and variance of each parameter of 11 different network parameters of all 20 teams in La Liga are calculated by the offensive network of shots and goals and the offensive network of shots that have not been entered.

Divide the average and variance of the parameters of the goal network and the goal failure network (as shown in Table 2), and use these two ratios as the basic difference analysis index for evaluating the offensive network that causes the two shooting results.

Through the results of the ratio, it can be seen that the average of each parameter has a significant difference. Since scoring a goal is a small probability event in a football game, the number of shots and missed shots is much higher than the number of goals scored. Therefore, the size of the shot without entering the offensive network is much larger than the shot on the offensive network. Among the 11 network parameters studied in this paper, the total number of nodes, total connections, and network density are three indicators that are directly determined by the size of the network. Therefore, these 3 parameters should be eliminated when comparing network differences based on average values.

After excluding these three parameters, it is obvious that the averages of the other eight parameters have a huge difference. This result shows that there is a significant difference between the network characteristics of shooting goals and not shooting goals. This discovery lays a theoretical foundation for the subsequent correlation analysis. However, the variance of the network parameters of the two types of networks is difficult to find obvious rules. More than half of the network parameters have a lesser variance than the variance of the shot-to-goal network.

3.1.2 Correlation analysis results of the two types of passing networks before shooting and the team reco

Correlation coefficient 1 refers to the Pearson correlation coefficient between the network parameters of the forward pass network and the team record; correlation coefficient 2 refers to the Pearson correlation coefficient between the network parameters of the missed forward pass network and the team record.

Summarizing points are linearly positively correlated with team record, and the degree of correlation is moderate; the total number of connections is linearly positively correlated with team record, and the degree of correlation is strong; network density is linearly positively correlated with team record, and the degree of correlation is strong Correlation; the average clustering coefficient is linearly positively correlated with the team's record, and the degree of correlation is strong; the average reciprocity is linearly positively correlated with the team's record, and the correlation degree is strongly correlated; the average transitivity is linearly positively correlated with the team's record , The degree of correlation is a strong correlation; the global connectivity is linearly negatively correlated with the team's record, and the correlation degree is moderately correlated; the average degree centrality is linearly positively correlated with the team's record, and the correlation degree is very weakly correlated; the average neighbor degree center There is a linear positive correlation between the team's record and the team's record, and the degree of correlation is moderate; the average close centrality is linearly positively correlated with the team's record, and the correlation degree is weakly correlated; the average betweenness centrality is linearly positively correlated with the team's record, and the degree of correlation It is very weakly correlated or not correlated.

The relationship between the network parameters of the missed front pass network and the team record:

Summarizing points are linearly negatively correlated with team record, and the degree of correlation is moderate; the total number of connections is linearly negatively correlated with team record, and the degree of correlation is moderately correlated; network density is linearly positively correlated with team record, and the degree of correlation is weak Correlation; the average clustering coefficient is linearly positively correlated with the team record, and the correlation degree is weakly correlated; the average reciprocity is linearly positively correlated with the team record, and the correlation degree is very weakly correlated; the average transitivity is linearly positive with the team record Correlation, the degree of correlation is medium correlation; the global connectability is linearly positively correlated with the team record, and the correlation degree is medium correlation; the average degree centrality is linearly negatively correlated with the team record, and the correlation degree is very weakly correlated; average neighbor degree Centrality is linearly negatively correlated with team record, and the degree of correlation is weakly correlated; average close centrality is linearly positively correlated with team record, and correlation degree is extremely weak; average betweenness centrality is linearly negatively correlated with team record. The degree of correlation is medium.

4 Analysis and discussion

4.1 Theoretical and practical significance of the research results

4.1.1 Theoretical significance

At present, more and more scholars or sports enthusiasts have come into contact with or even devote themselves to the subject of sports performance analysis for various reasons. The data processing method used in this article, social network analysis, is one of the most commonly used sports performance analysis methods. One is the theory most often used to process passing data in team sports. Although social network analysis belongs to the social sciences, it was first used to explore theories of communication modes between people, groups and groups, or organizations and organizations. But with the development of technology, all fields are inseparable from the support of data, and data analysis has become an indispensable skill in any field, and the sports field is no exception. In the papers using social network analysis to analyze sports performance, scholars from all walks of life use this method to realize their unique ideas and express their unique insights. More and more ideas collide in this field, and the development of sports will be stimulated, and the development of the whole field will bring more scholars into contact with this discipline. This forms a virtuous circle.

4.2 Discussion of statistical analysis results

This paper uses two statistical methods to explore the differences in social network parameters between the front pass network and the missed front pass network, and the correlation between the social network parameters and the team record.

In the part of exploring the parameter differences of the two types of pre-shooting passing networks, the most basic statistics are calculated: average and variance and the two types of networks are compared. The comparison method is division. The calculation results show that the ratio of the average value of the parameters of the pass network before the shot is very small to the average value of the parameters of the pass network before the shot is missed, which means that the average value of the parameters of the pass network before the shot hits is significantly smaller than that of the missed shot. The average value of the parameters of the mid-to-forward pass network. According to the understanding of the two types of offensive networks extracted during the data processing, the reason for this phenomenon maybe since the number of missed shots in the actual game under normal circumstances is far more than the number of shots hit, that is, the shot missed the pass before the goal. The size of the network is much larger than the size of the pass network before the shot hits, resulting in the former's network complexity being much greater than the latter. Therefore, there will be a huge difference in network parameters. Based on this result, it can be boldly guessed that stable network performance will not be a key factor in winning the game. The reason for this phenomenon may be that the game situation is always changing during football games. Perhaps the unstable offensive characteristics are the offensive strategies that can help the team win.

In exploring the correlation between the parameters of the two types of pre-shooting passing networks and the team record, the Pearson correlation test is used. According to the research results of this article, two predictions can be made: first, in addition to the 11 social network parameters discussed in this article, there must be more social network parameters that are relevant to the team's record; second, use others The correlation test method must also be able to find multiple social network parameters related to the team's record. Based on these two conjectures, future research can focus on network parameters related to the team's record. First, a variety of correlation analysis methods can be used to find network parameters that are highly correlated with the team's record, and then multiple regressions can be used to establish a regression model of network parameters and the team winning percentage. It is even possible to use highly correlated network parameters like features, and build a football match result prediction model based on the passing network through machine learning algorithms.

5 Conclusion

There is a significant difference between the network parameter characteristics of the forward pass network and the network parameter characteristics of the forward pass network. The team can train according to the information feedback from the social network to improve the shooting percentage.

The linear correlation between the 11 social network parameters of the front pass network and the team record is stronger than the linear correlation between the network parameters of the missed front pass network and the team record, indicating that the pass before the shot hits The network has more research value in improving the overall sports performance of the team than the passing network of missed shots.

References

1. Clemente, F.M., Martins, F.M.L., & Mendes, R.S. (2016) Social network analysis applied to team sports analysis, Springer International Publishing, Cham.
2. Rein, R., Raabe, D., & Memmert, D. (2017) "Which pass is better?" Novel approaches to assess passing effectiveness in elite soccer. J. Human Movement Science, 55: 172–181.
3. Mclean, S., Salmon, P. M., Gorman, A. D., Stevens, N. J., & Solomon, C. A. (2018) social network analysis of the goal scoring passing networks of the 2016 European Football Championships. J. Human Movement Science, 57: 400–408.
4. Kubayi, A. (2020) Analysis of goal scoring patterns in the 2018 FIFA World Cup. J. Journal of Human Kinetics, 71, 205–210.
5. Hagberg A., Schult D., Swar P. (2021) NetworkX Reference, https://networkx.org/documentation/latest/_downloads/networkx_reference.pdf. 2021-05-19.
6. Ashenagar, B. & Hamzeh A. (2019) Forming a well-connected team of experts based on a social network graph: a novel weighting approach. J. Social Network Analysis and Mining, 9(1).
7. Fan, Y., Agyenim, B., Kim, L., & Jiang, Y. (2021) Are bonds blind? Board-CEO social networks and firm risk. J. Journal of Corporate Finance, 68.
8. Chang, C. (2016) Based on analytic hierarchy process model of big data analysis of the NBA schedule impact on team ranking. J. Journal of Computational and Theoretical Nanoscience, 13(12).

Sportswomen and Digital Media: Case Study of "Dream Crazier" Campaign

Xiaotong Liu*

The London School of Economics and Political Sciences, London, UK

Abstract. The article focuses on popular feminist movements about sportswomen with the example of Nike's movement "Dream Crazier". It firstly considers the features of popular feminism, the media's role as a carrier, and the increased visibility of feminist movements in the digital era. In addition, given the context of the Nike sportswomen campaign, the essay proposes to analyse the engagement of digital media forms, especially posters on sportswomen, with a broader and more effective circulation of feminist claims. The essay contends that by illustrating sportswomen in videos and posters and presenting the discourse of renowned sportswomen like Serena Williams, the campaign conveys feminist appeals on sexual sporting equality, presents female power, and also works with intersectionality that protects equal rights for black sportswomen. This reflects the characteristics of popular feminism in the contemporary period. The essay concludes with further consideration of popular feminism when it refers to the relationship between celebrity figures and commercialized companies.

1. INTRODUCTION

The tradition of Nike campaigns on sports can be traced back to 30 years ago, when the brand's iconic "Just Do It" was released. In 2018 - the 30th anniversary of this movement, Nike continues with the campaign "Dream Crazy". Nike's selection of the spokesperson Colin Kaepernick showed its devotion to individual potential and equal sporting opportunity: "So don't ask if your dreams are crazy. Ask if they're crazy enough (McPhillips, 2019)[1]."

One year later, Nike cast its attention on women and launched the new campaign "Dream Crazier". The spot narrated by Serena Williams is a compilation of female athletes who are dedicated to fighting against barriers of sportswomen. It also stars Olympic gymnast Simone Biles, fencer Ibtihaj Muhammad, snowboarder Chloe Kim and members of the U.S. Women's National Soccer Team (Sweeney, 2019)[2], to redefine "what it means to call a female athlete crazy" (Monllos, 2019)[3].

The campaign aimed to highlight women athletes' power to break barriers, show their energy, and inspire generations of sportswomen to pursue their dreams. The campaign was soon spread out and triggered the hashtag #DreamCrazier on Twitter, with a number of female athletes and sports fans responding and resonating. The campaign also reached a high level of attention and clicks on YouTube.

The quintessential brand campaign is an indication of how feminist ideas can be embodied and disseminated through online popular media forms. After theorizing popular feminism with digital media, the representative videos and posters in this movement will be presented and analysed on their feminist connotations.

2. POPULAR FEMINISM AND THE DIGITAL ERA

With the technological progress of new media, feminists are more likely to bring traditional feminism's commitment to popularity. As Banet-Weiser (2019)[4] indicates, popular feminism refers to feminism "practices and conditions that are accessible to a broad public, from organising marches to hashtag activism to commodities" (p.9). Unlike the past forms, popular feminism redefines the new relationship between media and feminism, the spreading of feminist politics in a "post-girl power" context (Keller, 2015)[5]. The digital media performs an effective method for self-branding or "girl entrepreneur", and also creates a discursive space for feminist politics. Hence, girls are able to challenge the hegemonic postfeminist femininity that defines girls as apolitical.

Online spaces have been significant to this process, providing anonymous space to "experiment" (Keller 2013, 2012) with feminist identities, and affording girls a chance to connect with other feminist girls and women. (Keller, 2015, p.278)

Popular feminism enhances the visibility of feminist critiques and allows them to be disseminated more broadly. As social media becomes the critical platform for circulating feminist movements, the rise of "hashtag feminism" (Banet-Weiser, 2018)[6] has come into sight. The hashtag campaigns, along with other movements on digital media, are trying to reveal gendered inequality,

*Corresponding author: Liuxt15@outlook.com

emphasize female power, and create an opportunity for women to show their own voices to the broader public. By presenting young women's tendency to express their political views and communicating with different others online, Harris (2010)[7] also effectively clarified the role of social media as a "safer place" (p.480) for women speaking out.

What is more, intersectionality with race is also an outstanding point of popular feminism. As Williams (2015)[8] suggests, social media hashtags have aroused people's attention to the situation of black women, when the mainstream media has been neglecting them for decades. According to Pew research on technology use, 96% of African American users aged 18-29 use at least one kind of social networking site; moreover, black women use Twitter the most, compared with any other demographic group (Smith, 2014)[9]. The black feminists are hence provided with an online community to voice their opinions and speak for equal rights and condense them into appealing hashtags to spread globally. Besides, social media's nature of real-time discussion will help to reduce mainstream media's problematic framing of race.

3. DREAM CRAZIER: EMPOWERMENT, EQUALITY AND INTERSECTIONALITY

With the slogan of pushing us for our dreams, the campaign released a series of inspiring advertisements with the narrator Serena Williams and the heroines of famous athletes from other sports fields. These ads are posted and disseminated through online websites, calling for women's empowerment, fair sporting rights, and also equality across gender and race.

3.1. Empowerment

Nike's movement encourages female athletes to be proud of being called "crazy" (Binlot, 2019)[10]. In one of the video products from the campaign posted on YouTube, the outstanding accomplishments of women athletes of different ages and fields are highlighted. Williams's narration is also a central component in conveying the spirit of female power.

A woman running a marathon was crazy. A woman boxing was crazy. A woman dunking? Crazy. Coaching an NBA team? Crazy. A woman competing in a hijab, changing her sport, landing a Frontside Double Cork 1080, or winning 23 Grand Slams, having a baby, and coming back for more? Crazy. Crazy. Crazy. Crazy. And Crazy. (Williams, 2019)[11]

In the 1:30 clip, Serena Williams talked about the scorn that women athletes often endure: athletes who have triumphed over their opponents and achieved great success, are labelled "madwomen". In the public eyes, women are usually "delicate" and meanwhile conform to rules and regulations; any woman who breaks this impression is regarded as madness. Krane (2001)'s argument[12] captured the contradiction between athleticism and femininity - women must show their sports ability and "the ability to be accepted by society" at the same time. In other words, many sportswomen are forced to exhibit their feminine characteristics to avoid prejudice and discrimination.

Figure 1. Poster with Williams about sportswoman empowerment. From Ella.Marketing. Retrieved from http://ella.marketing/news/nike-y-serena-williams-redefinen-lo-que-significa-llamar-loca-a-una-mujer-atleta/.

The poster is simple but influential, illustrating Serena's confidence and willingness of showing her strong and vigorous look. Gender discrimination against female athletes and their hegemonic representations in the mainstream media (Shugart, 2003)[13] are traditional ways of devaluing and marginalising women's sports. As Heinecken's (2015)'s research[14] on women's jockographies (sports autobiographies), traditional media consolidates heterosexual femininity while ignoring the sporting inequalities in their reports on female athletes. By showing that women should not be discriminated against for participating in fierce competitive sports, the poster's circulation on social media effectively strikes back the stereotype of women of a "standard feminine role".

3.2. Sporting Equality and Intersectionality

Serena Williams emphasized much on equal rights for women (and especially black sportswomen) as a black female athlete herself. As Carter-Francique and Flowers (2013)[15] argue, the sociohistorical construction of ethnic groups and women as "the others" has situated them out of the sports world. Except for three concentrated fields for coloured female athletes (Basketball, Track and Field), the disproportionate number of coloured women in sports participation at both collegiate and professional levels reveals evident opportunity inequality (Lapchick, 2012; Lapchick, Hoff, et al., 2010)[16][17].

Figure 2. Posted on equal sporting opportunities. From Nike.com. Retrieved from https://news.nike.com/featured_video/dream-crazier.

The poster focused on the situation of Compton girls and their life in sports. As the Childhood hometown of Serena Williams, Compton contains 40.31% of African Americans (according to the 2010 United States Census). The crime-ridden area fostered long-lasting hostility and hatred towards black sportswomen. Serena, along with her sister, has been "the targets of racism, sexism, and body shaming - from getting booed by unabating crowds as teenagers, to being referred to as 'brothers' by the head of the Russian Tennis Federation" (Shackelford, 2015)[18].

To ask for equal sporting right for black women, Serena Williams tried to challenge the so-called "noble sport for white elites". In 2016, she wrote an open letter to "All Incredible Women", where she addressed the unequal payment and sexism faced by sportswomen. She portrayed her life as a "question of resilience" (Williams, 2016)[19]: regarding what others think of as the shortcomings of an athlete - race and gender - as the driving force for her success.

After calling for sporting rights and equality on the poster, she also did a similar video on her Twitter for the "Dream Crazier" campaign, encouraging girls from her hometown to realize their sports dreams, which received more than 150,000 likes and 50,000 retweets. Female athletes and sports lovers were responding to the campaign, making the feminist idea more visible and influential.

Figure 3. Responses on social media. From Twitter. Retrieved from https://twitter.com/serenawilliams/status/1034093664996343809.

4. LIMITATIONS AND REFLECTION

The "Dream Crazier" campaign over females in the sports area has received great public attention. Especially, the Internet and social media are instrumental in the popularity and circulation of their assertions, making women's rights for sports an increasingly valued topic. However, although the campaign was initiated by a group of prestigious female athletes headed by Serena Williams, which aimed at breaking the imbalanced power relation in the white-man-dominated competitive sports field; the Nike company is substantially the title quotient of this campaign. Therefore, the claims of authentic feminism assertions from celebrities could be a commodified "brand culture" (Genz, 2015)[20] in a neoliberal society.

Shortly after the campaign, Nike company was soon receiving critics of its policy of reducing an athlete's sponsorship payments for any reason he/she didn't meet a specific achievement threshold, including the reason for pregnancies. Several people have been reversely criticizing Nike's inconsistent behaviour, which also formed a trend on online media. Montaño spoke for the rights of those pregnant female athletes in The New York Times with the headline "Nike Told Me to Dream Crazy, Until I Wanted a Baby", along with a large-scale public outcry on social media. Fortunately, on August 12th, Nike issued a new contract that guarantees maternity protections for their sponsored athletes.

Figure 4. Critics of Nike's policy. From The New York Times. Retrieved from https://www.nytimes.com/2019/05/12/opinion/nike-maternity-leave.html.

This contradiction with its claimed feminist movement indicates the ambiguity of the "Dream Crazer" campaign: nominated by a commercial company, is it an authentic feminist movement, or just like commercialized postfeminism with "commodity fetish" that spread across social media, while its political significance gets erased (Gill, 2006)[21]? Is Williams' voice for sportswomen merely a commercial strategy to improve Nike's sales performance and popularity? This point indicates that the motivation and efficacy of popular feminism could be ambiguous because so many of the described endeavours are wrapped up in commodified notions of the individuals or leading characters.

5. CONCLUSION

While sportswomen may not be the major force of contemporary feminist movements, social media plays a critical role in expanding the visibility and influence of their voices. The core value they convey in their online campaigns, such as empowerment as being a "strong" woman, equal opportunities to receive professional training and to stand on the competition stage, and anti-discrimination towards black sportswomen, is consistent with broader rights for women in various identity.

However, when linking the campaign with the Nike brand, the efficacy of the "Dream Crazier" movement is greatly reduced by Nike's deeds that do not accord with the campaign's assertion. The visibility brought by digital media technology fosters not only feminist support but also a branding or commodified story (Grow, 2008)[22], which should be regarded as a further consideration.

AUTHORS' CONTRIBUTIONS

Xiaotong Liu contributed to the conception of the study, selected the case, analysed the documents and wrote the paper.

ACKNOWLEDGMENTS

The author thanks Prof. Sarah Banet-Weiser for assistance with the essay and valuable discussion.

REFERENCES

1. McPhillips, K. (2019). In the follow-up to its internet-breaking Colin Kaepernick ad, Nike features an all-star cast of women. Retrieved from https://www.wellandgood.com/good-sweat/nike-dream-crazier-women/.
2. Sweeney. E. (2019). Nike's 'Dream Crazier' spot with Serena Williams celebrates female athletes. Retrieved from https://www.marketingdive.com/news/nikes-dream-crazier-spot-with-serena-williams-celebrates-female-athletes/549146/.
3. Monllos, K. (2019). Nike and Serena Williams Redefine What It Means to Call a Female Athlete Crazy. Retrieved from https://www.adweek.com/brand-marketing/following-its-kaepernick-ad-nike-taps-serena-williams-for-a-stirring-celebration-of-women/.
4. Banet-Weiser, S., Gill, R., & Rottenberg, C. (2019). Postfeminism, popular feminism and neoliberal feminism? Sarah Banet-Weiser, Rosalind Gill and Catherine Rottenberg in conversation. *Feminist Theory*, *21*(1), 3-24.
5. Keller, J. (2015). Girl power's last chance? Tavi Gevinson, feminism, and popular media culture. *Continuum*, *29*(2), 274-285.
6. Banet-Weiser, S. (2018). *Empowered: Popular feminism and popular misogyny*. Durham: Duke University Press.
7. Harris, A. (2010). Mind the gap: Attitudes and emergent feminist politics since the third wave. *Australian Feminist Studies*, *25*(66), 475-484.
8. Williams, S. (2015). Digital defence: Black feminists resist violence with hashtag activism. *Feminist Media Studies*, *15*(2), 341-344.
9. Smith. A. (2014). African Americans and Technology Use A demographic portrait. Retrieved from https://www.pewresearch.org/internet/2014/01/06/african-americans-and-technology-use/.
10. Binlot, A. (2019). Nike And Serena Williams Inspire Women To 'Dream Crazier' With New Campaign. Retrieved from https://www.forbes.com/sites/abinlot/2019/02/28/nike-and-serena-williams-inspire-women-to-dream-crazier-with-new-campaign/#77382cd920a3.
11. Nike (2019, February 24). Dream Crazier | Nike [YouTube]. Retrieved from https://www.youtube.com/watch?v=whpJ19RJ4JY
12. Krane, V. (2001). We can be athletic and feminine, but do we want to? Challenging hegemonic femininity in women's sport. *Quest*, *53*(1), 115-133.
13. Shugart, H. A. (2003). She shoots, she scores: Mediated constructions of contemporary female athletes in coverage of the 1999 US women's soccer team. *Western Journal of Communication (includes Communication Reports)*, *67*(1), 1-31.
14. Heinecken, D. (2016). Gender and jockography: post-feminism and resistance in female sports autobiographies. *Feminist Media Studies*, 16(2), 325-343.
15. Carter-Francique, A. R., & Flowers, C. L. (2013). Intersections of race, ethnicity, and gender in sport. In Roper, E. A. (Ed.), *Gender relations in sport* (pp. 73-93). Leiden: Brill Sense.
16. Lapchick, R., Hoff, B., & Kaiser, C. (2010). *The 2010 racial and gender report card: College sport*. Retrieved from the Institute for Diversity and Ethics in Sport.
17. Lapchick, R. (2012). *The 2012 Women's National Basketball Association Racial and Gender Report Card*. Retrieved from the Institute of Diversity and Ethics in Sport.
18. Shackelford, A. (2015). Straight Outta Compton: The Williams Sisters Story. Retrieved from http://blog.mytennislessons.com/straight-outta-compton-the-williams-sisters-story/.
19. Williams, S. (2016, November 29). 'We must continue to dream big': an open letter from Serena Williams. *The Guardian*, Retrieved from https://www.theguardian.com/.
20. Genz, S. (2015). My job is me: Postfeminist celebrity culture and the gendering of authenticity. *Feminist media studies*, 15(4), 545-561.
21. Gill, R. (2006). *Gender and the media*. Cambridge and New York: Polity.
22. Grow, J. M. (2008). The Gender of Branding: Early Nike Women's Advertising as a Feminist Antenarrative. *Women's Studies in Communication*, 31(3), 312-343.

What Predicts the Intention to Engage in Home-Based Exercise: The Theory of Planned Behavior

Zhe Xu*

Wenzhou Business College

ABSTRACT. Drawing on the theory of planned behavior, the current study aims to explore the mechanisms by which social media attention influences the constructs of the theory of planned behavior and then affects individual intentions. A total of 430 participants in China participated in an online survey. The results revealed that attention to information on social media about home-based exercise and health care significantly influenced attitude, subjective norms, descriptive norms, and perceived behavioral control. These, in turn, positively influenced the individual's behavior intention to engage in home-based exercise. Theoretical and practical contributions are discussed in this study as well.

1 INTRODUCTION

During the pandemic, an increasing number of people have paid more attention to health information and exercising to keep fit. Doing exercises can improve an individual's immune function and reduce the risk and severity of respiratory viral infection [1,2]. In addition, home-based exercise, which requires a minimum amount of space, can bolster one's immunity and so has received widespread attention among several kinds of exercise.

Exercise is one of the important measures for people to keep healthy and can be done outdoors or indoors. During the pandemic, indoor exercises, especially home-based exercise, has gradually become people's favorite type of exercise. Home-based exercise clearly represented a convenient and safe way to promote people's health during the pandemic [3]. Participating in home-based exercise can protect individuals from being influenced by disease and improve the condition of one's personal health. Doing home-based exercises is not restricted by the weather or space limitations and keeps individuals healthy. Given these advantages, we need to explore the factors that influence individuals' home-based exercise intention.

The majority of studies have mainly concentrated on the role of knowledge in home-based exercise and have neglected to adopt a strong theory to predict home-based exercise. The theory of planned behavior (TPB) is a well-known theory that applies to many fields, such as psychology and communication. The theory proposes that people's attitudes, subjective norms and perceived behavioral control (PBC) can predict individuals' behavior. Previous research found that attitudes, subjective norms, and PBC could predict people's intentions toward sports participation [4]. Besides, an individual's descriptive norms could influence their behavioral intention unconsciously. Hence, this study has adopted the theory of planned behavior as the main theoretical framework to explore how these theoretical constructs influence individuals' behavioral intention.

In addition, social media attention has influenced individuals' perception of home-based exercise during the pandemic. Nowadays, social media platforms serve as the main source of most updated information for people. To some degree, an individual's social media attention may influence their attitude, subjective norms, and PBC to perform behaviors. Hence, this study also seeks to explore how social media attention influences individuals' behaviors via attitude, subjective norms, and PBC.

In a nutshell, the purpose of this paper is twofold. First, the study examines how individuals' attitudes, subjective norms, PBC, and descriptive norms influence their intentions to perform home-based exercise. Second, the study explores how social media attention indirectly influences people's behavioral intention to do home-based exercises via attitude, subjective norms, descriptive norms and PBC.

2 LITERATURE REVIEW

2.1 Theory of Planned Behavior

The theory of planned behavior (TPB) is a well-established theory for predicting exercise behavior, since it integrates diverse social cognitive variables that serve as robust predictors of exercise behavior [5,6]. TPB is extended and based on the theory of reasoned action, which is able to explain almost all human behaviors and has been proven effective in predicting and explaining human behavior in diverse contexts [7]. TPB includes three constructs, namely attitude toward the specific behavior, subjective norms regarding specific behavior, and PBC over engaging in the behavior. These three

*Corresponding author: 990269310@qq.com

constructs, in turn, influence intention [8]. For example, when individuals hold a positive attitude toward home-based exercise, perceive a high social expectation of exercise participation, and have a high level of behavioral control, they will be more likely to engage in home-based exercise.

2.1.1. Attitude.

Attitude can be broadly defined as "the degree of a person's favorable or unfavorable evaluation or appraisal of the behavior in question"[9]. It is also explained as a predictor of behavior [10]. Individuals with a positive attitude often perform related behaviors. Findings have demonstrated that a positive attitude can motivate individuals to engage in exercises during a pandemic [11]. Similarly, Liao found attitude was positively associated with home-based exercise. Thus, individuals with a positive attitude have a stronger intention to engage in home-based exercise[12]. Based on the literature, this study proposes a hypothesis as follows:

H1: Attitude is positively associated with home-based exercise intention.

2.1.2. Perceived behavioral control.

Perceived behavioral control (PBC) refers to 'perceived behavioral control refers to people's perception of the ease or difficulty of performing the behavior of interest [13]. In other words, the degree of a person's belief in his or her ability to perform behaviors may promote or hinder their behavioral performance [12]. It may encompass two components [13,14]. The first component reflects the availability of resources required to participate in the behavior [15]. Similar to self-efficacy in social cognitive theory [16], PBC is an assessment of an individual's ability to perform current behavior [13]. The second component reflects a person's self-confidence in their ability to accomplish the task [15]. When individuals possess a high degree of behavioral control, they tend to have a higher behavioral intention to engage in home-based exercise. Therefore, we propose a second hypothesis, as follows:

H2: Perceived behavioral control is positively associated with home-based exercise intention.

2.1.3. Subjective norms.

Subjective norms refer to "the perceived social pressure to perform or not to perform the behavior" [13]. In other words, the significant others, such as family, friends, and colleagues, exert their expectation to influence peoples' decisions. Two subsequent experiments focusing on healthy behaviors showed that subjective norms had a positive influence on intention [17.18].

H3: Subjective norms are positively associated with home-based exercise intention.

2.1.4. Descriptive norms.

Descriptive norms refer to "rules and standards that are understood by members of a group, and that guide and/or constrain social behavior without the force of laws"[19]. Smith found that descriptive norms were positively associated with home-based exercise [20]. Descriptive norms also serve as social functions that can help individuals coordinate their social action to achieve desired outcomes [21]. Therefore, we propose two further hypotheses:

H4: Descriptive norms are positively associated with home-based exercise intention.

2.2 Social media attention

Social media play an important role in shaping public awareness and opinions of participation in public activities [22]. During the COVID-19 pandemic, social media became the main source for individuals to obtain health information [23]. With the number of professionals and organizations involved in social media platforms constantly increasing in recent years, individuals now tend to rely on social media to get information about health matters. Seeking health information via social media can aid members of the public to take stock of their health status and take preventive measures to make sure their health is stable [24]. In addition, studies have shown that social media information functions as a significant factor influencing changes of intentions and modifying public lifestyles and habits [25]. Most of the existing research has focused on the direct impact of social concerns on individual intentions, ignoring the influence of TPB's own variables.

Social media attention also influences attitude, subjective norms, descriptive norms and PBC, and consequently influences individual intention. Media attention is a precondition for a media effect to occur, which can be illustrated by McGuire's information-processing model [26] and Bandura's social cognitive theory [27].

Individuals formulate an attitude based on their evaluation of the expected outcome of a behavior, which may be acquired from individual social media attention [28]. People are prone to get information from social media or participate in group discussions, and they thus realize the benefit of the behavior[29,30]. In addition, on social networks, the more information individuals gain, the more positive the attitude is that they have. Thus, people with stronger social media attention to exercise and health care are likely to hold positive attitudes toward them.

Individuals' PBC during the COVID-19 pandemic has largely depended on the information and knowledge they obtained from social media. People evaluate their individual ability through information they obtain online [29]. In this case, when individuals pay more attention to social media information, they tend to acquire more relevant and up-to-date knowledge, which could enhance their self-confidence and ability toward performing related behavior. Hence, a higher social media attention leads to a higher PBC.

When people pay more attention to information about a specific behavior, their parents, relatives, and friends may think they are fond of the behavior and are likely to expect people to do it [31]. In other words, individuals' social media attention to home-based exercise and health could influence subjective norms.

People's social media attention usually determines people's social circle, and friends around them have the same or similar interests and hobbies and partake of the same behaviors. In other words, social media attention is associated with descriptive norms. However, few studies have explored social media attention as a predecessor variable in TPB. Thus, it is necessary to examine social media attention as a predecessor variable in attitude, subjective norms, descriptive norms, and PBC. We thus propose the following research questions:

RQ1: How does the degree of social media attention relate to individuals' attitude to engage in home-based exercise?

RQ2: How does social media attention relate to individuals' PBC to engage in home-based exercise?

RQ3: How does social media attention relate to subjective norms to engage in home-based exercise?

RQ4: How does social media attention relate to descriptive norms to engage in home-based exercise?

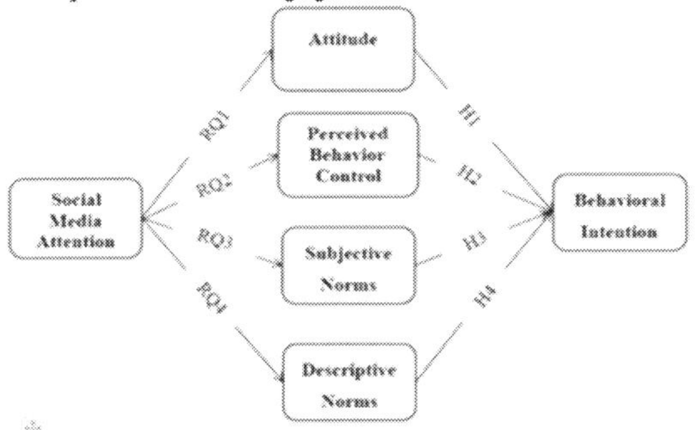

Fig. 1. Conceptual Model (The model is done by draw my own)

3 Method

3.1 Data collection

An online survey was conducted in July 2022. The participants were residents of southern, northern, eastern, and western China. A total of 477 responses were returned, of which 430 were valid responses, excluding incomplete or invalid responses. Of the respondents, 46.1% were male and 53.9% were female. More than half of the participants were undergraduates (53.4%), and 14.4% were graduates.

3.2 Measure

Our research team developed the questionnaire by adapting existing reliable and valid questionnaire scales to the study's context (see Table 1). First, we adapted three items from Skoric and Poor's (2013)attention to social media to fit the context of health and sport information [32]. Four items were adapted from the attitude scale of Kaplanidou and Gibson (2011), as well as a subjective norms scale (consisting of three items) to fit into our current context. In addition, we included Ho et al.'s (2017) descriptive norms scale (consisting of three items)[33,34]. Additionally, we used Young-Jae Kim and E-Sack Kim's (2021) perceived behavioral control scale (consisting of four items) [35] . Finally, we adapted Lim's (2017) corrective actions scale (consisting of five items) to the current research context. All of the adopted scales achieved a good level of reliability and validity [36].

Social media attention to health and sport information. For social media attention to health and sports, we concentrated on information concerning health, sports, exercise, and home-based exercise, including home-based exercise webcast. We used the items from Skoric and Poor (2013) with a seven-point Likert scale (1=no attention at all and 7=very close attention) to measure attention to health and sport information (M = 4.37, SD = 1.49, Cronbach's α=0.94) [32].

Attitude to home-based exercise. Based on previous research, the study adapted the attitude scale from Kaplanidou and Gibson (2011) to measure individual attitude to home-based exercise. We used a seven-point Likert scale (1=strongly disagree and 7=strongly agree). A higher score indicated greater attitude influence on home-based exercise (M=4.58, SD=1.49, Cronbach's α=0.94) [33].

Perceived behavioral control to home-based exercise. For measuring PBC, we modified three items using a seven-point Likert scale (1=strongly disagree and 7=strongly agree) [35]. A higher score indicated strong PBC (M=4.61, SD=1.49, Cronbach's α=0.93).

Subjective norms. Three items from a previous study were adapted to measure subjective norms using a seven-point scale (1=strongly disagree and 7=strongly agree)[33]. Respondents were asked whether their parents, relatives, or friends expected them to perform home-based exercise. A higher score indicated strong subjective norms (M=4.26, SD=1.51, Cronbach's α=0.95).

Descriptive norms. Three items were modified to measure descriptive norms using a seven-point scale (1=strongly disagree and 7=strongly agree) from Ho et al. (2017) [34]. Respondents were asked to indicate whether their parents, relatives, or friends engaged in home-based exercise. A higher score indicated strong descriptive norms (M=4.30, SD=1.43, Cronbach's α=0.91).

Behavioral intention to participate in home-based exercise. Behavioral intention to participate in home-based exercise was measured by three items using a seven-point Likert scale (1=strongly disagree and 7=strongly agree). The respondents were asked to indicate how likely they were to participate in home-based exercise. A higher score indicated a strong intention to participate in home-based exercise (M=4.55, SD =1.60, Cronbach's α=0.96).

3.3. Preliminary data analysis.

First, we used a question that served to screen those who did not carefully fill out the questionnaire and then we collected the data. Then, we excluded 47 participants who did not correctly answer the screening question. Next, we tested the collected data for the absence of outliers and whether all measured items were normally distributed. The results demonstrated that there were no outliers and all variables were normally distributed, as kurtosis and skewness were within the range of ±1 [37].

4 Results.

First, we ran a path analysis using the maximum likelihood tests for the hypotheses and answers to the research questions. The goodness-of-fit statistics of the model indicated a good fit to the data (CMIN/DF=2.27; CFI=0.99; TLI=0.98, and RMSEA=0.05). The results showed that social media's attention to health and sport was positively associated with attitude to participate in home-based exercise (β=0.81 p < .00), subjective norms (β=0.78, p < .00), descriptive norms (β=0.68, p < .00), perceived behavior control (β=0.78 p < .00), answering RQ1 – RQ4. In addition, attitude (β =0.29, p<0,001), behavioral control (β= 0.24, p < 0.001), subjective norms (β=0.20, p < .00), and descriptive norms (β = 0.26, p < .00) were positively associated with the behavioral intention to participate in home-based exercise, thus supporting H1, H2 , H3, and H4, respectively (see Figure 2).

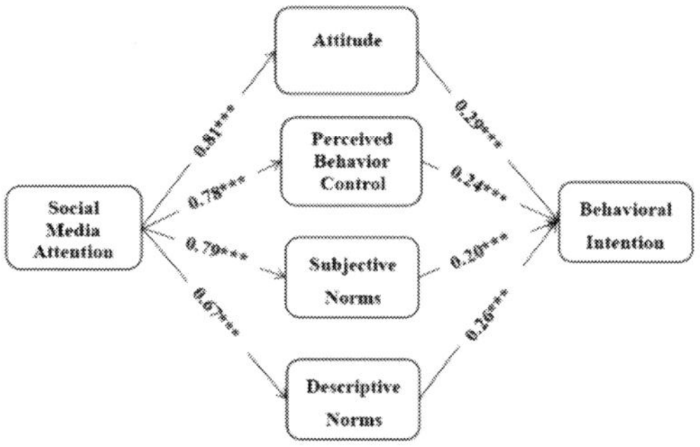

Fig. 2. TPB model to Predict home-based exercises intention (The model is done by draw my own)

Table 1. Item Scale (The model is done by draw my own)

Study measure	Survey items	Cronbach's α	M	SD
Social Media attention (Skoric & Poor, 2013)	(1) Attention paid to information about health topics.	0.94	4.40	1.63
	(2) Attention paid to information about sports and exercise.		4.35	1.57
	(3) Attention paid to information about home-based exercise .		4.32	1.56
Attitude (Kaplanidou & Gibson,2011)	(1) I think participating in home-based exercise is pleasant.	0.94	4.58	1.61
	(2) I think participating in home-based exercise is exciting.		4.41	1.58
	(3) I think participating in home-based exercise is entertaining.		4.49	1.61
	(4) I think participating in home-based exercise is worthwhile.		4.79	1.65
Perceived behavioral control (Young-Jae Kim & E-Sack Kim, 2021)	(1) I can participate in home-based exercise at any time I want.	0.93	4.37	1.55
	(2) Whether I participate in home-based exercise or not depends on myself.		4.82	1.71
	(3) I have time to participate in home-based exercise.		4.60	1.64
	I have the ability to do home-based exercise.		4.68	1.67
Subjective norms (Kaplanidou & Gibson, 2011)	(1) My friends or family members think that it is good for me to participate in home-based exercise.	0.95	4.23	1.57
	(2) My friends or family members support me in participating in home-based exercise.		4.30	1.56
	My friends or family members are likely to participate in indoor sports.		4.20	1.64
Descriptive norms (Ho et al., 2017)	(1) My family members do some home-based exercise.	0.91	4.30	1.59
	(2) My relatives do some home-based exercise.		4.14	1.54
	(3)My friends do some home-based exercise.		4.40	1.55
Behavioral intention (Young-Jae Kim & E-Sack Kim, 2021)	(1) I intend to participate home-based exercise in the next week.	0.96	4.55	1.65
	(2) I plan to participate home-based exercise in the next week.		4.54	1.64
	(3) I will to participate home-based exercise in the next week.		4.48	1.70

5 DISCUSSION

In line with previous TPB research, individual attitude is positively associated with home-based exercise intention. People with a positive attitude are more likely to have the intention to participate in home-based exercise. Perceived behavioral control is positively correlated to home-based exercise intention. It is possible that more active participation is a result of high perceived behavior control. The high perceived behavior control makes people realize the benefits and the shortcomings of engaging in home-based exercise and evaluate the self-condition. Subjective norms are positively correlated to home-based exercise intention. Individuals are prone to participate in home-based exercise after acquiring the support of important people. Descriptive norms are positively correlated to home-based exercise intention. Exercising with friends and family around creates a good atmosphere and stimulates them to devote themselves to engaging in it.

The current study adds social media attention not only as an antecedent variable to extend the TPB mechanism but also in the individual dimension and social dimension to observe the individual participating in home-based exercise. Social media attention through positively influencing the individual attitude, perceived behavioral control, subjective norms, and descriptive norms indirectly affect the individual behavior intention [28]. To some degree, the outcomes reflect the situation that people living in the social media era seek information and communicate with other people without social media.

Descriptive norms as a predictor of behavior intention make the social factors more reasonable [21]. This also indicates that the factor in the social dimension is truly vital for individual behavioral intention. Moreover, due to their great attention to health and sport information to keep healthy during the COVID-19 pandemic, they tend to communicate with people and encourage each other to engage in home-based exercise.

6 IMPLICATION AND LIMITATION

This research has both theoretical and practical implications. In regard to the theoretical implications, the current study focused on the background of COVID-19 and concentrated on promoting individuals to do some home-based exercises to stay healthy. Some scholars have researched home-based exercise for older adults [38]. However, few studies have focused on all ages of people and how to influence their intention to engage in home-based exercise.

Second, with social media development, social media attention gradually becomes an important variable to measure and predict individual behavior [34, 39]. Most studies have viewed social media attention as an independent variable directly influencing individual behavior or as a part of the independent variables of TPB. They neglected social media attention as an antecedent variable focusing on its influence on other factors and the relationship of TPB. This study is based on the theory of TPB adding social media attention as an independent variable, exploring the mediating effect of attitude, perceived behavioral control, subjective norms, and descriptive norms, and revealing the inner mechanism of TPB.

Finally, this study uses the descriptive norms as a societal factor to predict behavioral intention and to enhance social factors in the theory of planned behavior. Descriptive norms mostly focus on the social relationship formed by social media. Individuals can establish new positive relationships by making friends according to their hobbies.

In terms of practical implications, the findings indicate that improving individual social media attention to health and sport would change individual attitudes and prompt individual engagement. The more attention to health and sport that people pay in social media, the more chance for others to engage in home-based exercise. Additionally, professionals in health care should use intervention strategies to expose people to more social media health information and call for people to concentrate on social media attention.

However, the positive and negative nature of information is not considered in the attention of social media in this study. In addition, the current study collected cross-sectional data that just estimated that social media attention positively influenced the mediators, but the study does not explain specific causality. Lastly, age is an important demographic variable, as previous studies concentrated more on older people. But this study does not break down the effects of different age groups on dependent variables. Future research should pay attention to individual health concerns and be divided into different age groups to research, which would make the research more persuasive.

7 CONCLUSION

The study seeks to investigate how media attention influence attitude, subjective norms, descriptive norms, perceived behavioral control which in turn influence individuals' behavioral intention. The results showed that social media attention is positively associated with attitude, subjective norms, descriptive norms, perceived behavioral control. If participants pay more attention to the information about home-based exercises on social media, they are more likely to have a positive attitude, strong subjective norms and descriptive norms and positive perceived behavioral control, which motivate individuals to engage in home-based exercise.

REFERENCES

1. Nieman, D. C., & Pedersen, B. K. (1999). Exercise and immune function. Sports Medicine, 27(2), 73-80.
2. Martin, S. A., Pence, B. D., & Woods, J. A. (2009). Exercise and respiratory tract viral infections. Exercise and Sport Sciences Reviews, 37(4), 157. DOI: 10.1097/JES.0b013e3181b7b57b
3. Yang, Y., & Koenigstorfer, J. (2020). Determinants of physical activity maintenance during the Covid-19 pandemic: a focus on fitness apps. Translational behavioral medicine, 10(4), 835-842. DOI: https://doi.org/10.1093/tbm/ibaa086
4. Mohiyeddini, C., Pauli, R., & Bauer, S. (2009). The role of emotion in bridging the intention - behaviour gap: The case of sports participation. Psychology of Sport and Exercise, 10(2), 226-234. DOI: https://doi.org/10.1016/j.psychsport.2008.08.005
5. Ajzen, I. (1991). The theory of planned behavior. Organizational Behavior and Human decision processes, 50(2), 179-211.
6. Godin, G., & Kok, G. (1996). The theory of planned behavior: a review of its applications to health-related behaviors. American journal of health promotion, 11(2), 87-98.

7. Davis, F. D., Bagozzi, R. P., & Warshaw, P. R. (1989). User acceptance of computer technology: A comparison of two theoretical models. Management science, 35(8), 982-1003.
8. George, J. F. (2004). The theory of planned behavior and Internet purchasing. Internet research.
9. Ajzen, I., & Fishbein, M. (1975). A Bayesian analysis of attribution processes. Psychological Bulletin, 82(2), 261.
10. De Leeuw, A., Valois, P., Ajzen, I., & Schmidt, P. (2015). Using the theory of planned behavior to identify key beliefs underlying pro-environmental behavior in high-school students: Implications for educational interventions. Journal of Environmental Psychology, 42, 128-138. DOI: https://doi.org/10.1016/j.jenvp.2015.03.005
11. Cheng, O. Y., Yam, C. L. Y., Cheung, N. S., Lee, P. L. P., Ngai, M. C., & Lin, C. Y. (2019). Extended theory of planned behavior on eating and physical activity. American Journal of Health Behavior, 43(3), 569-581. DOI: https://doi.org/10.5993/AJHB.43.3.11
12. Liao, C., Chen, J. L., & Yen, D. C. (2007). Theory of planning behavior (TPB) and customer satisfaction in the continued use of e-service: An integrated model. Computers in Human Behavior, 23(6), 2804-2822. DOI: https://doi.org/10.1016/j.chb.2006.05.006
13. Ajzen, I. (1991). The theory of planned behavior. Organizational Behavior and Human decision processes, 50(2), 179-211.
14. Taylor, S., & Todd, P. (1995). Decomposition and crossover effects in the theory of planned behavior: A study of consumer adoption intentions. International journal of research in marketing, 12(2), 137-155. DOI: https://doi.org/10.1016/0167-8116(94)00019-K
15. Chiou, J. S. (1998). The effects of attitude, subjective norm, and perceived behavioral control on consumers' purchase intentions: The moderating effects of product knowledge and attention to social comparison information. Proc. Natl. Sci. Counc. ROC (C), 9(2), 298-308.
16. Bandura, A. (1982). The assessment and predictive generality of self-percepts of efficacy. Journal of behavior Therapy and Experimental Psychiatry, 13(3), 195-199.
17. Finlay, K. A., Trafimow, D., & Jones, D. (1997). Predicting health behaviors from attitudes and subjective norms: Between-subjects and within-subjects analyses. Journal of Applied Social Psychology, 27(22), 2015-2031.
18. Finlay, K. A., Trafimow, D., & Moroi, E. (1999). The importance of subjective norms on intentions to perform health behaviors. Journal of Applied Social Psychology, 29(11), 2381-2393.
19. Cialdini, R. B., & Trost, M. R. (1998). Social influence: Social norms, conformity and compliance. In Gilbert, D. T., Fiske, S. T., & Lindzey, G. (Eds.), The Handbook of Social Psychology, 151-192. New York, NY, US: McGraw-Hill.
20. Smith, J. R., Louis, W. R., Terry, D. J., Greenaway, K. H., Clarke, M. R., & Cheng, X. (2012). Congruent or conflicted? The impact of injunctive and descriptive norms on environmental intentions. Journal of Environmental Psychology, 32(4), 353-361. DOI: https://doi.org/10.1016/j.jenvp.2012.06.001
21. Gelfand, M. J., & Harrington, J. R. (2015). The motivational force of descriptive norms: For whom and when are descriptive norms most predictive of behavior?. Journal of Cross-Cultural Psychology, 46(10), 1273-1278.
22. Varshney, D., & Vishwakarma, D. K. (2021). A review on rumour prediction and veracity assessment in online social network. Expert Systems with Applications, 168, 114208. DOI: https://doi.org/10.1016/j.eswa.2020.114208
23. Wong, A., Ho, S., Olusanya, O., Antonini, M. V., & Lyness, D. (2021). The use of social media and online communications in times of pandemic COVID-19. Journal of the Intensive Care Society, 22(3), 255-260. DOI: https://doi.org/10.1177/1751143720966280
24. Friedman, D. B., Gibson, A., Torres, W., Irizarry, J., Rodriguez, J., Tang, W., & Kannaley, K. (2016). Increasing community awareness about Alzheimer' s disease in Puerto Rico through coffee shop education and social media. Journal of Community Health, 41(5), 1006-1012.
25. Bicen, H., & Arnavut, A. (2015). Determining the effects of technological tool use habits on social lives. Computers in Human Behavior, 48, 457-462. DOI: https://doi.org/10.1016/j.chb.2015.02.012
26. McGuire, W. J. (1989). Theoretical foundations of campaigns. In R. E. Rice, & C. K. Atkin (Eds.),Public communication campaigns (2nd ed., pp. 43 – 65). Newbury Park, CA: Sage.
27. Bandura, A. (2002). Social cognitive theory of mass communication. In J. Bryant & M. B. Oliver (Eds.), Media effects: Advances in theory and research (pp. 94 – 124). New York, NY: Routledge.
28. Chen, L., & Fu, L. (2022). Let's fight the infodemic: the third-person effect process of misinformation during public health emergencies. Internet Research. DOI: https://doi.org/10.1108/INTR-03-2021-0194
29. Dunne, A., Lawlor, M. A., & Rowley, J. (2010). Young people's use of online social networking sites – a uses and gratifications perspective. Journal of Research in interactive Marketing. DOI: https://doi.org/10.1108/17505931011033551
30. Park, N., Kee, K. F., & Valenzuela, S. (2009). Being immersed in social networking environment: Facebook groups, uses and gratifications, and social outcomes. Cyberpsychology & Behavior, 12(6), 729-733. DOI: https://doi.org/10.1089/cpb.2009.0003
31. Sun, Y., Shao, X., Li, X., Guo, Y., & Nie, K. (2020). A 2020 perspective on " How live streaming influences purchase intentions in social commerce: An IT affordance perspective". Electronic Commerce

Research and Applications, 40, 100958. DOI: https://doi.org/10.1016/j.elerap.2020.100958

32. Skoric, M. M., & Poor, N. (2013). Youth engagement in Singapore: The interplay of social and traditional media. Journal of Broadcasting & Electronic Media, 57(2), 187-204. DOI: https://doi.org/10.1080/08838151.2013.787076
33. Kaplanidou, K., & Gibson, H. J. (2012). Event image and traveling parents' intentions to attend youth sport events: a test of the reasoned action model. European Sport Management Quarterly, 12(1), 3-18. DOI: https://doi.org/10.1080/16184742.2011.637173
34. Ho, S. S., Chen, L., & Ng, A. P. (2017). Comparing cyberbullying perpetration on social media between primary and secondary school students. Computers & Education, 109, 74-84. DOI: https://doi.org/10.1016/j.compedu.2017.02.004
35. Kim, Y. J., & Kim, E. S. (2021). Analysis of Korean Fencing Club Members' Participation Intention Using the TPB Model. International Journal of Environmental Research and Public Health, 18(6), 2813. DOI: https://doi.org/10.3390/ijerph18062813
36. Lim, B., Son, S., Kim, H., Nah, S., & Mu Lee, K. (2017). Enhanced deep residual networks for single image super-resolution. In Proceedings of the IEEE conference on computer vision and pattern recognition workshops (pp. 136-144).
37. Talwar, S., Dhir, A., Kaur, P., Zafar, N., & Alrasheedy, M. (2019). Why do people share fake news? Associations between the dark side of social media use and fake news sharing behavior. Journal of Retailing and Consumer Services, 51, 72-82. DOI: https://doi.org/10.1016/j.jretconser.2019.05.026
38. Jette, A. M., Harris, B. A., Sleeper, L., Lachman, M. E., Heislein, D., Giorgetti, M., & Levenson, C. (1996). A home-based exercise program for nondisabled older adults. Journal of the American Geriatrics society, 44(6), 644-649.
39. Ho, S. S., & Chuah, A. S. (2022). Thinking, not talking, predicts knowledge level: Effects of media attention and reflective integration on public knowledge of nuclear energy. Public Understanding of Science, 31(5), 572-589. DOI: https://doi.org/10.1177/09636625211070786

Research and Exploration on the Optimization of the Completion System of the History Library Group of Colleges and University

Ming Lu[1,*], Qi Wang[2], Yali Liu[2] and Chen Cao[2]

[1] Air Force Early Warning Academy, Teaching and Research Support Center, Wuhan, Hubei, China
[2] Air Force Early Warning Academy, Teaching and Research Support Center, Wuhan, Hubei, China

Abstract. This paper first analyzes the commonness and current problems in the systematic construction of the university history museum, and then, combining with the experience and practice in the systematic optimization construction of the university history museum in our college, focuses on the conception of the content setting and spatial hierarchy optimization of the history museum, as well as the specific construction organization and working mechanism. It provides an example and reference for the systematic optimization and construction of history museums in other universities.

1 Introduction

From rudiment to perfection, from singleness to pluralism, from isolation to community, the history library system of colleges and universities is an important carrier to witness the course of running a school, to contain the experience of educating people, and to place the feelings of teachers and students, which has created an important cultural position for inheriting the red gene and deepening ideological and political education. At present, all kinds of special exhibition halls, science and technology museums, natural museums and other new ideas, new technologies and new achievements are constantly emerging, and various exhibition forms such as network history hall, cloud exhibition, virtual exhibition hall emerge in an endless stream, which are widely praised.

Faced with these new orientations, new developments and new changes, how to better adapt to the new era of education, build up the four beams and eight pillars of cultural education elements, and build a spiritual baptism space with complementary functions, interoperable systems and multi-dimensional interaction is an urgent problem to be solved. Therefore, the only way to get out of the current predicament is to conform to the trend of the times, carry out digital reconstruction and upgrading, build a digital education base, and optimize the construction of the system.

2 Analysis of the Current Situation of University History Museum [1]

Faced with the rapid development of the new era, the renewal of the Museum of History, which is in line with the pace of the construction and development of colleges and universities, is lagging behind.

On the one hand, the old building space restricts the content follow-up update. Because the exhibition space of the Academy History Museum is fixed, it is difficult to solve the drawbacks caused by the compression of time and space. The exhibition in the form of old exhibition boards and cabinets is difficult to increase or replace the new exhibition content in time. The theme of key landscape and physical exhibition are limited. It is difficult to expand the space, update the content and adjust the layout. With the evolution of time, the superposition of events and the increase of collections, it will be more and more difficult to adjust in the later period.

On the other hand, the situation of boundary dissolution affects the quality and effectiveness of education. With the increase of modern information channels, the broadening of knowledge horizons, the diversification of aesthetic tastes, cognitive differences and value conflicts show the characteristics of "cultural violence". The trend of ideological boundary digestion is becoming more and more obvious, and the traditional narrative mode of "I speak and you listen" is difficult to achieve results. The teacher-like preaching and cramming of information transmission inadvertently strengthen the dull display and the image of being superior. If we want to break the deadlock, we must seek to have a dialogue in an equal way.

Moreover, the vigorous development trend of the Network History Museum is forcing the pace of digital innovation to accelerate. With the continuous innovation and development of the network history museum under the background of "Internet +", the use of digital media, especially network platform technology, has become the mainstream trend, and all kinds of history museums and memorials at all levels have launched online terminals such as network version and cloud exhibition, and twin, derivative and independent exhibition cases have

*Corresponding author: 85074942@qq.com

emerged one after another. It has effectively realized the demand expectation of "closing the museum without closing the exhibition and visiting without restriction".[7]

Based on the above analysis of the current situation, the demand for systematic optimization and upgrading of university history libraries can not be ignored, and it is urgent to re-plan and construct to meet the requirements of new development.

3 Content setting and spatial hierarchy optimization of the History Museum of our Institute[3][4][5]

The History Museum of our Institute was completed in 2012, with an exhibition area of about 1000 square meters, 22 exhibition halls and more than 1000 physical exhibits. Since its establishment, it has undertaken thousands of practical teaching, freshmen enrollment, graduation education and various kinds of visits at all levels, showing the college's history of continuous struggle, the spirit of characteristic snow lotus and the achievements of talent phalanx construction, and has become a "living textbook" to enhance the sense of honor, belonging and pride of officers and soldiers, and a "small window" to show the new development and achievements of the college. It plays an important role in educating people, teaching and promoting education.

"Reform" is highlighted in the optimization of the construction of the History Museum of our Institute, focusing on the comparison of historical development and evolution, fixing the source of development, delimiting the stage of development, and determining major events. "Three to one long-term" thematic education highlights the "soul", focusing on condensing the spirit, and forming a positive guiding effect of value. The construction of online history libraries focuses on "wisdom" and strengthens the application of digital intelligent technology. Through two libraries and one hall, the layout of history libraries with different emphasis, complementarity and clear context is constructed and perfected by means of online and offline integration, historical facts and technology interaction.

3.1 The renovation of the Academy History Museum breaks the restrictions of the pattern, and the digital technology assists the physical exhibition.

In accordance with the principles of "respecting history, being well-founded, running through the main line, focusing on history, not leaking major events, highlighting highlights, excavating connotations, gathering souls, avoiding sensitivity and keeping secrets carefully" and the ideas of "based on facts, based on history, exhibiting features, highlighting representatives and highlighting spirit", we should break through the existing pattern of the Academy History Museum and introduce it. Assist in displaying the brilliant course of the college.

One is the transformation from historical data to outline. On the basis of careful and solid study of historical materials, we should keep the existing structure roughly unchanged, merge, compress, expand and update some of the contents, adopt the structure of time sequence as the main part and special topics as the supplement, and divide the historical stages according to the 100th anniversary of the founding of the Communist Party of China. The main line is the three historical nodes of "the initial stage of birth and preparation (1952-1977)", "comprehensive construction and development (1978-2012)" and "striving for a new era (2013-present)". The special exhibition areas of "historical honor", "teaching achievements", "scientific research achievements" and "three to one long-term" educational achievements are set up in the plate, and the long volume of history is spread out layer by layer.

The second is the transformation from exhibition hall to space. Optimize the visiting streamline, relocate the exhibition hall on the west side of the second floor to the east side of the first floor, and use the stairs on both sides to construct the closed-loop of the exhibition line, so as to avoid the repetition of loops and reduce the density of local people. It exchanges space for time extension, reflects the close connection and rhythm of space, and expands the vertical depth that traditional display can not achieve by atmosphere rendering.

Third, the transformation from text to vision. With pictures, images and objects as the main content, we should distinguish major national activities, key military scientific research projects, teaching and training achievements at or above the air force level, and major events in the development of the academy, and do a good job of "addition and subtraction", highlighting valuable historical materials and cultural relics, pruning branches and tendrils, simplifying them, developing limited space with auxiliary applications, seizing visual attraction and promoting memory. The history museum provides visitors with the best viewing experience and multi-cultural edification, so that the spirit of the inheritance of the flesh and soul, and enhance the sense of identity of officers and soldiers.

3.2 The network history museum is free to roam 360 degrees and create virtual space through multi-dimensional interaction.

Following the principle of "practicability, flexibility, fluency and easy operation", using virtual reality technology and combining three-dimensional animation to create a simulation space of realistic environment, the physical space is extended to any corner where intelligent terminals can be used, and the convenience, interaction and experience of visits are enhanced. They complement and promote each other in giving full play to the functions of propaganda and education.

First, self-guided tours enhance participation. Set up 360-degree free roaming, virtual explanation, location positioning, autonomous interaction and other functions, which can not only enter and exit freely according to the default route of the system, but also switch the exhibition hall independently for sightseeing. Fully consider the flexibility of content update, reasonably design the module layout structure, reserve the expansion interface, and meet the needs of dynamic display of hospital history.

Second, the co-construction of libraries and classes enhances plasticity. As a carrier of communication and teaching for participating in the Red Education and Ideological and Political Course, it is embedded in practical cases to tell the story of guarding the frontier and guarding the customs, to implement the "co-construction of library courses", to provide "the first lesson of entry" and "the first lesson of departure" for new entrants, especially freshmen and graduates, to cultivate the value of young officers and soldiers, and to establish a correct world outlook. Strengthen the backbone and ambition of guarding the frontier and guarding the frontier.

Third, cross-interaction enhances permeability. We should adhere to the unity of politics, ideology and artistry, and enhance the expressive force, dissemination and influence. With "three natures" and "three forces" as the traction target, an interactive comment and speech area is established, visitors' feelings are submitted at any time, and the "bullet screen" is displayed in real time, so as to enhance the interest of the tour, and at the same time, the brain space is penetrated by humanized intervention. By clicking on the hot label, the interest points of the exhibits on the exhibition board can be interactively operated, and the positive information can be displayed with amplification effect, so as to increase the cross-influence between people and achieve positive guidance.

3.3 "Three to one long-term" thematic immersion interaction, virtual and real blending expansion and mixed reality.

The special topic of "three to one long-term" education is an important spiritual link between the army and the Academy, and has become an important part of the Academy History Museum. The special exhibition hall can be used as an independent exhibition hall to organize visits and learning, as well as a new window for "academies linked to the army".

First is to reproduce the scene and display the activation of historical memory. The renovation of the "three to one long-term" special exhibition hall is divided into the corridor part and the main part. The corridor plans to use the form of space-time tunnel to interact with the image materials of the students guarding the frontier, and use the deep embedding technology to realize the "transfer of consciousness and cognition", so as to produce a sense of dedication and substitution wrapped in "going to the frontier, to the hard places, to the places where the motherland needs most, and serving the army for preparing for war for a long time"; The main part breaks the traditional physical barriers and chronological order of exhibition, releases the restrictions of visitor route and visiting methods, establishes the exhibition conception of full freedom, self-excavation and multi-directional interaction, and constructs the cultural empowerment in the process of spiritual baptism of mountains, islands, deserts, Gobi, snow areas and plateaus.

Second, technology touch constructs the interactive space of role transformation. Create an expanded mixed space of content and technology integration, reality and virtual integration, audio-visual materials and interactive devices, take all-media content as the standard, adapt to the development of "immersion communication" orientation and "ubiquitous communication" media, through the deconstruction and recasting of special exhibits in the frontier such as camel thorn, Populus euphratica, Gobi stone, etc. It integrates the significance of devoting to the frontier into the narrative voice of the scene, combines and crosses the simple vision, hearing and touch, effectively implants the value concept of defending the country and guarding the frontier, so that the education of "three loves and three dares" can penetrate from static to dynamic, from plane to three-dimensional, from single to multiple and omni-directional, and create a theme spiritual palace with feeling and temperature.

3.4 Duality shows deep and subtle influence, and internal and external forces produce common values.

The so-called dual display emphasizes that the individual feelings and public concepts of officers and soldiers can be strengthened in the process of visiting, and ultimately achieve the goal of strengthening individual feelings and forming collective concepts, so as to achieve the purpose of educating people.

First is to strengthen individual feelings. Control the psychology of the audience, let the officers and soldiers participate in the discussion of the history and spiritual culture of the college from various angles, respect the "basic right" of "personality feeling" which young officers and soldiers attach great importance to the greatest extent, promote the officers and soldiers to enter history and think actively, and better approach, infect and cohere the officers.

The second is to produce common ideas. Strengthen the role of the public sphere of the Museum of History, promote the development of the Museum of History community from one-way ideological indoctrination to spiritual sharing nodes, so that officers and soldiers can find themselves and discover others here, so that individual concepts and ideas can be gathered, exchanged and collided here, forming a "self-organized, growing spiritual resonance and ideological education system".

Third, to achieve value recognition. While fully stimulating the interest of officers and soldiers and effectively transmitting historical and cultural knowledge, the radiation affects the daily work and life of officers and soldiers, promotes their mutual influence, achieves the highest level of recognition in the value space, produces a higher cultural and spiritual value resonance, and ultimately forms a reaction to people's ideological interest and value recognition.

4 Suggestions on the Construction and Upgrading of the Museum of History[6] [7]

In the process of promoting the systematic optimization of university history libraries, a special preparatory working group for reconstruction should be set up, and a corresponding liaison mechanism should be established, which should be carried out simultaneously in three directions: data collation, design demonstration and liaison textual research.

4.1 Deeply excavate historical materials and establish exhibition archives.

Adhering to the principle of "systematically combing, exploring bright spots, seeking truth from facts and accurately identifying", we should sort out the main vein of discipline development and the main line of education culture and spirit, expand the collection of photos, objects and text descriptions of major events, contest awards and other activities, classify, file and register one by one, and establish a database containing existing pictures, picture albums, audio and video and cultural relics. High-definition transformation of basic historical materials, the use of plane scanning, three-dimensional modeling, virtual construction and other means, technology embedded exhibition outline, for the optimization and upgrading of the College History Museum to lay a solid foundation for information and rich spiritual support.

4.2 Integration of online and offline to carry out design demonstration.

As the systematic upgrading and transformation inevitably involves the integration of interior decoration, real sculpture, equipment disassembly and assembly, information integration, film and television production and other cross-domain systems, the construction work is complicated and needs to be promoted as a whole, and the work of project cost, procurement inquiry, consultation and demonstration should be started simultaneously, and professional cost companies should be selected to meet the needs of the transformation according to the market price. Around the idea and demand of transformation, we should communicate in advance and promote the research, demonstration and design of the scheme as a whole.

4.3 Multi-party contact and textual research to confirm historical events.

Establish a network covering all departments, departments, teaching and research offices, student unions and other liaison work, cooperate with data collection, historical textual research, project bidding and other liaison work, collect and summarize all kinds of historical data, sort out all kinds of honorary achievements, and determine the honor ownership and spiritual inheritance of the school on the principle of "exclusive source, honor sharing and EPL, **84** (2008)spiritual integration". Some veteran leaders and comrades are invited to check and verify historical events and put forward suggestions.

5 Conclusion

In the long run, in order to meet the needs of diversity and mobility of information in colleges and universities, we must optimize the construction of history museum community in a systematic way. The integration of various resources such as Academy History Museum and Network History Museum is the trend of future development. This paper takes the practice of the systematic optimization and transformation of the history museum of our college as an example, elaborates the transformation ideas of the content setting and spatial hierarchy optimization of the history museum in detail, and creatively puts forward a series of feasible transformation means, which provides an example and reference for the digital transformation of other history museums of colleges and universities.

Believing that practice is the only criterion for testing truth, I sincerely hope that all colleges and universities can find their own way of resource integration on the way forward, make full use of their own resource advantages, tap the value of resources, and improve the work of colleges and universities to a new and higher platform.

References

1. Miao Zengfu. Investigation and Reflection on Promoting the Construction and Use of the Digital Military History Museum of Qiangjun Network. Journal of Political Engineering, 06 (2021).
2. Zhong Guihua. A Study on the "Return Rate" of Museum Audiences. Shanghai: Shanghai University, (2020).
3. Jane Kidd. Museums in the New Media Environment: Cross Media, Participation and Ethics. Translated by Hu Fang. Shanghai: Shanghai Science and Technology Education Press, (2017).
4. Zhu Shaorui. Research on the Theme Narrative Design of the Exhibition Space of the School History Museum —— Taking the School History Museum of Dalian University of Technology as an Example.Liaoning: Dalian University of Technology, (2019).
5. Chen Zhiyin. Research on Narrative Expression of Exhibition Design of University History Museum. Hunan: Hunan Normal University, (2020) .
6. Qian Kai. Research on Museum Audience Autonomy in Cross-media Space. China Press,05 (2020).
7. Nie Siyu. The Media Practice of Museum Interpretation: Taking the Palace Museum as an Example. Jiangsu: Nanjing University,(2020).

Reasons Why the Ageing Population Stands Out on TikTok and the Analysis of Its Commercial Potential

Aolin Cheng*

Univeristy of Sheffield, Sheffield S1 4DH, UK

Abstract. Even in an era of information explosion, many older people still miss out on important information due to technical limitations. But TikTok, a global short video platform, has attracted more and more senior users in recent years through its creative and entertaining content, diverse community interactions, and easy-to-use methods. It is even creating senior netizens with millions of followers. These older consumers and content contributors are showing their energy and vitality through TikTok to rewrite the way people expect older people to behave on social media. This paper investigates the accounts of older celebrities on Chinese TikTok platform to understand how older people are expanding their reach by making TikTok videos and how they can engage with the rapidly developing society through this online platform, which is regarded as a virtual playground for teenagers. In addition, this paper also discusses the undervalued commercial potential of an aging society. The data analysis reveals that older people may challenge stereotypes brought about by ageism and conclusions can be drawn that the aging group on TikTok is standing out on the platform because of its positive image and optimistic attitude towards life. Moreover, because of the increased digitalisation and the changing attitudes of older adults, their purchasing power is continuing to grow, and the aging industry is booming.

1 Introduction

Despite the widespread belief that older adults are technophobic [1], the number of older people who are tech users has increased significantly over the past decade. The number of people aged 65 and older with a social media presence has more than doubled since 2010 [2]. TikTok, a mobile app launched in September 2016, allows users to create and share short videos with a community of like-minded individuals. It utilizes various marketing techniques such as 15-second clips to engage its users and has been installed on over 300 million devices globally. In February 2019, it reached one billion users; in mid-2021, it got a total of three billion [3]. Most people know that older adults are more likely to be passive consumers of active digital content producers [4]. The coronavirus pandemic in 2019 has led to widespread negative attitudes toward older adults on Facebook. The social media hashtag #BoomerRemover has also sparked a debate about the harmful effects of ageism [5]. In an environment where both ageism and intergenerational tensions have proliferated [6], Generation Z and Millennials, the primary users on TikTok, are having challenges from older users with millions of followers [7]. To analyse videos from elderly celebrities on TikTok is because the platform has received little attention among researchers in the field of gerontology. Also, TikTok is more suitable for young people than any other social media application [8], so their reactions and interactions indicate a trend where stereotypes resulting from ageism are being challenged.

This paper analyzes three elderly Internet celebrities on Tiktok through case studies and explores the reasons why the elderly group stands out on this platform. The author further analyzes the commercial potential of the elderly group under the background of China's aging, which is helpful to enrich the related research on the elderly group on new media platforms.

2 Reasons for the Aging Group Standing Out on TikTok

Although TikTok is commonly referred to as the virtual playground of teenagers [9], some of TikTok's older users have not only become content creators, but also amassed millions or even tens of millions of followers, debunking the notion that older people are passive users of social media [10].

2.1 A Positive Image

Most of the short videos posted by older adults on the platform and widely shared tend to promote healthy, active, and happy lifestyles. And some of the older TikTok celebrities have also profited from the platform [11]. In the past, the image of the elderly was mainly rigid, conservative, monotonous, outdated, etc. But the "silver old celebrities" on TikTok convey a completely different feeling. They are stylish, vivacious, and optimistic, with a

*Corresponding author: acheng2@sheffield.ac.uk

unique temperament endowed by age. It is this contrast of image that brings them a lot of fans.

For example, "Grandma Wang, who only wears high heels", writes on her TikTok page that age is just a number and her amazing life is just beginning. The 80-year-old Grandma Wang has nearly 15 million followers on TikTok. She is always dressed in slim-fitting clothes and high heels in her TikTok videos and often displays high standards of living and positivity. Grandma Wang calls her fans "little girlfriends" and often gently instructs them to discipline and love themselves. According to the data released by the TikTok short-video live-streaming data analysis platform "XinDou", the age group of Grandma Wang's fans is mainly distributed between 31-40 years old, followed by users aged 18-30. In March 2020, Grandma Wang officially started live-streaming with the goods. Grandma Wang's selection of products greatly caters to the needs of her young fans, with 42.43% of beauty and skin care products, followed by household products and food. The first live broadcast met expectations, with more than 8 million total viewers and sales of 4.7 million yuan. According to relevant media reports, as of June 2020, Grandma Wang's 30-day "Yinlang" (Virtual Currency on TikTok) revenue exceeded 84.47% of the live-streamers on the TikTok platform, and merchandise sales exceeded 99.53% of the platform's live-streamers. Fans of these older celebrities or consumers who buy these products tend to be people who recognise and appreciate a healthy ageing lifestyle. And these older adults, who are appreciated for their videos for staying active, elegant, and maintaining a high standard of living, are one of the groups that can stand out on TikTok. People tend to comment on this group by saying things like "I wish I could grow old like that" or "I wish I could be like her when I'm old".

At the same time, the amusing, entertaining, and down-to-the-earth older adults can also gain a lot of fans on TikTok and "I am Grandma Tian" is one of the most representative characters. During the epidemic, short videos once became a tool for most people to spend their time. In this context, Tian's grandson shared clips of Grandma Tian's daily life on TikTok. Due to the funny content and Tian's very personalised laughter, her videos went viral, and now she has more than 36 million followers on TikTok. In Grandma Tian's video, the funny daily routine between the mischievous grandson and the funny nagging grandmother brings a lot of joy to people. Unlike Grandma Wang's videos, Grandma Tian is more like a real older person. She does not have a slim figure and she wears loose clothes, likes to gossip with neighbours and friends, and is even a little grumpy and stingy. However, it is because of these imperfect characteristics of Grandma Tian that the audience can feel the vividness of the characters. Additionally, the real-life scenes make people feel warm and familiar to resonate with Grandma Tian. In the usual TikTok video, Tian's grandson will insert some brand advertisements through oral narration. And in terms of live streaming with products, Grandma Tian and her team will choose food, clothing, household items, etc. On June 21, 2020, Grandma Tian also cooperated with the local community to conduct live streaming to help farmers - using her TikTok account's popularity to help sell the local agricultural products. And she was awarded the "2020 Qianshan Area Network Charity Person". Grandma Tian's unique personality is making more and more people love her, and her fans often comment, "Grandma is really cute!" "It's so funny".

Elegant or funny older people on TikTok can indeed gain a lot of fans, but at the same time, another group of older people is getting more and more attention on TikTok. They usually are low-income families and want to improve their living conditions by selling goods through TikTok live-streaming. They live-stream in their home, with no complicated equipment, rehearsed words, or even little expression. Throughout the process, they usually sit up straight and introduce the goods in less-than-standard Mandarin. But they are gaining more and more fans because of their sincerity and earnestness.

2.2 The Establishment of Intimate Relationships with Strangers

When those older adults post their work on TikTok to bring positive energy to people, they also feel the warmth that the community brings to them. Many supportive words and emojis like roses or thumbs up can always be seen in their comments area. Their interaction suggests [12] a concept of 'stranger intimacy'. They define stranger intimacy as "an open, conditional relationship, albeit a very brief one. It is possible for two people who do not know each other to establish the emotional structure of understanding and even friendship or love through this relationship," The "stranger intimacy" seniors can obtain on TikTok creates a caring atmosphere, and its presence fosters their social participation. TikTok helps seniors transcend their physical location and get emotional support by providing them with new forms of convenient face-to-face communication and the opportunity to chat in real-time [11].

The fact that more and more older people can be active on the TikTok platform shows people a trend that the rapidly developing society has not abandoned the elderly. There is a group of older adults who are willing to accept and try new things, lead the lives of young people with their unique experiences, and infect young people with an optimistic attitude towards life.

3 Analysis of the Business Potential of the Ageing Population

In China, the official definition of middle-aged is over 40 years old, while the definition of elderly is over 60 years old. However, in a research reported by Wavemaker, an international marketing agency, "Seniors in China are the Hidden Treasure", it was found that if middle-aged and older people were to define their own age boundaries, they would consider people to be middle-aged until they are 49 and older until they are 66. This is a phenomenon that Wavemaker has named 'ageing procrastination'.

3.1 The Increased Digitalisation

In the report, middle-aged and elderly Chinese are divided into three generations according to their different social backgrounds during adolescence. Those born between 1934 and 1950, who experienced the first 15 years of the founding of New China during their youth, are known as "WITNESS OF REVOLUTION" and have a strong sense of collective values. The generation born between 1951 and 1963 experienced the turmoil of the Cultural Revolution when they were young but resuming their journey. They are known as the generation that "TOUGHENED NEW-CHINA 1st GENERATION". The generation born after 1963 experienced reform and opening up during adolescence and is known as the "THE OPENING UP NEWBORNS".

If these three generations are used to redefine 'middle-aged and older', then a number of judgements about this group will have to be re-examined. For example, it is often said that middle-aged and older people are slow to accept new things, such as new technology, new ways of consuming online, etc. But even in the generation of "WITNESS OF REVOLUTION", the digitalisation rate is around 30%. In the "TOUGHENED NEW-CHINA 1st GENERATION", the level of digitalisation is already close to 50%. And in the generation of the "THE OPENING UP NEWBORNS", the level of digitalisation is already more than 70%. Looking at the overall figures across these three generations, it can be found that they are not less exposed to new technology and new ways of consuming. For example, 40% of people usually go to Taobao, Pinduoduo and other websites for online shopping. The number of people who usually use Alipay is 48%. The percentage of those who use wearable devices is 20%. Also, more than 3/4 of the middle-aged and elderly group usually use websites such as Tencent and IQIYI to watch videos online. In addition, there has been a great controversy about the lifestyle of the elderly - for example, the stereotype like loneliness. The Wavemaker report shows nearly 70% of older adults say they "rarely" feel lonely. On the contrary, their lives are colourful, including walking, watching TV, going online, listening to music, and drinking tea. Many people also like gardening, shopping, and dancing. Regarding lifestyle, many older adults are not lonely and hope to have more independent time and arrange their own life.

3.2 A Continued Increase in Purchasing Power

According to data released by the National Bureau of Statistics, China's population over the age of 60 will double to 400 million around 2033 and reach 1/3 of the country's population by around 2050. It is predicted that between 2014 and 2050, the scale of China's elderly industry will grow from 4.1 trillion yuan to about 106 trillion yuan, and the proportion of GDP will increase from 8% to around 33%, making China's elderly industry with the most significant market potential. According to Wavemaker's calculations, in the present day, the annual spending power of the ageing population in China's first to third-tier cities is as high as RMB6.64 trillion per year, already far exceeding the assessments of different institutions.

3.3 Vigorous Development of Ageing Industry

In addition, many industries have been severely challenged by the sudden arrival of COVID-19 in 2021, and the public's health perception has risen to a new high. Wavemaker's survey of three generations of silver-haired people in China's first- to third-tier cities shows that the whole ageing industry demonstrates a vigorous development under the "healthy China" strategy and China's unique advantages. China's ageing industry presents three major changes in policy trends, industrial structure, and driving force. In addition, the number of new startups related to the ageing industry with registered capital (the total amount of capital contributions subscribed by the shareholders of a Chinese company, which must be registered with the business registration authority, that is, a competent office of the State Administration for Market Regulation) of more than 10 million yuan reached more than 1,900 in 2020. 2021 is destined to be a milestone starting point for developing China's ageing industry. The report also points out that the health industry, as the core of China's ageing industry, will develop towards greater segmentation and diversification. Among other things, the development of intelligent technology will further give rise to innovative practices in the ageing industry. The number of intelligent ageing products and services tripled from 2018 to 2020. The penetration of emerging medical devices and purification and innovative wellness products in households increases further.

Whether retiring from work and returning to life, building interest-driven social networks, or moving to a new city or community, old age opens up "new" roles and networks. Growing up in a collective-oriented social environment, they seek the sense of belonging and identity of the group, show increasing social needs and are keen on group action. This trend means a huge market space for the consumption of hobbies and social activities of the elderly.

In addition, for the elderly population with a large span of generations and age levels, the younger consumption attitude is also reflected in their choice of products. Moreover, their willingness to consume high-quality and imported products is increasing. This "new" elderly population, with more decent, comfortable, quality of second life expectations, will undoubtedly become a huge consumer market in the future.

4 Conclusion

The phenomenon of ageing is a social issue that is fraught with difficulties and opportunities at the same time. The importance of the huge consumer market lies not only in its commercial value but also in its social significance. This study explores the reason why the elderly adults can stand out on TikTok and the commercial potential of the aging industry. To conclude, TikTok's elderly adults have a unique affinity with the world, and their ability to bring in goods and business value is something worth tapping into. Although the elderly are not as responsive as they could be in terms of mobile phone operation, their good spirits, confidence, and happiness have influenced this audience. Also, through their inexpensive and good quality products, they are more likely to win the goodwill of this audience and strengthen the adhesion of both parties. This not only strengthens their sense of self-development, self-fulfillment, and self-improvement, but also updates the knowledge and skills of the elderly and improves their social adaptability. Meanwhile, the rapid growth of the elderly population provides an important opportunity for the development of the ageing industry. Vigorously developing the ageing industry can not only meet the needs of the elderly and improve their well-being, but also become a positive driving force to expand domestic demand and promote economic growth.

References

1. Jenkins, J. A.: An ageing workforce isn't a burden. It's an opportunity. World Economic Forum (2019). from https://www.weforum.org/agenda/2019/01/an-aging-workforce-isnt-a-burden-its-an-opportunity/.
2. Faverio, M.: Share of those 65 and older who are tech users has grown in the past decade. Pew Research Center (2022). from https://www.pewresearch.org/fact-tank/2022/01/13/share-of-those-65-and-older-who-are-tech-users-has-grown-in-the-past-decade/.
3. Surabhi, S., Shah, B., Washington, P., Mutlu, O., Leblanc, E., & Mohite, P. et al.: TikTok for Good: Creating a Diverse Emotion Expression Database (2022). Retrieved 13 August 2022, from https://openaccess.thecvf.com/content/CVPR2022W/ABAW/html/Surabhi_TikTok_for_Good_Creating_a_Diverse_Emotion_Expression_Database_CVPRW_2022_paper.html.
4. Jenny, W.: Older adults as digital content producers | Proceedings of the SIGCHI Conference on Human Factors in Computing Systems (2013). Retrieved 13 August 2022, from https://dl.acm.org/doi/abs/10.1145/2470654.2470662?casa_token=WZjorSgwyJ8AAAAA:nxDEr-OgXnHHLAKvJioQ1qPiOSa6I-JcuYuhsRHPAbY94wkkL8h4HYMDP-OSvxwflczphOTQglV0Ow.
5. Fraser, S., Lagacé, M., Bongué, B., Ndeye, N., Guyot, J., & Bechard, L. et al.: Ageism and COVID-19: what does our society's response say about us? (2020). Retrieved 13 August 2022, from. https://academic.oup.com/ageing/article/49/5/692/5831206.
6. Ayalon, L.: There is nothing new under the sun: ageism and intergenerational tension in the age of the COVID-19 outbreak (2020). Retrieved 13 August 2022, from https://www.cambridge.org/core/journals/international-psychogeriatrics/article/there-is-nothing-new-under-the-sun-ageism-and-intergenerational-tension-in-the-age-of-the-covid19-outbreak/E14BB8E757B3861FFB198E8C0CDB38DA.
7. Dellatto, M.: Meet the famous boomers and grannies of TikTok. Nypost.com. (2019). Retrieved 13 August 2022, from https://nypost.com/2019/12/18/meet-the-famous-boomers-and-grannies-of-tiktok/.
8. Muliadi, B. Council Post: What The Rise Of TikTok Says About Generation Z. Forbes (2020). Retrieved 13 August 2022, from https://www.forbes.com/sites/forbestechcouncil/2020/07/07/what-the-rise-of-tiktok-says-about-generation-z/?sh=4592748a6549.
9. Kurzrock, E.: Intensified Play: Cinematic study of TikTok mobile app (2022). Retrieved 13 August 2022, from https://www.researchgate.net/publication/335570557_Intensified_Play_Cinematic_study_of_TikTok_mobile_app.
10. A.K., L.: Social Media Use of Older Adults: A Mini-Review (2013). Retrieved 13 August 2022, from https://www.researchgate.net/publication/236224221_Social_Media_Use_of_Older_Adults_A_Mini-Review.
11. Yu, Y.: Can TikTok promote a healthier ageing paradigm? A case study of older digital celebrities from China (2022). Rgs-ibg-onlinelibrary-wiley-com.sheffield.idm.oclc.org. Retrieved 13 August 2022, from https://rgs-ibg-onlinelibrary-wiley-com.sheffield.idm.oclc.org/doi/full/10.1111/area.12777.
12. Koch, R., & Miles, S.: Inviting the stranger in: Intimacy, digital technology and new geographies of encounter (2020). Journals-sagepub-com.sheffield.idm.oclc.org. Retrieved 13 August 2022, from https://journals-sagepub-com.sheffield.idm.oclc.org/doi/10.1177/0309132520961881.

Barriers to sustainable development in digital communication of intangible cultural heritage: emotion drain under the technology dominance

Jiayi Wang*

University of Warwick, Coventry, United British

Abstract. With the surge of digital tide, the digital use of intangible cultural heritage is increasingly common. In recent years, scholars have focused on digital technology as an important force in enabling the development of intangible cultural heritage communication. However, in specific practical explorations, the media characteristics of digital technology and the remote participation of inherited community and other drawbacks are constantly exposed, resulting in the conflict between technology dominance and emotion loss in the digitalisation of intangible cultural heritage. Therefore, there is a need to reflect on the technology dominance of digitisation of intangible cultural heritage and to analyse the causes and effects of emotion drain. The paper also attempts to further propose measures that respect the cultural subjectivity of the inheriting community and focus on the emotional impact of digitisation, aiming to better realise the sustainable development of digitisation of intangible cultural heritage.

1 Introduction

As technology advances by leaps and bounds, there has been an increased recognition by scholars and organisations that digital technologies can largely help to preserve and communicate the intangible cultural heritage of nations and countries. From previous face-to-face communication to today's use of technology to communicate across time and space, technological developments have heightened the need for the digitalisation of intangible cultural heritage. However, this concept has recently been challenged by some studies demonstrating that as technology gradually dominates, participants might engage more with technology than with intangible cultural heritage. This phenomenon tends to cut off the most direct human experience and feeling of intangible heritage culture, thus losing the emotion it conveys. In the 2003 Convention for the Safeguarding of the Intangible Cultural Heritage, United Nations Educational, Scientific and Cultural Organization defined intangible cultural heritage as having social and emotional significance that is passed down from generation to generation in addition to the cultural expression itself (UNESCO, 2003)[19]. Therefore, it may be said that protecting its emotional value is also a way of preserving the diversity and uniqueness of human civilisation.

In the pages that follow, it will be argued the advantages and disadvantages of technology in the digital transmission of intangible cultural heritage and then analysed the probable causes and effects of emotional loss under technology domination. Based on proposing humanistic emotion as the core concept of communication, the aim of this paper is to explore the way in which to promote the sustainable development of the digital communication of intangible cultural heritage.

2 Advantages of digital technology

One of the advantages of being technology-driven is that it provides an efficient mode of preservation and presentation for the digital communication of intangible cultural heritage. Research in recent years has suggested that evolving 3D technologies have enabled significant progress in various digital intangible cultural heritage projects (Skublewska-Paszkowska et al., 2022)[13].

For example, Rongdong Xie's study (2021)[22] was based on VR virtual visualisation technology to construct an interactive virtual environment for 3D digital simulation models of Chinese Dai ceramic art. Through computer rendering and processing software, VR, an emerging technology that offers a human-computer interaction experience, restores the authenticity of Chinese Dai pottery while offering a panoramic 3D viewpoint visual experience in the digital scene generated. Through a computer interface, users may engage in interactive digital random browsing by entering the 3D display world of the digital Dai pottery.

This emerging technology is generally assumed to play a role in permanently storing intangible cultural heritage resources and enhancing the artistic effect of mobile displays (Sorea and Csesznek, 2020)[14], Moreover, it obviously height the user's multi-sensory experience

* Corresponding author: wangjiayi0615@gmail.com

and depth of thought within the virtual space (Sang et al., 2018)[12], see Figure 1.

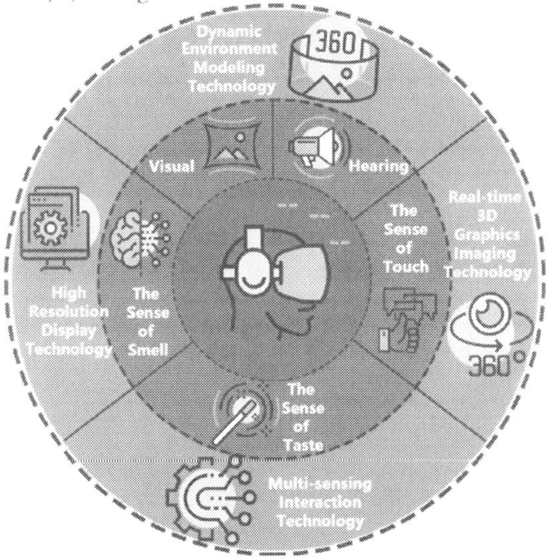

Figure 1 The realistic virtual world built by VR technology

The application of the five main features of VR experiences in the digitisation of intangible cultural heritage complements existing modes of presentation and communication, and provides new digital experience projects that are applicable to various domains within intangible cultural heritage. Figure 2 illustrates the characteristics of VR technology and VR experience projects applicable to the five major domains of intangible cultural heritage.

Figure 2 Applications of VR technology in intangible cultural heritage

The ambience and sense of space offered by conventional static displays are superior. However, they are inaccessible to visitors, and both exhibitors and visitors have to take into account the security of their exhibits and the time and financial expenses. By comparison, Digital technologies completely illustrate the intangible cultural heritage features in a straightforward real-time interactive manner, improving mass communication's publicity, fluidity, and virtual reality (Xie, 2021)[22].

In the more complex area of traditional folklore performing arts, the Terpsichore project's researchers (Ziagkas et al., 2019)[25] utilised the Vicon3D recording technology (Nexus Vicon, Oxford, UK) to recognise and catch two regionally specific traditional dances from Greece. They also improved the identification of subtle differences and the level of visual presentation through the data analysis of 3D kinematics. The "Syrtos Kalamatianos" and the "Syrtos Makedonikos" are two of the most popular folk dances throughout Greece and have the same rhythmic pattern. In their work to identify these two similar intangible cultural heritage traditional dances, they found that analysing the average displacement of the dancers' left toe reflex marker trajectories was a powerful tool.

It is a ground-breaking discovery that investigates various folklore performances in Europe. Furthermore, it subsequently supports numerous initiatives to preserve and present intangible cultural heritage by offering cost-efficient and semantically enhanced digital content of intangible traditional dances to various related industries and user groups. These positive results, therefore, seem to support further that digital technology, led by advanced information concepts, has continuously expanded the conservation and presentation of intangible cultural heritage into a higher dimensional digital space (Sang et al., 2018)[12].

Although it is almost certain that technology has dictated changes in the digital sphere, giving digital communication of intangible cultural heritage an increasingly high-tech power (e.g. augmented reality, virtual reality, artificial intelligence, blockchain) (Pérez-Seijo and Vicente, 2022)[10], it has accordingly come to dominate the paradigm shift (Manovich, 2013)[8] of digital intangible cultural heritage.

3 Disadvantages of technology: technological domination

The media imperative of technology in the digital transmission of intangible cultural heritage is constantly being amplified, leading to a tendency for digital technology to shift from technology-mediated to technology-led. Yang Gao (2020)[23] has argued that technology currently dominates intangible cultural heritage's digital production and dissemination, as it permeates the entire process from the collection and storage of intangible cultural heritage cultural elements to the design and dissemination of the displays.

The digitisation process of intangible cultural heritage can be divided into four stages: information collection, information storage, processing and dissemination, with each stage of intangible cultural heritage undergoing a different degree of digital transformation. Technicians frequently gather pieces of intangible cultural heritage from the regions where they originated and preserved them primarily as audio-visual works and written materials through interviews and recordings. Then, they are digitally processed for a variety of uses, including the development of open and interactive intangible cultural heritage digital museums, online games, documentaries, movies, and industrial design goods. The ability to digitise, network, and virtualise of intangible cultural heritage elements being exploited are necessary for these goals. So that web-based dissemination and offline expansion into an industrial chain for industrial production and administration are made possible by these capabilities. The specific digitalization process is shown in Figure 3.

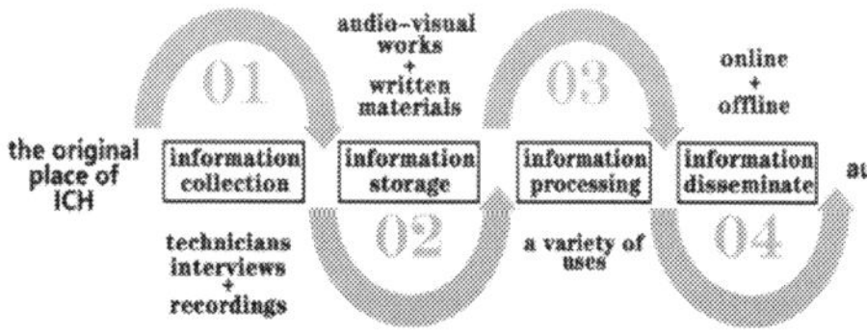

Figure 3 The digitisation process of intangible cultural heritage

An implication of this is the possibility that intangible cultural heritage elements need to be filtered, organised and reproduced by technology so that they can be digitally disseminated in the end, which creates a contradiction between technological dominance and cultural dominance in a given technical space. The single digital collection and storage, the processing and dissemination for another purpose, the technical teams that are not heirs, each of these links are mostly detached from and ignore the specific cultural spaces and groups of heirs on which intangible cultural heritage depends.

Similarly, Wang (2009)[17] found that because of the inherent nature of technological means to re-present information texts, this allows technology to act as a medium that not only disseminates culture but also creates new forms of cultural texts.

Regarding communication theory, the media pessimists followed the Frankfurt School's critique of technological rationality (Meng, 2012)[9]. In his book 'Amusing Ourselves to Death', the American scholar Neil Postman has suggested that every medium reinvents cultural content. It provides a new orientation to the ways of thinking, and expressing ideas and emotions, thus creating unique semiotics of discourse.

The Canadian scholar Marshall McLuhan also calls the "medium-as-message" theory, in which he argues that the medium influences our habits of understanding and thinking. On this basis, the digital technology medium transposes "content" to various smart devices and concepts. Each person views it becomes a network interface in the information society, acting as a neural mediator in the ecological web of the information network.

Neil Postman (2011)[11], as a media pessimist, argued that media technology should be treated with deep caution and that every technological innovation is a product of both advantages and disadvantages. It can be said that in the process of digitizing intangible cultural heritage, even the same textual information has undergone a series of effects of decoding and encoding by technical teams, re-creation effects by technical media and differential interpretation by end-users. It is also difficult to keep its original discourse power again.

In the Marxist theory of spatial studies, Henri Lefebvre considered the technical domination of a given space as an "abstract space with no substance of life, a solitary and closed space without interactive subjectivity or the possibility of interaction" (Liu, 2006)[7]. During the presentation of intangible cultural heritage, the space where the participant lives are no longer the outside space in which we typically live but rather a space intertwined with one or more technical methods and organically dominated by technology.

Thus, considering all of this evidence from various fields, it seems that the tendency towards technological dominance in the digital dissemination of intangible cultural heritage raises the question of touching the essence of intangible cultural heritage culture.

4 Technological dominance leads to emotional drain

Technological dominance has led to the loss of cultural sensibility in the digital communication of intangible cultural heritage. As intangible cultural heritage transitioned from being dominated by its native environment and inherited community to being overwhelmed by digital technology and technology users, it also became a technologically reproduced culture that could be stored forever and played back at any time (Wang, 2009)[17]. This means that under the domination of a specific technological space, digital intangible cultural heritage greatly loses its link to the original culture and its vitality and emotional value (Gao, 2020)[23].

For example, the performing arts of intangible cultural heritage, such as dance, song, drama, and gesture, rely on the ever-changing body language of the performer and the act of interaction with the audience to convey the emotions outwardly and cultural values contained therein. In turn, the performers activate their bodies and stimulate their emotions through live performance, again and again, conveying a sense of presence and gaining inner vitality to maintain their behaviour. These performative arts can be digitally recorded for mass distribution and cultural circulation, as the Figure 4 shows.

 https://doi.org/10.1051/shsconf/202315501008

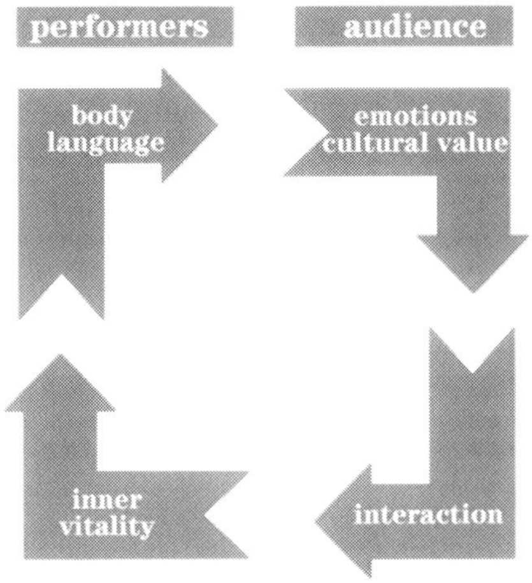

Figure 4 The interaction effect between the performers and audience

However, under the limitation of technical space and the dislocation of time and space, both audiences and performers lose the real interactive experience brought by the scene, which makes culture correspondingly lose its dynamic development.

Other scholars (Hou et al., 2022)[4] have also found that some digital display technologies that reproduce intangible cultural heritage seem to emphasise surface forms and compositions more closely, ignoring their human connotations. For example, China's Wuhan "Han Show" theatre, currently China's highest investment tourist performing arts project, but because of the excessive worship of technology and the overuse of high technology, the cultural connotation of the stage show is insufficient, and the audience's sense of experience and emotional interaction with local traditional culture is not satisfied.

In this regard, Adorno has argued that the technology that dominates the discourse is, on the one hand, trying to conceal the contradictions between subject and object in society and culture. In modern capitalist societies, the cultural industry turns all works of art into commodities, replacing use-value with exchange-value. On the other hand, most people could appreciate mass-producing models of cultural products, solidifying ideas and emotions and stifling individuality, and ultimately submerging culture in increasingly elaborate technological forms, superficial aesthetic tastes, and fashionable consumer behaviour (Wang, 2009)[17]. It allows people to enjoy a pleasant cultural sensation while covertly manipulating their physical, mental, and even subconscious activities to repeatedly engage in the familiar and safe consumption of the same products.

As a result, in the face of technological bullying, it is increasingly difficult for people to feel the emotional impact of intangible cultural heritage, which often poses problems for developing digital intangible heritage products.

5 Emotional drain is detrimental to the sustainability of intangible cultural heritage

Emotional loss is not conducive to the sustainability of digital communication of intangible cultural heritage. This is still a source of debate, but the generally accepted theory is that the essence of intangible cultural heritage is a people-oriented living culture (Li et al., 2016)[24]. It relies on the existence of people and has a unique way of transmitting continuity. It is handed down in verbal form and remembered in mind in life, passed on from generation to generation through the language and behaviour of people in a specific space and time. It is the accumulation and accumulation of the long-term life of the people of a region, reflecting a regional spiritual world, spiritual map, and values. It has a special humanistic significance for the communities and groups concerned, i.e., it provides them with the emotional value of 'identity and continuity. This emotional quality frequently transcends its aesthetics, artistry and other externally perceivable qualities and is a crucial motivation passed on and transmitted for generations.

Therefore, when intangible cultural heritage digital products are novel but not culturally emotive, it would not be easy to convey the cultural connotation to the outside world, thus resulting in cultural aphasia. Without cultural appeal, intangible cultural heritage products will conceivably not impress the viewers and potential inheritors (Fanzhuo et al., 2021)[3], which might not favour the sustainable development of digital communication of intangible cultural heritage. Due to the uniqueness of intangible cultural heritage, their existence inevitably depends on the community of inheritors, without which there would be no intangible cultural heritage. In turn, the inheriting community is more concerned with intangible cultural heritage's spiritual and emotional values (Li, 2012)[6].

In spite of the fact that digital technology is already being used in many places to preserve intangible cultural heritage, the inheritors of traditional crafts frequently lack the education necessary to comprehend the digital technology involved, as their level of education is usually not high. Besides, it is difficult for them to adapt to the fast pace of modern life. This often leads to a gradual loss of voice and participation in the digitisation of intangible cultural heritage by the community of heirs, which is detrimental to the digital dissemination of intangible cultural heritage.

According to scholars in Experimental Museology at the Swiss federal Institute of Technology in Lausanne (Hou et al., 2022)[4] have found that, in contrast to tangible heritage, the manifestations of intangible cultural heritage are generally defined by specific human practices and that current digital tools might be doubted to make little effort to demonstrate the 'living' nature of intangible cultural heritage fully. Tangible cultural heritage generally refers to all traces of human activity in our physical environment, such as buildings, monuments, artworks, and natural heritage (Swensen et al., 2013)[16] . Physical displays are the primary method of displaying physical cultural material and are typically seen in museum galleries (Wang, 2011)[18]. Traditional or digital static displays provide the spectator a face-to-face, even tactile, experience with their forms, colours, and features. This contributes to the materiality of tangible cultural heritage.

While intangible cultural heritage is a culture that 'lives' in a particular context, with characteristics such as vividness of performance, environmental originality, and territorial practice. They primarily include a variety of socio-historical and cultural practices, performing arts, festive customs and other cultural life systems and skills. Different from the tangible heritage which emphasizes the display and dissemination of materiality, intangible cultural heritage focuses on the presentation and communication of emotional and spiritual values. It is essential for researchers to understand that intangible cultural heritage integration with technology and media is only a way of presentation. Many intangible cultural heritage digital initiatives have become formal and superficial as a result of the drive for integration with new technologies and the overemphasis on this shiny technical veneer to the cost of their maintenance and sustainable development.

Taken together, recovering the substantial knowledge and intangible feelings of intangible cultural heritage in a meaningful digital communication environment is now a significant challenge for the sustainable development of digital intangible cultural heritage.

6 Measures to promote the sustainable development of intangible cultural heritage

The sustainable development of digital communication in the intangible cultural heritage should be centred on communicating cultural emotions supported by technology. The most crucial should be respected for the intellectual ownership of the heir community. The theoretical starting point for proposing this measure is that the person is the leading vehicle of inheritance, so reflecting on and acknowledging the fundamental role played by the person is key to grasping the specificity of intangible cultural heritage communication (Wulf, 2020)[20]. By interacting with their surroundings, people build skills and knowledge for a variety of circumstances, whereas movements and gestures are common forms of communication for expressing and processing wisdom (Hou et al., 2022) [4]. It could be argued that the accurate contact of these relies on displaying static decorative patterns and unique craftsmanship, more significantly, on sensory experiences such as smell, hearing, touch and even temperature. Suppose these people-related emotional elements of intangible cultural heritage are preserved for the most part. In that case, the drawbacks of traditional digital design, which overemphasises technical expression, could be overcome, thus most possibly achieving an organic blend of rationality and emotion and effectively reducing the distance between products and people (Huang, 2020)[5].

In line with this idea, many scholars are using motion capture (Mocap) technology to document the movements and performances of performative cultures. Beyond the frame and inside the volume, mocap enables neutral and synchronous data collection (Delbridge, 2015)[2], thus making the human presence and the emotions expressed by the person more salient in a virtual digital environment, as confirmed by Chalmers et al. (2021)[1] in their study.

Another example is Stavrakis et al. (2012)[15] conducted. They used mocap motion capture technology to record and archive high-quality movement data from Cypriot folk dances, creating a publicly accessible folk-dance archive. The implications of this research not only preserve this intangible cultural heritage through digital technology but also raise awareness of its dance heritage among local communities, teaching it to younger generations additionally. The project team has also developed a 3D video game as a novel means of teaching dance that anyone can use at home. The game uses 3D virtual reality technology to guide the user through the necessary body postures and movements, using pre-recorded templates of Cypriot folk-dance movements tied to the virtual character using motion capture technology. Users can select the dance they want to learn and perform along with the virtual teacher, while the user can also see visual feedback of their movements, which is done using Kinect real-time capture technology.

These studies have tried to consider the interactivity between people and objects in intangible cultural heritage culture, the powerful forces that emerge from human interaction, and the incorporation of environmental elements. Arguably, this allows the community of inheritors to express ideas and feelings effectively while simultaneously making the individuality and style of the intangible cultural heritage culture more distinct (Ziagkas et al., 2019)[25]. The inheritors, as one of the core cultural elements of the intangible cultural heritage, would like to accept the logic of digital technology and move from the native cultural environment into the digital technological display space because they expect to enhance the cultural vitality in the outside socio-cultural area.

Therefore, it may be essential to clarify the subjectivity of the inheritors in the digital communication of intangible cultural heritage, use digital technology as a means, and insist on communicating cultural emotions as the core appeal.

7 Conclusion

In conclusion, this paper has considered technology's positive and negative roles in the digital transmission of intangible cultural heritage. It has also indicated that technological dominance is detrimental to the transmission of cultural emotions when technology is over-relied upon. The sustainable development of the digital communication of intangible cultural heritage may not favour this scenario. In response to the role of technology in digital communication, this paper has argued that it might be possible to facilitate the digital preservation and presentation of intangible cultural heritage. However, current technology's limitations have led to technology dominating intangible cultural heritage communication, divorcing culture from people and its native environment. This slows emotion transmission and new cultural forms' healthy, timely growth.

This paper, therefore, attempts to suggest ways to squarely face the status of inheritors as "knowledge subjects" and to make the communication of human emotions the primary objective of digital technology. In this way, there is hope that the disadvantages of digital technology could be discarded while simultaneously affirming its value. Combining the two allows for a more robust and accurate expression of cultural emotions in digital communication, aiming to promote the sustainable development of intangible cultural heritage.

This paper is still at the stage of theoretical exploration and needs to explore in depth in a multi-disciplinary context. Further research should focus on the living essence of intangible cultural heritage, the interaction between people and the environment, and emotional factors of intangible cultural heritage. This would greatly help to examine the links between its sustainable development and technology more closely.

References

1. Chalmers, A. et al. (2021) "Realistic humans in virtual cultural heritage," *Emerging Technologies and the Digital Transformation of Museums and Heritage Sites*, pp. 156–165.
2. Delbridge, M., (2015). Motion capture in performance: an introduction. Springer.
3. Fanzhuo, D. et al. (2021) "Emotional drain in digital design for inheritance of Intangible Cultural Heritage," *E3S Web of Conferences*, 236, p. 05029.
4. Hou, Y. et al. (2022) "Digitizing intangible cultural heritage embodied: State of the art," *Journal on Computing and Cultural Heritage*, 15(3), pp. 1–20.
5. Huang Hui. (2020) "On the Design Programming Method of Interactive Products under Emotional Design [J] ". *Popular Literature and Art*, (14): 106-107
6. Li, L., (2012) "Does intangible cultural heritage law resolve everything in China," *J. Int'l Com. L. & Tech.*, 7, p.355.
7. Liu Huaiyu. (2006). The Mediocrity and Miracle of Modernity: A Textual Interpretation of Lefebvre's Critical Philosophy of Daily Life. Central Compilation Press, Beijing.
8. Manovich, L., (2013). Software takes command (Vol. 5). A&C Black.
9. Meng Jun. (2012). "From techno-rational critique, technological supremacy to technological dominance - On the impact of technology on media power." *Contemporary Communication* (1), 3
10. Pérez-Seijo, S. and Vicente, P.N. (2022) "After the hype: How hi-tech is reshaping journalism," *Studies in Big Data*, pp. 41–52.
11. Postman, N. (2011). Technopoly: The surrender of culture to technology. Vintage.
12. Sang, X. et al. (2018) "Interactive floating full-parallax digital three-dimensional light-field display based on wavefront recomposing," *Optics Express*, 26(7), pp. 8883–8889.
13. Skublewska-Paszkowska, M. et al. (2022) "3D technologies for intangible cultural heritage preservation—literature review for selected databases, " *Heritage Science*, 10(1), p. 3.
14. Sorea, D. and Csesznek, C. (2020) "Intangible Cultural Heritage of the Făgăraş Land. Priority Dimensions," *Bulletin of the Transilvania University of Braşov. Series VII: Social Sciences • Law*, pp. 155–164.
15. Stavrakis, E. et al. (2012) "Digitization of Cypriot Folk Dances," *Progress in Cultural Heritage Preservation*, pp. 404–413.
16. Swensen, G. et al. (2013) "Capturing the intangible and tangible aspects of heritage: Personal versus official perspectives in Cultural Heritage Management," *Landscape Research*, 38(2), pp. 203–221.
17. Wang Ailing. (2009). "Media culture: The paradigm of mass culture dominated by 'technological rationality," *Journal of Culture* (2), 5.
18. Wang Weihua. (2011). "Digital and Physical Exhibition of Cultural Heritage in Museums," *Southeast Culture* (05), pp. 91-95.
19. *What is Intangible Cultural Heritage?* [online]. Available at: https://intangible cultural heritage.unesco.org/en/what-is-intangible-heritage-00003. (Accessed 10 Sep 2022)
20. Wulf, C. (2020) "Performativity and dynamics of Intangible Cultural Heritage," *Ritual, Heritage and Identity*, pp. 76–94.
21. Xiao Bo, & Qian Shan. (2018). "Technology Cult and Cultural Dissonance in the Tourism Performing Arts Industry - The Case of Wuhan's Hanshow," *Journal of Tongji University: Social Science Edition*, 29(1), 10.
22. Xie, R. (2021) "Intangible Cultural Heritage High-Definition Digital Mobile Display Technology Based on VR Virtual Visualization," *Mobile Information Systems*, p. e4034729.
23. Yang Gao, Peng Chen. (2020) "Technology Dominance and Emotional Zero: Reflections on the Digital Technology Ethics of 'intangible cultural Heritage," [J]. *Guangxi Social Science*, No.301 07 p,140-145.
24. Zhang Li, Zhu Rong, Niu Chao and Shao Siyu, (2016) "Research on the Protection and Communication of Intangible Cultural Heritage in 3D Digitalization: A Case Study of Huishan Clay Figurines[J]," *Decoration*, no.280, pp.126-127.
25. Ziagkas, E. et al. (2019) "Greek Traditional Dances Capturing and a Kinematic Analysis Approach of the Greek Traditional Dance "Syrtos" (Terpsichore Project)," *In Interactive Mobile Communication, Technologies and Learning*, pp. 514–523.

The Development and Application of Digital Media Interaction in Artistic Experience

Yu Huang*

300 Garrison Forest Rd, Owings Mills, MD

Abstract: In this era of scientific and technological progress, digital media has made great contributions to the world and modern art and design works. The Internet has helped many artists make their work more visible and accessible to audiences around the world. Advanced technology has also enabled artists to transform and manipulate their work, thus becoming an important artistic medium. This paper summarizes the characteristics of interactive technology and new media, and analyzes the development and application of new media interactive art in experience.

1. Introduction

Interaction design is a kind of new media design for interactive products. In this kind of design, the designer's concern should include not only the product in development, but also the way users interact with the product and experience the effect. Therefore, interaction design needs to pay close attention to the user's needs, limitations, and environment. The background of the information age promotes the rapid development of new media interactive art. New media interactive art combines modern photography with the Internet and other technologies. The emergence of new art interacting with media not only reflects the progress and development of media technology, but also reflects the great progress of media art, which provides more ways for the development of art and is conducive to improving the diversity of people's artistic and cultural activities.[1] Interactive art is a virtual space that provides interactive experiences for artists and engineers through technology platforms. Interaction takes place in different areas, leading people to create it in different ways. This form of embodiment can not only be applied to many commercial fields in modern society, as an important means of dissemination of traditional culture in the cultural field, but also can bring people a sense of life experience full of science and technology.

Young people's entertainment programs can often reflect the development trend of new media in today's society. The three most popular new media experiential entertainment items among young people in the 21th century are games, animated generation, and virtual wear. These three main entertainment items can not only provide people with a variety of virtual experiences, can let people follow their daily life, but also can bring a lot of convenience to life.[2]

*Corresponding author: yuhuang@gfs.org

2. Interactive game experience

2.1. Introduction

Interaction design determines how people interact with computers and communications. It determines the value of the media to the user and the quality of the experience the user will have when using it. Young people love and even become addicted to games not only because of the relaxing animation and interesting mechanics, but also because of the sense of unlimited freedom gained in games. The eleven key game design factors include game goals, game mechanics, game fantasy, game value, interaction, freedom, narrative, feel, challenge, sociability, and mystery; Interaction design should consider all the above factors and combine them effectively.[3]

2.2. Research Status

Highly interactive games have many of the same complexities as the real world, while allowing us as researchers to conduct tightly controlled scientific research. Because games are fundamentally an interactive medium, we see them as tools for studying people, cognitive processes, decision-making, and behavior. We seek to understand how people behave in games, how the actions they take in games relate to cognitive and behavioral processes, and how players' personal characteristics influence the games they choose to play (and how they play). Much of the work in this area draws on concepts from social and cognitive psychology as inspiration, but our main tools are still AI and ML techniques.[4]

In addition, because games are a low-cost approach to behavioral research, and because the analytical data collected in games is often very extensive, there is a unique opportunity to apply machine learning and data mining tools. As the number of players reaches 100,000 or more, commercial game companies can collect vast amounts of data that describe the player experience. One of the ways we do research in this area is to use these massive data sets to gain insight into how players behave in games. To this end, we design new data mining and machine learning techniques. These algorithms can provide automated reasoning tools. Recently, we developed a "semantically interpretable" technique that not only makes accurate inferences about players, but also provides insight into why those inferences are made. This semantic interpretability can help us understand the game qualities that make a player unique, that predict cognitive processes, or that explain player motivation.[4]

2.3. Typical Case Analysis

On the basis of in-depth analysis and research of the key features of motion sensing game and Flash AS programming technology, the designer team has developed a motion sensing game based on camera sensing. They use the camera to accurately capture the user's body movements, collision detection, control functions, scoring functions and exit mechanisms. During the game, users can get rid of the limitations of mouse, gamepad and keyboard, operate the game through their body movements, and interact with the game content in an immersive way. It's a technology that allows players to use their bodies, free of gamepads and all the extra finishing tools, to have a fully immersive experience. The technology and principle of the system are respectively using the camera to obtain the user's body action part through the use of Flash AS file to realize the interaction between the camera and the user's body action, for the camera video motion detection and other related processing can be created accurately.[5]

2.4. Application Value

The combination of game design and interaction design generally changes in two ways. The first is convenience: whether the player knows how to play the game, whether it takes too long to wait, whether the game mechanics are hard to learn, or whether the game objects are awkward to use are crucial to the experience. The second is accessibility: can the player understand the content of the cutscene, or can they hear the sounds of other characters interacting (for example, can a hearing-impaired person feel the sound of footsteps creeping up behind them during the experience), Or whether the game supports the use of specific input devices (such as some one-handed controllers or joysticks that allow players with severe physical disabilities to play). In order to improve the user experience, the highly interactive motion-sensing games have come into being and attracted more and more attention. Whereas traditional games are played with a mouse, gamepad, or keyboard, motion-sensing games with interactive design interact with the game through the user's body movements. This not only breaks through the traditional interaction mode, but also creates a new form of game based on the development of interaction technology. The realization method is to capture the user's body movement through the camera to complete the motion control of the game. It gets rid of the traditional interaction mode, and gives the player a new interactive experience. For the user, this category of motion-sensing games not only improves the efficiency of interaction in the game, but also greatly improves the immersion of the game. This is a very attractive point for modern young people.[6]

3. Animation image generation

3.1. Introduction

Functional animation is a subtle animation that has a clear logical purpose. It reduces cognitive load, prevents change blindness, and builds better recall of spatial relationships. This animation technique brings the user interface to life. In the profession, it is an art form capable of telling stories and conveying multiple emotions and messages. Animation is used almost everywhere: from advertising to movies to games. There are many different types and styles of animation generation :2D, 3D, motion graphics, visual effects. Each category may have multiple workers using their own special skills to make the final product.[7]

3.2. Research Status

Animation generation has developed as early as the last century. The first 2D animation was created by Emil Cole in 1908 and was called Fantasy Orie. According to Kieran Kane's lecture "The History of Animation", the animated film is 70 seconds, 700 images, 24 frames per second to make the ending. In 1928, Walt Disney created the first Mickey Mouse cartoon, Steamboat Willie, in 2D animation. It was also the first animated film to feature History as its soundtrack.[8]

3.3. Typical Case Analysis

The most famous character in animation must be Walt Disney's Mickey Mouse. As the opening animation of many movies today, Disney's Steamboat Willie is a landmark in the history of animation. It was the first film about the character Mickey Mouse, and the first animated feature with synchronized sound. It made silent animation obsolete and ushered in a whole new era of empire. The film opened at the Colonial Theater in New York City on November 18, 1928, which became known as Mickey Mouse's birthday. The earliest technology was not enough to make animation smooth; Instead, a short film is made by designing the movements of an animated character frame by frame.[8]

3.4. Application Value

Animation is important because it allows us to tell stories and communicate emotions and ideas in a unique, palpable way that both children and adults can understand. Animation helps connect people around the world in a way that no other field can. Nowadays, anyone can pick up a drawing board and present their ideas to the world, or have machines produce their own images. Characters can have a variety of styles to make the audience feel more comfortable. As an animated character, this character feels like their own existence.[8]

4. Virtual Wear (VR)

4.1. Introduction

As the diversity of interaction design changes, there has been a flurry of controversy on the Internet about virtual wear. Virtual wear, as the name implies, is to use online digital design and 3D technology to create some goods to meet some people's imagination of the future. Artists will sell their futuristic garments and trade them across platforms; Users who buy them are then given a stylish virtual "look". Virtual fashion is the new trend, and it goes beyond the online fashion shows that quickly became the norm during the pandemic. Virtual fashion is not physically created streaming content -- including fashion shows, interviews and the clothes themselves -- but emanates from the digital space.

4.2. Research Status

Fashion design is about planning, thinking, and planning. It also includes graphics, drawing, and modeling. Virtual clothing design is to use the combination of computer technology and mathematical engineering. It is a wide range of fringe disciplines, and literature, art, history, philosophy, religion, aesthetics, psychology, physiology, ergonomics and other social and natural sciences are closely related.[8]

4.3. Typical Case Analysis

The most famous virtual clothing stores will be Carlings and The Fabricant. The Norwegian fashion house Carlings is the most advanced fashion brand among them. In November 2018, when virtual fashion was first gaining popularity online, the company launched and sold a fully virtual collection called "Neo X" . The collection contains 19 fashionable clothing and the cost only ranges between \$9 and \$30. Their imaginative designs have been warmly welcomed by fashion lovers and Internet celebrities alike. Founded in 2018, The Fabricant is another important frontrunner in The virtual fashion space. Although the price is much more expensive. However, of the creative design and outstanding concept, it is still very popular among young people. The brand gained recognition in May 2019 when it sold its rainbow-colored dress for \$9,500. Since then, the brand has gone from strength to strength through successful partnerships with industry giants such as Tommy Hilfiger and Soorty.[9]

4.4. Application Value

Virtual fashion is a trend these days. Many wealthy people choose to pay modelers to design their own virtual fashion. But beyond that, virtual fashion gives people a new way to choose clothes. In the apparel industry as a whole, electronic tailoring is bound to take off because it's hard to find clothes that fit you on the shelf. However, nowadays, many clothing stores on the online platform provide more and more limited services to people. It is difficult for consumers to buy clothes that fit them, and the situation of returning goods is even more serious. So many consumers still prefer to buy clothes in physical stores; If we can use the virtual reality environment of 3 d garment design, consumers can online clothing stores the freedom to choose their favorite clothes and try it on, and then select the dress according to the size of the virtual human body model of consumers consider recommending dress, size, let the consumer can choose the more suitable for their own clothes.[9]

5. Conclusion

The biggest difference between interaction design and the general design profession is that it is more focused on the external elements of outcome, environment, action, and delivery of a specified message. The main purpose of interaction design is first to plan and specify how information will work, and then to provide effective ways and methods to facilitate the public. From the perspective of the experiencer, interaction design is a project to facilitate and understand the user's goals. It's a much more user-friendly, comfortable technology.

Reference

1. G. N. Yannakakis and J. Togelius, "A Panorama of Artificial and Computational Intelligence in Games," in *IEEE Transactions on Computational Intelligence and AI in Games*, vol. 7, no. 4, pp. 317-335, Dec. 2015, doi: 10.1109/TCIAIG. 2014.2339221.
2. Liu Junjie, Ding Siqi, and Lin Dongmin." Design and Development of motion sensing Games Based on camera Sensing." Wireless Internet Technology 5 (2020).
3. Babich, Nick. "Functional Animation In UX Design: What Makes a Good Transition? 2017 (accessed December 31, 2017)." *URL: https://uxplanet. org/functional-animation-in-ux-design-what-makes-agood-transition-d6e7b4344e5e.*
4. Lieberman, Debra. (2006). What Can We Learn From Playing Interactive Games?.
5. Marx, Christy. *Write your way into animation and games: create a writing career in animation and games*. Routledge, 2012.

6. Bell, Desmond. "Practice makes perfect? Film and media studies and the challenge of creative practice." *Media, Culture & Society* 26, no. 5 (2004): 737-749.
7. Liu, Huabai. "Computer 5G Virtual Reality Environment 3D Clothing Design." *Mobile Information Systems* 2022 (2022).
8. Jung, Jaesuk, Jihye Yu, Yuri Seo, and Eunju Ko. "Consumer experiences of virtual reality: Insights from VR luxury brand fashion shows." *Journal of Business Research* 130 (2021): 517-524.

Research on the Path of Promoting the High-quality Development of State-owned Enterprises in the New Era

Xin Li*

Beijing Normal University, Beijing, China

Abstract. High quality development is an important proposition put forward at the 19th National Congress of the Communist Party of China and a new requirement for all-round economic development. As an important material and political foundation of socialism with Chinese characteristics, whether state-owned enterprises can make substantial steps in high-quality development is of great significance for building a modern economic system, promoting high-quality development of regional and national economies, promoting the transformation of growth rate and quality status, replacing scale with efficiency, and upgrading power.

1. Introduction

The new normal of China's economy is the symmetrical state of economic structure and sustainable economic development based on the symmetrical state of economic structure, including sustainable and stable economic growth. The new economic normal is an economy that emphasizes structural and steady growth, not an aggregate economy; it focuses on the symmetry of economic structure and sustainable development based on the symmetry, not just GDP, per capita GDP growth and maximization of economic scale. The new economic normal is to use growth to promote development, and use development to promote growth. The new economic normal is not the absence of GDP, but the absence of the GDP growth mode; not the absence of growth, but the positioning of GDP growth in the development model, so that GDP growth becomes part of the regenerative growth mode, productivity development mode.

The transition from high-speed growth to high-quality development is a major feature of economic development in the new era.

Since China became the second largest economy in the world in 2010, the Chinese economy has obviously taken on different characteristics. China has entered the middle and late stage of industrialization, and its economic growth rate is below 7%. This is determined by the cycle of the Chinese economy. This feature can also be verified from the industrialization process of developed countries.

The second feature of the new era of economic development is the strategic adjustment of the economic structure

Establishing the basic socialist economic system and promoting socialist economic construction have laid the institutional and economic foundation for China's leapfrog economic development. They have also provided valuable experience, theoretical preparation and material foundation for how to build a socialist economy with Chinese characteristics in the new historical period. Reform and opening up are integrated and mutually reinforcing. Development is the absolute principle and economic development is the inevitable requirement of development. We have thus established and developed a socialist market economy and implemented a series of all-round and three-dimensional reform and opening up measures, which have greatly liberated and developed the productive forces and become the most important driving force behind China's economic miracle.

In the new era, the Chinese economy has shifted from a stage of high-speed growth to a stage of high-quality development. Guided by the new development philosophy, we have followed the theme of high-quality development, deepened supply-side structural reform, accelerated the building of an innovative country, implemented the strategy of rural revitalization, implemented the strategy of coordinated development among regions, and accelerated the improvement of the socialist market economy. We will push forward major measures such as opening up on all fronts and winning the three key battles, shift economic growth from quantitative expansion to qualitative improvement, and maintain the general trend of steady and long-term growth.

State owned enterprises are an important force in building a well-off society in an all-round way, an important pillar of socialism with Chinese characteristics, and an important basis for our Party's governance. To develop the productive forces and realize the industrialization and modernization of the country, we must always rely on and give play to the important role of state-owned enterprises. Under the situation of economic globalization and accelerating scientific and

*Corresponding author: 1280538273@qq.com

https://doi.org/10.1051/shsconf/202315501010

technological progress, state-owned enterprises shoulder the responsibility and mission of leading high-quality economic development, and must actively practice and take the lead. Over the past 40 years of reform and opening up, the development of state-owned enterprises has made brilliant achievements with the rise of the country. State owned enterprises have become an important part of the national economy, an important support for key construction projects and play a leading role in leading and driving the development of key industries and key areas [1]. The 19th National Congress of the Communist Party of China put forward the requirement of high-quality development, which is the requirement of all-round economic development. For state-owned enterprises undertaking heavy responsibilities, they must deeply understand and actively practice the connotation and actions of high-quality development, draw impetus from deepening the reform of state-owned enterprises, accelerate the pace of transformation and development, optimize the industrial layout, improve investment efficiency and production efficiency, we will accelerate the development of state-owned enterprises in a new way.

2. The Importance of Promoting the High-Quality Development of State -owned Enterprises in the New Era

First of all, promoting the high-quality development of state-owned enterprises is a development that can well meet the growing needs of the people for a better life. It is a development that reflects the new development concept. It is a development in which innovation becomes the first driving force, coordination becomes an endogenous feature, green becomes a universal form, openness becomes a must, and sharing becomes a fundamental goal [3]. To be more specific, high-quality development means that economic development changes from "whether there is" to "whether there is". Since the 18th National Congress of the Communist Party of China (CPC), the understanding of the phased characteristics of economic development has been deepening. In 2013, the CPC Central Committee made a judgment that China's economic development is in the "three phases superposition" stage of growth rate shift, structural adjustment pain and early stimulus policy digestion. In 2014, it was proposed that China's economic development enter a new normal. The new normal is an objective state that will inevitably occur when China's economy reaches a certain stage of development. The great logic of China's economic development is to adapt to the new normal, grasp the new normal, and lead the new normal. Under the new normal, new changes have taken place in the environment, conditions, tasks, requirements, etc. of China's economic development. The growth rate should be changed from high speed to medium high speed, the development mode should be changed from scale speed to quality efficiency, the adjustment of economic structure should be changed from the focus on incremental capacity expansion to the focus on adjusting the stock and improving the growth simultaneously, and the driving force for development should be changed from mainly relying on the input of resources and low-cost labor and other factors to innovation driven [3]. The 19th National Congress of the Communist Party of China further made it clear that China's economy has shifted from a stage of rapid growth to a stage of high-quality development. These changes are the necessary process for China to promote the evolution of state-owned enterprises to a higher form, a more optimized division of labor, and a more rational structure.

Secondly, promoting high-quality development of state-owned enterprises is the fundamental requirement for determining development ideas, formulating economic policies and implementing macro-control at present and in the future. We must adhere to the principle of quality first and efficiency first, promote the quality, efficiency, and power changes of economic development, and constantly enhance economic innovation and competitiveness. We will accelerate the formation of an indicator system, a policy, standard system, statistical system, performance evaluation system, and performance appraisal system to promote high-quality development, to create and improve the institutional environment, and promote China's economy to make new progress in achieving high-quality development. Over the past 40 years of reform and opening up, state-owned enterprises have experienced the development process of market economy from the decentralization of power and interest within the planned economy system to the establishment of a modern enterprise system, and have made remarkable achievements and accumulated rich experience. However, at present, the development of state-owned enterprises still faces many problems. The outstanding point is that the overall industrial layout is "heavy", and the enterprises are lack of innovation motivation, vitality, and reform. They are still in the middle and low quality development stage. Therefore, we must firmly promote the key tasks of strengthening the main business, optimizing the layout, innovating the power, improving efficiency, deepening reform, and accelerate the implementation of the new development concept of state-owned enterprises, adapt to the development requirements of the new era, and achieve high-quality development.

Thirdly, there is the inevitable choice of high-quality development of enterprises in the new development environment, conditions and stages. At present, socialism with Chinese characteristics has entered a new era. China's economy has changed from high-speed growth to high-quality development, and it is important to focus on economic construction to rejuvenate the country [4]. Promoting high-quality development is an inevitable requirement for maintaining sustained and healthy economic development, adapting to changes in major social contradictions, building a moderately prosperous society in an all-round way and building a socialist modern country in an all-round way, and complying with economic laws. On September 7, 2018, The State-owned Assets Supervision and Administration Commission of the State Council pointed out in the report "Promoting the high-quality development of state-owned enterprises in a down-to-earth manner": China's economy has shifted from a stage of rapid growth to a stage of high-quality development. As the pillar of the socialist economy with Chinese characteristics, state-owned enterprises must strengthen their sense of responsibility and urgency, deeply understand the rich connotation and practical requirements of high-quality development, completely abandon the complex of scale and speed, and solidly promote the high-quality development of state-owned enterprises to achieve tangible results. We must unswervingly consolidate and develop the public ownership economy, adhere to the dominant position of the public ownership, give play to the leading role of the state-owned economy, deepen the reform of state-owned enterprises, develop a mixed ownership economy, cultivate world-class enterprises with global competitiveness, and promote state-owned capital to become stronger, better, and larger.

3. Key Problems to Be Solved in Promoting High-quality Development of State-owned Enterprises in the New Era

In the process of promoting the development of height and quality in state-owned enterprises, the following major problems have emerged or always existed in recent years:

First of all, the problem of state-owned enterprises becoming stronger, better, larger and let the market play a decisive role. This problem mainly exists in two aspects:

(1) The contradiction between state-owned enterprises becoming stronger, better, larger and protecting the market vitality of private enterprises. The state-owned enterprises are superior at the starting point, and a fair and reasonable competitive environment is far from being formed;

(2) The evaluation of core competitiveness should come from the market, not from the protection of the government. Data shows that in 2019, the total assets of state-owned enterprises were 233.9 trillion yuan, with a year-on-year growth of 11.1%. While in 2008 it was only 34.7 trillion yuan, with the scale expanding more than six times in just 11 years.

This report card can almost "crush" most non-state-owned enterprises. Although the relevant departments have explained for many times that making state-owned enterprises bigger and stronger does not mean that the country will advance and the people will retreat. But the so-called "big, strong and excellent" is a relative concept, and its benchmarking can only be non-state-owned enterprises. Making state-owned enterprises bigger and stronger means that private enterprises in this field will be smaller, weaker and worse to some extent. The essence of making state-owned enterprises bigger and stronger is to compete with the people for money." In addition, state-owned enterprises (especially central enterprises) have three major responsibilities: political responsibility, economic responsibility and social responsibility. One of the three is indispensable. On the three major responsibilities, the difficult points for state-owned enterprises to grasp are: first, state-owned enterprises should not only undertake economic tasks, but also fulfill social and political responsibilities. Second, the social responsibility of many state-owned enterprises is to assume the responsibility of the government to help the government serve enterprises and people. Third, it is necessary to clarify the function of state-owned enterprises in the national economy and give them a correct positioning and development environment.

Secondly, the problem between peeling off non main businesses and the extensive development of enterprises. At present, the reform of state-owned enterprises requires improving quality and efficiency, and the merger and reorganization of central enterprises presents a trend of stripping off non main businesses and focusing on main businesses. On the surface, it is conducive to getting rid of the historical burden and going into battle lightly. The boundary is clear and can be followed, but it runs counter to the general trend of increasingly extensive development of enterprises. According to the law of enterprise development, the super large enterprises have reached a certain stage, and the enterprise stage they are pursuing is the extensive development of enterprises, that is, they are beyond the existence of classic enterprises, form is scattered but spirit is not scattered, the organizational form is vague, there is no obvious boundary, and the function is generalized rather than weakened. In this regard, it is difficult to define which is the main business and which is the non main business. Stripping off the non main business and enterprise generalization may belong to different development stages, but the size and status of central enterprises will have to accommodate these two contradictory stages at the same time.

Third, the management incentive of state-owned enterprises. Entrepreneurs are the main body of the market. Both resource allocation and technological progress come from the exertion and application of entrepreneurship. Unfortunately, under the state-owned enterprise system, it is almost impossible to create entrepreneurs. Governments cannot replace entrepreneurs. How the state-owned enterprise leaders position and transform between the two specific roles of entrepreneurs and leading cadres of state-owned enterprises is a problem that every state-owned enterprise leader needs to face. Because the characteristics of the two are different or even completely opposite. Zhang Weiying, a famous economist, once said: "It is unlikely that the leaders of state-owned enterprises will consider the issue three years later. If you want to make him innovative, I think it is impossible". "The leaders of state-owned enterprises cannot be regarded as entrepreneurs, because they do not have to bear the final business risks, and their positions are always changing." From the comparison between the two, the leaders of state-owned enterprises can be clear: First, the fundamental attribute of the leaders of state-owned enterprises is the administrative attribute, the relationship with the party and government departments,

https://doi.org/10.1051/shsconf/202315501010

not the market attribute. Second, state-owned enterprises incentive mechanism is first the party and government incentive, from the deputy to the principal, to the department level and even the deputy ministerial level. In addition, there are the following three points: (1) personnel mobility and equity lock; (2) The contradiction between the tenure system of state-owned enterprise leaders and the medium and long-term incentive mechanism; (3) People flow and focus on innovation. Innovation needs a period of five years or twenty years to finally produce results. These three problems are also the pain points that need to be solved urgently in the management incentive of high-quality development of state-owned enterprises in the future.

Fourth, the market-oriented assessment of state-owned enterprises. At present, most of the listed companies of state-owned enterprises have not implemented truly market-oriented assessment. In the capital market, more attention is paid to net profit, earnings per share, operating cash flow, return on equity, industry attributes and development prospects. However, the assessment of listed companies of state-owned enterprises still focuses on operating income, total profits, economic added value, two funds, and the proportion of costs, and requires continuous optimization.

If the state-owned enterprises fail to effectively solve these four major problems, in order to truly promote the high-quality development of the state-owned enterprises in the future, they need to break through the shackles of the existing system and mechanism. The deep-seated, systematic and long-term contradictions still have a long way to go.

4. The Practical Path of Promoting the High-Quality Development of State -owned Enterprises in the New Era

1. We should promote the formation of an investment optimization system for building high-quality projects and high-quality investment.

The high-quality development of real enterprises starts from investment and depends on the quality of investment. Effective industrial investment is the main way for enterprises to grow and enhance their competitiveness. Therefore, it is necessary to strengthen the combination of enterprise investment and demand, and innovation, expand effective and innovative investment, increase increment, make living stock, improve quality, and add new strength. First, we should strengthen the industrial planning and project planning of state-owned enterprises, and plan a number of new leading growth points such as modern energy, major infrastructure, modern logistics, finance, and credit services around the city's industrial transformation and upgrading direction. Second, we should focus on effective investment, optimize the comprehensive investment layout of enterprises, reduce investment in backward industries and backward production capacity, orderly promote the construction process of projects under construction as planned, and ensure efficient capital operation.

2. We should promote the formation of a power transformation system for technological innovation, industrial innovation and business type innovation.

Innovation needs long-term accumulation. The majority of state-owned enterprises are traditional industry led development models, and the foundation for innovative development is weak. From now on, we need to establish the concept of innovation driven development, accelerate the cultivation of new drivers, and gradually get rid of dependence on traditional, resource-based, and basic industries. Among the new driving forces identified in the report of the 19th National Congress of the Communist Party of China, green low-carbon and modern supply chain are the key breakthrough directions in line with the actual development of state-owned enterprises. Focusing on green and low-carbon, we should do a good job in modern energy economy, such as clean coal, green thermal power, photovoltaic, wind energy, conventional and unconventional oil and gas resources development and pipe network facilities; It is necessary to accelerate the development of railway collection and distribution system, integrated logistics base, logistics information platform, big data application and other industries around the modern supply chain.

3. We should improve the formation of a production management innovation system for enterprises with strong quality, brand and credit.

Engineering and product quality are important manifestations of high-quality development, which determine whether an enterprise has market competitiveness. The brand determines whether the product has added value and whether the enterprise is in the leading position in the industry. Credit rating is an important embodiment of the comprehensive strength of state-owned enterprises, and an important guarantee for state-owned enterprises to expand cooperation, finance credit, deepen reform, and build corporate culture. Therefore, we should focus on the above objectives, accelerate the establishment of enterprise quality standardization system, build state-owned enterprise brand mainly by improving quality, increasing category and creating brand, and implement the requirements of social credit construction to shape the image of enterprise integrity and improve the credit rating.

4. We should enhance the formation of a system to improve the production factors of resources, technology and talents.

For state-owned enterprises, the key point of high quality development of total factor productivity is to recruit talents, cultivate high-level enterprise management teams, technical talent teams, and party building work teams; We should eliminate inefficient, backward and overcapacity, eliminate sunset industries, sick enterprises and zombie enterprises, and "vacate the cage for birds" for the development of emerging industries.

5. We should increase the formation of an institutional guarantee system focusing on the reform of state-owned enterprises.

Economic restructuring is an important guarantee for high-quality development. The reform of state-owned enterprises is the inherent requirement of high-quality development of state-owned enterprises. We should further promote the separation of state-owned enterprises from government and enterprises, between government and capital, and between ownership and management, implement enterprise autonomy in accordance with the law, improve the market-oriented operation mechanism of enterprises, and promote state-owned enterprises to truly become independent market players. We will optimize the overall institutional arrangements for state-owned enterprises, actively implement the top-level design of classified reform of state-owned enterprises, differentiate and promote the reform of the property rights system of public welfare and commercial state-owned enterprises, the reform of corporate governance, and the reform of three systems within enterprises, steadily promote the reform of mixed ownership of state-owned enterprises, and stimulate the internal vitality of state-owned enterprises. We will continue to optimize the distribution of state-owned capital and the structure of the state-owned economy, improve the management system of state-owned assets, accelerate the transformation of state-owned assets supervision from focusing on managing enterprises to focusing on managing capital, improve the efficiency of supervision, and ensure that state-owned assets maintain and increase their value.

6. To promote the high-quality development of state-owned enterprises in the new era, we must improve the management level of state-owned enterprises in an all-round way.

First, realizing power conversion. Power transformation is the starting point of high-quality development of state-owned enterprises. To realize the power transformation, it is necessary to change the enterprise development mode, and make its development power shift from mainly relying on resources, low-cost labor and other factors to innovation driven. Second, promoting efficiency reform. Efficiency reform is the key to achieve high-quality development of state-owned enterprises. To promote efficiency reform, state-owned enterprises should adhere to efficiency orientation and value orientation, make efforts in product production and service provision, enterprise operation, etc., and constantly improve the technical efficiency, resource allocation efficiency and management efficiency of enterprises, so as to achieve the overall efficiency improvement of enterprises. Third, improving innovation ability. Schumpeter, the American economist who first put forward the innovation theory, said in the Economic Development Theory that from one carriage to 10000 carriages is growth, and only from carriage to locomotive is development. Innovation is the driving force of development, and entrepreneurs lead innovation.

Improving the innovation ability is the basis of realizing the high-quality development of state-owned enterprises. To improve innovation ability, we should focus on cultivating high-quality innovative talents, increase R&D investment aiming at the world's scientific and technological frontier, and build a good enterprise innovation ecosystem. Be brave and good at carrying out management reform, eliminate the shortcomings of system, mechanism and management, vigorously promote the reform of enterprise organizational form, improve corporate governance, and build a flexible and efficient enterprise management model that conforms to industry characteristics and is conducive to enhancing core competitiveness. Fourth, improving the corporate image. Good corporate image is an important intangible asset for enterprises to achieve high-quality development. State owned enterprises should make greater contributions to safeguarding national security and the operation of the national economy, and developing forward-looking strategic industries. They should play a leading and exemplary role in promoting high-quality development, enhancing independent innovation capabilities, protecting resources and the environment, accelerating transformation and upgrading, and fulfilling social responsibilities.

5. Conclusion

This paper focuses on the analysis of the characteristics of the new normal of China's economic development, and on this basis, puts forward suggestions for the adaptation of State-owned Enterprises to China's economic development. The suggestions includes,

1.We should promote the formation of an investment optimization system for building high-quality projects and high-quality investment.

2.We should promote the formation of a power transformation system for technological innovation, industrial innovation and business type innovation.

3.We should improve the formation of a production management innovation system for enterprises with strong quality, brand and credit.

4.We should enhance the formation of a system to improve the production factors of resources, technology and talents.

5.We should increase the formation of an institutional guarantee system focusing on the reform of state-owned enterprises.

6.To promote the high-quality development of state-owned enterprises in the new era, we must improve the management level of state-owned enterprises in an all-round way.

References

1. Liu, Q. (2018) Research on the Nature and Role of State-owned Enterprises in the Reform of State-owned Enterprises. J. World of Labor and Social Security (Theoretical Edition), 8: 53-55.
2. Publicity Department of the CPC Central Committee. (2019) Study Outline of the Thought on Socialism with Chinese Characteristics for a New Era. People's Daily Press: 6.
3. Guangming Daily. (2015) Guiding Opinions of the CPC Central Committee and the State Council on

https://doi.org/10.1051/shsconf/202315501010

Deepening the Reform of State-owned Enterprises. J. 10: 5-10.

4. Li, Q. (2012) Research on the Corporate Governance Structure of State-owned Enterprises. J. Enterprise Guide, 22: 68.
5. An Yuhong（2014) The New Normal of Economy. J. Macroeconomic Management.@;23-34.
6. CAI Yuelan.(2016) Supply-side Structural reform: Adapting to and leading the new normal of China's economy. J. Finance and Economics, 10:12-23.
7. Hu Angang, Zhou Shaojie, Ren Hao. Supply-side structural reform: Adapting to and leading the new Normal of China's economy. J. Journal of Tsinghua University, 07:22-45.
8. Xi Jinping. Adhere to the New Development Concept to Fight the "Three Major Battles" and Strive to Write a New Chapter in the Development of Hubei in the New Era. N. People's Daily, 2018-04-29.
9. Zhang Peng. To Recognize the Development Characteristic of the New Era by the Characteristic of Economic Law. T. Finance and trade Economics.
10. Pan Yunliang. Grasp the Three Fundamental Features of the New Era of Economic Development. T. Small and Medium-sized Enterprises in China.

 https://doi.org/10.1051/shsconf/202315501011

Have Housing Prices Affected Residents' Level of Social Integration?

Yiping Li*

Faculty of Economics, Chulalongkorn University, Bangkok, Thailand

Abstract. In recent decades, most cities in China have experienced a large number of population movements, and the local integration of mobile populations has become a new challenge, not only in relation to their own quality of life, but also in relation to the development potential of the cities. While urban housing prices have soared, real estate has gradually become the main economic industry in China, and high housing prices have brought enormous housing pressure to this group. This paper analyzes the negative impact of house prices on social integration by combining labor market discrimination theory.

1. Introduction

Since 1978, China's urbanization process has undergone a steady rise and a rapid rise (Zhang & Song, 2003)[1]. According to the World Bank, China's urbanization rate has increased significantly from 17.9% in 1978 to 61.43% in 2020, and Kojima (1995)[2] argues that China's urbanization development is hyper-urbanization, especially in the 1980s, when China was even the fastest urbanizing country in the world. However, such urbanization inevitably leads to a series of problems, one of which is the rapid increase in housing prices. Figure 1 shows the price increases for different uses of houses in China over the past 20 years. We are more concerned about the price of residential housing for China's mobile populations, and the graph shows that although residential prices are lower than business and office housing, they still have increased significantly, with the average residential house price in China rising from RMB 1,948/m^2 in 2000 to RMB 9,980/m^2 in 2020. The real estate industry is booming, with its added value rising from 414.1 billion yuan (2000) to 7.5 trillion yuan (2020) and its share in GDP rising from 4.13% to 7.34%, making as one of the most important reasons for pulling China's economy. The mega-cities formed as a result of policy support and urbanization, such as Beijing and Shanghai, have seen their average residential prices rise 8 to 10 times in 20 years.

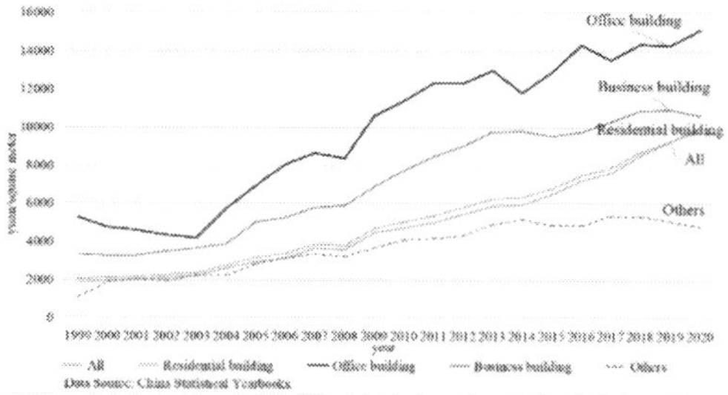

Fig. 1. Average house price by use in China. *Data Source*: China Statistical Yearbooks.

The other problem that comes with urbanization is the massive movement of people, especially those with rural household registration, to the urban areas. For decades, China's household registration system (hukou) has been deeply entrenched, and even after a comprehensive reform of the household registration system in 1990, it is still very difficult to switch and relocate hukou, and wage discrimination, employment discrimination, and education discrimination based on hukou are common (Song, 2014)[3]. Wu (2011)[4] demonstrates that hukou status affects educational achievement, with rural hukou inhabitants receiving fewer years of education than urban hukou residents. Liu (2005)[5] argues that rural populations have difficulty accessing schools in urban areas or they are unable to access the high cost of education to pay for urban education because the system becomes less accessible to urban employment.

The social bonds of human interaction are essential for maintaining individual psychological well-being (Turner & Turner, 1999)[6], and the economic, social and cultural exclusion of mobile populations is not only painful for them, but also a waste of social resources for the society as a whole, as the talents and strengths of these mobile populations are not fully utilized, and group conflicts are

* Corresponding author: 113212569055@gmail.com

caused. As a result, urban development will be slowed down or suffer from problems (Madanipour, 2015)[7]. Eraydin et al. (2010)[8] argue that immigrants contribute to production and services, bringing cheap labor and new social relations. The discrimination brought by household registration is a manifestation of economic and social exclusion, and such inequitable treatment and outcomes clearly reduce social integration significantly.

Soaring housing prices in China have highlighted the issue of social integration, which plays a significant role in urban development, and cities with higher levels of social integration have greater urban development potential. Therefore, this paper investigates whether house prices have an effect on social integration. Since existing studies related to house prices are mainly related to the economy, this finding fills the existing gap in regard to house prices and social integration and provides theoretical support for future exploration of the field.

2. Labor market discrimination theory

The existence of different degrees of discrimination between different groups in the labor market is a common phenomenon in all countries. In addition, compared to the discrimination that exists in other countries, household registration discrimination is a special but very common discrimination phenomenon in China. Discrimination theory provides a theoretical basis for studying urban settlement and willingness to social integration.

The discussion of discrimination can be traced back to the early 20th century when Fawcett (1918)[9] inspires subsequent research in this area by discussing the phenomenon of unequal pay between men and women for equal work. He believes that the cause of gender discrimination is due to social conventions and employer bias. Women are perceived to be incapable of performing technical work, even though they can actually do it well. Myrdal (1996)[10] studies the problem of discrimination against blacks by whites in the United States, which led to a vicious cycle of discrimination against blacks, and argued that if discrimination against blacks by whites was reduced, the situation of blacks would be improved, and he also proposed the principle of cumulative causation to explain this vicious circle, that is, the stronger the discrimination of whites against blacks, the lower the standard of living of blacks, and the lower the standard of living of blacks, which in turn will strengthen the discrimination of whites against blacks.

The modern theory of discrimination was developed by Becker. Becker (2010)[11] delves into unexplored areas of early discrimination economics, suggesting that the roots of discrimination stem from personal biases and developing an economic model that can be quantitatively analyzed. The individual bias model includes discrimination by employers, discrimination by employees, and discrimination by customers. Different employers may prefer to hire groups with certain types of characteristics, they prefer to pay higher employment costs and prices, and perhaps even the employees in that category are less productive or profitable than employees in other categories. Employee discrimination occurs when employees with certain characteristics exclude another group of employees they do not like, causing employers to pay them more or to segregate them professionally in order to retain them. Customer discrimination refers to the preference of customers for service workers, which causes those discriminated against to lower their wage demands or work harder.

China's household registration system is based on geographic and family member relationships and is directly linked to land and housing. The household registration disparity in China's labor market is an economic discrimination based on the household registration system, and it is the unequal treatment suffered by the disadvantaged groups of household registration in relation to labor market related employment resources, social security, etc. Although the household registration system is being reformed, the negative effects have not been eliminated on a large scale, such as the wage discrimination generated in the labor market. The Analysis and Forecast of Social Situation in Guangzhou, China (2011) reports that, according to a citywide sample survey of rural household registration migrant workers in 2010, 79.1% of migrant workers earned less than RMB 2,000 per month, with 16% of the respondents even earning an average monthly salary of less than RMB 1,030. More than half of the rural-named workers said they received unfair treatment, including discrimination in the labor market and discrimination in welfare resources, such as different pay for the same work, unreasonable fees for labor groups coming to the city from outside, and Companies do not pay social insurance for them. According to Fan (2020)[12], peasants who migrate to cities and towns but find it difficult to obtain urban hukou are considered as outsiders in the urban society, disadvantaged in the urban labor market, working in undesirable or even dangerous jobs, but with low or even below-average wages and not enjoying the benefits they deserve.

3. How do housing prices affect social integration?

a. High housing prices increase regional discrimination and thus reduces social integration. Regional discrimination is mainly due to the wealth gap, and high housing prices exacerbate this wealth gap.

In fact, the common residents rarely have the extra money to afford the cost of purchasing a property. According to China Statistical Yearbook 2021, the national per capita annual disposable income in 2020 is only 32,188.8 yuan, and the per capita consumption expenditure alone reaches 21,210 yuan. People who cannot afford housing keep their money in the bank, take out loans or rent, but the low interest rates on savings do not outweigh the price of housing, and the price to wage ratio is getting larger, shrinking wealth in comparison.

Fig.2 shows residential prices by region, and cities in the south have higher average prices and faster rising house prices, at RMB 13,085.8/m^2 in 2020. The high average house prices are due to higher house prices in Xiamen, Guangzhou and Shenzhen. Cities in the

 https://doi.org/10.1051/shsconf/202315501011

southwest and northwest regions have lower average house prices, at RMB 5512.5/m^2 and RMB 5343.4/m^2 in 2020, respectively. Thus, we can see that the regional appreciation of house prices varies.

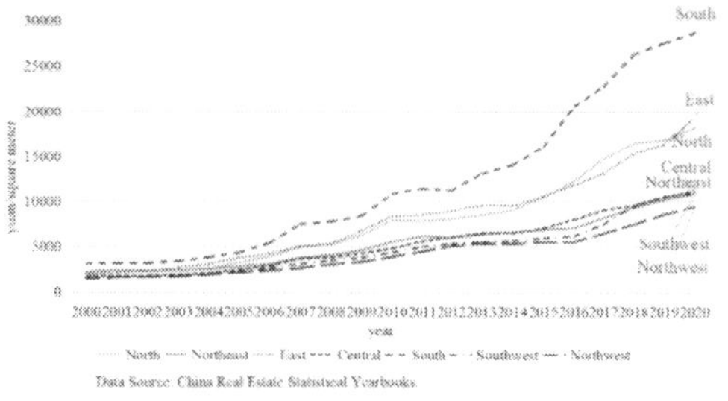

Fig. 2. Average house price (residential building) by region in China. *Data Source*: China Real Estate Statistical Yearbooks.

Wealthy people buy property because they have sufficient funds, and accumulate more wealth through rent collection, housing appreciation and resale, and reinvestment, the property premium will bring solidification of interests. In addition, there are already different degrees of regional discrimination in different cities, especially in some provinces and cities with advanced economic development, such as Jiangsu and Shanghai. And the rising property prices have become the material basis for their bragging rights, creating a sense of superiority among the locals. In these cities, the aborigines have an inherent advantage, not only the house prices have risen much more than most other cities, but also the wealth has skyrocketed due to housing demolition and other reasons. Then, the wealth gap has increased, regional discrimination has become more serious, and it has become more difficult for mobile people to integrate locally.

b. High housing prices make household registration conversion more difficult, thereby reducing social integration.

Due to the scarcity of social resources, the household registration system has become a screening mechanism for resource allocation in China. Policies vary from city to city; in some areas, people can only register in the city if they buy local property, and in others, they can only buy local property if they have a local household registration. Regardless of the policy regime, the bundling of property and household registration means that mobile populations must endure unfair treatment, even though they may have lived there for a long time. Sato (2006)[13] finds that rural populations come to the city and a new housing poverty emerges. After the Chinese government issued a policy of privatizing housing and providing subsidies to purchase housing, it was still difficult for them to own a house of their own, even though they had migrated to settle and work in the city for a long time. This is not only because of their low wages, but also because these people have difficulty qualifying for low-cost housing programs and subsidized home purchases, from which they are excluded. Inequality in social resources relates to maternity, education, health care, social protection welfare, etc. For example, children of mobile people who are not registered in the area cannot attend nearby quality schools because they are not "locals" with local registration, and there are schools of relatively low quality or for migrant or migrant workers, and they may only be allowed to attend these schools or return to their place of registration, or they may have to pay higher tuition fees in advance (Song, 2014)[14].

4. Conclusions

This paper analyzes the negative impact of housing prices on social integration in the context of labor market discrimination theory, both in terms of regional discrimination and household registration. Housing prices exacerbate the gap between rich and poor in society and the difficulty of switching household registration, further making the mobile populations unfairly treated, which comes from wages, social resources, etc. China has been committed to urbanization, and the migration of the population has led to the existence of a large number of mobile populations. Although the urbanization rate has increased significantly, the urbanization rate of the household registration population is much lower than the urbanization rate of the resident population against the background of high housing prices, and a large number of mobile people do not choose to settle in the area and cannot integrate into local life, which will also further affect the quality and efficiency of urban development. Therefore, the Chinese government should adopt a series of property price regulation measures or provide more government-subsidized housing projects to solve the housing problem of the mobile populations.

References

1. Zhang, K. H., & Shunfeng, S. O. N. G. (2003). Rural–urban migration and urbanization in China: Evidence from time-series and cross-section analyses. China Economic Re-view, 14(4), 386-400.
2. Kojima, R. (1995). Urbanization in China. The Developing Economies, 33(2), 151-154.
3. Song, Y. (2014). What should economists know about the current Chinese hukou sys-tem?. China Economic Review, 29, 200-212.
4. Wu, X. (2011). The household registration system and rural-urban educational ine-quality in contemporary China. Chinese Sociological Review, 44(2), 31-51.
5. Liu, Z. (2005). Institution and inequality: the hukou system in China. Journal of com-parative economics, 33(1), 133-157.
6. Turner, R. J., & Turner, J. B. (1999). Social integration and support. In Handbook of the sociology of mental health (pp. 301-319). Springer, Boston, MA.
7. Madanipour, A. (2015). Social exclusion and space. In The city reader (pp. 237-245). Routledge.
8. Eraydin, A., Tasan-Kok, T., & Vranken, J. (2010). Diversity matters: Immigrant entre-preneurship and contribution of different forms of social integration in economic per-formance of cities. European Planning Studies, 18(4), 521-543.
9. Fawcett, M. G. (1918). Equal pay for equal work. The Economic Journal, 28(109), 1-6.

10. Myrdal, G. (1996). An American Dilemma, Volume 2: The Negro Problem and Modern Democracy (Vol. 2). Transaction Publishers.
11. Becker, G. S. (2010). The economics of discrimination. University of Chicago press.
12. Fan, C. C. (2002). The elite, the natives, and the outsiders: Migration and labor market segmentation in urban China. Annals of the association of American geographers, 92(1), 103-124.
13. Sato, H. (2006). Housing inequality and housing poverty in urban China in the late 1990s. China Economic Review, 17(1), 37–50.
14. Song, Y. (2014). What should economists know about the current Chinese hukou system? China Economic Review, 29, 200-212.

Analysis of Urban Ecological Vulnerability and prospects under the Impact of Urban Expansion

Xingyu Yang*

Forestry Institute, Hebei Agricultural University, Baoding, Hebei Province, China

ABSTRACT. Currently, urban expansion has been an important means of urban economic development, but the coercive impact of its rapid expansion on the ecological vulnerability of the region has become an increasingly important issue that restricts sustainable urban development. This paper uses a SWOT analysis method to analyze this phenomenon, and then proposes solutions based on the problem in order to better grasp the future and lead the development of a new type of city that is low-carbon and ecological.

1 BACKGROUND (INTRODUCTION)

The transformation of Chinese cities in the process of urbanization is the general background of this era [1]. Urbanization is an important initiative to promote regional socio-economic development, and is a necessary path for China to achieve modernization. Since entering the 21st century, China has made remarkable progress in the level of urbanization, which has greatly contributed to the development of regional socio-economics [2]. The expansion of land for construction is one of the main signs of rapid urbanization [3]. The ecological vulnerability of China's urban sprawl is not a new topic, from the initial cries of stifling the blind expansion of cities to the present sighs of helplessness over the loss of ecological territories, the criticism of which is sometimes strong and sometimes weak, accompanying the main theme of China's rapid urbanization and becoming a hot topic for urban and architectural workers to discuss urban development issues. According to statistics, between 1978 and 2019, China's urbanization rate increased from 17.92% to 60.60%. And the urban built-up area expanded from 7438 km2 in 1981 to 146,102 km2 in 2017, an increase of 19.64 times.

In the past decades, human activities have caused a large amount of habitat loss, habitat fragmentation and habitat quality degradation with various adverse consequences on human well-being [4-5]. Since the early 20th century, when the famous American ecologist Clement first proposed the"ecotone" [6], this theory has evolved into the study of ecological vulnerability [7], especially in the 1980s. Since the 1980s, in the wave of reform and opening up, the level of urbanization in China has been increasing, and the resulting ecological and environmental problems have become increasingly prominent, and the public has gradually become aware of the fragility of the ecological environment. For example, the Outline of the Beijing-Tianjin-Hebei Cooperative Development Plan released by the state in April 2015 clearly states that how to balance the contradiction between social and economic development and ecological environmental protection is a key issue in realizing the Beijing-Tianjin-Hebei Cooperative Development Plan [8].

*Corresponding author: 3329325421@qq.com

2 Analysis of the current situation of urban expansion (SWOT analysis method)

2.1 Advantages

2.1.1 Development of economic level

The large-scale construction carried out during urban expansion absorbs surplus rural labor to participate in urbanization and makes it transfer to secondary and tertiary industries, which better drives the development of our economy. Urban expansion can also drive the development of rural areas in all aspects, improve the regional industrial structure, facilitate urban-rural exchanges, and narrow the gap between urban and rural development. When the economy is developed and people's material and spiritual life is satisfied, they will have more energy and money to pay attention to and improve the ecological problems caused by urban expansion.

2.1.2 Mass quality improvement

As far as education is concerned, China's rural population base is large, the total number of people is more than the urban population, the urbanization process has increased the connection between rural areas and towns, even the direct transformation of rural areas into towns, the education of children is guaranteed [9]. Urban development has led to the spread of education, raising the quality level of the general public, making it easier for the public to accept the ecological problems caused by urban expansion and will be more concerned about the ecological problems of the city.

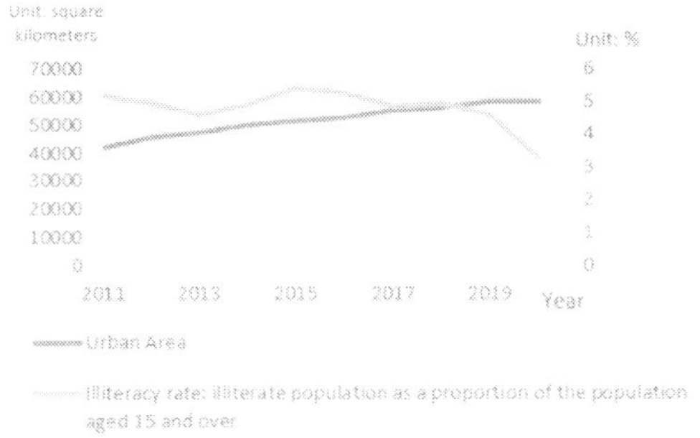

Fig. 1. Urban built-up area and illiteracy rate [owner-draw]

2.2 Disadvantages

2.2.1 Urban sprawl creates stress on ecological service functions

Wind and sand control is one of the major ecosystem services, which is mainly influenced by vegetation cover, local climate factors (wind, soil moisture), and soil erodibility.

Soil conservation is one of the important regulating service functions provided by ecosystems and is influenced by local vegetation cover, climatic conditions, topographic factors, etc[8].

The water-holding function of ecosystems is mainly manifested in intercepting precipitation, regulating runoff, influencing precipitation, and purifying water quality, and also plays a key role in improving hydrological conditions, preventing silting of rivers and reservoirs, and regulating the regional water cycle [10].

2.2.2 Urban expansion leads to weakening of ecological territories

The first thing caused by the urban expansion process is the reduction of natural habitat area. Due to the reduction of habitat area, the migration of many animals will be affected, while the competition between plants and animals will be enhanced, which can easily lead to the disappearance and extinction of species. For example, Erdao Baihe in Antu County, Jilin, which was also a lush temperate forest, has now become a prosperous town in the Changbai Mountain area, with no native forest left. Another example is the development of the city of Wuhai in Yimeng, Inner Mongolia, in the midst of endless grasslands, which resulted in the endangerment of the endemic tetraena mongolia [11].

Secondly, urban expansion has facilitated the construction and expansion of urban infrastructure, resulting in the isolation of many natural green spaces and urban water habitats into "islands" in artificial buildings, resulting in severe fragmentation and islanding of urban habitats. For animals, according to the species-area principle, it is difficult for many large animals or animals requiring large activity areas to survive in urban patchy habitats; for plants, islanded habitats are not conducive to pollen and seed dispersal, causing genetic decline or even extinction of many species. Therefore, islanded habitats are very unfavorable to the survival of organisms and can easily cause a decrease in biodiversity.

2.2.3 The urban artificial system causes heat island effect and pollution

Due to the low heterogeneity of urban environment, there is little space for the development of biodiversity, which leads to its simple food chain composition and small share of participation in biogeochemical cycles [11], and a large number of urban artificial systems, such as factories, buildings, businesses, roads, etc., are incomplete artificial systems and fragile systems.

Moreover, with the conversion of a large amount of vegetation into impervious surfaces of artificial construction such as asphalt and cement, the green areas such as grass and woodland are greatly reduced, which inevitably leads to an increase in surface heat capacity, a decrease in heat absorption rate and evapotranspiration, and triggers the heat island effect (UHI), which in turn hinders the sustainable development of cities and the normal life and health of the public [3]. Many studies have shown that urban sprawl significantly enhances urban surface heat island intensity, but its variation varies significantly from day to night and from season to season [12].

According to the seventh national census, the urbanization rate of China's resident population reached 63.89% in 2020, compared to less than 50% (49.68%) in the sixth national census in 2010 and 36.09% in 2000, maintaining a high rate of growth in China's urbanization rate over a 20-year period. This horizontal spread of "big pie" and the expansion of low-density cities have brought about a very serious "big city disease". Cities can not control and control pollution in time, thus causing a series of environmental problems [13].

2.3 Opportunities

2.3.1. Strong support from government policies

As early as 2010, the National Development and Reform Commission (NDRC) launched the pilot low-carbon provinces and cities, national low-carbon cities (towns), the construction of ecological civilization demonstration zones, the creation of circular economy demonstration cities (counties), and the pilot kitchen waste resource utilization and harmless treatment cities. Electricity demand-side management city comprehensive pilot.

The Ministry of Housing and Urban-Rural Development (MOHURD) has also carried out the demonstration of renewable energy building application cities, the construction of green ecological urban areas, the continuous assessment of water-saving cities, and the cooperation with Guangdong, Hebei, Shenzhen and Wuxi to carry out a number of low-carbon ecological city demonstrations.

China is paying more attention to this area and there are endless possibilities.

2.3.2. Development requirements of major national strategies

The Proposal of the "Central Committee of the Communist Party of China on Formulating the Fourteenth Five-Year Plan for National Economic and Social Development and the Visionary Goals for 2035 "sets out the main goals of achieving new progress in ecological civilization construction, specifically including optimization of the pattern of territorial space development and protection, significant green transformation of production and lifestyle, more rational allocation of energy resources and significant improvement in utilization efficiency, sustained reduction in total emissions of major pollutants, and The ecological environment will continue to improve, the ecological security barrier will be more solid, and the urban and rural living environment will be significantly improved.

2.4 Threats

2.4.1. Weak concept of residents' initiative to protect ecology

Although the country has put the ecological environment on the same level as the economy and society, the public has not established the consciousness of protecting the environment for a long time, and still ignores the fragility of urban ecology and destroys the ecological environment with reckless pollution. In China's cities in general, the public's ecological concept is weak, and domestic sewage is discharged at will, and the sewage is not filtered and processed and flows into the urban water ecosystem.

2.4.2. Urban ecological environment problems have a long way to go

To really develop the city well, urban ecological and environmental problems will become a huge challenge for the government and society as a whole. How to effectively achieve the coordinated development of urban economy and urban resources and environment, as well as how to strengthen urban resources and environmental protection and take strong measures are the current challenges for the development of each city. The old urban areas are dependent on resources for development, and this process will inevitably cause irreversible damage to the urban ecosystem. When building and expanding new urban areas, ecological issues will directly become the first element that cities need to consider.

3 SWOT matrix analysis

Combined with the above SWOT analysis of the impact of urban expansion on urban ecological vulnerability, this study uses the SWOT matrix to make further analysis and recommendations for urban ecological development planning.

① "Strengths-Opportunities" (SO) strategy: urban expansion is guided by policies and financial support, and more financial resources are available for environmental restoration and pollution control

②"Disadvantage-Opportunity" (WO) strategy: to regulate the contradiction between urban construction land and ecological land, the "cluster" development mode of cluster-type multi-center can be adopted, and the coordinated development of the main urban area and the new area should be strengthened, following The city should make efforts to promote the transformation of resource-based cities, improve the long-term mechanism of sustainable development of resource-based cities, cultivate and strengthen succession replacement industries, actively develop new industries, promote the optimization and upgrading of industrial structure, and improve the capacity of sustainable economic development [15].

③"Strength-Threat" (ST) strategy: the state and government establish and improve laws and regulations, and make efforts to build people's awareness.

④"Disadvantage-threat" (WT) strategy: according to the spatial pattern of urban space, economic industry, land resources, water resources, transportation network system, etc., guide the construction of polycentric compact ecological cities [16]; in the direction of smart, low-carbon, sponge, and forest cities, make efforts to promote green buildings, the development of three-dimensional farms, in the face of drastic changes and transformations, urban space also faces the fate of survival and death, therefore, urban space must also be sustainable, like the growth of an organism, both to maintain the essential characteristics and to have the ability of self-renewal and adaptability [17].

4 Summary and Prospect

Sacrificing much or more land for construction does not necessarily achieve the goal of ecological protection, nor does it have to be at the expense of the economic value of land use. Instead, it can be ensured by using as little land as possible through scientific and reasonable spatial pattern design. The basic urban ecosystem services and their security patterns, while still meeting the requirements of urban development; The spatial pattern and scale of a city should be defined by the ecological bottom line of the city, and should not be planned and defined by an image of the urban space [18]. Therefore, when we expand the city, we should also consider the ecological problems brought about by it, not to develop the economy at the expense of ecology, and truly implement the idea of "green water and green hills are the silver mountain of gold" of General Secretary Xi Jinping, and strive to establish a low-carbon city, a forest city, and a resilient and sustainable ecological city. We will truly realize the harmony between human and nature.

In the process of expanding new urban areas, it is necessary to continuously explore ecologically compatible development models and develop new environmentally friendly urban ecosystems in order to fit in with the ecosystems native to urban expansion. Turning around, the formation of new ecosystems in the process of urban expansion is something that can contribute to the improvement of old urban ecosystems. The new city and the old city ecosystem will form a positive and negative relationship with each other, reversing the symbiosis.

As Patrick Geddes, an urbanist, said, the relationship between nature and man is not a decorative backdrop for the human stage, but needs to be maintained as a source of life, a social environment, a teacher and a sacred place.

REFERENCES

1. Shen Shaojie. Discrimination and Reflection on Criticism Re-cognition of "All Cities with the Same Features [J]. ARCHITECTURAL JOURNAL, 2013(06):96-98.
2. Wang J, Zhou W Q, Xu K P, Yah J L. Spatiotemporal pattern of vegetation cover and its relationship with urbanization in Beijing-Tianjin-Hebei megaregion from 2000 to 2010. Acta Ecologica Sinica, 2017, 37(21): 7019-7029.
3. Zhao Z A, Pei T ,Cao S ,Zhang A B, Fan Q Q, Wang J J. Impacts of urbanization on vegetation growth and surface urban heat island intensity in the Beijing-Tianjin-Hebei[J]. China Environmental Science,2020,40(04):1825-1833.
4. Haddad N M, Brudvig L A, Clobert J, et al. Habitat fragmentation and its lasting impact on Earth's ecosystems[J]. Science advances, 2015, 1(2): e1500052..
5. Newbold T, Hudson L N, Hill S L L, et al. Global effects of land use on local terrestrial biodiversity[J]. Nature, 2015, 520(7545): 45-50..
6. Nelson R, Kokic P, Crimp S et al. The vulnearability of Australian rural communities to climate variability and change :Part II—Integrating impacts with adaptive capacity[J]. Environmental Science & Polity,2010,13(1):18-27.
7. Bradley M P ,Smith E. Using science to assess environmental vulnerabilities[J].Environmental Monitoring and Assessment, 2004,94(1-3):1-7.
8. Liu J H, Li W F, Zhou W Q, Han L J, Qian Y G. Prediction of expansion pattern and impact on regional ecological security of Beijing-Tianjin-Hebei Metropolitan Region. Acta Ecologica Sinica, 2018, 38(5): 1650-1660.
9. Chen xu. China's urbanization and its pros and cons[J]. Global Market Information Guide,2014(11):13.
10. Tallis H T, Ricketts T, Guerry A D et al. In VEST 2.5.3 User's Guide[M]. Stanford: The Natural Capital Project, 2003
11. Li J S, Gao G X, Zhang X L, Zheng X M. Effects of urbanization on biodiversity [J]. Chinese Journal of Ecology,2005(08):953-957.
12. Chen B H, Feng Y, Yuan J G, Zhou Y M, Zhao X Y. Spatiotemporal Difference of Urban Heat Island in Jing—Jin-Ji Area Based on MODIS Land Surface Temperature [J]. Acta Scientiarum Naturalium Universitatis Pekinensis,2016,52(06):1134-1140.
13. Liu X H. Study on the effect of urban sprawl on haze pollution [J]. Urban Problem,2020(04):90-96.
14. Xu M R, Zhang J, Li Z H, Zhang M, Mo Y. Delineation of Urban Growth Boundary Based on Ecosystem Services Tradeoffs: A Case Study of Dalian, China [J].Geography and Ger-Information Science,2022,38(02):89-95.
15. Zhao L, Wang W, Zhang Y S, Li R, Wu D T. The Spatio-Temporal Evolution about Urban Vulnerability on the Northeast Region since Northeast Revitalization [J].Economic Geography ,2014,34(12):69-77.DOI:10.15957/j.cnki.jjdl.2014.12.011.
16. Hang Y B ,Chen M H, Li H B.The Return of Reason: From Rapid Urban Expansion to Ecological Awakening[J].Surveying and Mapping Bulletin,2016(S2):298-304.
17. David Harvey , Paris, Capital of Modernity[M], 2005
18. Yu K J; Wang S S; Li D H; Qiao Q. Ecological Baseline For BeiJing's Urban Sprawl: Basic Ecosystem Service And Their Security Patterns [J].City Planning Review,2010,34(02):19-24.

The Impact Of Douyin (Chinese TikTok) On The Socialization Of Chinese Youth

Xinting Gao*

The School of Business Administration, South China University of Technology, Guangzhou, China

Abstract. The more efficient network transmission rate and increasingly popular hardware technology brought by 5G technology have driven the rapid development of online HD short videos. As one of the most popular online short video mediums in China, Douyin, its international version known as TikTok, has a large youth user base, and its influence on youth growth has attracted widespread attention from families, society, and academia. Therefore, it is important to study the influence of Douyin on youth socialization and propose a media environment that is conducive to building a healthy socialization environment for youth.

1 INTRODUCTION

China's economy has soared in recent years, and the Internet has advanced quickly. According to the white paper "Chinese Youth in a New Era," China will have 180 million Internet users between the ages of 6 and 18 by the end of 2020, and the percentage of children who use the Internet will have reached 94.9%. There is a vast base of Internet users as well as the developed low latency 5G technology that has given birth to online short video media like Douyin. Douyin, one of the most well-known short-video social media platforms in China, was formally launched in 2016 and received 45.8 million downloads from the Apple App Store in 2018, exceeding other famous apps like Facebook, YouTube, and Instagram.

The United Nations defines "Youth" as those persons between the ages of 15 and 24 years, and users aged 18-23 account for 22% of the total users of Douyin, as a short video media with a large number of youth users. As a short video platform with a sizable youth user base, Douyin offers a wealth of engaging material for young people, including news, movies, animation, celebrity entertainment, technology, animals, music, dance, and more. For Chinese youth, it has grown into a significant media outlet that helps them expand their horizons, have fun, interact with others on a regular basis, and define their social responsibilities. It also subtly affects how they interact with others. Traditional socialization is a slow, gradual and orderly process, relying on intergenerational transmission and influence through language, text, body and expressions [1]. In contrast, social media such as Douyin accelerates the process of youth socialization and shortens the time of socialization.

Youth socialization is a process in which individuals shed their infancy and mature development and continuously adapt to social rules. American scholar S. Koenig defines socialization from the perspective of social norms. He holds that socialization is the process by which people grow up to be a member of the community from which they originate, and that every action a person takes should be in accordance with the civic and moral norms of that society. Talcott Parsons, an American sociologist of the structural-functionalist school, sees socialization as a process of role learning [2]. Adolescents' socialization is impacted by their cultural surroundings, families, schools, the media, and their peer groups, with internet media having a growing influence on the media. Bozeman claims that the "childhood vanishing" phenomena has been going on since the 1970s. From early childhood until maturity, the media has been influencing how individuals think and act, and to some extent, how they see many things. This study takes Douyin as an example, and analyzes the main issue posed by China short-video social media for young people, focusing on the positive and negative effects of Douyin on the socialization of youth based on the large number of youth users of Douyin and the fact that Douyin has a significant impact on youth growth. We hope that the discussion in this paper will make the society pay more attention to the influence of Douyin on the growth of youth and help create a good media environment.

*Corresponding author: ellaxgo@gmail.com

2 THE POSITIVE IMPACT OF DOUYIN ON YOUTH SOCIALIZATION

2.1 Enriching the way of learning for youth

Youth no longer prefer the old one-way passive information-gathering methods including classroom instruction, instructional films, and textbook reading. The rise of Douyin has improved the methods that young people study. The videos on Douyin cover a wide range of topics, including astronomy, geography, gaming, music, movies, food, and more, so young people can always find

https://doi.org/10.1051/shsconf/202315501013

something interesting to watch there. Additionally, there are bloggers who live all over the world in Douyin; their films allow you to learn about regional traditions and people without traveling, substantially expanding your horizons. Moreover, 15 seconds-1-minute-long videos with popular music and extra voice introduction, this way of spreading information makes youth more interested in learning than traditional teaching, thus turning passivity into initiative. According to the statistics, 44.3% of young people learn new knowledge on Douyin, and interest becomes the biggest reason for them to learn on Douyin. (Data source: Ocean engine – Douyin Youth Research (N=6171), June 2022)

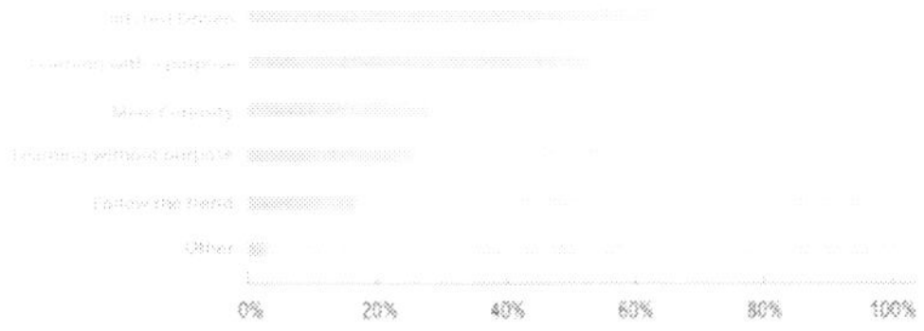

Fig. 1. Motivation of young groups to learn in Douyin

At the same time, in this kind of short video, the content presented is short and precise, hitting the key points directly, which is easier for youth to remember deeply.

2.2Enriching youth's social style

The centrality of social networks in the adolescent environment may also accelerate over time. As specific platforms gain influence among adolescent peer groups, social networks become more significant for daily interactions. Social networks are not only modalities that facilitate interactions with unfamiliar peers but also essential tools for intensifying connections to the existing peer group [3]. The 2022 Douyin Youth Observation Report demonstrates the tremendous desire for sharing among young people. The Douyin youth group's members frequently post the pictures and videos they shoot on social media. They are not afraid to express themselves, record their lives on social platforms and create quality content, thus establishing their persona on the network and initially building their social status.

Fig. 2. Frequency of sharing pictures and videos on the social platform by young people on Douyin

Douyin has enriched the way of self-expression and socialization for youth. Youth have more methods to engage with friends and strangers because to Douyin's variety of interaction options, including likes, retweets, comments, private messages, collaborative movies, etc. At the same time, Douyin video shooting threshold is very low, no complicated technical operation, you can make the video more interesting through Douyin's own video effects, music, and Douyin's parent company, Toutiao, has also developed a software for Douyin video editing cutting, so that you can complete the video editing through the mobile App. The interesting video content is pushed to interest users through Douyin's algorithm, which helps to form a social circle with unique attributes, satisfying young people's social and sharing desires.

2.3Accelerating the socialization of youth politics

As social media today are a key method by which young people engage with political activism [4]. Douyin is also raising the attention of Chinese youth to political and state authorities. The administration and evaluation of China's new government media began in 2019, and for the first time, new government media have been included in a quantitative assessment of government activity. State Council, National Development and Reform Commission, State-owned Assets Supervision and Administration Commission of the State Council, Ministry of Foreign Affairs, Ministry of Transport, public security agencies, and others have offices in Douyin. The presence of government agencies in Douyin can shorten the distance between government departments and youth, enhance the status of the network, help youth understand the latest government information, grasp national political trends, learn political knowledge, form political attitudes and build political personalities, while youth can also interact with these agencies through Douyin and communicate directly between the two sides.

Besides, Douyin as an entertaining short video media, government civil servants have formed an IP phenomenon in Douyin. By way of illustration, the "Diplomatic Troupe," a spokesman for the Ministry of Foreign Affairs, has amassed a large following of young fans on Douyin because to its impressive appearance. Many of these young fans have created movies of the "Diplomatic Troupe" and published them there. A number of young admirers have created films of the "Diplomatic Troupe" and posted them on Douyin, drawing a fanbase. The development of the "Diplomatic Troupe" has altered the traditional media's serious and stereotypical portrayal of government employees. On Douyin, they now have a more personable charisma that may persuade some youth to pay attention to foreign policy and increase their favorability to the Ministry of Foreign Affairs.

3 NEGATIVE EFFECTS OF DOUYIN ON YOUTH SO-CIALIZATION

3.1 False information misguides youth's value judgment

Although Douyin serves as a channel for youth to get the latest information, Douyin is also a platform created by individuals, filled with a large number of third-party personal accounts, and the information released by the authors lacks the guaranteed mechanism of news authenticity supervision and screening [5], and some short video authors release false information to earn traffic, which triggers public opinion.

Li Min, vice president of Didi, questioned the Douyin short film authors for defaming Douyin drivers in a post on her personal microblog in May 2021. In her example, the texts of these short video authors making up stories were all almost identical. The creator's name is concealed on Douyin, but the conversation can be amplified to the maximum. These short film creators create false stories about Douyin and stir up social mood to attack a group's beliefs just to gain viewers.

3.2 Youth produce irrational herding behavior

Many young people exhibit a pretty strong herding mentality at this period, which is influenced by how each person has been socialized. Information herding and rule herding, according to researchers, are two types of herding. Rule-following is a way to gain social approval, to establish good social relationships by forming friendly alliances with others, and to some extent, to pursue a sense of group identity and belonging, fearing rejection by the group as an alien (Deutsch & Gerard, 1955) [6]. Adolescents tend to be more affected by their classmates and adopt a herd mentality as a result. It is clear that herd mentality influences how teenagers view their own values and behavioral inclinations. Compared with traditional formal education, the potential herd mentality will have a relatively stronger impact on many teenagers, and the creators of short videos on Douyin come from different backgrounds in various industries. Some Douyin weblebrities broadcast everyday life films on the site that are filled with fancy automobiles, designer goods, upscale resorts, and other things in an effort to attract attention. Youth are prone to blind imitation when brushing Douyin videos, and are proud of such following, all kinds of unskilled following, imitating what segments, teasing children, family members The phenomenon is endless, as well as the phenomenon of elementary school students learning to wear heavy makeup and pretend to be mature. The user's tastes will be determined by Douyin's big data algorithm, which will intensify the pushing of such filthy movies. When kids are exposed to low-quality videos with terrible sexual values, they are likely to copy or have their lives and worldviews influenced. [7].

3.3 Douyin algorithm creates information cocoon

In "Information Cocoons", Robert Walmsley University Professor Cass R. Sunstein pointed out that when people are confronted with a plethora of information online, they tend to pay attention to what they want to see, and the algorism will select their preferred information to them, eventually narrowing their field of vision, much like a silkworm making a cocoon for itself[8].

"User-centered design" is the core of Douyin's prototype. Simply put, Douyin will only recommend content that the current user likes, and will continue to track and reinforce this user preference from the cold start of the app. If you tap on a skiing video, the system will initially customize your preferences to sports, and then continue to track your behavior for further analysis, eventually providing you with highly tailored and accurate recommendations.

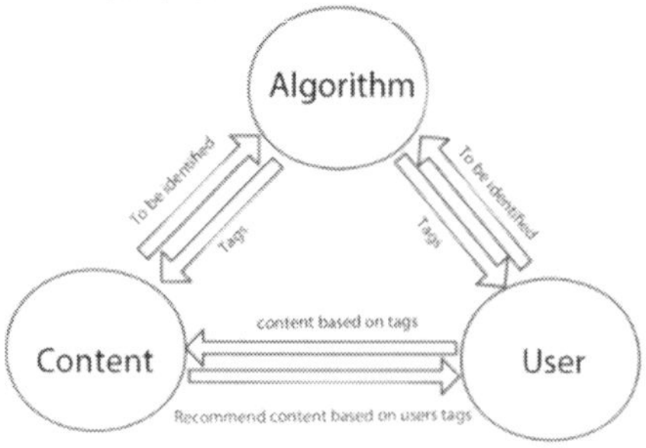

Fig. 3. Douyin algorithm prototype,

Source: Short Video Popular Research Institute

Although this algorithm can increase the viscosity of users to the platform, the convergence of content to a certain extent exacerbates the formation of the information cocoon, limiting the breadth of adolescents' horizon and knowledge. Being caught in the information cocoon also tends to give adolescents a one-sided view of things, which can gradually aggravate, making adolescents blindly believe and follow, which is eventually difficult to correct and seriously hinders the normal socialization process of individuals [9].

4 MEASURES TO DEAL WITH THE NEGATIVE IMPACT OF DOUYIN ON YOUTH

4.1 Strengthen the audit of the authenticity of the information content on Douyin and the punishment

Douyin has established a comprehensive procedure for rumor detection, identification, and countering mechanisms, established a "rumor prevention zone," and established close ties with authoritative organizations and inventors in many fields of expertise to carry out rumor prevention activities. The establishment of a "rumor database" for automatic machine review, new warning labels for suspected rumors, big data and artificial intelligence for "accurate disinformation". However, despite these measures, there are still many authors on Douyin who publish false, vulgar and other content that distort the values of youth. In terms of safety mechanism for youth using Douyin, Douyin also launched Douyin teenager mode, the default length of use is 40 minutes; unable to publish content, watch and open live; unable to use private messages, comments, recharge reward function; the same city and hot search portal is closed, the content is selected by the platform, and only selected content can be searched. Although this addresses the issue of adolescents being misled by material, it also lessens young usage of Douyin and diminishes Douyin's contribution to the development of youth social skills, making it a short-term fix. Douyin needs to join hands with government agencies to impose strong legal punishment on authors who publish bad content, and invest more supervisors to increase the review and punishment. The review and punishment efforts, rather than just be video deletion, blocking and other treatment. The law should be passed to regulate the online environment of Douyin.

4.2Families strengthen the socialization guidance for youth

Parents should make clear what bad impacts the Douyin environment would bring to youngsters in order to favorably lead them, even if the media progressively outweighs family as the key influential force in juvenile socialization. According to social learning theory, youth's important life decisions, such as school and career choices, are linked to the influence of parental role models [10]. According to the social learning hypothesis, learning and observation might have an impact on teenage personality development. Instead of allowing children to get excessively hooked to the Internet, parents should pay greater attention to the social realities of children's lives, offer advice on the right values in all facets of everyday life, and set an example for children to follow.

5 CONCLUSION

According to Herbert Marshall McLuhan, "the scale and form of interpersonal interactions and actions" are shaped and controlled by the media. Douyin, as one of the most popular social media for Chinese youth, plays a role in advancing youth socialization, opening up their horizons and allowing them to establish their social status more easily. Youth may access a broad range of material from Douyin that does not just depend on teens' online conduct as they develop their online social skills. Back in real life, the immediate and effective information resources obtained in cyberspace are reflected in youth daily family life, school studies, and peer relationships [11]. However, this effect is not all favorable; it also poses certain obstacles to young people's socialization. As a result, society, families, and schools should pay attention to this and enhance the control over Douyin public opinion while also supporting actual socialization guidance.

REFERENCE

1. Han Mei, Social media profoundly affects adolescent socialization, Ethnic Education of China 2018,(06)
2. Yu Binbin, A Psychological Perspective on New Development of Adolescent Socialization in the Internet Age, Research Report on Youth and Youth Work in the Internet Age - The 6th China Youth Development Forum and the Outstanding Paper Collection of China Youth Research Association (2010)
3. Schwartz D, Fritz H L, Kelleghan A R, et al. Adolescents who are nonusers of fashionable social networking platforms[J]. Cyberpsychology: Journal of Psychosocial Research on Cyberspace, 2021, 15(3).
4. Grasso M, Giugni M. Youth doing politics in times of increasing inequalities[J]. Politics, 2022, 42(1): 3-12.
5. Zhu Jiangli, New Media Boosting Citizen Participation in Society Governance: Mechanism, Problems and Solutions, Chinese Public Administration, 2017,(06)
6. WEI Zhenyu, DENG Xiangshu , ZHAO Zhiying, The effect of conformity tendency on prosocial behaviors, Advances in Psychological Science
7. Zeng Ying、Liu Yang , Research on the negative effects of Douyin short video content and its guidance countermeasures, Media Forum, 2021,4(10)
8. Sunstein C R. Infotopia: How many minds produce knowledge[M]. Oxford University Press, 2006.
9. Wang Zhizhao, Pros and cons of short video on youth and its governance[J].Journalism Lover,2019(11)
10. Hou Yihua, Wang Zijia, Who influences adolescent socialization - A comparative analysis of the role of parents versus the role of peers, Education Research Monthly, 2021,(07)
11. Yu Binbin, A Psychological Perspective on New Development of Adolescent Socialization in the Internet Age, Research Report on Youth and Youth Work in the Internet Age - The 6th China Youth Development Forum and the Outstanding Paper Collection of China Youth Research Association (2010)

Marketing Strategies of Huangshan Hotel Homestay Industry in the Post-Epidemic Period

Ziyi Li*

Huangshan University

Abstract. In an increasingly competitive business environment, marketing strategy plays a very important role, providing the basis for marketing plans and indicating future courses of action. A systematic marketing strategy includes the organization's goals, policies, and actions to conduct business effectively. With the passage of time and the improvement of the transportation system, the tourism B&B industry in Huangshan City has made great progress and has played a vital role in enhancing the image of tourism in Huangshan. Tourism is one of the fastest growing and oldest industries; the marketing of available hotel services is critical at all levels and destinations. The new crown epidemic in recent years will also promote the role of innovative ideas that help to enhance the hospitality industry is developed and effective decision making strategies are followed and implemented by the industry in general and prominent hotel groups in particular to attract potential travelers. This paper highlights the challenges faced by the Huangshan hotel B&B industry in the post-epidemic period in terms of hardware facilities and services and future strategies to address these challenges through literature analysis and a study of the current situation of small and medium-sized hotels and B&Bs in Huangshan. It also proposes measures to improve operational efficiency to make this prestigious industry practical and feasible.

1 INTRODUCTION

With the growing demand of people's tourism consumption, more and more B&Bs continue to go hot. At present, B&Bs in China's mainland are concentrated in areas with developed tourism such as the southeast and west, while in Anhui, the development of B&Bs is still in its initial stage. Huangshan, as a 5A tourist attraction in China, is influenced by the strong Hui culture and has a bright future for B&Bs. However, there are still problems such as low visibility and immature development of B&B in Huangshan. The purpose of this thesis is to analyze the current situation of small and medium-sized hotels and B&B in Huangshan and create a new strategic marketing plan for them, which can solve the current problems such as lack of characteristics and single marketing channel for small and medium-sized wine industry and B&B, improve the hardware facilities while making full use of the Internet to improve their competitiveness and adapt to the post-epidemic period, leading to a more successful competitive business. The theoretical framework of the thesis contains definitions and general information about marketing and marketing analytics, especially in the hospitality industry. The theoretical analysis of the thesis uses SWOT and PEST as tools to create a clear picture of the current situation and to set the basic direction for the new marketing strategy.

*Corresponding author: ziyili1130@163.com

2 Analysis of the current situation of the B&B industry in Huangshan

2.1 Macro environment PEST

Table 1. PEST analyses of the Homestay industry (edited by the author)

Politics: Government regulations on services, hygiene, food, and encouragement of B&B development.	Economy: Rising prices and higher taxes Decline in Gross Domestic Product (GDP)
Social : People's Thirst for Rural Leisure Vacation villagers' use of their own idle resources. Lack of business travelers in this area	Technology : Possibility of online booking Possibility of accessing hotel information through social networks and booking portals No wifi in the room

The second is to improve the water and electricity infrastructure, change the rural often stopped water, electricity, stop the network of the problem, strengthen the treatment of rural household garbage, sewage, and equip with rural special cleaning staff, full-time, to improve the environmental health of rural areas; third is to implement the traditional village protection project in each village, repairing collapsed, broken ancient dwellings, repairing the stone road, maintaining the construction of the top 100 photography points, the implementation of village lighting, greening, to create a harmonious appearance, beautiful environment of B&B points. Fourth, the introduction of relevant policies and norms to stimulate the development of B&B [1].

The financial and economic aspects have not been good in recent years due to the occasional outbreak of the New Crown epidemic. In particular, unemployment among young people has risen sharply and taxes have increased. All these factors have affected all aspects of people's lives and people have started to travel and spend less. At the same time, local financial institutions in Huangshan played a forward-looking role in the development of B&Bs. For Yixian County's B&B development boom and development direction, increase the injection of credit funds for B&B development. Although the B&B is small, it is the lever fulcrum to pry the second leap in tourism and rural tourism development in Yixian County. Because of the small, light, beautiful, it is easier to attract a large number of small and medium-sized social capital, a large number of investment and business crowd, revitalize a large number of idle resources in society, creating a fine, characteristic multi-experience products [1].

New technologies occupy an important place in contemporary life. The quality of WI-FI has a direct impact on the guest's impression of the hotel. Nowadays, the easiest way to get any information is to search for it on the Internet. Therefore, it is crucial for the hospitality industry to monitor travel and booking portals, as this may influence the guest's choice. Online marketing efforts and careful monitoring of portals may be an easy and inexpensive way to increase the number of customers.

2.2 SWOT analyses of the Homestay industry

SWOT analysis is a combination of internal and external factors involved in situation analysis. Strengths are usually considered as internal factors and are the company's resources and capabilities that can be used as a basis for competitive advantage. For example, a company's strengths may be a strong brand, a good reputation among customers and favorable access to the distribution network. The lack of certain strengths can be seen as weaknesses, and these weaknesses are also seen as internal factors. Not all weaknesses can be eliminated, but once identified, a company should try to minimize their impact or eliminate them altogether. Strengths and weaknesses are often a matter of perception, which makes them sometimes difficult to identify. Consumer research is a good way to understand these internal factors, so customer feedback should be understood and thoroughly addressed [2].

Table 2. SWOT analyses of the Homestay industry (edited by the author)

Strengths: Location (near the scenic spot) Unique decoration style (Huizhou-style architectural features) Private Parking High evaluation scores on booking portals The presence of regulars	Weaknesses: Small area, only a dozen rooms Location (walking distance from downtown, but no walking path) Limited range of additional services Most of the B&Bs have not yet made good use of Internet marketing tools, and the publicity channels are relatively single
Opportunities: Increase the variety of additional services Define the target market Improve the policy of improving the advertising campaign Establish a clear customer loyalty system Build a team of experienced and motivated employees	Threats: Price war with local competitors Lack of activities and entertainment in this area

Location is both an advantage and a disadvantage, as the hotel complex is located in the Hongcun scenic area, 10 km from the center of Huangshan City. It is a good location, but there is no convenient road to the city center, which causes some guests who want to visit the city to not choose this place as their accommodation [3].

The size of the hotel is also a point of inconsistency, as smaller hotels offer personalized service to each guest, but the limited number of rooms limits profits and during peak seasons, rooms can easily fill up. The hotel has a relatively regular customer base, but not a large one. Getting a high score on booking portals is an advantage.

Threats are important aspects that can be controlled. Control over competitors, price policy, market situation, domestic political and economic situation, booking sources, etc. are all very important.

3 Consumer analysis

In order to attract visitors, it is vital to understand who the current guests of the hotel are. Once the customers and their needs are identified, it is possible to create a target market. Once the hotel's target market has been identified, special pages can be designed on the hotel's website specifically for them. It is important to focus on the parts of the hotel's marketing mix that appeal to each specific segment and to create packages and special offers for that specific segment [3].

Defining a target customer means identifying specific characteristics of an individual or business, such as age, gender, income level, buying habits, occupation, family status, geographic location, ethnicity, hobbies and interests.

Couples, seniors, families with children, and nationals or foreigners who are mainly interested in Huizhou culture and Huangshan scenic vacations can be categorized as customers of Huangshan Yixian B&B. It is logical and efficient to focus on these people and create some points to make them our customers for a long time.

4 NEW MARKETING STRATEGIES

4.1 Objectives

The main goals of the new marketing plan are to increase profits, expand the number of regular customers and make the area a thriving business. The opportunities identified in the SWOT analysis have been taken into account in the marketing plan.

4.2 Marketing mix

Marketing Mix is an important part of a company's marketing strategy and refers to a holistic activity that combines the basic marketing measures under the company's control. The main purpose of marketing is to satisfy the needs of consumers. The marketing mix is defined by Kotler as a controlled set of tactical marketing tools that a company mixes in order to generate the response it wants in its target market. There are many such tools, but they are traditionally grouped into four sets of variables 4Ps - product, price, place, and promotion [3].

It is the basis for developing a company's marketing strategy, and a good marketing mix ensures that the company meets the needs of consumers as a whole.

It is logical to start the marketing mix with product analysis. The question that can be asked is what would be of interest to the target customer. All products have both tangible and intangible aspects. However, in service industries, such as hospitality and tourism, the product is largely intangible. The fact that products in the hospitality industry are primarily in the form of services makes the industry's offerings very unique relative to other industries.

Therefore, for a hotel to successfully implement its marketing strategy, it must develop a marketing mix [4].

4.2.1 Designing the Product:

The survey found that Huangshan B&Bs lack personalization and do not have their own business philosophy. Although B&Bs were initially developed slowly in the form of farmhouses in response to market changes, they are a concentrated expression of nostalgia culture and artisan spirit[5]. Most of the B&Bs in Huangshan are just ordinary hotels with good infrastructure but no sentiment. Considering the marketing strategy of differentiation, homogeneity limits the further development of B&B in Huangshan. Huangshan B&Bs could work on creating a unique service, such as offering a spa or spa services. There are currently no spa or spa-themed B&Bs in the area. Add a spa or hot spring service and open it all year round. It would be able to balance the traffic between the low and high tourist seasons.

In addition, Huangshan B&Bs can also integrate sightseeing, food, and accommodation to create a one-stop tourism service, changing the previous fatiguing way of travel and focusing on getting close to nature and enjoying life.

At the same time, Huangshan B&B and the hotel industry should realize the digital operation of hotel check-in and basic services as soon as possible. For example, with the WeChat sweep function, various touch points such as user check-in, free borrowing of rechargeable batteries, free going to the gym, hotel screen casting, and opening electronic invoices after leaving the hotel can be integrated and completed online through one cell phone [6]; stores can also be equipped with robots that can provide services such as meal delivery, delivery, cleaning and disinfection. In the post-epidemic era, people are more concerned about hygiene, safety and health, so equipped with hotel robots can basically realize the whole process from check-in to departure without contact, which can ensure both privacy and health and safety, and further upgrade the service experience.

4.2.2 Selection of Place

Location is another important aspect of marketing distribution channels. Currently there are two ways to sell hotel accommodation: directly through the hotel phone or through three booking portals C.trip, Meituan and Feizhu. By searching for Huangshan B&Bs on different travel software, a total of over 700 were searched on Ctrip, over 200 on Meituan and over 500 on Feizhu, among which there are a small number of high-end B&Bs and the rest are mostly The rest are mostly economical and ordinary B&Bs. Most of the high-end B&Bs online have complete infrastructure and provide many services such as transportation and sightseeing tours. Most of the other online B&Bs are privately owned, similar to the business model of farmhouses. According to the research on "Hongcun", "Xidi", "Tachuan", "Huangshan Scenic Area" and other places, it was found that these The existence of the B&B group in the scenic area, has not yet been well used the Internet marketing tools. In Hongcun, for example, only 60% of the stores use Internet publicity methods, and another 40% of the B&Bs do not use Internet booking, but still use tourists' local search, telephone booking, and friends' recommendations [5].

Therefore, Huangshan B&B owners can open their own WeChat and Weibo public numbers, write illustrated scenic travel guides and travelogues, or post well-produced B&B videos on major video websites. In addition to the popular marketing number publicity methods, Huangshan B&B can also be implanted in TV programs to promote the introduction of Huangshan B&B, or sponsor the filming location of variety shows. In response to the low coverage of online booking, Huangshan B&Bs should try their best to improve the information description and make full use of the app platform for introduction and promotion.

4.2.3 Strategies for Promotion:

It is not enough for a company to have a good product; it will not generate any sales unless the consumers are aware of the product's existence and also have access to the product.

The promotion here includes promotion, advertisement, public relations cooperation, discount, product experience activities, etc. With the development of mobile internet and internet of things, consumers are paying more and more attention to the quality and experience of products, and the way of experience will become more and more efficient, so products will become the best connector between enterprises and consumer users. Through a more catchy and deep-rooted slogan for consumers to clearly perceive, it also establishes the core competitive advantage of the enterprise. In the era of the mobile Internet, it is imperative to pay attention to and utilize the connection function of the product itself.

4.2.4 Decision about Price

Price is not only a reflection of product positioning, but also a guarantee of corporate profitability, and moreover, a competitive tool. So it is important to make price adjustments based on strategic intent and to make good progress and retreat. B&B operators should consider how pricing can maximize profits when there is an oversupply, tough competition and a market downturn, as well as how to build a competitive advantage and establish a price standard. Be good at responding to price wars forcefully and focus on building value wars.

5 Conclusion

The purpose of this paper is to develop a new marketing plan for small and medium-sized hotels and B&Bs in Huangshan City. This paper successfully develops a theoretical framework for marketing, including the concepts of marketing and consumer analysis. Different sources of information such as books, articles and websites are used in the process. The current state of marketing is accurately analyzed. SWOT, PEST analysis and target group analysis were conducted. The hotel was evaluated in two internal categories such as strengths and weaknesses, and two external categories such as opportunities and threats.

Based on the market research, a new marketing strategy is proposed. The portfolio and marketing communication mix have been improved by correcting the current deficiencies in marketing for small and medium-sized hotels and B&Bs in Huangshan, such as how to use the Internet to enhance their competitiveness. The implementation of the new strategy is a dynamic and long-term process. During the implementation process, it is very important to pay close attention to the market conditions, trends and spontaneous leapfrogging changes in the hotel industry. Further future research will test, validate and improve the marketing strategy proposed in this paper. The results obtained will be presented in future studies.

References

1. Fang Jinhua, (2018), Survey on the Development of B&B in South Anhui Tourism, Chinese and foreign entrepreneurs, https://www.meipian.cn/37nn42fh
2. Maarit Karppinen, (2011), STRATEGIC MARKETING PLAN FOR A HOTEL, pp.32-34
3. Elizaveta Nazarova, (2014), A NEW STRATEGIC MARKETING PLAN FOR A HOTEL, pp.4-27
4. Mohd. Yameen, (2013), MARKETING STRATEGIES OF HOTEL INDUSTRY IN INDIA, Global J. of Arts & Mgmt., pp.147-150
5. Wu Haoyue, (2018), Research on Enhancing the Competitiveness of Huangshan B&B in the Context of "Internet +", pp.28-29
6. Li Monan, (2022) Where will the hotel industry go in the post-epidemic era? https://www.zgswcn.com/article/202204/202204010947181021.html

The Impact of Taobao's Negative Comments on Consumer Willingness

Liyuxi Sun*

The University of Newcastle, school of business and management, 039594, Singapore

Abstract. Based on the theory of trust orientation and perceived risk, this study explores the impact of Taobao's negative evaluation on consumer behaviour, and then conducts research. According to the theory of trust tendency, this paper discusses how consumers view the negative evaluation of goods. Through data analysis, it is found that consumers will have lower purchase intention when they browse negative comments. The psychological mechanism of this is that consumers are affected by perceived risks, resulting in lower brand identity. The subjects of this study are 70 Taobao users, and the hypothesis proposed is tested through correlation analysis.

1 Introduction

With the development of the Internet era, more and more consumers are shopping online. As with traditional WOM communications, online consumer reviews are important for driving the actions of consumers. (Jumin Lee a, Do-Hyung Park, Ingoo Han, 2006) [1] Product reviews can influence consumers' consumption decisions. (Haoxing Tang and Hui Cao, 2020) [2] Most of the studies on the negative evaluation of goods focus on Amazon and other overseas retail brands. However, China's shopping software Taobao is also an indispensable and important research object.

1.1 Research background

Gao Bo and Song Yanping[3] introduced the consumer behaviour model in the analysis of consumer behaviour, that is, consumer behaviour is one of the most common and important activities of human beings. In his article on the discussion of cross-border e-commerce consumer behaviour mode and marketing strategy, Liu Li mentioned that the consumer behaviour process is understood from both psychological and behavioural perspectives, including the psychological activities and actual purchase behaviour related to purchase decisions. Through consulting the domestic and foreign literature, we know that most of the domestic academic research focuses on the relationship between advertising and consumer behaviour. For example, Zhuo Jun's article[4] on online marketing studies the impact of online advertising on consumers' external behaviour. Zhang Peipei[5] wrote many similar studies on the impact of short video ads on women's consumer behaviour. However, the research on consumer behaviour from the perspective of commodity evaluation is less involved. For example, Bai Li's[6] analysis of the impact of commodity customization on consumption evaluation behaviour in e-commerce platform and Wang Hui's[7] empirical analysis of the impact of social factors on online shopping consumption behaviour - taking online evaluation of digital goods as an example were mentioned in two articles. The articles published by domestic scholars in foreign journals also have some relevant research, such as the reference groups and their language styles in the research of Jinhua Zhang and Yongsheng Huang's[8] positive online comment brand attitude towards consumers, and the summary of research on counterfeit goods detection by Hao Xingtang and Hui Cao.[2] The research on this aspect by foreign scholars has increased significantly compared with that in China, mainly focusing on foreign shopping software such as Amazon. Boban Melovi ć aDamir Š ehovi ć aVesna Karad ž i ć aMarina Dabi ć b cDragana Ć irovi ć A.[9] The determinants of millennial online shopping behavior - the impact on consumer satisfaction and e-commerce development has been mentioned a little. Agostino Vollero, Domenico Sardanelli, Alfonso Siano[10] also explored the role of Amazon effect on customer expectations from the perspective of merchants: analysis of user generated content in consumer e-retailing, and Farhad Khoshbakht, Atena Shiranzaei, S. M. K[11] Comparison performance evaluation of search items and experience items in the Amazon customer reviews of Quadri. Based on the above information, I found that few scholars at home and abroad have studied the negative evaluation of Taobao, a shopping software product. Therefore, I will take Taobao, a Chinese e-commerce platform, as an example to study the impact of negative evaluation on consumer behaviour.

1.2 Theoretical development

With the development of the Internet era, more and more consumers are shopping online. Product reviews can

* Corresponding author: 3224830863@qq.com

influence consumers' consumption decisions. (Haoxing Tang and Hui Cao, 2020[2]) Since most scholars at home and abroad focus on positive evaluation, negative evaluation is the key to affect consumers' purchase intention. Therefore, this paper will start from the negative evaluation of consumer behaviour. There are two theoretical bases to support this paper. First, consumers tend to trust other consumers. The tendency of trust is a general rather than a specific tendency, that is, showing belief in human nature and taking a trusting attitude towards others (McKnight et al., 1998) [12]. Second, consumers often have perceived risks before purchasing goods. Glover and Benbasat (2010) [13] believed that perceived risk was an important determinant of consumers' unwillingness to buy online. When consumers take action under uncertainty, trust will play a role (Kim, D., Ferrin, D., & Rao, H. 2008[14]). Then they will choose to trust some comments left by consumers, especially some negative comments. Based on these two theories, this paper conducts a more in-depth and comprehensive study of consumer behaviour and consumer psychology, hoping to fill the gap in this field of academia through research.

2 Method

2.1 Participants and Study Design

I created an online questionnaire using a questionnaire star to enable me to collect online data. this

The questionnaire is opened and filled by scanning the QR code. Participants are 70 Taobao users.

Almost everyone in China uses Taobao for online shopping, so you can choose convenient sampling for research.

2.2 Measures

In order to evaluate the trust tendency (McKnight et al., 1998), the answer is based on the 5-point scale, 1 point=very inconsistent, 5=very consistent, and answer the question.

To determine the browsing order, use sorting for evaluation.

In order to understand the perceived risk (Kim, D., Ferrin, D., & Rao, H. 2008), the written answers are used for further evaluation

3 Data

3.1 Reliability Analysis

The data obtained from the questionnaire survey has a certain stability, and can give consistent results. This stability is the reliability of the satisfaction survey results, also known as reliability. The higher the reliability is, the more successful the survey is. That is, the data reflects the true opinion of the survey object. In general, the Kronbach coefficient is between 0 and 1. When the coefficient is less than 0.6, the reliability is not enough, while when it is greater than 0.8, it indicates that the reliability is good. This paper uses SPSS22.0 to analyse the reliability of the 3.2questionnaire:

Table 1. Reliability Statistics

Cronbach's Alpha	Item
0.711	8

According to the table, the Kronbaha Alpha coefficient of the survey results is 0.711, indicating that the reliability of the questionnaire is higher than the standard, and the internal reliability is reasonable.

3.2 validity analysis

validity is to measure the internal consistency of the questionnaire. When the internal consistency is high, it means that the views represented by the questionnaire are more unified. In this paper, SPSS26.0 is used to test the KMO and Bartlett sphericity of the questionnaire structural variables: according to the research analysis, when the KMO value is less than 0.6, it is not suitable for factor analysis, while when it is greater than 0.6, it is suitable for factor analysis. From the following table, it can be seen that the KMO value is 0.723, The questionnaire can be used for subsequent analysis

Table 2. KMO and Bartlett's Test

Kaiser-Meyer-Olkin Measure of Sampling Adequacy.		0.723
Bartlett's Test of Sphericity	Approx. Chi-Square	103.334
	df	16
	Sig.	0.004

3.3 Frequency analysis

Table 3. Frequency analysis result

name	item	frequency	Percentage (%)	Cumulative percentage(%)
Age	18 years old and below	16	22.86	22.86
	19-25 years old	48	68.57	91.43
	26 - 40 years old	6	8.57	100.00
Monthly income	3000 yuan and below	10	14.29	14.29
	3001RMB-5000RMB	22	31.43	45.71
	5001RMB-10,000RMB	27	38.57	84.29
	10,001-$20,000	8	11.43	95.71
	RMB 20001 and above	3	4.29	100.00
Gender	Male	2	2.86	2.86
	Female	68	97.14	100.00
Education	High school and below	3	4.29	4.29
	Bachelor's degree/college	63	90.00	94.29
	Graduate and above	4	5.71	100.00
Annual consumption amount	5000 yuan and below	46	65.71	65.71
	5001RMB-	20	28.57	94.29

	10,000RMB			
	10,001-$20,000	1	1.43	95.71
	20,001 - 30,000	2	2.86	98.57
	RMB 30,000 and above	1	1.43	100.00
Total		70	100.0	100.0

A total of 70 final samples were selected as research objects in this survey. Among them, there are 16 people under 18 years old, accounting for 22.86%, 48 people between 19 and 25 years old, accounting for 68.57%, and 6 people between 26 and 40 years old, accounting for 8.57%. In the distribution of monthly income, there are 10 people with income of less than 3000 yuan, accounting for 14.29%, 22 people with income of 3001-5000 yuan, accounting for 31.43%, 27 people with income of 5001-10000 yuan, accounting for 38.57%, 8 people with income of 10001-20000 yuan, accounting for 11.43%, and 3 people with income of 20001 yuan and above, accounting for 4.29%. In the distribution of different genders, there are 2 males, accounting for 2.86%, and 68 females, accounting for 97.14%. In the distribution of educational background, there are 3 persons in senior high school and below, accounting for 4.29%, 63 persons in undergraduate/junior college, accounting for 90%, and 4 persons in graduate and above, accounting for 5.71%. The distribution summary of annual consumption amount, the highest proportion is below 5000 yuan, accounting for 65.71%, followed by 5001-10000 yuan, accounting for 28.57%.

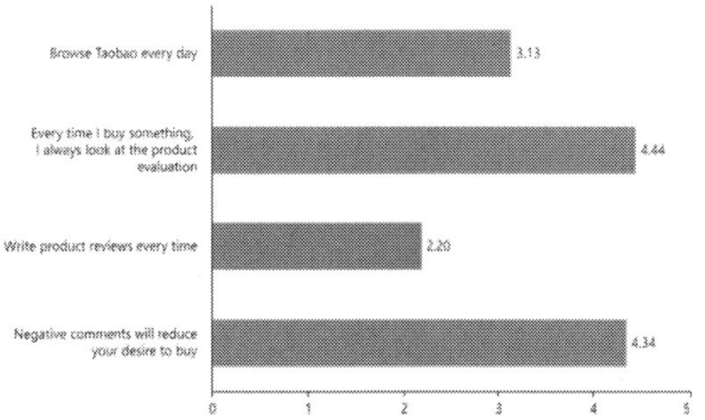

Figure 1. Occurrence of the above events

According to the description of the use of Taobao, the rating of browsing Taobao every day is 3.13, which means that the daily browsing of Taobao is not very obvious. In the survey where you will see product reviews and negative reviews will reduce your purchasing desire every time you buy something, the score is above 4 points. This indicates that evaluation and negative evaluation in shopping are the direction of great concern for consumers. Every time a product evaluation is written, the score is 2.20, which means that consumers do not often write comments.

When to write comments

Among those who can write comments, 48.57% are uncertain, 34.29% are satisfied, and 17.14% are dissatisfied.

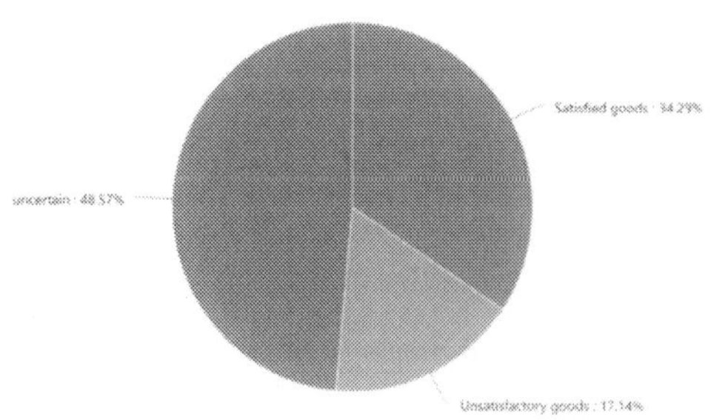

Figure 2. When to write a review

There is a significant difference between different genders in each evaluation ($p<0.05$), and there is no difference in other dimensions. Gender shows 0.01 level significance ($t=-8.500$, $p=0.000$) for each commodity evaluation. The average value of men (1.00) is significantly lower than that of women (2.24). The frequency of evaluation of male groups will be significantly lower than that of female groups.

Table 4. Results of t-test analysis by gender

	gender(M±SD)		t	p
	Male (n=2)	Female (n=68)		
Read product reviews every time they buy something	3.50±0.71	4.47±0.98	-1.379	0.172
Write product reviews every time	1.00±0.00	2.24±1.20	-8.500	0.000**
Negative reviews will reduce your desire to buy	3.50±2.12	4.37±0.94	-1.244	0.218
Under what circumstances would you write a review	2.50±0.71	2.13±0.91	0.563	0.575

* p<0.05 ** p<0.01

3.4 Correlation analysis

The correlation analysis mainly uses the person coefficient to test the degree of correlation between the data. If the correlation coefficient of two variables is positive, it indicates that there is a positive correlation between the two variables; if it is negative, it indicates that there is a negative correlation between the two variables.

Table 5. Pearson correlation

	Prefer to trust merchant or product reviews	Prefer to believe in good or bad reviews	Whether to avoid negative reviews of products	Whether or not they will change their opinion of the product because of the reviews
Age is	-0.034	0.031	0.121	-0.003
Monthly income	0.120	-0.242*	-0.121	-0.087

	Prefer to trust merchant or product reviews	Prefer to believe in good or bad reviews	Whether to avoid negative reviews of products	Whether or not they will change their opinion of the product because of the reviews
Education	0.073	-0.005	-0.039	-0.072
Annual consumption amount	0.264*	-0.014	-0.042	-0.098

* $p<0.05$ ** $p<0.01$

In the correlation analysis, it was found that the monthly income and the tendency to believe in positive or negative reviews had significant effects. The annual consumption amount and whether it is more inclined to add up or commodity evaluation have a relevant impact. Among them, the monthly income can be explained as the higher the monthly income, the more likely the group is to believe in bad reviews. The higher the annual consumption amount, the more likely the group will believe in the evaluation content of the goods.

4 Conclusion

4.1 Results

My purpose in this study is to study the impact of negative evaluation on consumer behaviour. As predicted, negative evaluation is the focus of consumers. Consumers will reduce their purchase desire after seeing negative comments. However, there are too few male samples in the sample, so it is uncertain whether male samples can bring the same results. In addition, groups with higher monthly income tend to believe in poor reviews. In other words, high-income people have a stronger ability to perceive risks. Because high income buys life safety action (Daniel Kahneman and Angus Deaton, 2010)[15].

4.2 Research Significance, Limitations and Future Directions

Many studies have shown that word of mouth communication affects consumers' attitudes towards various products and services (Klement Podnar, Pina Javernik, 2012).[16] Consumers have a tendency to trust, and the probability of failure in online shopping is reduced. My findings show that negative comments will largely affect consumers' purchase desire. Therefore, many Taobao merchants have chosen to spend money to buy praise to attract more consumers. However, the data shows that 15.71% of consumers are still willing to believe the few bad comments. Therefore, the development direction of merchants should be changed from changing the direction of commodity evaluation to improving product quality.

References

1. Do-Hyung, P; Jumin, L; and Ingoo, Han "Information Overload and its Consequences in the Context of Online Consumer Reviews" (2006). *PACIS 2006 Proceedings*. 28.
2. Tang, H., & Cao, H. (2020). A review of research on detection of fake commodity reviews. Journal Of Physics: Conference Series, 1651(1), 012055. https://doi.org/10.1088/1742-6596/1651/1/012055
3. Gao, B., & Song, Y. (2012). Analysis of consumption behaviour.
4. Zhuo, J., 2008. Wang luo ying xiao. Beijing: Qing hua ta xue chu ban she.
5. Zhang, P. (no date) The Influence of Short Video Advertising on Women's Consumption Behavior, The Impact of Short Video Advertising on Women's Consumer Behavior -- Marketing Circle, Issue 05, 2021. Available at: https://www.cnki.com.cn/Article/CJFDTotal-YXJI202105095.htm (Accessed: September 28, 2022).
6. Bai, L. (2021). Analysis of the impact of customization on consumption evaluation behaviour in e-commerce platform. Retrieved 12 September 2022, from.
7. Wang, H. (2021). An Empirical Analysis of the Influence of Social Factors on Online Shopping Consumption Behaviour -- Taking the Online Evaluation of Digital Commodities as an Example. Retrieved 13 September 2022, from.
8. Zhang, J. and Huang, Y., 2020. Good Reviews Go out the Door, Bad News Travels Fast?—Research on the Influence of Reference Groups and Their Language Styles in Positive Online Reviews on Consumers' Brand Attitudes. Open Journal of Social Sciences, 08(02), pp.271-295.
9. Melović, B., Šehović, D., Karadžić, V., Dabić, M. and Ćirović, D., 2021. Determinants of Millennials' behaviour in online shopping – Implications on consumers' satisfaction and e-business development. Technology in Society, 65, p.101561.
10. Vollero, A., Sardanelli, D., & Siano, A. (2021). Exploring the role of the Amazon effect on customer expectations: An analysis of user-generated content in consumer electronics retailing. Journal Of Consumer Behaviour. https://doi.org/10.1002/cb.1969
11. Farhad K, Atena S, S. M. K, A Comparative Performance Evaluation of Search Items and Experience Items in Amazon Customer Reviews. Retrieved 16 September 2022, from.
12. McKnight, D.H., Choudhury, V. and Kacmar, C. (2002) "Developing and validating trust measures for e-commerce: An integrative typology," Information Systems Research, 13(3), pp. 334–359. Available at: https://doi.org/10.1287/isre.13.3.334.81.

13. Glover, S., & Benbasat, I. (2010). A Comprehensive Model of Perceived Risk of E-Commerce Transactions. International Journal Of Electronic Commerce, 15(2), 47-78. https://doi.org/10.2753/jec1086-4415150202
14. Kim, D., Ferrin, D., & Rao, H. (2008). A trust-based consumer decision-making model in electronic commerce: The role of trust, perceived risk, and their antecedents. Decision Support Systems, 44(2), 544-564. https://doi.org/10.1016/j.dss.2007.07.001
15. Kahneman, D., & Deaton, A. (2010). High income improves evaluation of life but not emotional well-being. Proceedings Of The National Academy Of Sciences, 107(38), 16489-16493. https://doi.org/10.1073/pnas.1011492107
16. Podnar, K. and Javernik, P. (2012) "The effect of word of mouth on consumers' attitudes toward products and their purchase probability," Journal of Promotion Management, 18(2), pp. 145–168. Available at: https://doi.org/10.1080/10496491.2012.668426.

SHS Web of Conferences 155, 01016 (2023)
SDMC 2022
https://doi.org/10.1051/shsconf/202315501016

"Cultural and Creative IP" Empowerment Model for Red Culture under the New Media Environment

Xiaoling Ye[1], Chenya Liao[2], Xiaofan Cai[3], Zixin Wu[4] and Lin Xue[5,*]

[1]Department of Journalism and Communication, Minjiang University,350108 Fuzhou, China
[2]Department of Business Management, Minjiang University,350108 Fuzhou, China
[3]Department of Physics and Electronic Information Engineering, Minjiang University,350108 Fuzhou, China
[4]Department of Law, Minjiang University,350108 Fuzhou, China
[5]School of Innovation, Entrepreneurship and Creation, Minjiang University,350108 Fuzhou, China

Abstract. Boasting exemplary national qualities and high educational value, Red Culture is a unique spiritual wealth that make the Chinese nation stand out from the rest nations. Under the new media environment, however, the collision between the disparate characteristics of Red Culture and new media has given rise to the predicament plaguing the communication of Red Culture. With the advent of the 100th anniversary of the founding of the Communist Party of China, both tangible and intangible cultural heritage and resources are being valued, tapped and cultivated across the country. With the cultural and creative industry in the ascendant, the concept of intellectual property ("IP") has seeped into all aspects of our everyday lives. Against that backdrop, this study attempts to explore novel ways to link up the Red Culture resources scattered across China, to combine education with cultural tourism through the "4+5+2" cultural and creative IP empowerment model, and to pass on and spread Red Culture and add new value to the industry on the strength of influential IPs.

1 Introduction

In our days, the widely accepted Red Culture refers to "a characteristic, advanced culture brought into existence through integration and absorption by the Chinese people under the leadership of the Communist Party of China amid the historic struggles for the liberation and freedom of the Chinese nation and during the three pivotal periods of socialist transformation after the founding of New China"[1]. It's also called the "Revolution Culture", and the revolutionary spirit is generally hailed as the "Red Spirit".

On the occasion of the centenary of the founding of the Communist Party of China, a sweeping wave of learning the party history and carrying out red ideological education has been set off across the country. Availing themselves of their historical connections with the red revolution, provinces, cities and districts have been keen on tapping resources available at hand to build Red Culture memorial halls and education bases, in a drive to boost the effectiveness of publicity and education. However, in the context of the on-going epidemic control and the unstoppable rise of new media, the publicity and education work in many places is seemingly missing the point and has become a mere formality, not to mention the poor collaboration among publicity agencies, who tend to roll out similar and homogeneous cultural and creative products that instead curb the audience's enthusiasm for learning. Given the communication plight consequent on the divide between Red Culture and new media, this study will draw on some cases relating to the cultural and creative industry and IP creation to shed light on the basics and connotations of the "Cultural and Creative IP" model, and to explore effective ways to better materialize the communication value and economic value of Red Culture, in an effort to empower the cultural industries in addition to fulfilling the mission of educating people.

* Corresponding author: 2184@mju.edu.cn

2 Pain Points and Predicament of Red Culture Communication under the New Media Environment

As a new form of media entirely different from the traditional media, new media are best known for their instant, efficient, interactive and experience-centric ways of communication and the diversified, massive, fragmented and entertaining contents of communication. In the old days, Red Culture was mainly communicated through then important and popular traditional media such as newspapers, radio, etc. The collision between the disparate characteristics of traditional media and new media has given rise to the predicament plaguing the communication of Red Culture. Drawing on the classic "5W" model of communication proposed by Harold Lasswell[2], the problem can be examined from the following five dimensions:

2.1 Who (communicator): lack of voices

The communication of Red Culture in the past relied heavily on the traditional media, and the communicators were mainly journalists working for traditional paper media. Although they have gained deep expertise and rich experience in the sector of traditional media, most journalists fail to acclimatize themselves to the new media era, and some even encounter career crises. It takes time to bring in and foster talent proficient in the use of new media for communicating Red Culture. For the time being, the uneven competency and professional quality of practitioners will inevitably lead to the lack of voices from Red Culture communicators for a certain period of time.

2.2 Says what (message): trapped in the three nets of "amusing ourselves to death", "decentralization" and "information cocoons"

As mentioned earlier, "amusing ourselves to death", "decentralization" and "information cocoons" are unarguably the three nets impeding the communication of Red Culture. Fast-paced life and ever-mounting pressure make people favor those amusing and entertaining contents, the viral spread of which can be best explained by the "Meme" theory[3] — that is, people amuse and entertain themselves through the constant imitation and reproduction of the so-called popular culture. "Decentralization" allows each individual to take part in the entertainment on the internet, thereby weakening the voice of influencers or even misleading others. Driven by big data algorithm, the "personalized recommendation" is virtually narrowing the scope of information received by individuals in an imperceptible way, thus creating "information cocoons". These three barriers are marginalizing the Red Culture by trapping it in the predicament of being unable to reach out to the people, or to increase its visibility, or to make its voice.

2.3 In which channel (medium): fragmented contents are imperiling the integrity and weight of Red Culture

It is undeniable that compared with traditional media, new media are more multi-dimensional in terms of media presentation, breaking through the plane boundaries of graphics and texts and playing to their strengths in sound, image, video and real-time interaction. However, if their fragmented forms of presentation are not artfully handled, the integrity and weight of Red Culture itself will be imperiled. If a story is pieced up with one-sided elements and plain information, it would be difficult for the audience to gain a comprehensive understanding of Red Culture. Such kind of superficial information, lacking a good sense of atmosphere and substitution, may leave the audience with a "boring" stereotype, prompting them to resist receiving such information and even the akin culture.

2.4 To whom (receiver): gradually waning attachment to and dependency on the contents

Amid its long-run development, traditional media have acquired a fixed set of audiences and developed relatively fixed forms of expression. The new media, however, have even wider audiences and tend to go after headlines and hot topics. Compared with the audiences of traditional media, audiences of new media are less attached to and dependent on the contents of Red Culture.

2.5 With what effect: lack of appeal and credibility

Diversified cultural collisions has sparked public opinion, the network discourse system impacts the voice of mainstream values, and the guiding mechanism is becoming weak under the massive fragmentation of information and the profit-seeking big data algorithm. Under the influence of these factors, the red ideological and cultural communication is submerged in a large number of entertainment, one-sided and even wrong information, which appears to be unattractive, and the credibility will inevitably decline, bit by bit.

3 Cultural and Creative Design and IP Exploitation of Red Culture

Although the rise of new media is impeding the spread of Red Culture, it is also bringing about new technologies and new opportunities for the cultural industries, especially for the cultural and creative industry. According to the China Culture Consumer Index 2019, "the Internet has emerged as the most important conduit for cultural consumption, and the applications of 5G, VR, blockchain and other technologies are beefing up this trend ... short video related activities have become the most popular online cultural services."[4] On top of that, with 75 cities listed as the first batch of demonstration cities and pilot cities in cultural and tourism consumption by the Ministry of Culture and Tourism in December 2020, provinces and cities are seizing on this very opportunity to fully tap their cultural consumption potential and to make constructive attempts in expanding cultural tourism consumption, providing cultural services and designing cultural and creative products. Without question, all these efforts will lay a strong technological foundation and provide new consumption patterns for cultural and creative design, product consumption and cultural communication.

Improving both the content and expression of cultural products is also a key direction for development in China's cultural and creative industry in recent years. A number of outstanding creative cultural enterprises and brands have carried out "IP exploitation" around the most representative cultural relics, characters and even animals. They also further combine with popular things through the Internet for secondary creation, creating all kinds of cultural and creative lifestyle products including tableware, stationery, decorations, clothing and even film and television works loved by consumers[5]. For example, Forbidden City Taobao has built on the historical and cultural heritage of the Forbidden City to launch a stunning collection of cultural and creative products that are both artistic and useful in daily life. By integrating the royal culture people used to look up to with everyday items, Forbidden City Taobao not only effectively drives up the added value of its products but also successfully spreads the unique charm of traditional Chinese culture with cute and trendy designs[6]. The ever-mounting influence and customer base of Forbidden City Taobao have not only buoyed the IP exploitation efforts of the Palace Museum, but also provided a proven and replicable business model for other cultural and creative brands and practitioners and laid a strong industrial foundation for sustainable development.

Aside from the splendid IP products created by Forbidden City Taobao, there are also a wide array of shoddy and imitative products churned out by others. Those products are hardly distinct from each other. The imbalanced development of cultural and creative industry in various parts of China has led to significant differences among provinces. Against the macro backdrop of recalling history and advocating the protection and mining of red resources, and given the high spirit of patriotism among people, we must, building on the existing technological and industrial basis, exploit featured IP contents while passing on and carrying forward the Red Culture, and devise characteristic cultural and creative products tailored to local conditions. To this end, the "Cultural and Creative IP" model tailored to the promotion of Red Culture is proposed below to assist in the development of cultural and creative industry in respective provinces, including cultural tourism.

4 Construction of the "Cultural and Creative IP" Model for Red Culture

4.1 Put emphasis on the unique characteristics and turn disadvantages into advantages

Under the new media environment, the fragmented communication must be rebuilt by supplementing and introducing extracted contents rather than looking at matters in isolation and interpreting out of context. Under this premise, the fragmented spread of Red Culture is conducive to its acceptance among the audience, who will gradually supplement their own red thoughts, deepen their ideological and political understanding, and eventually develop a full-fledged ideological and political mindset.

Moreover, big data analysis also allows accurate and dynamic demand forecasting, thus pointing the way forward for product design. By collecting feedback, designers can also make timely adjustments to products and the forms of cultural communication.

4.2 Make explicit orientations and engage fans in product design

In addition to identifying the target audiences and consumers, market-oriented red-themed cultural and creative products must also take into account their emotional and demand orientations.

In terms of target audiences and consumers, it is imperative to discern the different preferences of teenagers and adults. Serious, obscure sermons cannot touch the hearts of young people. To them, short stories narrated by cartoon characters might be more appealing, and that's why IPs such as "Red Star Boy" and "Little Solder A-rong" are gaining currency among teenagers. For middle-aged adults and elders who share strong feelings for the red history, "cute" cartoon characters are obviously inappropriate. Instead, plain but graceful products would better cater to their practical or gift-giving needs.

From the perspective of meeting the diversified needs of consumers, cultural and creative products can be divided into two broad categories for meeting either the practical needs or cultural needs of consumers, such as the accessories, furniture, stationery and other products launched by the British Museum. Merchants sometimes recommend different products according to different gift recipients, a reflection of their keen insight into consumers' emotional needs. The overarching value the Red Culture holds is education and emotional identity. Bearing this in mind, we can devise some parent-child interactive products or educational toys aimed at presenting certain cultural images. In this way, not only can parents tell red stories to educate their children in spending time with their kids, but parents themselves can also deepen their understanding of and find ideological resonance and spiritual identity in Red Culture, thereby strengthening their national sentiment and national pride.

Furthermore, when a specific IP has a lot of fans, online social platforms and offline facilities such as cultural centers and museums can be utilized to engage fans in the design of cultural and creative products, such as by indirectly collecting fans' opinions and suggestions on product design.

4.3 How the model works

4.3.1 Element extraction and design

4.3.1.1 Extract cultural elements from the four aspects of "people, things, stories and spirit" through meticulous investigation

"People" refer to those who fought bravely and relentlessly for the victory of the revolution, such as the well-known revolutionary heroes, martyrs, and veterans. "Things" refer to the various objects used during the revolution, such as red scarves, red armbands, military uniforms, military caps, grass roots, tree bark, rickshaws, radio stations, rifles, etc., which are not only the witnesses of the revolutionary history but also the embodiment of the revolutionary spirit. "Stories" refer to those major historical events during the revolution, such as important meetings, important battles, important milestones, etc., which jointly made up the great revolutionary history. "Spirit" refers to the revolutionary spirit, such as Hongyan spirit, Jinggangshan spirit, Yan'an spirit, etc. The revolution witnessed the birth of many revolutionary heroes who were indomitable, selfless, self-denying and innovative[7]. The lofty outlook on life and values and the holistic views that these heroes represent should be fully extracted, as they still have great educational significance for the current and even future generations.

"People, things, stories and spirit" are independent of each other, yet they also unite as a whole through progressive reinforcement, thus contributing to the in-depth interpretation and all-round transformation of the connotations of Red Culture. Cultural and creative products allow Red Culture to seep into all aspects of people's everyday lives through figurative objects adored by people, who will strengthen their socialist ideals and beliefs after realizing that today's happy life has been earned hard. By extracting cultural elements from the four aspects of "people, things, stories and spirit", we can make sure that each cultural and creative product has a rich and profound heritage of Red Culture, and that these products will remind people of the ideological connotations and historical value of the revolutionary spirit and increase people's awareness of potential dangers and patriotism[8].

4.3.1.2 Enhance product experiences from the five senses of "sight, hearing, smell, taste and touch" through scientific design

In taking information from the external environment through sensory organs such as ears, eyes, mouth, nose and skin, people consciously observe, analyze, perceive and process things surrounding them, followed by the deconstruction and integration of the information obtained and then higher-level rational construction. Five-sense design aims to enhance the consumers' sense of experience and participation, to enrich the forms of products, and to add more fun to products. On the strength of the five-sense design techniques and concepts, different cultures and stories can be interpreted through human senses such as sight, hearing, smell, taste and touch. By adding immersive, interactive and experience-rich design elements, consumers will find emotional resonance in the products and develop deeper feelings, which can in turn help pass on the Red Culture.

For example, Henan Museum has recently rolled out an archaeological blind box which contains different kinds of exquisite imitations of cultural relics with rich historical and cultural connotations. The blind box also comes with a Luoyang shovel, the archetype of Chinese archaeological tools. The archaeological blind box not only highlights the unique characteristics of Henan Province, but also conveys a sense of immersion, enhances the interactivity of cultural and creative products, enable consumers to empathize with the difficulties facing archaeological exploration and cultural relics preservation, and helps today's young people develop their interest in and enthusiasm for archaeology[9].

4.3.2 Construction of "online + offline" IP matrix

We can harness various social media platforms to open official accounts for IPs and to launch special columns introducing local Red Culture. According to the styles and characteristics of different platforms, we can design interactive adventure-themed mini-games based on various short stories, publish interesting short videos about playing the red hero for one day, properly adjust the approaches to video editing or the length/depth of articles, spread Red Culture in the form of special topics, create adorable and captivating IPs, and design unique cultural and creative products. Furthermore, we can also team up with revolutionary scenic spots to help boost the visibility of their IPs. We can combine local unique IP characters with location IPs and red story IPs so that IP characters can tell about Red Culture in an easy-to-understand way for teenagers, thus bring into being a vertical matrix of "locations, stories and people", which can be further expanded to include book publishing, online fun courses and cultural tourism. After building a vertical matrix for each IP, different vertical matrices can be bound together and expanded to form a greater horizontal matrix, thereby inaugurating a grand ecosystem of socialist Red Culture with Chinese characteristics. By promoting the close linkage between online and offline communication efforts, IP characters and Red Culture will become deeply rooted in the hearts of the people.

5 Value and Significance of the "Cultural and Creative IP" Model for Red Culture

5.1 Add momentum to industrial development

From the perspective of promoting economic development, the IP exploitation and the cultural and creative design of Red Culture seem to be the breakthrough point for the development of cultural and creative products in many places. Building on the collections of cultural facilities and museums, and harnessing the revolutionary stories and heroes in these places, we can roll out diversified cultural and creative products and kick off the relevant commemorative activities, which would contribute to the creative transformation and innovative development of Chinese culture. These efforts can also be combined with services such as cultural tourism related hospitality and catering to extend the industry chain, thereby preparing the ground for the Red Culture industry to thrive and undergirding local economic growth and high-quality development of regional and national economies[10].

5.2 Give full play to the value of Red Culture in boosting patriotism and educating people

In the current era of ubiquitous entertainment, "eyeball economy" and "traffic mania" have spawned a flurry of chaos in the cyberspace. People who have long been bothered by vulgar tastes and insane groupies are eager for more high-quality works and products that spread positivity. The popularity of The Age of Awakening well justifies the charm of Red Culture. Meticulously produced and history-respecting film and television works with intriguing storylines and appealing characters will never go unnoticed. In recent years, many patriotic films were not only greeted with roars of approval, but also became box office hits, suggesting that the successful works of Red Culture are by no means scant. Keeping up with the trend of the times, these works are leading the way of thinking and have won the support and recognition of a vast number of audiences. In such a context, IP-oriented cultural and creative designs can immerse our next generation in Red Culture since childhood, encourage them to follow the example of heroes, and help them develop a sense of patriotism[11].

On the strength of new media, we can break temporal and spatial boundaries and spread Red Culture in various forms across the whole country. Through online and offline endeavors, we will take Red Culture to the hearts of more people, help them find resonance in Red Culture, and shoulder the paramount responsibility of boosting patriotism and educating people.

6 Conclusion

"The color of the People's Republic of China is red, and we shall never let it fade." Red Culture represents those fine qualities of the Chinese nation, the Chinese communists and the general public. It contains the spirit of adhering to the revolutionary ideals, the spirit of unity and cooperation, and the indomitable spirit of hard work and innovation, which fits the Chinese nation 's personality and value pursuit of enduring hardship, advocating justice and daring to fight. Consequently, red culture is an important part of Chinese values. In order to entrench Red Culture in the hearts of the people and take ideological and political education to a whole-new level, we must spread Red Culture in forms favored by the people. The "4+5+2" cultural and creative IP model is ideal for Red Culture communication, as it allows the adjustment of target audiences and orientations to cover multiple spheres and age groups in addition to promoting the coordinated development of cultural industries. Aside from creating economic value, we must stay true to our original aspiration of boosting patriotism and educating people, and play a greater role in passing on, spreading and transforming Red Culture.

Acknowledgements

The research is supported by the Minjiang University Principal Fund Project in 2021(Project name:" Cultural and Creative IP" Empowerment Model for Red Culture under the New Media Environment; Subject No.:103952021060).

References

1. Wang, E.L. (2012) Feature: Red Net in Red Culture. http://www.xibaipo.com/news2007/news/2012HSWH/121112304831578 7G6DDCFIB4A3A5E.html
2. Lasswell, H.D. (2013) The Structure and Function of Communication in Society. Beijing Broadcasting University Press, Beijing.
3. Dawkins, R. (1990) The Selfish Gene. Oxford University Press, New York.
4. The Beijing News, (2019) China 's Cultural Industry Index 2019 Released : China 's Popularity Overtakes Japan. https://ishare.ifeng.com/c/s/7sikU1c3wuD
5. Deng, C.J. (2019) Design of Cultural and Creative Products of Museums under ' IP Remodeling '. J. Grand View of Fine Arts, 2019 (04) : 126-127.
6. Dong, Y.Q. (2021) Soft Cultural Communication in Cross-Border Marketing from the Perspective of New Media - A Case Study of the Palace Museum. J. New Media Research, 2021, 7 (11) : 50-52.
7. Wang, J.T., Liu, W.F. (2021) Red Revolutionary Spirit Cultural and Creative Product Design and Development Research. J. Design, 2021, 34 (09) : 8-10.
8. Jiang, X.W., Li, J., Zhang, S.H. (2020) Research on Cultural and Creative IP Construction of Red Culture Museum. J. Beauty and the Times (City Edition), 2020 (09) : 93-94.
9. Li, M.Z. (2021) Blind box makes museum cultural and creative trend. J. Cultural industry, 2021 (19) : 136-137.
10. Yuan, J.L., Miao, D.W. (2020) The Symbolic Strategy of Constructing Red Tourism Cultural and Creative Products from the Perspective of IP. J. Journal of Hechi University, 2020, 40 (01) : 49-53.
11. Wang, H.L. (2019) Opportunities, Problems and Strategies of Red Culture Communication under the Background of New Media. J. Journal of Nanjing Xiaozhuang University, 2019, 35 (05) : 80-83 + 124.

https://doi.org/10.1051/shsconf/202315501017

Exploring the Communication Strategy of Intangible Cultural Heritage in the New Media Context

Mei Meng*

Fujian Normal University, Radio and TV Editing Department, 350100 Fuzhou City, Fujian Province, China

Abstract: As the pearl of China's intangible cultural heritage, there are still many difficulties in the development of opera. This paper takes Meilin opera of Sanming City, Fujian Province, as an example, analyzes the current situation of the inheritance of Meilin opera in the context of "Internet+", and compares it with the development of similar operas, and finally discusses the new strategies for the protection and development of intangible cultural heritage such as opera from the perspective of communication path and communication mechanism.

1. Introduction

Originated from Hui opera, Taining Meilin opera has a long history of more than 300 years, retaining the artistic characteristics of Hui opera and is integrated with local folk arts in Fujian. In 2006, Taining Meilin opera was listed as the first batch of "National Intangible Cultural Heritage", which is a rare physical heritage resource for studying the art of Hui school and exploring the history of Chinese opera.

However, according to Mr. Li Xiuzhen and Ms. Chen Canxia, the non-genetic inheritors of Meilin opera, in today's digitized society, the troupe is still mainly in the rural touring performance, the overall economic situation of the troupe's capital investment is not optimistic, talent training is yellow, and is even facing the danger and challenge of making the troupe unsustainable and the extinction of the genre.

2. Literature review

From the available literature, studies on the development and dissemination of Taining Meilin opera have some focuses: introducing protection policies, changing the dialect to Mandarin to make it more worth-seeing, designing cultural products such as gift boxes and nurturing elites, etc. In recent years, scholars have mostly focused on the innovation of traditional programs such as the performance system and art modeling of Meilin opera and have systematically researched techniques such as the steps of Lao Sheng opera, the neck work of three-flower in the roles of Chou and the duck-shaped step, as well as the mostly lost masterpieces such as spitting fire and changing skirts (Feng Youcai, 2016).

At present, few scholars have conducted in-depth discussions on the media communication of Taining Meilin opera in the digital era. In this paper, we take the digital media communication of Meilin opera as the research object to explore the communication value, current situation and future communication methods of it in the context of "Internet+" full media era.

3. Communication dilemma of Meilin opera in the context of media change

The antiquity, rigidity, and ruggedness of the Taining Meilin opera make it both important in historical, cultural, aesthetic artistic and educational value (Xu Xiaoyan, 2015). As one of the first intangible cultural heritage projects, Meilin opera not only plays an important role in the study of Hui art and historical tracing, but also is a specimen of literary fusion where the cultures of the Central Plains and Min-Yue collide and combine and is an important literary resource for the study of Chinese folk aesthetics. In addition, the Meilin Drama Troupe of Taining was awarded the honor of "the first troupe in the world" in the activity of "unique local characteristics drama" by the Ministry of Culture of China. Its communication value is evident. However, 77.63% of the 219 respondents said they had never heard of Fujian Meilin Theatre, and 77.89% of them were Fujian household members.

3.1 Status of the dissemination of Taining Meilin opera

At present, the inheritance and dissemination of Meilin opera is mainly divided into two parts: online and offline,

*Corresponding author: mmvanessa@163.com

with offline as the main focus. First, the leading group for the protection of intangible cultural heritage of Taining County and the research group of Meilin opera are set up to hold regular protection seminars, excavating the characteristics of the genre, collecting ancient materials and rehearsing traditional plays. (Ding Xueqin, 2017) Second, a wide range of public performances are carried out in rural areas, communities and schools, and has been performed in other regions and countries like Hong Kong, Taiwan, Japan, and Korea. Third, young talents are needed. Only one official Weixin account "Taining Meilin opera" of Meilin Opera Art Heritage Protection Center was opened, releasing performance previews, introduction videos and fixing camera videos in a few related institutions media and self-media accounts.

But in today's "Internet+" era, all elements of communication are changing.

Regarding the dissemination of non-heritage culture represented by Meilin opera in the "Internet+ era", the author conducted an online questionnaire survey from October 3, 2022 to October 11, 2022, and a total of 219 valid questionnaires were collected. In terms of geography, it covers 17 provinces and cities in China, among which Fujian ranks first with 43.38%. In terms of gender, most respondents were female, accounting for 64.84%. In terms of age, most of them (73.97%) are in the age of 18 to 23. In terms of education and profession, those with bachelor's degree are the most numerous, accounting for 76.26%, followed by master's degree students (15.07%). Among them, those studying or engaged in humanities and social sciences accounted for 45.21%. Among the 219 valid questionnaires, 90.41% of the respondents said that intangible cultural heritage such as opera is currently facing modern communication difficulties.

With the establishment of the new media ecosystem, the absolute authority and leadership of traditional media has been weakened. Everything can be a "medium". Gradually, professional media (including the traditional and the new), institutional media, self-published media, platform and machine are coexisting in the communication pattern.

The audience has also changed from "audience" to "user". With the widespread use of the UGC model, individuals can edit, create and publish messages freely after receiving them. The secondary dissemination extends the reach of the message, and then a wider range of users again becomes the producer of the media product, and the message spreads exponentially.

Figure 1 Change of communication audience

The demand triggers a revolution in imaging. With the development of technology, the content of communication has also shifted from the traditional single text, picture and sound to both audio and visual images (Cui Baoguo, Xu Lijun, Ding Mai 2021). Many theater companies have chosen to perform in a "split version" to meet the public's preference for "short, frequently and fast". However, the emergence of too many folding plays has led to controversies about the overall watchability of opera books.

The questionnaire results show that 48.86% and 22.37% of the respondents generally learn about the non-heritage culture of opera through self-media user graphics and videos and social media. In the case of Taining Meilin opera, for example, the audience can only search and passively receive information on their own, and the opera propaganda is not a large-scale system, too single in form and channel and overly dependent on user autonomy, which is caught in a vicious circle of communication dilemma.

When the elements change, the previous means of communication are no longer applicable. Therefore, placing the Taining Meilin opera in the new media context has become an inevitable trend and a necessary path for the future development of non-heritage culture.

3.2 Conservation and dissemination measures of other drama genres of the same type

Along with the development of the Internet and digital media, local opera non-heritage cultural preservation units have gradually begun to modernize the new form of opera art communication.

Table 1 Protection and dissemination measures of other operas of the same genre of Taining Meilin opera

Sichuan opera	French-style Sichuan opera *Jing Hua Yuan*	Combining traditional vocal and skills like "face-changing" and "water-sleeving" in Sichuan opera with French-style choreography and nonsensical lines.
	Game-style Sichuan opera *Dream Jianghu*	The plot and costume combine classic elements of Sichuan opera, giving the game the connotation.
	Promotional video *Dream Jianghu*	Jin Yong's Jianghu world is interpreted by traditional forms of Sichuan opera.
Guangxi Wenchang, fishing drum opera, Cai diao, Guangxi opera	Digital Mass Art Gallery (digital album, virtual exhibition hall, etc.)	AI panoramic technology is adopted to experience the charm of opera performance online.
Qinqiang opera clubs: Jingju Theater Company of Beijing, Yi Su She, San Yi She	"Cloud live" payment model	Through the form of live broadcast, the transmission of the traditional stage is extended.
Peking opera *Spring Feast*	The first VR opera recorded in China	Breaking the boundary between virtue and reality, to build a sense of

		"immersive".
Gao Jia Opera *Sister A Da*, Peking Opera *Mu Guiying* and *Zhao's Orphan*	Theatrical film	Breaking the "fourth wall" and create modern audio-visual effects.
Hunan Huagu Opera *Tinkering*	(micro) Anime *Tinkering*	Speeding up and refining the original meaning, according with the new media's transmission pattern.

4. The two-track parallelism of the dissemination path of Taining Meilin opera

4.1 The means of dissemination of Taining Meilin Opera

CNNIC's 49th Statistical Report on the Development Status of the Internet in China shows that in December 2021, the size of China's internet users reached 1.032 billion, and the Internet penetration rate reached 73.0%. Among internet users, the scale of video users was 975 million, accounting for 94.5%. The scale of short video users was 934 million, reaching 90.5% (2022).

From the above data, new media communication in China is increasingly flourishing, showing four main communication trends: mobile as the mainstream of information dissemination, social media as an important channel for news acquisition, media to achieve intelligent, segmented and accurate communication, and short video has become the latest form of information dissemination. Traditional media still occupy a dominant position, but online digital media and platforms have become the main segment of the media industry. (Cui Baoguo, 2021)

The development of the opera genre depends on cultural policy and environment. Based on this, online communication should be the primary focus to expand the influence and radiation of Taining Meilin opera. The interactive media matrix needs to be established to expand the Internet publicity space and communication channels of Meilin Opera. Besides, two-track parallelism can dig offline entity depth exchange program, to spread both breadth and depth.

4.2 The dissemination effectiveness of Taining Meilin Opera

In *Personal Use of Mass Communication*, Katz et al. proposed a basic model of "Uses and Gratifications" in the process of contact between the public and the media, that is, the causal chain process of "social condition + personal characteristic → media expectations → satisfaction of demand".

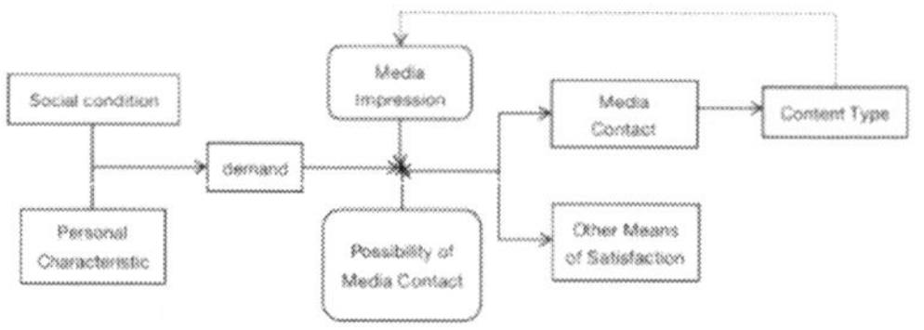

Figure 2 "Uses and Gratifications" model diagram

Immordino-Yang's research suggests that opera is not only a tool for education, but also an alternative to experiencing and exploring real-life situations in a fictional setting. (van de Water Manon.2021)

Opera enriches audience's experience. Therefore, in terms of communication effectiveness, we should also focus on humanistic emotions in communication, audience's psychological needs, possible access to the communication channels and the enhancement of the contact experience to explore new possibilities of digital communication of opera in the "Internet+ era".

5. Communication Strategy of Taining Meilin Opera

In *Culture and Commitment*, Margaret Mead categorized human cultural transmission mechanisms into three types: pre- inter- and post-metaphorical cultures (Mead M., 1987). WMU added an information literacy module to the original theatre curriculum to familiarize students with the sources of theatre, (Sachs Dianna E. & Duffy IV Michael J.. , 2022) which is a typical pre-metaphorical cultural setting.

The questionnaire shows that the access of respondents to information in the current internet era is mainly focused on video (62.56%), audio (62.1%), knowledge communities (54.79%), social (54.34%), and short self-published videos (53.42%). It shows that the current forms of cultural communication are more concentrated in the mutual metaphorical culture of internet communities among peers and the former metaphorical culture from descendants to predecessors.

Therefore, the new strategy of communicating Meilin opera should be based on community communication on all platforms, supplemented by traditional authoritative media news reports and offline participation at the same time.

5.1 Online Communication

5.1.1 IP image construction and application

Firstly, we should let the virtual IP "live" in the form of graphics and videos on the social media platform.

There are two ways to build the virtual image of the IP. One is to extract the characteristic elements and refine the spirit, character-modeling, moves and makeup of Meilin opera players to enrich the character connotation of the IP image. For example, the "Jiang Ricai" portrayed by the inheritor Ms. Li Xiuzhen is strong with great personality value. The performance of the artist Wang Yaokang in "Half of a Ruyi" is humorous and vivid with his neck work and duck walk. In addition, there is also a unique "three-dot red" makeup of Meilin opera. The IP image can also be made into expression packs containing the classic phrases in the play for social communication, so that more people can accept and spread the essence of Taining Meilin opera in a subtle way.

Secondly, we can create new IP stars by relying on classic IP genes. Based on the opera culture of Meilin opera, a series of virtual IP characters are formed with typical original ones, such as Cui Yunlong, Bian Yixiao, and Zhang Xuyu from "The Deprecated Official".

5.1.2 Community

Unlike traditional media such as professional media organizations, the main body of social media is users, which is a tool and platform for people to share their opinions and experiences based on network technology. That is, to stimulate users' desire to create, express and share, emphasizing interactivity, allowing users to enjoy the ability to create, to gather themselves into some kind of reading and listening community, and to realize the precise delivery of message preferences.

The establishment of the Taining Meilin opera community on various social media breaks the limit of geography, time and space, aggregating the scattered individual Meilin opera fans into a community with certain influence. They have an identity to share and recreate the opera with related peripheral, graphic and video resources within the community. In this way, more and more information about Meilin opera will be generated, which can form a secondary or even multiple dissemination with greater influence.

5.1.3 Being available in various video specifications, multi-platform linkage to produce a sensational effect

As a traditional opera genre, the modernization of Taining Meilin opera can be spread in both visual and auditory communication channels, choosing a combination of horizontal and vertical video screens or a combination of long and short videos, and developing output content strategies according to the characteristics of the platform.

Among the 100 most popular IOS apps, more than 90% are for vertical screen apps. Snapchat reports that the playback completion rate of full-screen vertical video ads on its platform is 9 times higher than that of horizontal ones. On Facebook, the use of vertical video can increase views by 130% and engagement by 4 times (Tencent Media Research Institute, 2019). This means that vertical video is the big trend of mobile terminals.

Compared with horizontal screens, vertical screens can bring a more natural and intimate feeling. Therefore, vertical videos with strong audio-visual impact should be the first thing to consider, which can be an overview and introduction of a theater company, with a simple structure, fast editing, high soundtrack and beautiful images, to play an "eye-catching" effect.

The global video creation platform can focus on favoring drama, documentary and science videos. Each video should have a complete story structure and a plot with twists and turns to meet the expectations of the audience. The use of strong interactive means such as pop-ups can bring a sense of simultaneous participation. The advantage of the large number of expatriates in Fujian Province has, to a certain extent, satisfied the audience base and dissemination conditions for the overseas dissemination of Meilin opera.

The social media platform is not only a promotion channel for cultural products, but also has strong interactive features and relies on KOLs who have certain influence. Therefore, Meilin opera can seek cooperation with relevant KOLs and release a series of participation and experience videos with "I" as the main body for dissemination.

On (paid) live streaming platforms, Meilin opera's stage is brought to the screen of mobile devices. The use of internet technology broadens the boundaries of traditional stages, reduces the cost of performance and theater-going, and gives audiences a different kind of theater-going experience.

Opera originated from the art of storytelling. Therefore, auditory communication should also be emphasized. Meilin opera plays, storytelling, historical commentary and song adaptations can be recorded through audio software such as podcasts, allowing users to hear Meilin opera anytime and anywhere.

5.1.4 Re-creation with animation, games, film integration

In the 1930s, Benjamin wrote in his book *Toward the Age of Fading Light* (2004) that "film can be analyzed in greater detail and reveal more dimensions of perspective than painting and theater can offer. Compared to stage theater, film is more amenable to analysis because it is easier to isolate the individual components." (P.97) Therefore, the classic plays of Meilin opera can be expressed through the film language, breaking the "fourth wall", presenting multi-view, multi-directional and three-dimensional audiovisual effects, increasing the sense of space and communication of modern opera, which allows for wider dissemination.

The results show that 90.41% of the respondents are willing to participate in opera games, among which 50.68% are strongly willing. Taining Meilin opera can rely on its own IP. With animation as the appearence and traditional stories as the core, it can refine the spiritual connotation of each character, and create roles rich in Meilin Opera characteristics. Gamification of the opera story is carried out, so that players can be transformed into opera characters and experience the beauty of Meilin opera in an immersive way.

5.1.5 Application of VR/AR technology

VR technology gives the power of viewing to audience, simulating various stimuli for people's senses, trying to create a "complete space" that "surrounds the individual", so that opera no longer stops at "imaginary reality". Users can freely choose the viewing types, to get a "private custom" stage opera viewing experience.

AR technology allows the AR experience to see both real and virtual images in a head-mounted device. Users can project the virtual opera stage to their own homes and outdoors to experience the "opera in front of them" without having to leave home.

Relying on the technical principle of VR and AR, the mobile device can create a 360-degree free viewing database of Taining Meilin opera, so that the cultural venues and resources database of opera can be visualized and freely presented.

5.2 Offline Communication

5.2.1 To research and develop peripheral cultural and creative products

The results show that 72.6% of the respondents have the willingness to purchase the cultural and creative peripheral of opera, but 52.51% of them think that the existing design is not impressive enough. The theater company can choose the currently coveted blind box and other practical peripheral products to cross the border co-branding, and carry out multi-dimensional and interactive promotion from production to sale on social self-media platforms, ultimately realizing the role of publicity and funds to feed Meilin opera.

5.2.2 Combination with emerging cultural industries

Currently, murder mystery games are very popular, which are of mixed quality, and excellent works rich in cultural connotation are urgently needed. The results show that 78.99% of the respondents are willing to participate in script killing or board games adapted from opera stories, and 38.81% of them said they are very willing. Therefore, the Taining Meilin Theatre Company can make new adaptations of opera stories into script-killing plays or tabletop games and provide traditional Meilin Theatre costumes to wear, allowing for an immersive experience.

5.2.3 Combining culture and tourism to enrich the reality experience

Fujian is famous for opera and tourism. The results of this study show that 93.15% of the respondents indicated that they are willing to experience non-heritage culture in person when traveling, among which 56.62% of them are strongly willing to do so. It is possible to reverse the road to the promotion of Meilin opera by promoting cultural tourism and non-heritage experience activities in and around Taining County, Sanming City, and releasing activities such as offline punching and on-site teaching of Meilin opera. In-depth study and hands-on experience of classic works can help regenerate the inheritance of Meilin opera.

6. Conclusion

China has a long history and a fruitful intangible cultural heritage. "A good wind sends 'intangible heritage' to the clouds". We are in the era of rapid development of "Internet+", we should make good use of new means of communication brought by the times to create a communication chain, which integrates front-end design, content output and integration of the media matrix. We should implement a dual-track communication mode with online and offline linkage to complement each other, so as to promote the Meilin opera of Taining and even all intangible cultural heritage on the road of succession and development.

References

1. CNNIC. (2022). The 49th China Statistical Report on Internet Development. Publishing House of Electronics Industry.
2. Cui BG, Xu LJ, Ding M. (2021). Report on Development of China's Media Industry (2021), social sciences academic press (CHINA).
3. Ding XQ. (2017). Research on the Protection and Development of Non-Foreign Heritage -Taking Sanming Daqiang Opera and Meilin Opera as Examples. Populer Literature (09),3-4.
4. Feng YC. (2016). The contemporary heritage of Meilin opera and the characteristics of its genre. Oriental Art(S1),83-84.
5. Margaret Mead. (1987). Culture and Commitment: a study on generation gap. Hebei People's publishing House.
6. Sachs Dianna E., Duffy IV Michael J.. The drama of information literacy: Collaborating to incorporate information literacy into a theatre history curriculum[J]. College & Undergraduate Libraries,2022,28(2).
7. Tencent Research Institute. (2019). Global vertical screen development trend report: six dimensions to explain the new vertical screen ecology.
8. van de Water Manon. (2021). Drama in education: why drama is necessary. SHS Web of Conferences. doi:10.1051/SHSCONF/20219802009.
9. Walter Benjamin. (2004). A Short History of Photography (The Work of Art in the Age of Mech anical Reproduc). Guangxi Normal University Press Group, P.97.
10. Xu XY. (2015). A Preliminary Study on the Artistic Characteristics of Meilin Opera and its Research Value. Appreciation (04),101.

Exploration of Rural Revitalization Planning in Guangdong under the Strategy of Rural Revitalization: An Example from Youshan Town of Nanxiong City

Youhuan Li [1,*], Songtao Shou [2]

[1] Social Cooperation Department, Guangdong University of Finance and Economic, Guangzhou 510320, Guangdong, China.
[2] College of Business Administration, Guangdong University of Finance and Economics, Guangzhou 510320, Guangdong, China.

Abstract. To achieve common prosperity is the fundamental goal of socialism, and the implementation of rural revitalization strategy is an inevitable requirement for achieving common prosperity. Based on the background of rural revitalization strategy, this paper summarizes the theories related to rural revitalization in line with China's national conditions, clarifies the main tasks of rural revitalization planning in Guangdong, further analyzes the ways to promote rural revitalization planning in Guangdong, and on this basis, takes the rural revitalization planning of Youshan Township in Nanxiong City, which was concluded and promulgated in July 2022 under the auspices of the author, as an example, and discusses the four aspects of planning background, general requirements, main tasks and guarantee measures. This is with a view to providing certain reference and reference for the relevant functional departments to prepare and improve the rural revitalization plan.

1 INTRODUCTION

In 2017, President Xi Jinping first proposed the rural revitalisation strategy in the report of the 19th Party Congress. Implementing the rural revitalisation strategy is not only an urgent requirement to resolve the imbalance between urban and rural development, but also an important grasp to resolve the contradictions in the operation of the market system, and an important element to address the development of modernised agriculture. In 2021, the Central Committee of the Communist Party of China formulated the 14th Five-Year Plan for National Economic and Social Development and 2035 Outline of Visionary Goals, which further clarifies the planning and leading role of strengthening rural construction. Planning is a reflection and consideration of future global, long-term and fundamental issues, a development vision that integrates the views of many elements and people, and a programme for designing a whole set of future actions. Rural revitalization should be guided by the rural revitalization plan to lead the comprehensive development of rural revitalization; and the rural revitalization plan should be used as a grasp to promote the high-quality development of agriculture and rural areas. Therefore, in order to revitalise the countryside, planning needs to come first.

Guangdong is one of the regions with the most dynamic economy and the highest level of openness in China, yet as of 2021, only 7 of the 21 prefecture-level cities in the province have a GDP per capita of over 70,000 yuan, which is higher than the national average; the rest are below the national average, and 10 cities even have a GDP per capita of less than 50,000 yuan, far below the national average, with a wide gap between the rich and the poor. For this reason, cities and counties in Guangdong have been actively preparing rural revitalization plans according to the actual local development situation, and are committed to promoting common prosperity in Guangdong in the comprehensive promotion of rural revitalization. Based on the above-mentioned background, this paper summarises the theories related to rural revitalisation in line with China's national conditions, clarifies the main tasks of Guangdong's rural revitalisation plan, further analyses the ways to promote Guangdong's rural revitalisation plan and, on this basis, takes the rural revitalisation plan of Youshan Township in Nanxiong City as an example to analyse the planning background, general requirements, main tasks and safeguard measures, with a view to providing some reference for the relevant functional departments to prepare and improve the rural revitalisation plan. It is intended to provide certain reference and reference for the relevant functional departments to prepare and improve the rural revitalization plan.

2 THEORY OF RURAL REVITALIZATION PLANNING

If we want to prepare the rural revitalization plan scientifically, we must first grasp accurately the relevant

* Corresponding author: 469280272@qq.com

theories that are highly compatible with the essential characteristics and basic laws of China's countryside. The following mainly introduces and elaborates two theories that are more closely related to China's countryside.

2.1 New Urbanisation Theory

The theoretical basis of new urbanisation originates from Marx's theory of urban-rural integration, and scholars have studied the connotation of new urbanisation from different perspectives. Peng Hongbi and Yang Feng (2010) focus on the scientific connotation of new-type urbanisation, i.e. the development goal of urban-rural integration, led by the scientific concept of development, with intensification and ecology as the development mode, and with functional diversification and system rationalisation as the basic content [1]. After the 18th Party Congress first proposed to adhere to the new type of urbanisation path with Chinese characteristics, the academic community has gradually enriched the research on the connotation of new type of urbanisation, with the following representative views: Shan Zhuoran and Huang Yaping (2013) proposed that new type of urbanisation has three core connotations of both people's livelihood, sustainable development and quality [2]. Song Liansheng and Jin Yuehua (2016) considered new urbanisation as the urbanisation of lifestyles, employment patterns, public services, spatial areas, social governance and the beautification of the human living environment [3]. Dong et al. (2017) point out that the connotation of new urbanization is the path of urbanization in the new era of facing urban problems, facing deep reforms, and seeking transformation and development around the coordinated development of quality and chemistry [4]. According to Chen Mingxing et al. (2019), the connotation of new urbanisation includes four aspects of human-centredness, synergy, inclusiveness and sustainability [5]. Ren Shine et al. (2022) outlined four features of the evolution of the connotation of new urbanization, which are people-centred, integrated urban and rural areas, coordinated development of towns and cities, and ecological civilization constraints [6].

In summary, although there is no unified definition of the connotation of new-type urbanisation in the academic community, a theoretical system has generally been formed to adhere to the concepts of people-oriented, sustainable development, integrated urban-rural development and green development, so as to achieve high-quality development of new-type urbanisation.

2.2 Urban-rural integration theory

Marx and Engels have long expounded on the relationship between urban and rural areas, which has provided the theoretical cornerstones and methodological principles for Chinese Marxists (Zhao Xiuling, 2015) [7]. At present, foreign scholars have not yet explicitly proposed the concept of urban-rural integration, and domestic scholars have their own different understanding of the connotation of urban-rural integration. According to Fang Liling (2006), urban-rural integration is a systematic project, and the premise and foundation of urban-rural integration is to grasp the development status and degree of connection between urban and rural areas in the region, and on this basis to formulate corresponding countermeasures for integrated urban-rural development [8]. Ge Dandong and Hua Chen (2010) suggest that urban-rural integration is the interactive development of city and township, which depends on the drive of city and even on township itself [9]. Song Jinsong (2012) points out that urban-rural integration is an important way to truly achieve integrated urban-rural economic and social development [10]. Chen Xiaofei et al. (2016) outline that the purpose of urban-rural integration is to integrate both urban and rural areas into the socio-economic system, improve the separation between urban and rural areas, gradually eliminate the urban-rural dual structure, establish a new urban-rural relationship and achieve an optimal and rational allocation of urban and rural resources [11].

In summary, although there is no uniform definition of the connotation of urban-rural integration in the academic community, the general idea is to achieve smooth flow of production factors between urban and rural areas and to coordinate urban-rural relations.

3 MAIN TASKS OF GUANGDONG RURA L REVITALISATION PLAN

The implementation of the rural revitalization strategy in Guangdong should comprehensively implement the spirit of the 19th CPC National Congress and carry out the five-in-one construction of economic, political, cultural, social and ecological civilization in accordance with the general requirements of the rural revitalization strategy of prosperous industry, pleasant ecology, civilized countryside, effective governance and prosperous life. The construction of the Five-in-One economic, political, cultural, social and ecological civilization.

3.1 Rural industrial planning under the goal of flourishing industries

In the revitalisation of the countryside and the construction of beautiful villages, the prosperity of industry is the focus. The specific planning content includes: firstly, the planning of efficient and ecological modern agriculture; secondly, the design of transformation of agriculture into a new business mode combining agriculture, tourism and culture; thirdly, the programme of cultivating new agricultural business subjects; fourthly, the planning of the development of traditional rural special industries.

3.2 Ecological construction planning under the objective of eco-livability

Ecological livability is the key to the revitalization of the countryside and the construction of beautiful villages. The specific planning contents include: firstly, to continuously improve the living environment in rural areas. Secondly, to promote the comprehensive environmental improvement in urban and rural areas; thirdly, to protect and preserve the rural style and carry out the demonstration of field architecture; fourthly, to implement the action of greening, beautifying, sequencing and brightening the countryside.

3.3 Rural cultural planning under the objective of civilization of the countryside

In the revitalization of the countryside and the construction of beautiful villages, the civilization of the countryside is a guarantee. The specific planning content includes: firstly, inherit and develop the excellent traditional culture in rural areas; secondly, strengthen the construction of rural public culture and improve the rural public cultural service system; thirdly, the construction of cultural halls and enrich the rural cultural industry; fourthly, the planning of rural public cultural product development and service supply.

3.4 Rural cultural planning under the objective of civilization of the countryside

Effective governance is the foundation for the revitalisation of the countryside and the construction of beautiful villages. The specific planning includes: firstly, strengthening the construction of rural grassroots party organisations; secondly, deepening the practice of villagers' self-governance; thirdly, improving the level of moral governance in the countryside; fourthly, building a countryside under the rule of law and a peaceful countryside.

3.5 Rural livelihood planning under the goal of living well

Wealthy living is fundamental to the revitalisation of the countryside and the construction of beautiful villages. The specific planning includes: firstly, the design of income generation channels for farmers; secondly, the construction of a platform for farmers' entrepreneurship and innovation; thirdly, the promotion of upgrading rural infrastructure; fourthly, the promotion of the development and use of renewable energy in rural areas.

4 WAYS TO PROMOTE RURAL REVITAL ISATION PLANNING IN GUANGDONG

Guangdong's strategy to promote rural revitalization should be based on the theoretical foundation of new urbanization and urban-rural integration that conforms to the essential characteristics and basic laws of China's countryside, and guide the orderly and sustainable implementation of related projects.

4.1 New Rural Urbanisation Project

Small rural towns are key nodes in the development and revitalisation of rural areas, and are also growth poles for the economic growth of rural areas. Promoting the new type of rural urbanisation can effectively drive the social, economic and cultural revitalisation of rural regions.

4.2 Urban and Rural Integration Project

The solution to rural problems should not be limited to the internal perspective of the village. By addressing the three rural issues from a new perspective of urban-rural relations, many problems will be solved. The key to urban-rural integration is the extension of urban infrastructure to the countryside and the coverage of urban public services to the countryside. Therefore, promoting urban-rural integration and urban-rural integration is a strategic way to achieve rural revitalisation.

4.3 Agricultural industry structure transformatio n and upgrading project

Traditional agriculture is built on the soil of a smallholder society, with low commodity rates, small scale and the goal of self-sufficiency. It is obvious that the small-holder production model cannot improve the economic development of villages. Therefore, traditional agriculture should be optimised and upgraded in terms of industrial organisation, spatial layout, leading industries and value chains. Modern agriculture, ecological agriculture, recycling agriculture, integrated agriculture of three industries and other advanced agricultural modes are the targets for the optimisation and upgrading of rural industrial structures.

4.4 Rural Cultural Development Project

Rural culture is the most important component of traditional Chinese culture, and China's excellent traditional culture is mainly rooted in the countryside. The culture of farming and reading in the countryside has nurtured the excellent national qualities of the Chinese nation. Remembering nostalgia and retaining the charm of the countryside is the main content of rural revitalization. The construction of rural cultural projects is equally important as the transmission of rural culture and the cultivation of a rural cultural atmosphere, and the undesirable tendency of emphasizing the construction of facilities at the expense of cultivating an atmosphere should be prevented.

4.5 Rural grassroots organisation building proje cts

Rural grassroots organisations have a political function but are different from political organisations. They play a major role in the social stability of villages, the optimisation of economic order, the transmission of public opinion and the implementation of local guidelines and policies. Office space, funding and the cultivation of mobility are the three main aspects of planning for the construction of rural grassroots organisations.

5 EXPLORING THE PREPARATION OF A RURAL REVITALISATION PLAN FOR Y OUSHAN TOWNSHIP

Based on the theory of new urbanization and urban-rural integration, this paper takes the rural revitalization planning practice of Youshan Township in Nanxiong City, which the author presided over, as an example, and discusses four aspects: planning background, general requirements, main tasks, and guarantee measures .

5.1 Planning Background

Youshan Town belongs to Shaoguan Nanxiong City, Guangdong Province, and is located in the north of Guangdong Province. It is adjacent to Wuchun Town in the east, Huangkeng Town in the south, Dengfang Town in the west and Chijiang Town in Dayu County, Jiangxi Province in the north, 33 kilometres away from Nanxiong City. The township covers an area of 146.78 square kilometres, of which 48,000 mu is arable land and 147,000 mu is forest land, making it a typical mountainous town in Nanxiong. Youshan Town has 1 community under the jurisdiction of Youshan Community, 17 administrative villages including Datang, Pingtian O, Lianshan, Dalan, Pinglin, Huangdi, Aijing, Shangjiao, Paste Tian, Kongcun, Yancun, Jinpi, Shangshuo, Jiehekou, Huangtian, Hahui and Gucheng, and 90 natural villages, of which the town government is located in Datang Village. In 2021, the per capita disposable income of rural residents in the town was RMB 18,512, lower than the average level of RMB 44,993 in Guangdong Province; the financial expenditure was basically the same, both being RMB 25,572,200.

5.2 General requirements

Adhering to the guidance of Xi Jinping's thought of socialism with Chinese characteristics in the new era, comprehensively implementing the requirements of the 19th Party Congress and the 19th Plenary Session on rural revitalization, implementing the spirit of General Secretary Xi Jinping's important instructions and instructions on Guangdong's work, implementing the spirit of Secretary Li Xi's speech during his research in Shaoguan, deeply implementing the new development concept of innovation, coordination, green, openness and sharing, and building a rural revitalization pattern for the township based on the general requirements of prosperous industry, pleasant ecology, civilized countryside, effective governance and prosperous living. The town will build a rural revitalization pattern, strengthen and enlarge traditional industries, cultivate new industries such as green energy, actively develop cultural and tourism industries, strengthen the construction of rural e-commerce industries, build a modern industrial system with Yushan characteristics, and promote the high-quality development of Yushan economy. At the same time, we will promote the construction of industry, talent, culture, ecology and organisation, and accelerate the development of high quality and high efficiency agriculture, livable and livable countryside, and rich and prosperous farmers.

5.3 Main tasks

This plan is based on the core concepts of urban-rural integration, industrial foundation, characteristic protection, sectoral coordination and multiplicity of plans. The premise is to strengthen the spatial use control of rural areas, to balance efficiency and equity, to coordinate the construction of urban and rural infrastructure and public facilities, and to create a distinctive rural landscape with the goal of rural scenery.

5.3.1 Shaping a new pattern of rural revitalisation

We will adhere to the double-wheel drive of rural revitalisation and new urbanisation, implement the urban and rural territorial spatial planning pattern, construct the overall layout of Youshan, optimise the production, living and ecological space of Youshan countryside, realise rural revitalisation in a classified and orderly manner, and co-ordinate the balance of resources and environment and the unity of economic, social and ecological benefits.

5.3.2 Build a modern industrial system with in-depth integration of one, two and three industries

Actively promote the integrated development of one, two and three industries, and enhance the modernisation of the whole industrial chain. Deeply explore the special agricultural resources such as high-quality rice, navel oranges and watermelons in Youshan Town, mountain and forestry industry resources such as honey and moso bamboo, red resources such as the ruins of the Shangshuo Soviet Government and the People's Hall, as well as the transportation location advantages of the national highway and expressway in northern Guangdong, develop special agriculture, rural e-commerce and cultural tourism, and accelerate the transformation and upgrading of traditional industries.

5.3.3 Promote ecological improvement of the countr yside

The creation of the human environment in Youshan will be oriented towards the construction of beautiful and livable villages, with the treatment of rural rubbish and sewage, the improvement of village appearance and ecological protection as the main directions, to carry out rural habitat improvement actions, to comprehensively improve the quality of rural habitat, to improve the stability of the ecosystem and the quality of the ecological environment, to promote the value-added of rural ecological capital, and to achieve the unity of the people's wealth and ecological beauty.

5.3.4 Improve infrastructure support

The construction of rural revitalization in Youshan Township should be combined with the construction of roads, water conservancy, water supply, electricity supply, logistics and information in rural areas, so as to accelerate the improvement of the quality of good living of the residents in the township and villages, promote the livability and livability of the countryside, and provide strong support for the rural revitalization of Youshan Township.

5.3.5 Strengthening public service construction

Insist on the organic connection between rural revitalization and the consolidation and expansion of the results of poverty eradication, and make the consolidation of the results of poverty eradication a priority task in the implementation of the rural revitalization strategy. In conjunction with the construction of beautiful pueblos, promote the improvement of the basic public service system in the countryside, optimize education, medical and health care, and social security resources in rural areas, and gradually improve the comprehensive functions of the farmers' living circle centred on the Youshan township.

5.3.6 Prosperity and prosperity of rural culture

With the core socialist values as the leader, inherit and carry forward the excellent traditional culture, actively cultivate a new era of civilized countryside, good family style and simple folk style, promote cultural prosperity, enhance the spirit of farmers, continuously improve the level of civilization in rural society, accelerate the formation of a new style of rural civilization, and create a rural culture with Youshan characteristics.

5.3.7 Strengthen rural talent support

Focusing on who will cultivate the land and who will revitalise the countryside, we will adapt to the demand for high-quality talents in the modernisation of agriculture and rural areas, insist on a combination of local cultivation and importation, and promote the revitalisation of rural talents in Youshan.

5.3.8 Strengthen the governance of grassroots orga nisations

Fully carrying forward the fine traditions of the historical revolution in Youshan, we are solidly promoting the grasping of party building for rural revitalisation and constantly strengthening the standardisation of grassroots party organisations. Focusing on the standardized construction of four party group service centers in Yau Shan neighborhood committee, Ping Tian Au village, Huang Di village and Ai Jing village, we concentrate on building a rural governance system that combines self-governance, rule of law and moral governance led by the Party's grass-roots organizations as the core and sound Party organizations.

5.4 New Rural Urbanisation Project

We will understand the importance of promoting the implementation of the rural revitalization strategy, to comply with the objectives and tasks, it grasps the pace and puts forward sustained efforts to ensure achieving the rural revitalization in Youshan on schedule.

5.4.1 Strengthen organizational leadership

Adhere to the overall leadership of the Party. Strengthen the Party's leadership of rural revitalization work, give full play to the Party Committee's leading role of overseeing the overall situation and coordinating all parties, and provide a strong guarantee for the realization of the township's rural revitalization plan. Insist on the mechanism that major matters, important issues and important work of rural revitalization are discussed and decided by the party organization, so as to grasp the work of the three rural areas and give full play to the role of the grass-roots party organization as a fighting base and the pioneering role of party members, so as to lead the masses to join the great cause of rural revitalization.

5.4.2 Strengthen coordination and linkage

Strengthen the co-ordination of major matters, enhance coordination and cooperation, and form a more efficient mechanism for promoting work. The Party Committee of the town takes the lead in promoting the implementation of the plan and the implementation of relevant policies, and is responsible for coordinating various relevant departments to promote it in a solid and orderly manner. The town's Agricultural Office and all relevant departments strengthen coordination and cooperation, formulate implementation plans for the main objectives and tasks in the areas characterized by rural revitalization undertaken, and clarify the responsible subjects, timetables and roadmaps. Ensure that the plan is coordinated with provincial and municipal tasks in terms of development ideas, main objectives and key tasks.

5.4.3 Strengthening financial security

Establish and improve the financial input guarantee system for the implementation of the rural revitalization strategy, implement various subsidies from the governments of Guangdong Province, Shaoguan City and Nanxiong City in strict accordance with the relevant regulations, tilt public finance more strongly towards the three rural areas, ensure that financial input is compatible with the objectives and tasks of rural revitalization, increase the co-ordination and integration of financial agricultural-related funds, and make reasonable use of We will also make greater efforts to integrate financial resources related to agriculture and make rational use of special funds to improve the level of financial services in towns and villages. Take key construction projects as a carrier, actively make use of national credit policies to expand domestic demand, and actively seek bond funds for rural public welfare construction projects. Strengthen the supervision and management of funds to support agriculture and improve the efficiency of the use of financial support for agriculture. Give full play to the role of finance and finance in driving economic growth and optimising economic structure.

5.4.4 Strengthening assessment and supervision

According to the main tasks and responsibility requirements determined by the plan, and based on the principle of scientific setting and reducing the burden on the grassroots, a list of tasks and construction targets will be formulated each year, and the binding indicators determined by the plan will be incorporated into the assessment and evaluation index system of the town government to implement the management of planning targets and responsibilities. Improve the system of annual assessment and mid-term evaluation of the plan, check the implementation of the plan, analyse the effect of the plan's implementation, identify problems and propose improvements in a timely manner, and revise the content of the plan in an appropriate manner. Strengthen the comprehensive evaluation and assessment of the completion of target tasks such as structural optimisation, facility upgrading, livelihood improvement, resource conservation and environmental protection. Departments and villages (dwellings) should effectively carry out the main responsibility, co-ordinate and promote the implementation of the tasks, and strive to create a number of demonstration projects, strengthen demonstration and leadership, to ensure the effectiveness of the work. Units and individuals with outstanding work achievements will be commended and rewarded.

5.4.5 Reserved land adjustment

Strengthen the collation and planning of rural land use, and report construction land use plan indicators to protect rural land use needs. Coordinate all land use activities in agriculture and rural areas, and the town's overall land use plan may set aside a certain proportion of planned construction land indicators for agricultural and rural development. On the premise of ensuring agricultural land for agricultural use, the demand for land for the construction of agricultural industrial parks is reasonably guaranteed. On the premise of conforming to the overall land use plan, the town-level government can prepare village land use plans by reporting to the Nanxiong Municipal People's Government, adjust and optimise the layout of land for arable land protection, village construction, industrial development and ecological protection, and arrange for a certain proportion of annual land use plans to be used specifically to support the development of new rural industries, new business models and industrial integration. Sound land use policy for facility agriculture, under the premise of not occupying permanent basic farmland, land for various production facilities and ancillary facilities needed in the process of agricultural production, and land for ancillary facilities that must be built for large-scale agricultural operations are included in the management of land for facility agriculture and are implemented for record.

6 CONCLUSION

Focusing on the twenty-word strategy of the rural revitalization strategy, this paper explores the relevant solution path towards in Guangdong while highlighting the implementation of the strategic objectives and requirements, it takes the five key elements in Youshan Town of Nanxiong City as the research objects. In general, the practice of rural revitalization planning in Youshan Township is somewhat cutting-edge, but under the new normal of rapid knowledge innovation, new progress in policies and concepts may be made at any time, and this paper is written to exchange and provide some ideas.

The implementation of the rural revitalisation strategy is an inevitable choice to achieve common prosperity for all people and a necessary way to realise the Chinese dream. Therefore, it is of practical and academic significance to explore the planning for the revitalization of Guangdong's countryside under the rural revitalization strategy, and it should be expanded and improved in the future with the planning practice, so as to provide more comprehensive support and guarantee at the planning level for the realization of Guangdong's rural revitalization strategy.

ACKNOWLEDGMENTS

This work was supported by the 2022 annual project of Nanxiong Youshan Township, Nanxiong City, entitled "Rural Revitalization Development Plan of Nanxiong City Yushan Township" [No. GDCH- 21YB072].

References

1. Peng Hongbi,Yang Feng. The scientific connotation of the new urbanization road [J]. Theoretical exploration,2010(04):75-78. M. Ben Rabha, M.F. Boujmil, M. Saadoun, B. Bessaïs, Eur. Phys. J. Appl. Phys. (to be published)
2. Shan Zhuoran, Huang Yaping. Analysis of the connotation, target content, planning strategy and cognitive misconceptions of "new urbanization"[J]. Journal of Urban Planning,2013(02):16-22.
3. Song Liansheng,Jin Yuehua. On the essential connotation of new-type urbanization [J]. Shandong Social Science,2016(04):47-51.
4. Dong Xiaofeng,Yang Chunzhi,Liu Xingguang. Expl oring the theory of new urbanization in China[J]. Ur ban Development Research,2017,24(01):26-34.
5. Chen Mingxing,Ye Chao,Lu Da Dao,Sui Yuwen,Gu o Shasha. The perception and construction of the the oretical connotation of new urbanization with Chine se characteristics[J]. Journal of Geography,2019,74 (04):633-647.
6. Ren Gao, Song Yingchang, Zhang Liao. The connot ation characteristics, level measurement and promoti on path of new urbanization in China[J]. Ningxia So cial Science, 2022(05):102-110.

7. Zhao Xiuling. The process of Chineseization of Marxist urban-rural integration theory [J]. Fujian Forum (Humanities and Social Sciences Edition),2015(09): 5-10.
8. Fang Liling. Urban-rural integration:An analysis of urban-rural linkage perspective[J]. Research on Finance and Economics, 2006(06):16-21.
9. Ge Dandong,Hua Chen. New directions of rural planning in the integrated development of urban and rural areas [J]. Journal of Zhejiang University (Humanities and Social Sciences Edition), 2010,40(03):148-155.
10. Song Jinsong. Three stages of urban-rural integration[J]. Urban planning,2012,36(01):33-38+45.
11. Chen Xiaofei,Yao Shimou,Zhang Chengcheng. Theoretical and practical issues of urban-rural integration in China in the context of new urbanization[J]. Geoscience,2016,36(02):188-195.DOI:10.13249/j.cnki.sgs.2016.02.004.

Investigation of the post-05 adolescents' families' attitudes toward fertility——A case study of Changshu

Junze Xie*

United World College Changshu China

Abstract: Under the influence of social, economic, cultural, and other factors in the new era, young people, who are responsible for the country's future, have different attitudes toward fertility from those of their predecessors. It is of great practical significance to study their attitudes toward fertility. Based on the analysis of the questionnaire survey on post-05 adolescents' families in Changshu City, this paper indicates:

1. The family income has a certain degree of influence on the number of children, but there is no positive correlation;
2. post-05 adolescents are more inclined to have brothers or sisters who have the same biological genders as they do;
3. Compared with their parents, who have low reproduction willingness, adolescents' reproductive willingness gradually increases.

Based on these conclusions, the following policy suggestions are put forward:

1. More opportunities should be created for practitioners to accompany their families.
2. Schools should avoid transmitting extreme gender normative education.
3. People should advocate the sustainability of adolescents' positive attitudes toward childbearing under the inclusive and diverse concept of fertility.

1 Introduction

People's perspectives on fertility refer to their relatively stable attitudes and proposals regarding fertility issues. It is closely related to the region, society, and period as a group value. From a micro perspective, a society's attitudes toward fertility directly impact a family's fertility willingness. In addition, a family's fertility willingness is also influenced by the social economy, regional customs, social culture, etcetera. The attitudes toward fertility can be interpreted by five aspects as mentioned in the Dictionary of Population Science: the motivation and purpose of fertility, the choice of the number and quality of children, the preference for female children, the requirement of childbearing age, and the mode of childbearing. [1]

Adolescents and their families, who are responsible for building the country's future, have different attitudes toward fertility than their predecessors due to the influence of social, economic, cultural, and other factors in the new era. It is of practical significance to study their attitudes toward fertility. However, there is currently a lack of systematic research on adolescents' and their families' attitudes toward fertility in Chinese academia. The lack of systematic analysis makes understanding the current situation of teenagers' views on family fertility impossible. It also leaves a blank study on its influencing factors, which is not conducive to the fertility policy's sustainability and effectiveness. Using a questionnaire, we analyze the post-05s generation family group in Changshu's attitudes toward fertility and their influencing factors. We propose policy suggestions to fill gaps in previous research and enrich the content of this research field.

*Corresponding author: xgjunze@gmail.com

2 Literature review

The study of people's attitudes toward fertility started at the end of the twentieth century in China, with a clear division of the research object's geographical category instead of the research object's age level. It primarily combined with factors such as fertility policy and economic system to study the influence of fertility attitudes in a specific area. For example, Cao Jingchun (1985) investigated the impact of commodity economy development on rural fertility [2]; Gu Shengzu (1986) discussed the relationship between birth policy and birth view [3]. By the beginning of the twenty-first century, academic circles had analyzed the various perspectives on the fertility of middle-aged, young, and older people. The research primarily focused on non-adolescent groups such as adult women and adult men in rural or urban areas. Wang Xu et al. (2019) stated that the change in rural women's fertility outlook is primarily due to improved family status and emphasis on quality of life [4]. Yuan Ming

(2019) interviewed 22-40-year-old women in Jinan following the implementation of the "universal two-child" policy and stated that the core of their fertility view lies in "self-value judgment and maternal value judgment." Their fertility view is influenced by "family status." [5] There have been few studies on teenagers' perspectives on fertility and few direct studies on teenagers. Most of them are presented indirectly through the mouths of elders, books, and movies. Wu Chengchao (2019), for example, used the film How to Be At Home to highlight the difference between parents' and offspring's perspectives on fertility [6]. The investigation of adolescents' and their families' fertility attitudes has received insufficient attention in existing research. In this paper, adolescent families in Changshu City, Jiangsu Province, are chosen as the subjects of the investigation on the attitudes toward fertility. The study not only compensates for the lack of existing research on the research object but also provides the theoretical value of the research and makes targeted suggestions for China's fertility policy. Therefore, the investigation has practical significance.

3 Questionnaire survey and analysis

3.1 Basic information of children's questionnaire

This questionnaire collected data on parents' and children's identities. A total of 152 questionnaires were distributed to children, and 150 questionnaires were effectively collected. Regarding the age distribution among 150 post-05s adolescents, 137 were born in 2005-2007, accounting for 91%; 1 was born in 2008-2010, accounting for 7.3%; Very few samples were born after 2010.

3.2 Survey status of adolescents' questionnaire

Table 1. Basic information of children's questionnaire survey (the table is created by the author)

Title	Option	Quantity	Proportion (%)
Year of birth	2005-2007	137	91.33
	2008-2010	11	7.33
	2011-2013	1	0.67
	2014-2016	1	0.67
Number of children	0	62	42.76
	1	57	39.31
	2	18	12.41
	3	2	1.38
	4	2	1.38

62 (41.3%) of the 150 valid survey samples have one child, 57 (38%) have two children, 18 (12%) have three children, and very few families have more than three children. Approximately 80% of families have one or two children.

According to whether post-05 adolescents and their families want to have younger siblings, 107 post-05s (70.67%) desired younger brothers and sisters, while 45 did not. When asking for reasons, 61 (30.2%) said: "my younger brothers and sisters can accompany me when I am lonely." 59 (30.2%) said: "I want to experience being an elder brother/sister," and 60 (30.2%) said: "younger brothers and sisters are adorable." Only 19 (9.41%) people "support their parents' desire for a second or third child."

When asked why they did not want younger siblings, 20 adolescents (24.7%) said: "my parents are too busy to take care of younger siblings." 7 (21.0%) said: "the financial burden of my family is too heavy." 13 (16.0%) said: "I want to be my parents' only child." The love for younger brothers and sisters, a sense of accomplishment as a big brother/sister, and the joy of life brought by younger brothers and sisters are the main reasons why some post-05 adolescents want younger brothers and sisters. In contrast, some post-05 adolescents do not want younger siblings because they do not have enough time or money to take care of them or because investing too much energy in younger siblings causes them to love themselves less.

When asked if losing parental care due to having a younger sibling is acceptable, more than half of the respondents said the loss was utterly acceptable. 12% said they would not have negative emotions, and 10% said their loneliness would be enhanced significantly. The polarization was evident.

According to adolescents' preference for their children's gender, 41 (30%) of the 136 valid samples have no explicit preference for their children's gender. Women are preferred by 53 (39%) of the population, while men are preferred by 42 (31%). The most common reason adolescents expect their gender to be male is: "I am a boy; my younger brother and I will have a common interest." Adolescents who expect their gender to be female gave two main reasons: "because a younger sister is easier to take care of than a younger brother" and "I am a girl; my younger sister and I will have a common interest." From the research, respondents' gender identities heavily influence their choice of gender for younger siblings.

We collected 93 valid answers from adolescents and their parents regarding childbearing age. The average childbearing age expected by people was 28.6 years, the median was 29 years, the maximum childbearing age was 40 years, and the minimum childbearing age was 24 years. There is a tendency to procreate later based on the age of childbearing.

3.3 Basic information of parents' questionnaire

We distributed 101 questionnaires for parents and recovered 100 effectively. The age of children is evenly distributed among 6-17 years old, of which 9-11 years old are the majority, accounting for 32%. Most respondents have two children, accounting for more than half, followed by one child, accounting for 39%. 8% of the sample have three children, and some samples have four children and no children.

3.4 Survey status of parents' questionnaire

Table 2. Basic information of parents' questionnaire survey (the table is created by the author)

Title	Option	Quantity	Proportion (%)
Year of birth	2005-2007	21	21
	2008-2010	24	24
	2011-2013	32	32
	2014-2016	23	23
Number of children	0	1	1
	1	39	39
	2	51	51
	3	8	8
	4	1	1
Monthly income	More than 10000 yuan	82	82
	5000-10000 yuan	12	12
	4000-5000 yuan	2	2
	3000-4000 yuan	1	1
	2000-3000 yuan	1	1
	Less than 2000 yuan	2	2

Regarding whether families want a second or third child, 47 parents want a second or third child, and 52 do not, with little difference. When asked why some parents want a second/third child, 29 (61.7%) said: "I do not want my children to be lonely." 18 people (38.3%) said: "I hope my children can accompany and take care of me when I become old." 14 people (29.8%) said: "I think babies are adorable," and 11 people (23.4%) said: "my children want to have younger siblings, so I support them." From the responses, we can conclude that parents want a second/third child, accounting for 61.7%, mainly to alleviate their children's loneliness. It is consistent with the post-2005 generation's response, "Younger brothers and sisters can accompany me and relieve my loneliness."

When asked why some parents do not want a second/third child, 33 people (62.3%) said: "they have no time to take care of their children." 19 people (35.8%) said: "their career development will be hampered." 17 people (32.1%) said: "they have a heavy financial burden," and 15 people (28.3%) said: "their physical health limits their fertility." Other reasons (20.8%) included "children will be under too much educational pressure" and "I am satisfied with the current number of children." Therefore, parents are hesitant to have a second/third child because they do not believe they have enough energy and time to care for another child simultaneously. The answer is consistent with the post-05s generation's response that "parents care less about themselves due to the birth of younger siblings."

When asked, "If you have a second or third child, do you think your child will lose parental care?" 39% of the parents said: "yes, but it is acceptable," while 26% said: "not at all." The distribution of parental responses to this question is similar to children's answers, but the responses are generally more positive.

According to parents' preference for their children's gender, 62 respondents (62%) have no specific preference for their children's gender. Female children are preferred by 24 people (24%), while male children are preferred by 12 people (12%). There are more than twice as many girls as boys among the answers. The first reason parents expect their children to be male is that boys are easier to take care of, and the second is that they do not want their children to suffer from childbirth. In contrast, parents who mainly expect their child's gender to be female think girls will spend more time with them and have a better parent-child relationship, followed by less financial pressure to have a girl.

The increasing number of parents who want their children to be female demonstrates that the patriarchal phenomenon is gradually fading in today's society. However, some parents have given stereotyped reasons for expecting their children to be male, such as "boys are emotionally insensitive" and "boys can protect their sisters," demonstrating the existence of gender stereotypes.

According to the respondents' monthly income, the overall level is high, with more than 80% of the samples earning more than 10000 yuan, 12% earning 5000-10000 yuan, and only a few samples earning less than 5000 yuan. We used the Chi-square test to investigate the relationship between the monthly income of parents and the number of children. We assume H_0 is "there exists no correlation between parents' monthly income and the number of children," and H_1 is "there is a correlation between parents' monthly income and the number of children." According to Table 3, the monthly income of different families significantly affects the number of children ($p<0.001$). The p-value is very small, indicating that the H_0 hypothesis can be rejected. Therefore, family income has an impact on the number of children.

Table 3. Chi-square test of the relationship between monthly income and the number of children (the table is created by the author)

Title	Option	Monthly income						Total	χ^2	p
		10000 yuan or more	5000-10000 yuan	4000-5000 yuan	3000-4000 yuan	2000-1000 yuan	Less than 2000 yuan			
Number of Children	0	0(0.00)	0(0.00)	0(0.00)	0(0.00)	0(0.00)	1(50.00)	1(1.00)	54.926	0.000**
	1	31(37.80)	7(58.33)	1(50.00)	0(0.00)	0(0.00)	0(0.00)	39(39.00)		
	2	42(51.22)	5(41.67)	1(50.00)	1(100.00)	1(100.00)	1(50.00)	51(51.00)		
	3	8(9.76)	0(0.00)	0(0.00)	0(0.00)	0(0.00)	0(0.00)	8(8.00)		
	4	1(1.22)	0(0.00)	0(0.00)	0(0.00)	0(0.00)	0(0.00)	1(1.00)		
Total		82	12	2	1	1	2	100		

* p<0.05 ** p<0.01 *** p<0.001

The monthly household income showed a significant level of 0.001 for the number of children (χ^2=54.926, p=0.000<0.001). The percentage comparison shows that the proportion of two children and three children for families whose incomes are more than 10000 yuan is 51.22% and 9.76%, respectively, slightly higher than the comparative value of 51% and 8%. 58.33% of parents whose income is 5000-10000 yuan choose to have one child, which is higher than the comparative value of 39.00%. The ratio of parents with 4000-5000 yuan income and one child is 50.00%, more significant than the comparative value of 39.00%. The proportion of two children for parents with both 2000-3000 and 3000-4000 yuan income is 100.00%, significantly higher than the comparative value of 51.00%. The proportion of zero children of parents whose income is less than 2000 yuan is 50%, considerably higher than the comparative value of 1.00%.

4 Conclusion & Policy Suggestions

As for post-05 adolescents, some desire younger siblings and sisters for various reasons, including their love for them, a sense of accomplishment in being a big brother or sister, and the joy they bring to life. On the other hand, some post-05 teenagers do not want younger siblings because they do not have the time or resources to care for them or because spending too much energy on taking care of their siblings makes them feel less satisfied.

Similarly, most parents desire a second or third child to lessen their kids' loneliness. Most parents are hesitant to have a second/third child because they do not believe they have enough energy and time to care for another child simultaneously. The answers are consistent with the post-05s generation's responses.

As for childbearing age, post-05 adolescents generally expect to reproduce later than their parents did. Parents have more optimistic views than adolescents regarding whether children will lose parental care if their parents have a second or third kid.

Eventually, the relationship between income and the number of children in a family is analyzed. Parents' decisions to have a second/third child are not only influenced by financial costs, but also factors such as time, energy, and their children's feelings.

Based on the above questionnaire survey, the following policy suggestions are proposed.

1. Although the chi-square test indicates that family income influences the number of children, this relationship is not positively correlated. Parents' decisions to have a second/third child are influenced by factors such as time, energy, and their children's feelings, not simply financial costs. The employment pressure of practitioners can be appropriately alleviated by increasing holidays or compressing working hours, allowing practitioners to spend more time with their families and rationally arranging maternity and paternity leave. As a result, parents will have sufficient time to take care of their children and thus will be more willing to have a second/third child.

2. Teachers and parents should reach a consensus to avoid transmitting extreme gender norms through school and social education to reduce the degree of gender stereotypes.

3. Teenagers' positive fertility willingness is gradually increasing compared to their parents. To advocate for the concept of social inclusion and diversity, we should support the long-term viability of teenagers' positive attitudes toward fertility.

References

1. Wu Zhongguan, Zhou Junyu, etc. Dictionary of Population Science, Southwest University of Finance and Economics Press, December 1997.
2. Cao Jingchun, On the Influence of the Development of Commodity Economy on Rural Fertility Situation, Population Research, No.4, 1985, pp.16-19.
3. Gu Shengzu: On Fertility View and Fertility Policy, Journal of Central South University of Political Science and Law, No.4, 1986, pp. 74-75.
4. Wang Xu and Liu Li: A Study on the Factors Affecting the Change of Rural Women's Fertility Outlook, Society and Public Welfare, 11th issue, 2019, pp. 68-70.
5. Yuan Ming: Research on the Fertility View of Urban Women Based on the Implementation of the "Universal Two-Child" Policy, Shandong University of Finance and Economics, 2017.
6. Wu Chengchao: Reflections on How to Be At Home, Popular Literature and Art, 17 (2019), pp. 196-197.

A study on the implementation of precise poverty alleviation policies in Luotan Village, Zhangjiajie City, Xiangxi Province —— A perspective based on the Mitt-Horn policy implementation system model

Dingke Yuan*

Civil Aviation University of China, School of Economics and Management, Department of Public Service Management

Abstract: Since General Secretary Xi Jinping proposed the policy of precise poverty alleviation in 2014, localities have made sufficient preparations to actively promote the implementation of poverty alleviation policies. China gained a comprehensive victory in poverty eradication in 2020, and the experience process is worth studying as each region has developed poverty eradication policies with local characteristics and achieved good implementation results. This article uses the Mitt-Horn system model of the policy implementation process as a framework to study the implementation of the precision poverty reduction policy in Luotan Village, Zhangjiajie City, Xiangxi Province. The study shows that the implementation of poverty alleviation policies in Luotan village of Zhangjiajie City is closely related to the degree of coordination between the respective functions and interactions of six factors: policy goals and standards, policy resources, implementation methods, characteristics of policy implementation agencies, system environment, and values of policy implementation personnel. Therefore, with the help of this model, we can better understand how poverty alleviation policies are effectively implemented in Luotan village, and encourage the participation of multiple parties to broaden the path of rural economic development and stimulate the endogenous power of rural construction.

1 Introduction

In 2020, China achieved a comprehensive victory in the battle against poverty, with 98.99 million rural poor people under the current standard all out of poverty, 832 poor counties all removed from the hat, regional overall poverty was solved, and the arduous task of eliminating absolute poverty was completed. In order to achieve this goal, the country has promulgated a series of "precise poverty alleviation policies". In the process of goal achievement, policy formulation is important, but policy implementation is also a key factor. The American scholar Alison has clearly stated that the program identification function accounts for only 10% of the process of achieving policy goals, while the remaining 90% depends on the effective implementation of the policy.[1]In their book "Policy Implementation Process - A Conceptual Framework", policy scholars Van Miette and Van Horn argue that there are many influencing factors between policy decisions and the transformation of policy effects, including the influence of the policy system itself and the policy environment. Therefore, the two scholars propose the famous Miette-Horn policy implementation system model to achieve further understanding of the whole process and elements of policy implementation in conjunction with practical research. Considering that the implementation of poverty alleviation policies involves a wide range of subjects and is influenced by many factors, and its implementation process is relatively complex, this paper attempts to introduce the Mitt-Horn policy implementation system model in the study of precise poverty alleviation policy implementation, to explore the applicability of the system model and to enhance the explanatory power and persuasive power of some practical issues such as the future development orientation of the model[2]

2 Analytical framework of policy implementation system model and its applicability in China

A theoretical model to explain the effectiveness of policy implementation was proposed by Mitt and Horn in the article "The Policy Implementation Process: a Conceptual Framework". The model includes six factors that influence policy implementation. Among the factors are: first, policy goals and standards. That is, the ultimate goal to be achieved by formulating policies and taking corresponding actions to solve public problems, and a series of indicators and other measurement bases to test

* Corresponding author: 1143303851@qq.com

whether the goal is achieved are specified in the relevant documents. Compared with general laws and regulations, the policy objectives and targets are more detailed and specific, and more operable and practical. Second, policy resources, i.e., the sum of the various resources needed to ensure the effective implementation of policies[3], The specific modalities can be the provision of technical advice and assistance, site visits, project reviews, feedback mechanisms, etc. Fourth, the characteristics of the policy implementation agency. This factor describes the attributes of the policy institution, such as the size of the institution, the capacity of the internal staff, the level of implementation, and the formal and informal links of the implementing agency with other institutions. Fifth, the system environment, most commonly the economic, political and social environment. For example, the overall development trend of the country, the level of economic development, the political climate, strategic opportunities, and the participation of society at large. It is important to emphasize here that the environmental factors in the model are not absolutely favorable, but also include unfavorable factors. Sixth, the value orientation of policy implementers. Policy implementers may be influenced in their ability and willingness to implement policies by their level of knowledge (knowledge or understanding) and attitude (acceptance, neutrality, or opposition) towards the policy. Implementers with positive attitudes toward policy will significantly increase the likelihood of successful implementation, while implementers with negative preferences for policy may try to surreptitiously divert attention from or even reject policy implementation, resulting in poor or even unimplementable policy implementation[4]The policy implementation model is shown in Figure 1.

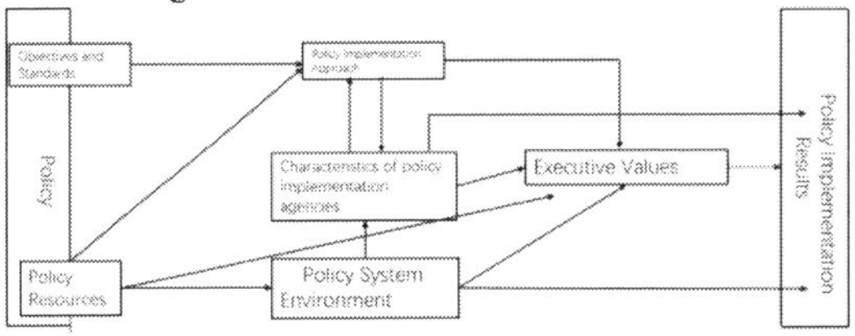

Figure 1 Mitt-Horn Policy Implementation Process Model

As an emerging discipline in China, public policy has been developed in China for a relatively short period of time. And the Mitt-Horn policy implementation system model was also proposed by Western academics and then introduced to China, and its research in the field of policy implementation in China is still immature. Although it has had success in applications in Western countries, But is it applicable to China, where the political system, economic system and even cultural habits are vastly different? In this regard, the author makes the following analysis. First, the model emphasizes the "top-down" leadership approach, which is fully consistent with the implementation of China's poverty alleviation policy, which was first proposed by the Party Central Committee and the State Council. The special position of the Communist Party of China in China has created a public policy implementation mechanism that is "party-led" and "embedded" between the party and the state, showing the characteristics of "high level promotion".[5] In addition to this, there is a distinct hierarchy of government departments at all levels, and as administrative organs, all levels of government, under the leadership of the State Council, promote the implementation of policies at all levels, creating a top-down political system. Therefore, the model is relevant for the study of precise poverty alleviation policy implementation in terms of the prerequisites used in the model. Second, from the domestic research in recent years, many scholars have applied the Mitt-Horn policy implementation system model in the fields of education, healthcare, artificial intelligence, and other policy implementation areas, and obtained better application conclusions or made adjustment suggestions to make the model more suitable for China's policy implementation characteristics. In summary, the Mitt-Horn policy implementation system model has a good fit for the analysis of pro-poor policy implementation performance in this case.

3 Analysis of the Mitt-Horn policy implementation system model

In the analysis of precise poverty alleviation policy implementation, Luotan village, as a typical representative of a mountainous area, inhabited by ethnic minority people and once a deep poverty village, has the ability to complete the task of poverty eradication in 2019, achieve "self-blood" and withstand the test of the epidemic without returning to poverty, its poverty alleviation policy implementation efficiency is undoubtedly worthy of our study.

3.1 Overview of Luotan Village

Luotan village is located in Zhangjiajie City, Hunan Province, Cili County, Sanguanshi Tujia Township, is located in Wulingyuan, Huanglong Cave and Zhangjiajie Grand Canyon, the intersection of three major scenic spots, is a typical remote mountainous areas, because of the closed terrain and closed information, the local people difficult to contact with the outside world, most of the low level of cultural quality, self-development ability is weak. At the same time, because most of the local villagers are ethnic minorities with unique ethnic customs and languages, it is difficult to popularize science and culture, the level of regional economic development is low, and the motivation is insufficient, making it a key poverty-stricken village in the province.

In order to achieve a well-off society, win the battle against poverty and gnaw the "hard bones", the Zhangjiajie Municipal Government even sent a task force across three levels to help the village directly to Luotan Village for point-to-point, one-to-one precision poverty alleviation. After the development of "Luotan Village Poverty Alleviation Plan", "Luotan Village Rural Tourism Industry Development Plan" and other documents to determine the development ideas, the municipal government has widely received funds to help alleviate poverty. By the end of 2019, a total of 24.05 million yuan has been invested in the construction of the countryside, and 10.2 kilometers of new and improved asphalt tourism roads and other roads have been built to fully promote the tasks of poverty alleviation. After 4 years of precise poverty alleviation, Luotan Village completed the removal of poverty and passed the national acceptance in 2019, and successfully realized the perfect transformation from a provincial key poverty-stricken village to a national rural tourism key village. The success of poverty alleviation policy implementation in Luotan Village is not a coincidence, it is the inevitable result of the interconnection and interaction of various factors such as policy objectives, policy environment, implementing agencies, personnel and policy environment, and its successful experience is worth exploring.

In order to further systematically study and summarize the experience of Luotan village's poverty eradication, this paper uses the Mitt-Horn policy implementation system model to conduct a Chinese study, with a view to preserving the valuable experience of China's poverty eradication and for other regions to learn from.

3.2 Specific analysis of the model

3.2.1 Policy goals and standards: hierarchical policy development goals lay down criteria for poverty alleviation effectiveness

Whether public policies can be effectively implemented cannot be separated from the rational formulation of their goals and standards. The American policy scientist Smith has highly valued the significance of policy objectives and standards, and even more directly regarded rational policy objectives and standards as the primary factors affecting policy implementation. The goal of public policy is the expected result of the implementation of the policy formulated by the government, which is the main line and direction of the whole policy implementation process. The policy criteria are the conditions for assessing whether the policy meets the pre-defined requirements, which are more objective and can be quantified.

Around 2013, the Central Committee of the Communist Party of China (CPC) and the State Council issued a series of national policy documents, such as "Implementation Plan for Establishing a Precise Poverty Alleviation Mechanism", "Opinions on Innovative Mechanisms to Promote Poverty Alleviation and Development in Rural Areas" and "Decision on Winning the Battle against Poverty", which clarified the poverty alleviation standard of "two no worries and three guarantees", with the ultimate goal of poverty alleviation, so as to achieve the ultimate goal of building a moderately prosperous society by 2020.National policies have laid a solid foundation for poverty alleviation in poor areas, but the poverty situation and the government's financial disposition vary from place to place. Therefore, based on national policies and consideration of its own actual situation, the Hunan Provincial Government adopted the Regulations on Rural Poverty Alleviation and Development in Hunan Province at the meeting of the Standing Committee of the Hunan Provincial People's Congress in December 2015 to further refine and standardize the work of rural poverty alleviation and development in five major aspects, from the determination of poverty alleviation targets to poverty alleviation planning measures to supervision and assessment, in order to achieve the goal of increasing rural poverty alleviation and development, accelerating the pace of poverty alleviation and promoting the economic and social development of poor areas .Immediately afterwards, Zhangjiajie City, rooted in the actual local situation suitable for the development of tourism, formulated the "3+1 guarantee" and other related policies for precise poverty alleviation, forming a hierarchy of clear policy objectives and standards. After the central, provincial and municipal governments formulated relevant poverty alleviation policies, the people's government of Cili County further refined the policies based on the superior documents, and really different villages formulated to give for specific policies to get rid of poverty. Although the subjects of poverty eradication policies are different, the goals at all levels are highly consistent, and they are all committed to achieving the century-long goal of winning the battle against poverty and achieving the overall building of a moderately prosperous society. Thus, it can be seen that in the process of implementing precise poverty alleviation policies in Luotan Village of Cili County, the pioneering policy objectives and hard policy standards provide a leading role in the implementation of realistic poverty alleviation and point out the direction for all subsequent development, which greatly affects the development of poverty alleviation work in poor areas and the effectiveness of poverty alleviation policy implementation. Some of the relevant policies are listed in Table 1.

Table 1 Relevant documents and policies introduced

Issuance of documents	Generation time	Issuing unit
Implementation Plan for the Establishment of a Precise Poverty Alleviation Work Mechanism	May 2014	State Council Poverty Alleviation Office
Decision on Winning the War on Poverty Alleviation	November 2015	The State Council of the Central Committee of the Communist Party of China
Performance Management Approach for Poverty Alleviation Project Funds	May 2018	Ministry of Finance, State Council Poverty Alleviation Office, National Development and Reform Commission
Opinions on Realizing the Effective Interface between Consolidating and Expanding the Results of Poverty Alleviation and Rural Revitalization	March 2021	The State Council of the Central Committee of the Communist Party of China
Regulations on Rural Poverty Alleviation and Development in Hunan Province	December 2015	Hunan Provincial Government
Resolution on Implementing Precise Poverty Alleviation and Accelerating Poverty Alleviation and Development	August 2015	Hunan Provincial Committee of the Communist Party of China
Regulations on Rural Poverty Alleviation and Development in Hunan	July 2020	Hunan Provincial Government

Province		
The Action Plan of "Five Base Projects", the Action Plan of "Sunshine Three Guarantee Projects" and the Action Plan of "Vigorously Promoting Tourism Precise Poverty Alleviation" "1+3" series of documents	March 2016	Zhangjiajie Municipal Government
Cili County rural low income security and social security underwriting to get rid of poverty objects identified clean-up and rectification work program	June 2017	Cili County People's Government Office
Implementation Measures for the Relief and Support of Special Hardship Cases in Cili County	September 2017	Cili County People's Government Office
Cili County 2019 Implementation Plan for Consolidation and Enhancement of Poverty Alleviation in Chinese Herbal Medicine Industry	July 2020	Cili County People's Government Office

3.2.2 Policy resources: participation of diverse elites amplifies the effectiveness of poverty alleviation

elites add to Luotan Village's poverty alleviation with outside streng According to Smith scholars' interpretation of policy resources, the supporting resources for successful policy implementation include: human resources, economic resources, organizational resources, intellectual resources, and information resources. It is true that any policy is just a piece of paper if it is not supported by certain supporting resources after designating policy objectives and standards.

From the poverty alleviation process of Luotan village, the advantages of human resources, economic resources and organizational resources are its most prominent features. The so-called human resources should include the number of people who implement policies, their education level, and their ability to be creative and practical. In the case of Luotan village, the unique composition of its people has led the author to divide its human resources into different categories in order to better distinguish and describe them. Of course, in the following different divisions, factors such as the quality of personnel and the number of people involved in poverty alleviation also play a crucial role, and these factors are all interlinked and complement each other thus maximizing the advantages of human resources. The types of human resources in Luotan village can be roughly divided into three categories: political elite, village elite, and social elite. There is no doubt that the first category of political elite is often referred to as the staff of government agencies and institutions. In the road to poverty alleviation the staff of the agency government is instrumental in leading and demonstrating the role. The Zhangjiajie Municipal Government assigned three-year resident poverty alleviation cadres to provide one-to-one, point-to-point assistance. Among the cadres stationed in the village, there is no shortage of old, middle-aged and young power, in addition to this, there is a constant stream of "mobile power" to supplement the power of various aspects. Zhangjiajie Municipal Government, together with Zhangjiajie People's Hospital, Zhangjiajie Township Primary School, tour guide qualification training institutions and relevant technical institutes have provided medical, educational and technical support to Luotan Village, focusing on the construction of the "five bases" of village organization positions, promoting the construction of spiritual civilization and improving the comprehensive quality of farmers. During the poverty alleviation period, a number of leading cadres visited the village, talked with Du Xiufu, the secretary of the village party branch, and Yang Chunchu, the first secretary of the municipal government office in the village, and guided the poverty alleviation work, which greatly encouraged the work of the "political elite". The second category, "village elite", "village elite" is the characteristic human resource of Luotan village. As a village inhabited by ethnic minorities, the village has very dignified elders as well as young and forward-thinking villagers. Diverse village elites occupy specific advantageous resources and have certain influence. The elders have authority within the village and have a role in unifying the villagers' thinking and encouraging and supervising them to cooperate in the poverty alleviation work, which is difficult to be replaced by government agencies. At the same time, the village elders acted as a "softening agent" in some cases of cultural conflicts, mediating the conflicts and making the whole policy implementation smooth. The younger villagers are better at accepting new ideas and learning abilities than the older ones due to their education, so they have become the main force in promoting cultural tourism in Luotan Village and even in Cili County. In addition, young villagers have played a significant role in creating a special cross-shaped communication channel with upward dialogue with the government, downward communication with villagers, and horizontal use of social media software to promote the local area to foreign travelers. The last category: "social elites", which refers to external forces that are not directly involved in the village's poverty alleviation efforts, but contribute greatly to the development of the local economy. Based on the geographical advantage of Luotan village scenic spot "crossroads", the superior decided to develop tourism to alleviate poverty and launched a series of policies to attract business elites into the village. They grasp the highlights of rural leisure farming and ethnic culture experience, and develop a special tourism program of "climbing mountains, watching water and learning culture". Social th, undertake policy poverty alleviation projects with capital advantages, drive the development of special industries based on local characteristics, enhance the countryside's endogenous power, and achieve sustainable poverty alleviation development. The reality of the relevant data: the village's net per capita income of farmers in 2014 was 2,140 yuan, and by 2016 the village's net per capita income of farmers reached 6,580 yuan. Only the Hetianju Resort Hotel project and the Hetianju Valley Tourism Industry Chain project have increased villagers' employment income by nearly 4,250 yuan. The pluralistic and elite human resources provide the impetus for the implementation of poverty alleviation policies in Luotan Village and stimulate the intrinsic vitality of poverty alleviation, and the related participants are shown in

Figure 2.

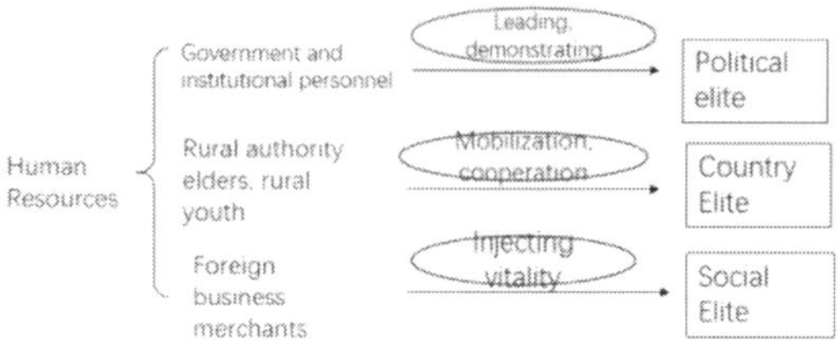

Figure 2 Human Resource Advantage of Multiple Subject Composition

3.2.3 Implementation approach: multi-subject communication

Poverty alleviation in poor areas has always been considered a "hard nut to crack", so a lot of human and material resources are often invested in poor areas. At this point, it also makes it possible for different implementing agencies and executives to work together to alleviate poverty. In order to make the poverty alleviation work go smoothly, close cooperation among agencies is especially important. Among them, timely communication and exchange becomes an important way to connect various subjects. In addition to the communication and collaboration between implementing organizations and organizations, it also includes the coordination and interaction between implementers and poverty alleviation targets, and the communication between external social forces and implementers as well as target audiences. The task of poverty alleviation and poverty eradication in Luotan Village is complex and arduous, and the implementation activities of precise poverty alleviation policies require the cooperation of various institutions such as the government, villagers, and enterprises, etc. Inter-organizational communication becomes the window linking the poverty alleviation policies with the target people, and its related links are shown in Figure 3.

(1) Interaction between implementing agencies. The Commune Party Committee actively develops party members, gradually establishes and perfects the organization of grassroots party committees and has developed dozens of party members. At the same time, the Zhangjiajie Municipal Government has sent a poverty alleviation team to the village, the Poverty Alleviation Office has visited the village from time to time for guidance and inspection, and other institutions and departments have participated in large projects to alleviate poverty, thus constituting a cross-sectoral and cross-level working linkage mechanism of "party committee leadership, government leadership, mass participation and social assistance", and striving to promote the development of employment poverty alleviation work in a concerted manner. The party government uses special poverty alleviation funds to increase investment in rural infrastructure construction and create public service projects in the direction of creating a modern leisure and vacation tourism industry. At the same time focus on attracting investment, through extensive publicity and tax relief preferential policies to attract foreign enterprises, especially the hotel resort type enterprises stationed in Luotan Village. The county government and the village committee of Luotan Village focus on improving villagers' skills and literacy, integrating a variety of training resources, inviting a number of experts to discuss and analyze, and organizing poor people to participate in pre-employment training and job skills upgrading training to improve the employment competitiveness of the poor. One of the exemplary examples of employment driven by enterprises is that Hotanju Group has created Hotanju Valley Leisure Experience Farm, introduced Zhangjiajie Runhui Agricultural Company, which operates and manages Peach Garden, Yangmei Garden and Wajiaobao Diner, etc. This has realized friendly communication between outside social forces and villagers internally, and greatly improved the income level of villagers.

(2) Interaction between the performer and the target population. Policy implementation requires not only inter-agency communication, but also communication between policy implementers and target groups. The understanding and support of the poor people directly determine the effectiveness of the implementation of precise poverty alleviation policies. To ensure that each villager understands the poverty alleviation policy, Luotan Village insists on taking the mass route, simplifies the villagers' feedback process, increases the frequency of communication between poverty alleviation cadres in the village, builds a file and card for the villagers, understands the villagers' family situation in depth, organizes learning technology if it is suitable for cultivating skills, provides funds and relevant training conditions if it is suitable for developing farm caravans, and does propaganda and work in different categories according to the actual situation of each family and the villagers' requirements to ensure that the villagers understand the help policy and help measures, and ensures the poor people's support and cooperation with the poverty alleviation work.

(3) Communication of external social forces with the executors and the target population. In order to open up the tourism market and broaden the source of funds for poverty alleviation, the municipal government has taken the initiative to take the lead in attracting investment and providing policy support for enterprises that carry out business activities in Luotan Village. Attracted by the government's tax exemption and good market prospect, foreign enterprises recruit local people to work in Luotan Village, providing them with education and training in the service industry and considerable income, while local villagers join the enterprises under the attraction of good working environment and income, and participate in poverty alleviation work invisibly while realizing their personal value. Since then, external social forces, implementing agencies and target people have formed a communication mechanism that can provide two-way feedback and response, building up strength for the development of the village's tertiary industry.

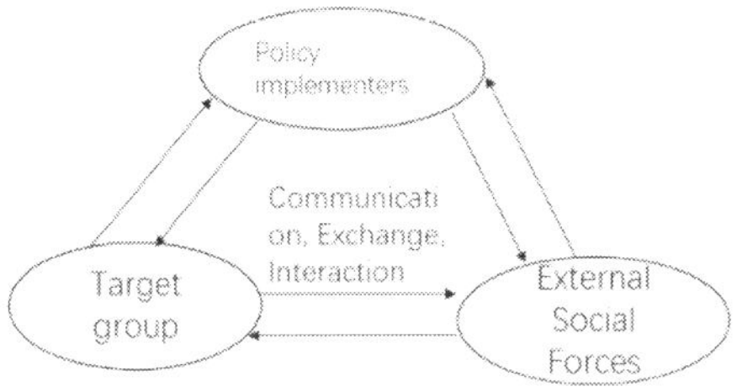

Figure 3 Three-party subject communication model

3.2.4 Policy implementation agency characteristics: distinct layers and interlocking

No policy can be implemented without people to implement it, and the best way to manage and dispatch people is for the organization to perform the implementation. The implementation of pro-poor policies cannot be separated from the implementation of organizational behavior, and the reasonable division of labor in the organization largely affects the efficiency and effectiveness of implementation. In the practice of poverty alleviation in Luotan Village, the government, grass-roots party committees, enterprises, etc. are both learning and summarizing their experiences in the practice of poverty alleviation nationwide, and exploring and innovating the paths suitable for their own development. In the process of poverty alleviation through tourism, the municipal party committees and governments take the lead in promoting the villages to attract capital, cooperate with banks to take out loans to build infrastructure, reasonably coordinate the "UNESCO" support work in the construction of rural civilization, transfer land to share dividends, and implement the decisions of higher-level organizations from top to bottom. Village level party committees and party members demonstrate leadership, take the initiative to participate in education and training efforts, and implement poverty eradication programs and frequent field visits to confirm each household. The tourism enterprises of the hotels support the "selection of enterprises and capital into the village", and strive to provide employment, which greatly improves the level of income of the villagers. Luotan Village's tourism poverty alleviation development model encourages a top-down, multi-body participation, internal and external cooperation poverty alleviation mechanism to ensure; stability and long-term implementation of precise poverty alleviation policies.

3.2.5 System environment: excellent environment to strengthen the foundation of poverty eradication

Policy implementation involves not only "policy insiders" but also links to the outside world. That is to say, the implementation of the policy in the social environment, the effective implementation of the policy can not be separated from the excellent policy environment. The systemic environment of a common policy in general contains social, economic, and political aspects. For the implementation of poverty alleviation policies in Luotan village, the role of economic and political environment is more direct and even plays a decisive role.

From the perspective of economic environment, at present and for a long period of time in the future, the financial investment of national governments at all levels for poverty alleviation policies will not be reduced, the market attractiveness of Luotan village is gradually growing, and the overall economic situation is favorable. From a macro point of view, China's rapid economic development, the total amount continues to break new highs, even during the epidemic economic growth has slowed down but still considerable, in 2017 and 2018 China's GDP exceeded 80 trillion yuan, 90 trillion yuan, the total economic volume for two consecutive years to reach a new level, and steadily ranked second in the world[6].The momentum of "good and fast" economic development has laid a solid foundation for guaranteeing financial expenditure for poverty alleviation, and the stable and positive domestic economic environment has enhanced the confidence of the masses to get out of poverty. From the micro environment, in recent years, Zhangjiajie City has spent more than 10 billion yuan in enhancing people's livelihood. Cili County "fine" poverty alleviation funds, and constantly improve the utilization rate of funds, poverty alleviation in the past five years, the county invested financial special poverty alleviation funds amounted to 230 million yuan. Since 2015, under the arrangement of the "four bases project" for poverty alleviation, an 8,000-meter tourism road to the village was opened, two famous enterprises were stationed in Luotan, and the special livestock breeding and rural tourism projects helped to alleviate poverty precisely. Compared with the end of 2 0 1 4 , the per capita net income of farmers in Luotan village increased from 2140 yuan to 3570 yuan in 2015, and the poverty-stricken population decreased from 253 people to 164 people[7].The internal and external economic environment has solidified the economic foundation for poverty alleviation in poor areas. The development of Luotan village further confirms that a positive economic environment has a direct impact on the implementation of poverty alleviation policies. The dynamics of the internal economy and the dynamics of the external economy provide fertile economic ground suitable for the development of Luotan Village.

As for the political environment, this paper has already mentioned several times the importance of the state to the cause of poverty alleviation and the care of the Hunan Provincial and Zhangjiajie Municipal governments for the poor areas of Cili County, so I will not repeat it here.

3.2.6 Policy implementers' value orientation: never return until poverty is broken

The implementation of the policy is ultimately down to the people, so the degree of understanding and perception of the policy, the responsibility and motivation of the task by the implementers all affect the effectiveness of the implementation of the poverty alleviation policy. In order to clarify the division of responsibilities and delineate uniform requirements, in October 2016 the CPC Central Committee and the State Council formulated the "Implementation Measures of the Responsibility System for Poverty Alleviation" to combat poverty in accordance with the central coordination, the province is responsible for the overall responsibility, cities and counties to grasp the implementation of the work mechanism, to build a clear responsibility, each responsible for their own, joint efforts to attack the responsibility system. According to the instructions of the state, departments at all levels will also divide the power and responsibility to individuals at each level.

The introduction of the approach document is a hard requirement for the executives in the poverty alleviation policy, while the spontaneous feelings of each participant in poverty alleviation is the source of motivation to promote the policy implementation. With the deepening of poverty eradication in Luotan Village, one group after another of grassroots workers speak their hearts and souls into it, taking leading villagers out of poverty and getting rich as their greatest pursuit. Since the poverty alleviation team has been stationed in the village, the team has been eating, living and working with the villagers, taking advantage of the location, finding the scientific path of tourism poverty alleviation, and formulating the "Luotan Village Poverty Alleviation Plan" and "Luotan Village Rural Tourism Industry Development Plan". Raising and integrating funds in many ways, repairing village roads, laying pipe networks and renovating dangerous houses not only solved the old problems of difficult travel, drinking water, electricity and housing for villagers, but also met the high standard demand of rural tourism industry for basic environment, which injected full vitality for the development of Luotan village. Poverty alleviation members actively develop new ideas of tourism, the local government and Hetianju Group jointly create "Hetianju Valley", taking the "characteristic park + famous scenic spot supporting industry" model has been recognized by the central, provincial and municipal governments. At the same time for Luotan village to create a new industrial development model, take the "company + village collective + farmers" industrial development model, farmers with land shares to participate in the dividend, the company accounted for 79% of the shares, the village collective shares accounted for 17%, farmers accounted for 4% of the shares. As participants, villagers have also taken the initiative to join in, actively participating in personal skills training, giving up their land for transfer, and even using online platforms to create a "net red Luotan" to cooperate with the government to promote their hometown. This new model of poverty alleviation and development, with the party organization as the core, farmers participating in management, and entrepreneurial operation, allows land to become equity and farmers to become shareholders, all reflecting the seriousness and responsibility of grassroots poverty alleviation personnel for their work and the high level of trust of villagers in the policy. There is no doubt that in the process of policy poverty alleviation in Luotan village, the executive staff is always full of enthusiasm, the villagers trust and cooperate, and the high recognition and conscious implementation of the policy continuously improves the level of Luotan's economic development.

4 Conclusion and Discussion

Comprehensive analysis of the above, the reason why Luotan Village in Cili County has always been able to lead the villagers out of poverty and become rich and guaranteeing the effective implementation of precise poverty alleviation policies and obtaining the expected benefits. Its success story fits highly with the inner logic of the Mitt-Horn policy implementation system model. The six factors in the model are all reasonably verified in the process of poverty alleviation policy implementation, which also proves the explanatory power of Mitt-Horn's system model for the study of poverty alleviation policy implementation in China. Although the policy implementation model originated in Western countries and was developed and first applied in an environment different from that of China, the essential connotations of the model still have important implications for policy implementation in China. The Mitt-Horn policy implementation system model integrates and analyzes the complex factors and multiple actors involved in the implementation of poverty alleviation policies in Luotan Village, Cili County, so that the whole process of policy implementation is not only shown as much as possible to provide reference for other poverty alleviation-related departments, but also plays an important role in the development of the "Chineseization" of the system model. The policy application in the field of poverty alleviation also provides more possibilities for the model to be studied in other fields, which is what the academic community is continuously working on.

References:

1. Chen Qingyun. Public Policy Analysis [M]. Beijing: Peking University Press, 2011: 171
2. Qin, L.F., Chen, Y.. Research on the effective implementation of precise poverty alleviation policies in Zhaomen Township, Yao'an County, Yunnan Province--a perspective based on the Mitt-Horn policy implementation system model[J]. Journal of Wenshan College,2020,33(06):41-46.
3. Guan Mengxue. A study on the heterogeneity of health poverty alleviation policy implementation--a perspective based on the Mitre-Horn system model[J]. Journal of Shanxi Youth Vocational College,2020,33(04):79-83
4. Wang Fashuo, Chen Ling. Why is it difficult to implement intelligent and innovative policies of social governance? --A case study based on Mitt-Horn model[J]. E-Government, 2020(05): 49-57. DOI:10.16582/j.cnki.dzzw.2020.05.005.
5. He Donghang, Kong Fanbin . The Chinese Experience of Public Policy Implementation [J]. China Social Science, 2011(5): 61-79
6. Data source: 2018 Statistical Bulletin on National Economic and Social Development of the People's Republic of China, published by the National Bureau of Statistics on February 28, 2019.

 https://doi.org/10.1051/shsconf/202315501020

7. Data source: extracted from "Hunan Cultural Tourism - Cili County Sanguanshi Tujia Township Luotan Village Help Documentary" on December 9, 2015

https://doi.org/10.1051/shsconf/202315501021

Study on construction quality Evaluation Index of prefabricated Buildings and Identification of Workers' occupational literacy Evaluation Index

Nana Liu, Wei Wang, Cheng Chen, Junxia Sun*

Chongqing College of Architecture and Technology, Chongqing, China.

Abstract. This study through reading to sort out quality management and quality inspection standards and specifications, identify five the prefabricated building quality specifications and workers professional quality of 16 factors, and then using the theory of quality management theory, labor productivity, literature research, identify quality evaluation specifications and general factors of professional workers, using the method of case analysis and expert interview to collect quality evaluation specification and workers professional quality characteristic factors, identify common workers professional quality influencing factors of 4 first-level indicators, 16 secondary indexes, prefabricated building quality five evaluation indexes, and further to a total of 21 indexes both generality and individual character classification, found 11, 10 personality. The extraction and analysis of the key factors affecting workers' occupational literacy on the quality of assembly construction provide a research basis.

1 Introduction

Construction labor productivity refers to the efficiency of construction workers producing construction products during the reporting period. It is expressed by the ratio of the output or value of construction products and the corresponding labor consumption. It is an important index to evaluate the improvement of production efficiency and labor saving of construction enterprises. Under the same input labor force, the more value the enterprise creates, the higher labor productivity, if not, it is lower. The commonly used labor productivity indicators of construction enterprises include the labor productivity calculated by the gross output value of construction industry and the added value of construction industry[1].

Relevant studies show that labor workers' expenses in the construction industry are only about 20% of the product value, and construction labor productivity is still at a low level.

With the continuous development of prefabricated buildings, the evaluation of the quality of prefabricated buildings y also needs very comprehensive indicators[2]. Among them, the professional quality of prefabricated construction personnel is the most important factor affecting the quality of prefabricated construction. How to evaluate the professional quality of an assembly construction personnel needs a comprehensive evaluation system. By extracting evaluation indexes of prefabricated building quality and professional quality of construction personnel, this study provides necessary basis for talent training in the transformation and development of the construction industry.

2 Factor identification method

2.1 Selection of factor identification methods

The influencing factors of workers' occupational literacy on the quality of prefabricated construction are mainly divided into two categories. One is the common factor, that is, the factor which coincides with the construction project management theory, and this factor is mainly identified by theoretical induction method and literature research method. The second type is personality factors, which are mainly identified by the existing prefabricated building case analysis method and expert interview method[3,4]. The factors extracted by case analysis method may not be universal, so it is necessary to conduct further verification by expert interview to form personality factors for the study of workers' occupational literacy in this study. Finally, the common factors and personality factors are combined to form the research factors of workers' occupational literacy in this study. The same prefabricated construction quality evaluation index is still extracted from the relevant standard books by theoretical induction, literature research, case analysis, expert interview method and other comprehensive extraction[5].

*Corresponding author: ldnnana@sina.com

2.2Step of factor identification method

This study under the perspective of professional workers to identify factors influencing the quality of prefabricated construction in the process of using a variety of methods, covers the theory of induction, literature research, case analysis, expert interviews and other methods, at last, through comprehensive analysis the influence factors of the final of this study, the role of the recognition step and the method as shown in figure 1.

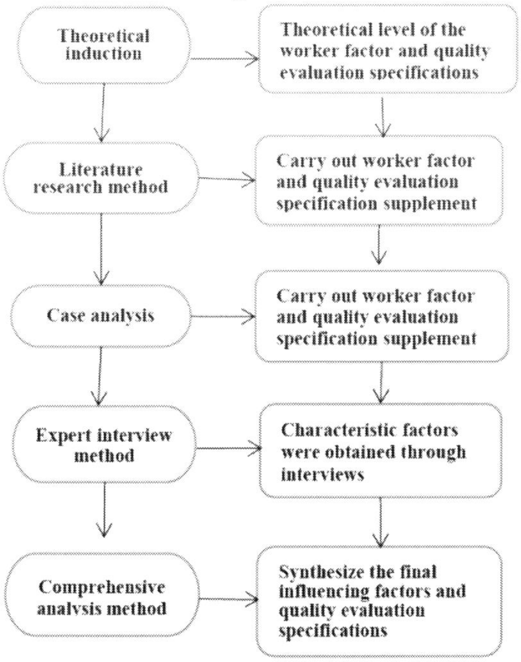

Figure1 Factor identification process

2.3. Prefabricated construction quality evaluation index identification

Based on the prefabricated construction theory, the influencing factors of quality were analyzed and supplemented from the aspects of standard specifications, books, literature retrieval and expert interview.

2.3.1. Standard specifications and books

In recent years, with the gradual warming of research and application in the field of prefabricated building in China, research and application of related technologies have been carried out. In order to meet the development needs of prefabricated building application, a series of national standards, industry standards and product standards have been compiled and revised[6]. Based on the fact that workers mainly work in the prefabricated construction stage, combined with relevant standards and specifications, the prefabricated construction quality evaluation specifications are extracted from production, transportation, construction and installation. The main reference standards and specifications and books are 7 national standards, 4 industrial standards and 1 other management regulations[7].

2.3.2. Literature retrieval

Based on the reference extraction of the existing standard specifications for prefabricated buildings, part of the quality management content can be extracted to evaluate specifications and understand the quality standards for prefabricated buildings[8]. However, due to the limitations of practice and application degree, it is difficult to ensure the comprehensiveness and pertinence of quality management content to evaluate specifications only by relying on the standard specifications. Combined with 148 relevant literature studies in recent ten years, the quality management contents that are not reflected in the standard specifications are adjusted and analyzed from four aspects of quality inspection, quality cost, quality safety and environmental quality[9].

This template, modified in MS Word 2007 and saved as a "Word 97-2003 Document" for the PC, provides authors with most of the formatting specifications needed for preparing electronic versions of their papers. All standard paper components have been specified for three reasons: (1) ease of use when formatting individual papers, (2) automatic compliance to electronic requirements that facilitate the concurrent or later production of electronic products, and (3) conformity of style throughout a conference proceedings.

2.3.3 Case analysis and expert interview

This study enumerates the common quality problems of prefabricated buildings through case analysis, among which the individual influencing factors are consistent with the quality management theory, which can be directly used as the quality management content to evaluate specifications, and the rest can be verified by expert interview method[10].

In addition, this study collected and understood the factors affecting the quality of prefabricated construction, including quality cost, construction schedule, environmental quality, on-time delivery rate, green building requirements and so on, through face-to-face interviews with professional and technical personnel or managers with years of work experience in construction enterprises.

In standard specification and books as the basic content of quality management as a benchmark, combined with the literature retrieval, and in case the empirical and expert interview considers unscrambles prefabricated construction quality from the following five aspects: F01 quality qualification, F02 quality cost, F03 quality safety accident frequency, F04 on time delivery rate, F05 clean environment quality.

2.4. Identification of influencing factors of workers' occupational literacy

2.4.1. Theoretical induction

Recognition for prefabricated construction quality evaluation specification factors at the same time, based on the theory of labor productivity, and cognitive biases, and increase the workers related policies, summarized, analyzed and summarized, because workers professional quality will affect the quality of the prefabricated building factors, these factors will be needed by regression analysis model is established in this study potential variable factors.

Referring to the relevant specifications, it can be seen that influential factors with high frequency will be used as observation variables of this study. As can be seen from the above table, "labor intensity" has a low frequency and is not listed as observation variables of this study for the time being. "Safety education", "safety training", "certificate entry", "operation standard", "professional skills", "vocational training" and "professional knowledge" appear frequently. Combining these influencing factors with the quality management of prefabricated buildings, through analysis and summary, can form part of the observed variables of this study.

2.4.2. Literature research method

Labor productivity level decides the comprehensive professional quality workers, plain is distinguish between work performance and the key to good work performance, whether to choose a scientific and reasonable professional characteristics, build effective and practical workers professional quality factors determine whether find workers professional quality cannot adapt to existing building industrialization, in order to find out the existing workers professional quality problem, Nearly a hundred articles were selected from cnKI, WIp, etc. in the form of subject search for "prefabricated construction quality" + "workers" and "migrant workers" respectively, taking into consideration the quality of research and journal quality. By analyzing the key factors affecting workers' professional quality in each article, 11 characteristics of workers' professional quality were extracted. Have a high level of education, professional knowledge, qualification certificate, have the awareness of the operation norms, have the sense of teamwork, have a responsible work attitude, have the awareness of quality assurance, through vocational training, familiar with the operation norms, rich work experience, with safety protection.

2.4.3Case analysis and expert interview

In this chapter, the case study on the quality problems of prefabricated construction shows that the influencing factors of workers' occupational literacy are consistent with the literature research method and can be directly used as the observed variables of this study, while the rest can be verified by the expert interview method. This study collected and understood the influencing factors of workers' occupational literacy by conducting face-to-face interviews with professional and technical personnel or managers with years of experience in production, construction and management of prefabricated buildings in construction enterprises. Including cost awareness and learning ability, hard-working quality, learning willingness and adaptability.

Can be determined through expert interview method integrated, professional workers for the construction of prefabricated construction quality is very important, at the same time building enterprise is faced with how to change now workers professional quality, to adapt it to the prefabricated construction development, the above investigation and interviews of experts are rich experience in the industry production and the construction technology and management personnel, has high credibility, In addition, some factors are the same as those pointed out in the case analysis in chapter 1 of this study, such as "workers have good cost consciousness", "workers have the quality of hard-working" and "workers need to have good learning ability", which can be directly used as the observation variables of this study. After sufficient investigation and research, the 16 influencing factors can be taken as the observational variables of this study.

3. Identification results and discussion

3.1 Identify the results

In this study, the identification of prefabricated building quality evaluation specifications and the identification of workers' occupational literacy factors were implemented through various methods. Based on the labor productivity theory and the competency model, a 4-dimensional index system was constructed to reflect workers' operational ability knowledge, attitude, professional skills and potential based on 16 influencing factors. Having a high level of education, professional knowledge, qualification certificates, operating norms cognition is divided into knowledge dimension, having the spirit of bearing hardships and standing hard work, having the consciousness of teamwork, having a responsible working attitude, having the consciousness of quality assurance, having the consciousness of cost control is divided into attitude dimension, Having undergone vocational training, being familiar with operation specifications, having rich working experience and having safety protection can be divided into professional skills dimension, and having strong learning ability, strong adaptability and strong learning willingness can be regarded as potential dimension.

3.2 The results discussed

Through the comprehensive application of a variety of methods, all the observed variables of this study are obtained, which are divided into two categories. The first category is the result identified by theoretical induction and literature research, which is a general factor of quality management and also applicable to the quality management of prefabricated buildings, belonging to the common factor. The standard and specification theory induction method and literature research method are consistent with professional knowledge, knowledge of the code of operation, qualification certificate, teamwork consciousness, familiarity with the code of operation and rich work experience. The focus of theoretical research and analysis is consistent; The second type is the factors identified by case analysis and expert interview, which are unique factors in the overall evaluation of workers' development into industrial workers and the quality of assembly construction, and belong to individual factors.

4 The research conclusion

The purpose of this study is to identify the influencing factors of workers' professional literacy and the evaluation indexes of prefabricated construction quality. By reading the finishing quality management and quality inspection standards and specifications, identify five the prefabricated building quality specifications and workers professional quality of 16 factors, by using the theory of quality management, labor productivity, the method of literature research, case analysis and expert interview collection quality evaluation standard and staff professional quality characteristic factors, Determine the factors affecting the professional quality of ordinary workers 4 first-level indicators, second-level indicators 16, assembly building quality 5 evaluation indicators, further to a total of 21 indicators, both common and personality classification, 11 items, 10 items of personality. These factors form a scale, and then a questionnaire is distributed to the public, which lays a foundation for data collection, and provides a research foundation for extracting the key factors that affect workers' professional literacy on the quality of prefabricated buildings and the effectiveness of the regression model.

References

1. Sun Yu-Fang, Wu Xia, HE Meng-lin, CONG Xu-hui (2021). Research on quality Management of Prefabricated Building Based on BIM+ Internet of Things Technology [J]. Building Economy, 42(05):58-61.
2. Sun Qianjin (2021). Construction Technology and Quality Management of Prefabricated Buildings [J]. China Housing Facilities, (05):8-9.
3. Fan jun, Yang sixin(2017). Thoughts and suggestions on realizing prefabricated building [J]. Building technology,48(02):118-122.
4. Xiaolong Gan, Ruidong Chang, Jian Zuo, Tao Wen, George Zillante (2018). Barriers to the transition towards off-site construction in China: An Interpretive structural modeling approach[J]. Journal of Cleaner Production,197: 8-18.
5. Duan Yadi, Li Guanlei (2020). Analysis on the Quality Problems and Preventive Measures of Prefabricated Building Construction[J]. Journal of Physics: Conference Series,1648(3):
6. Aynur Kazaz, M. Talat Birgonul (2005). Determination of Quality Level in Mass Housing Projects in Turkey[J]. Journal of Construction Engineering and Management,131(2): 195-202.
7. Gul Polat (2010). Precast concrete systems in developing vs. industrialized countries[J]. Journal of Civil Engineering and Management,16(1): 85-94.
8. Imane Krimi, Zoubeir Lafhaj, Laure Ducoulombier (2017). Prospective study on the integration of additive manufacturing to building industry—Case of a French construction company[J]. Additive Manufacturing,16: 107-114.
9. Ying Wang, Heng Li, Zezhou Wu (2019). Attitude of the Chinese public toward off-site construction: A text mining study[J]. Journal of Cleaner Production,238
10. Liu Xinghong, ZHANG Hanyu (2018). Research on prefabricated building quality management based on TQM theory and BIM [J]. Building economics,39(10):25-30.

A Study Media Coverage of College Crisis under Framing Theory

Haonan Luo*, Qianwen Ling*

Chongqing College of Architecture and Technology, Chongqing, China.

Abstract: Framing as an academic concept to examine human cognition andc ommunication behavior was first seen in anthropologist Bateson's paper "A Theory of Play and Fantasy" published in 1955. The most direct manifestation vehicle of framing effect in news communication is framing theory, which means that a certain framing effect is achieved by selecting and emphasizing a certain content. However, the media coverage of the college crises has been ignoring the framing theory and changes in the information technology era. The article will analyze the current research on media coverage of college crises based on the framing theory in the information era and try to innovate and apply it to this. The information environment has made the information that has been diluted and hidden under the traditional news media framework come to the surface, and more and more information has been focused under the spotlight to be examined by the public. Such a journalistic framework has undoubtedly had a crucial impact on the public crises that arise in today's universities.

1 Introduction

With the development of the times, public crises are getting more and more attention and concern from academia as well as the public, and the study of crisis events in colleges and universities has gradually become a hot spot. As a special group environment, the development and evolution of crisis events in colleges and universities have an important impact on society. In the information environment, with social media as the main communication channel, the communication carrier and communication mode of college crisis events become diversified. In addition, media reports and network public opinion spread faster and deeper,and the social impact of the college crisis is more extensive and will cause a series of social problems.

The U.S. Department of Education's 2003 Practical Resource for Crisis Planning: a Guide for Schools and Communities, issued to all levels of schools, defines a college or university crisis as a crisis in which school leaders must confront and make one or more important decisions under conditions of inadequate information, time and resources. Lichtenstein defines a school crisis (school crisis)[1] as "a sudden, unanticipated event in which the school in general or a significant part may be seriously and negatively affected, usually including serious injury or death." The actors of college crisis events are college students. The frequent college crisis events in recent years are the following: academic integrity, daily management, campus safety, teacher morality,and so on. Therefore,the news media need to select and process the information about the college crisis in what way, through what framework to highlight its connotation and meaning, and in what news picture to present the matter.

The powerful ability to guide public opinion and disseminate information is the most unique ability of the news media in society, and the exercise of this ability is crucial to and even determines the operation of any social organization. In recent years, the crisis in universities has attracted strong attention from the media and the public. In the age of information technology, if the crisis in colleges and universities is not handled properly, it will not only affect the normal teaching order, but also make the crisis management in colleges and universities out of control and produce negative social influence. The analysis of this paper focuses on the selection and highlighting of the study of media reports of college crises under the frame theory in the new media era, and exploring the value orientation of media reports and the social significance and impact produced according to the content gated by different media.

2 Theories related to news frames

The frames constructed by the news media provide audiences with the right to know about news events and constitute the "reality" on which we base our perception of society. Chinese scholar Pan Zhongdang also believes that frames exist in at least two places: one is implicit in the text;[2] the other is in human perception, that is, in human memory. Frames represent how social events are subjectively presented by journalists or media organizations on the one hand, and how audiences

*Corresponding author: 1469611942@qq.com, 1079549789@qq.com

subjectively interpret the symbolic system of the news media on the other.

2.1 News framework

In 1974, Irving Goffman published his book Framing Analysis, which first applied the concept of "framing" to communication situations. Goffman argued that framing is an important means by which people translate social reality into subjective thought, that is, the structure of people's or organizations' subjective interpretations and thinking about events [3]. Goffman's framing theory has been widely used to analyze the organizational principles of media coverage of news events and to explore the factors that influence the media's perception of a particular group's image, which is customarily referred to as "news framing".

Huang Dan, a communication scholar in China, summarizes news framing theory as follows: First, the central issue of framing theory is the production of media, that is, how media reflect reality and regulate people's understanding of it. Second, the media reflect reality, construct meaning and regulate people's understanding, which is ultimately realized through text or discourse-media products; therefore, text construction, interpretation, or discourse production analysis is the focus of frame theory[4]. Thirdly, frame theory focuses on media production, but does not regard production as a closed process, but places production and its products (texts) in a specific context - in various relations.

The Taiwanese scholar Zang Guoren further divides frames into high-level structure, medium-level structure, and low-level structure. The high-level structure of the frame refers to the characterization of a certain subject matter; the medium level structure of the frame includes the main events, previous events, history, results, and evaluation; and the low-level structure of the frame refers to the concrete representation of the frame through language or symbols.

According to Gitlin, "media framing is the principle of selection, emphasis, and expression ...... is the process of selecting, emphasizing and excluding news stories. Media framing ensures that journalists can process large amounts of information quickly and routinely, identify it into cognitive categories, and then package it for more effective presentation to the public." Specifically in news texts news framing mainly includes the presentation of text content, some keywords, stereotypical terms, sources, news graphic structure, etc. Therefore, facing a more specific topic like a college crisis, journalism researchers should consider, when studying its media coverage, how the news frames differ from media to media in response to the same college crisis. What are the different frames of media coverage at different stages of crisis development? How does the media frame the coverage of the university crisis?

2.2 Audience framework

The media selectively filters and communicates complex and diverse information to the audience through frames, constituting a mimetic environment that influences the audience's perception of the world. After years of life and cognitive accumulation, audiences have also formed their audience framework and cognitive system. In common parlance, it means that there are thousands of people, and for the college crisis, the audience will also be structured according to their audience framework. When the issue of the Xi'an University of Electronic Science and Technology students looking for a gunman to do their graduation design for them aroused public discussion, the focus of the audience also showed diversified characteristics, such as some views that the focus of this incident is on academic misconduct, some views that universities educate people but not morality, and some views that the phenomenon of gunmen in the academic world needs to be rectified urgently. And some audiences who do not know much about this matter, once they accept one side's view, they accept the other side's definition of this issue and reconstruct this issue.

Combining the above definitions of the frame, we can point out some common points, that is, "frame" is mainly a structure for perceiving and presenting things, highlighting specific connotations in selecting and processing things, and expressing ideas reflecting a specific cultural value [5].

3 Analysis of frames in college crisis reporting

3.1 Framework analysis of news production

The way news is disseminated in the information age has changed significantly, i.e., traditional media are no longer dominant and new media are taking over, both of which together form the overall framework of news production today. This reconfigured news framework, which has been reconstructed by the times and new conditions, provides new requirements for audiences and news producers.

News sources and news values, as important intrinsic elements of news production, are already framed and prescribed in the framework. The media and news agencies serve as the main sources of news, scattered to all corners of the world, but there is no shortage of general ideas and guidelines in line with the basic journalistic profession, which facts make news and which are just ordinary information, and news sources are filtered out layer by layer in this well-defined framework. News producers can then use the framework to find sources, the identify angles and interviewees for specific news content. As a result, some of the facts are selected for inclusion in the frame, and others are excluded due to some routine practices of journalists in obtaining news sources. In turn, news value determines the further selection of news, and the degree of importance of some news. This framing is also maintenance of balance in this process, and news values are determined by the agreement of journalists to maintain a fixed pattern of news values to maintain the balance between them.

https://doi.org/10.1051/shsconf/202315501022

News frames, as principles chosen in news production, are to some extent deliberately emphasized, interpreted, and presented as symbols, which involve selection and prominence, swaying the birth of news and influencing the audience's view of a specific event. In the age of information technology, news frames have their favorable aspects and require a higher level of media literacy on the part of audiences and news producers. Frames are ultimately a way and a tool, and only by taking advantage of the features and advantages can we better report news and disseminate information. The first time I saw the company, I had to go to the university to find out what it was doing. The two students involved have been suspended from defending their graduation designs. Public opinion swelled rapidly after the incident, and many media reported related information such as the Central Committee of the Communist Youth League, China News Weekly, Global Network, Red Star News, Surfing News, and Observer, and the microblogging topic #Xi'an University of Electronic Science and Technology responded to the incident of students' graduation design doing on behalf of students# was read by more than 200 million people and discussed by more than10,000 people, triggering widespread concern among netizens. It was not until May 30 that Xi'an University of Electronic Science and Technology officially issued a notice that the university's Academic Style Construction Committee confirmed the academic misconduct of Lei and Lu, and gave Lei and Lu two students one year's probation according to relevant regulations, during which they were not allowed to apply for a degree; and canceled Lu's qualification for graduate school exemption. The media coverage of such academic misconduct is bound to cause an uproar in the university community and expand the discussion of research ethics and academic misconduct to the public agenda.

In October 2018, more than one hundred papers by Liang Ying, a young Changjiang scholar, were withdrawn from major academic journals, and in 2019, Zhai Tianlin "did not know what the Internet was", and then this year, the gunman did the bibliography and other academic misconduct news reports, the majority of netizens jokingly called "Tianlin four years, Lu Lei The incident". To a certain extent, the news framework of the relevant events has been drawn up to allow the public to deliberately feel the specific impact of the public opinion environment, where academic misconduct is not a trivial matter and the environment of honest governance can take root in the news framework. Traditionally, the disclosure of academic misconduct and its treatment has been somewhat limited to the scientific community, and its impact has not entered the general public's view or generated much feedback on the public agenda. However, in recent times, media exposure and disclosure has become important way to report and expose academic misconduct[6]. The incident of Zhai Tianlin's not knowing "Zhiwang" has highlighted the importance of the public agenda or focus on exposing academic misconduct.

3.2 Framing theory in communication content research

In the field of communication content research news text research, it is not difficult to find being placed at the center of frame analysis. For the news media, the frame is a regular filtering device, and the frame is like a funnel, through which the facts that fit the size of the funnel are selected and left behind, processed, or reorganized by the symbolic and ideographic elements within the frame, becoming news with the "brand" of the media. Therefore, in a sense, the content of communication produced by the media may be mixed with personal subjective factors or the competition of various power institutions.

All major media have different frame settings based on different attributes, audiences,and positions. The development of education is very significant for a country, especially for a country in rapid development like China, it has special significance and the public has high expectations of education, and some crisis events in colleges and universities are often more likely to attract public attention and discussion. Therefore, colleges and universities as a special group environment, the report of college crisis, the monitoring of public opinion after the report, and the subsequent public opinion guidance and public reaction are especially important. In recent years, central-level mainstream media such as Xinhua News Agency and People's Daily play the role of official mainstream media in reporting college crisis events, while the reporting framework of some online media may have the characteristics of diversification, extremism, and entertainment based on the attributes of the audience groups they are facing[7]. Once these features take over the main role, they will mislead the audience and influence them to view and analyze the events rationally and fairly, and then affect the development of public opinion.

3.2.1 Information Sources

Information sources are seen as an important variable in news framing. At the beginning of the creation of a college crisis, information sources tend to be mainly conventional sources. And after the incident fermented and attracted widespread attention, information sources rapidly increased, and official sources and private sources expanded. Take Zhai Tianlin's academic misconduct incident as an example, in February 2019, Zhai Tianlin claimed in a live broadcast that he "did not know what Zhiwang was", and the authenticity of his doctorate caused netizens to question. Then he claimed on Weibo that he was "just joking", at which point the news media relied on conventional sources of information, namely the direct parties involved in the incident.

However, this statement did not reduce the heat of public discussion, and subsequently, Beijing Film Academy, Peking University Guanghua School of

Management, and Zhai Tianlin's academic misconduct were investigated in public statements. The sources of information reported by the media at this time were official sources and private sources. In this incident, the sources of news reports of People's Daily and Xinhua News Agency are official sources, which pay more attention to authenticity and authority. The sources of the online media were more extensive, which aroused public discussion on social phenomena such as "academic misconduct" and "social fairness" and expanded the controversy.

3.2.2 Selection of topics

As a special type of event, the coverage of the college crisis and the subsequent direction of public opinion have far-reaching effects on society. The topic of the college crisis is undoubtedly a hot topic for online media to arouse audiences' reading interest. When organizing reports, their topics tend to seek novel angles, while some of them maliciously cause confrontation and even worsen the impact of the event. And after the college crisis aroused attention, the reports of various online media further expanded public opinion, and the breadth and depth of discussion gradually expanded. When the incident rises to the level of academic integrity, daily management, campus safety, and teachers' morality, and causes adverse social influence, the official mainstream media will report on the topic, to provide good social guidance. Take Zhai Tianlin's case as an example, Xinhua published "Zhai Tianlin's case is not only academic misconduct", People's Daily published "The best persona is self", Global Daily published "I hope Zhai Tianlin's case will form a wide deterrent", and Beijing Daily commented on the case. " and "Beijing Daily" commented "I hope there will be less 'tofu-dreg' persona projects", which tracked and analyzed the incident and attracted wide public attention. Compared with the broad and diversified choice of topics of college crisis events by online media, the official mainstream media shows more characteristics of targeting and guiding.

3.3 Framing theory in communication effectiveness research

The framing effect is another significant manifestation of the framing effect in the field of news communication, where news production and communication content have specific media effects on particular content. Unlike agenda-setting theory, framing analysis focuses not only on what people think, but also on how they think.

3.3.1 Irrational framework results

Most of the frame clues in mass communication in the information era originate from self-media, which determines that the more open the self-media is, the greater the frame risk is. Another example is the case of Xi'an University of Electronic Science and Technology students doing their bibliography on behalf of the students. With the push of public opinion, more and more media organizations began to pay attention to this matter, ranging from the Central Committee of the Communist Youth League, Li Zhi News, Observer.com, and Peng Pang News to countless WeChat public numbers and opinion leaders of major platforms reprinting and reporting, making this event dominate the hot search list of major platforms at that time. At the same time, platform users were also discussing and analyzing the event as individuals, fully demonstrating the openness and rapidity of the information age. The timeliness of the platform means that more and more communicators can become recipients for the first time. In the era of self-media, the public can immediately post information sources related to themselves without rational processing to the self-media platform, turning the sources into risky framing clues in mass communication quickly, and the more open the framing clues are, the more framing risks continue to be superimposed. The existence of framing risks led to many extreme and unwise discussions of this incident at the time, with some arguing that the punishment was too lenient, and some netizens even steering it into gender antagonism and stirring up disputes.

3.3.2 Negativity of framework content

News communication is susceptible to the influence of the plurality of communication subjects, more subjective and tendentious, and negative and negative emotions will increase. In the information age, the subjects, objects, Methods, and channels in mass communication activities have undergone subversive changes. News sources evolve into opinion frames with certain communication directions through public dispersal. Then the negative attributes of frame cues will also turn into frame contents with negative emotional pointers[8].

In response to the #Xi'an University of Electronic Science and Technology's response to the student's graduation design incident, the audience showed different views, but subjectivity and tendency made the netizens' views mostly negative emotions: first, they thought that hiring someone to do the graduation design for them was academic misconduct and hoped that the university would make serious punishment for the students involved under the regulations; second, they thought that the students involved had poor conduct and questioned the existence of shady practices in their examinations; third, they called it "the first year of Lu Lei" and were worried that the graduation audit of universities would be tightened after this incident; fourth, they dug deeper into the information of the students involved and the follow-up of the incident, and expressed dissatisfaction with the university's indifferent attitude. The third one is that the university has called "the first year of Lu Lei" and is worried that the graduation audit of universities will be stricter after this incident; the fourth one is to dig deeper into the information of the students involved and the follow-up of the incident, and expresses dissatisfaction with the school's indifferent treatment attitude; the fifth one is related to the previous incident of the university's graduate school couple cheating the poor students' subsidy, and thinks that the school's enrollment work may be affected as a result; the sixth one is that some The netizens sarcastically called the students involved "fairies" and were criticized for stirring up the confrontation between men and women ...... The negativity of the frame attribute is specifically expressed in the new communication ecology as the negative emotion of the communication event, and this negative emotion is generated by the source of the frame cue, which gradually dissolves and deepens in the communication frame of the whole media, and finally affects the decision.

3.3.3 Irrationality of the framework results

In the information age, especially the non-traditional nature of information dissemination on the self-media platform makes framing cues with high risk and negative framing content spread geometrically, gathering more complex and irrational framing results of communication. Such framing results inevitably lead to irrational opinion tendencies, which are more likely to cause the collapse of the positive frame of audience psychology and induce public social behavior backlash. The student hired a "gunman" to do the graduation design for him and asked for a refund, and was reported by the "gunman", and the chat records that were revealed attracted numerous netizens to watch and troll. Academic falsification, student conduct issues, and school spirit construction issues have been criticized by all parties, bringing a large negative impact on the reputation of the school. After the incident triggered a heated debate, the school quickly responded to the matter, but the decision to "suspend two students from defending their bachelor's degree" did not satisfy public opinion. Some netizens continued to question, further digging and other negative events related to the school, which brought more pressure to the school. In addition, the irrational expressions such as the uncontrolled dissemination of private information about the students involved, the malicious mockery to provoke the confrontation between men and women, and even the defense of the "shooter" on behalf of the students, also put pressure on the guidance of public opinion.

The development of social media platforms in the information era has dissolved the framework structure of traditional news media and made the communication environment more complex. Especially in the reporting of sudden public crisis events in universities and the public relations processing of universities themselves, the riskiness of the whole media itself increases the uncertainty of information dissemination, which requires the media reporting and public relations processing to change its inherent discourse system and actively adapt to the media usage habits of Internet users, to win the trust of more audiences. At the same time, it is also necessary to speed up the pace of deep media integration development, build a communication platform for the integration of old and new media, and construct a space for guiding media opinion in the context of integration. Although the investigation of academic misconduct cannot rely on the media, the media can still play a proper role in the process of combating academic misconduct, such as following up the investigation of academic misconduct and announcing the progress on time, letting the public understand the serious consequences arising from academic misconduct, and reporting the results of the treatment of academic misconduct, to maintain a positive image of scientific research and public trust in science. In addition, the media can also assist the academic community to do a good job in this regard by reporting to preach scientific ethics and academic style building.

4 Conclusions.

The development of social media platforms in the information era has dissolved the framework structure of traditional news media and made the communication environment more complex. Especially in the reporting of sudden public crisis events in universities and the public relations processing of universities themselves, the riskiness of the whole media itself increases the uncertainty of information dissemination, which requires the media reporting and public relations processing to change its inherent discourse system and actively adapt to the media usage habits of Internet users, to win the trust of more audiences. At the same time, it is also necessary to speed up the pace of deep media integration and development, build a communication platform for the integration of old and new media, and construct a space for guiding media opinion in the context of integration.

Reference.

1. Lichetenstein, R, etal., (1994). School crisis response: Expecting the unexpected. *Educational Leadership*, 52
2. Wu Xiaobo, A Comparative Study of People's Livelihood News between Party Newspapers and Urban Newspapers from the Perspective of Frame Theory. [J] Master's Thesis of Anhui University, 2011
3. Erving Goffman F*raming Analysis*
4. Luo Yicheng, Chen Gang: On the framework of news media coverage of controversial events: A study of the "South China tiger photo" incident[J]. Contemporary Communication, 2008.
5. Sun Caiqin, A Literature Review of the Development of Frame Theory in the Past 35 Years -- Also discussing the problems and suggestions of the development of frame Theory in the past 11 years in Mainland China. [J]International Press. 2010
6. Wang Dapeng, Media should do something to punish academic misconduct. [J]Young Journalist,2019
7. Liu Xiaolin, Agenda Setting in the New Media Environment: A Case Study of Zhai Tianlin, [J] News Research Guide,2020
8. Guo Qingguang: Communication Courses [M]. China Renmin University Press,1999.

Research on the Countermeasures of Military Equipment Instructors' Teaching Ability Construction in the New Era

Wei Zhang[1,*], Chenggang Gu [2] ,Qiqing Fang [2] and Jingjing Wang [2]

[1] Air Force Early Warning Academy, Air-defense Early Warning Equipment Department, Wuhan, Hubei, China
[2] Air Force Early Warning Academy, Air-defense Early Warning Equipment Department, Wuhan, Hubei, China

Abstract. According to the requirements of military education policy in the new era, military equipment instructors must constantly strengthen the construction of teaching ability in order to provide professional equipment technical personnel for the new information war in the future. In view of how to comprehensively improve the teaching ability of military equipment instructors, this paper first analyses the main problems existing in the current teaching ability construction of equipment instructors in our Institute, and then puts forward some specific measures to strengthen the teaching ability construction of equipment instructors, which has certain guidance and reference for rapidly and efficiently improving the actual combat teaching ability of equipment instructors.

1 Introduction

President Xi clearly put forward the new era military education policy of "adhering to the absolute leadership of the Party over the army, serving for strengthening the country and rejuvenating the army, cultivating people by virtue, educating people for war, and cultivating high-quality and professional military personnel with both ability and political integrity", which put forward higher requirements for training military personnel and running military education well. How to run a direct battlefield education, a satisfactory education for the army, and a future-leading education is a heavy responsibility that falls on the shoulders of every military Academy teacher.

Facing the complex international situation, the transformation of quality-efficiency and technology-intensive is the direction of military equipment development. From the perspective of several local wars in recent years, the role of sophisticated weapons and equipment in the war has become increasingly prominent, and it has become a magic weapon to win future wars. The high-quality and professional new military talent phalanx is the talent advantage to support the strong army to win. How to train professional and technical personnel who understand equipment, understand the mechanism, and are good at using it is a new challenge for military equipment instructors. Therefore, military equipment instructors must constantly strengthen their own teaching capacity building, in order to continuously provide nutrition for the training of new military personnel, so that the supply of personnel training can accurately meet the needs of the future battlefield.[1] [2]

2 Main problems existing in the current construction of equipment instructors' teaching ability

In order to fully understand the quality of equipment personnel training in our Institute and further find out the new requirements and changes of the army for the professional competence and quality of equipment personnel posts, in recent years, our Institute has organized instructors to conduct targeted research in air force radar brigades in various war zones every year in a planned and targeted manner, and collected opinions and suggestions from the army on the requirements of personnel posts and the reform of actual combat teaching. From the investigation, discussion and analysis, some problems existing in the current construction of equipment instructors are reflected at different levels, specifically as follows:

(1) There is still a gap between the teaching ability of equipment instructors and the job requirements of students. In recent years, although the first post ability and the professional comprehensive quality of combat support of the cadets in the army Academy have been unanimously recognized, there is still a gap in the professional ability and quality of equipment of the cadets. Performance: a small number of students do not understand the meaning of equipment performance indicators, equipment operation is not familiar with the equipment, the ability to use the full function of equipment is weak, the ability to maintain and manage equipment is weak. All these reflect that we have shortcomings in teaching, and we have not done well in training objectives, teaching materials and teaching

*Corresponding author: 466673564@qq.com

content, teaching mode, practical teaching, classified guidance and other links.

(2) There are still shortcomings in the training objectives of equipment teachers' teaching ability and students' ability and quality. A very important aspect of equipment teaching is to cultivate the students' professional development ability and quality, including the thinking ability to analyse and solve problems independently, the strong ability to learn equipment independently, and the practical skills to use and support equipment, which were not effective enough in the past.

(3) The teaching ability of equipment instructors is not closely related to the actual combat of the army. With the iterative upgrading of the demand for equipment combat support, the army has made new changes in the requirements for the professional competence and quality of personnel at the three levels of command, engineering and sergeants. However, some equipment instructors are not closely connected with the army, and many new cases and combat examples have not been absorbed into the teaching content in time, resulting in a slight lag in the training of post personnel competence compared with the demand for equipment combat support.[3]

(4) The construction of equipment instructors is not compatible with the transformation of combat effectiveness generation mode. Young teachers still have some problems in varying degrees, such as the lack of solid equipment theory, the lack of actual combat teaching level, and the unskilled operation of full-function equipment. Teachers' follow-up training, information literacy and technical and tactical literacy training, exchanges between academies and troops, and service experience are not enough. Guided by post demand and operational demand, the realistic demand for the transformation of teachers' ability and the upgrading of their quality is constantly improving.[4]

3 Specific measures to strengthen the teaching capacity building of equipment teachers

3.1 Strengthen the foundation and seek a breakthrough from the basic teaching ability to the professional ability. [5]

1.Tamping the basic teaching skills of equipment teachers. To equip teachers to stand on the "three-foot platform", the most important thing is to have the basic teaching skills required by a qualified teacher. The so-called basic teaching skills of equipment teaching include the basic teaching abilities of "speaking, acting, writing, drawing and posture", as well as the production and application of teaching AIDS (such as the compilation of teaching plans, the production of multimedia courseware, the application of teaching AIDS, etc.), and the basic skills of organizing and arranging teaching contents. Through training, all teachers are equipped with the basic teaching skills required for the preparation of lessons.

2. Strengthen the ability of equipment teachers to organize classroom teaching and training. The remarkable feature of equipment preparation is "integration of theory and practice", which is the technical realization of basic theory to equipment, and also the bridge from the basic operation of equipment to the operational application of troops. The ability of equipment instructors to organize training in classroom teaching is particularly important. Through "demonstration class", pre-class trial lecture, in-class supervision and other ways and means, the ability of equipment instructors to tell the principle class is improved, so that they can sort out the equipment process clearly in the classroom, explain the basic principles thoroughly, and analyse and interpret comprehensively the introduction of operational use, the explanation of equipment technology system, and the anatomy of equipment realization principle. Through operation demonstration, situation creation and task drive, the equipment practice teaching is suitable for the actual situation of equipment maintenance and operational application.

3. Strengthen the ability of equipment operation and maintenance support of equipment instructors. It takes the early warning technology and command professional quality of the new era and the ability of three or four generations of equipment operation and maintenance support as the basic criteria. Continuously carry out the "nine one-pass" teaching ability of equipment instructors (mechanical and electrical structure inspection, electrical connection correspondence, principle circuit running, signal flow combing, sensitive point parameter testing, inspection and adjustment operation, equipment maintenance organization, operational application drill and typical fault repair). Through "post competence training, post deepening training and post development training", we can train a team of equipment instructors with the ability of equipment application and maintenance support, and realize the ability of equipment instructors to model experts and famous teachers.

3.2 With the help of training, broaden the way to train the actual combat teaching ability of equipment instructors. [6]

1. Enhance the awareness of actual combat teaching in the army (acting) posts. Equipment instructors should seize the rare post training opportunity of post exchange activities, go deep into front-line combat troops and forward combat areas, and strengthen their own awareness and concept of actual combat. By taking on daily combat duty and support tasks, equipment instructors have experienced and deeply felt the combat environment in which the front-line troops are facing heavy combat readiness tasks, great air pressure and rapidly changing enemy situations, thus imperceptibly building a strong sense of combat-oriented teaching ideologically.

2. Accumulate actual combat teaching strength in large-scale exercise tasks. The importance of large-scale

exercise tasks for the improvement of individual technical and tactical literacy, as well as the exercise and improvement of comprehensive ability is self-evident. On the one hand, it provides opportunities for young teachers to go to the army and the exercise ground; on the other hand, young teachers should consciously take the initiative to seek opportunities to improve their actual combat level, and strive to be qualified to participate in battlefield training. Young teachers and teaching backbones participate in the task of exercises and drills, which not only enables teachers to learn operation, planning, tactics and coordination, but also cultivates their system combat capability. It has also obtained a wealth of practical cases, laying a solid foundation for the teaching of case courses and seminars in the future.

3. In the normal teaching competition, the actual combat teaching ability is tempered. Young and middle-aged teachers should actively participate in the normal teaching competition, and the teaching team can also organize internal teaching plans, PPT courseware compilation, "teach a good lesson" and other evaluation activities to fully mobilize the enthusiasm and initiative of teaching.

4. The actual combat teaching content is extracted from the deep training and follow-up training of equipment. Every year, equipment manufacturers and research institutes organize activities such as centralized training, follow-up training and in-depth training for various types of equipment. However, during the training period, we should not only be satisfied with the study of general working principles, composition structure, function mode and maintenance support skills, but also dig deeper content from the perspective of equipment overall design, obtain more first-hand information and data, and enrich the teaching content of innovative equipment.

5. Carry out the actual combat teaching mode in the equipment type term class teaching. Equipment teaching is the only standard "main battlefield" to test the actual combat teaching ability of teachers. The research results of actual combat theory and practice should eventually be transformed into classroom teaching in order to bring its actual benefits into play and reflect its actual combat value. Therefore, we should pay attention to every detail in the process of equipment teaching management (from the selection of teaching content, the writing of teaching plans, the production of PPT, the teacher's trial lecture, the simulation trial, the scientific grouping of students, the support of students' computer time, and the test of students' actual combat ability), and explore how to integrate the concept and method of actual combat into classroom teaching. Strive to become excellent courses and case courses.

3.3 Build an environment and seek benefits in the resource construction of team leadership and teaching and research. [7]

1. Adhere to the system of collective lesson preparation to improve teaching efficiency and quality. As an effective form of teachers' cooperative discussion, collective lesson preparation plays an indispensable role in giving full play to teachers' team spirit, drawing on the wisdom of the masses and learning from each other's strengths. Centring on equipment teaching, through collective participation, mutual exchange, joint discussion, and mutual inspiration, it is not only conducive to increasing knowledge, but also conducive to individual teachers better grasping the "ruler" and "degree" of radar equipment theory and practice teaching in the process of preparing lessons, making full use of their strengths and avoiding their weaknesses, giving better play to their teaching level, and improving the quality of teaching.

2. Adhere to the system of "passing on, helping and leading" with the team as the core. Make full use of the advantages of team teaching and scientific research, pay attention to the backbone role, teach by words and deeds, care about the study, work and life of the training objects, help them get familiar with the working environment as soon as possible, understand the characteristics of equipment teaching and master the teaching methods of radar equipment preparation, and carry out centralized lesson preparation in the teaching group according to the situation of lesson preparation and trial lecture of the training objects, so as to improve their equipment teaching level.

3. Adhere to the resource sharing mechanism of teaching and research team. The integration of teaching and research should focus on improving the quality of education, organically combine teaching research with scientific research, take teaching and research as the carrier, take scientific research as the task, let teaching and research drive the development of scientific research, let scientific research promote the progress of teaching and study, and realize the improvement of individual and team comprehensive ability.

3.4 Strict standards and seek development in the implementation of equipment instructor assessment and acceptance mechanism.

The examination and acceptance of equipment instructors is the core link of the equipment instructor access system, and it is necessary to study and formulate the corresponding organization and implementation methods of the examination and acceptance of equipment instructors. In view of the training objects who are going to engage in the teaching of different types of equipment, an expert group composed of department and teaching and research section leaders, department teaching supervision group and teachers with rich experience in equipment teaching and excellent equipment use and support technology is organized to assess the training objects. When conditions permit, military experts can be invited to conduct assessment and guidance. We should adhere to the principle of fairness and justice in the acceptance assessment, put an end to any form of cheating, and ensures the authenticity of the assessment of the ability level of equipment instructors. The standards are directly linked to the mastery of new equipment, and it is clear that the evaluation results are directly linked to job evaluation, professional title evaluation, priority evaluation, promotion, and overseas selection, so as to encourage teachers to consciously touch equipment, repair equipment and understand equipment.

4 Conclusions

The policy of military education in the new era systematically expounds the objectives, tasks, ways, key grasps and basic requirements of the reform and development of military education under the new historical conditions, and points out for us the major practical issues such as where and how to go in military education in the new era. Efforts should be made to train professionals to meet the needs of army construction and future wars.

Faced with the issue of how to comprehensively improve the teaching ability of military equipment teachers in the new era, this paper first analyses the main problems existing in the current teaching ability construction of equipment teachers in our Institute, clarifies the shortcomings in the construction and teaching of equipment teachers, and then puts forward some specific measures to strengthen the teaching ability construction of equipment teachers. Emphasis is laid on strengthening the teaching ability in the improvement of the ability to strengthen the foundation, teamwork, equipment in-depth training and follow-up training, and equipment model class teaching, and through various activities such as large-scale exercise tasks, normalized teaching competition, teamwork, equipment in-depth training and follow-up training, and equipment model class teaching, to enhance the actual combat teaching ability of equipment teachers. [8]

In this paper, the ways and methods of faculty capacity building are put forward, which are combined with the achievements of the pilot practice of improving the actual combat teaching ability of equipment faculty in recent years, and are summarized and improved, which have certain guidance and reference for improving the actual combat teaching ability of equipment faculty and improving the conversion rate of actual combat teaching results.

References

1. Cheng Hua, Han Tong, Tang Shangqin, Du Haiwen. Thoughts on Accelerating Equipment Curriculum Construction in Military Academies. China Educational Technology and Equipment, 15(2019).
2. Dai Hongde, Jia Linsheng, Wang Xibin, Lu Jianhua, Liu Rui. Exploration of equipment teaching skills training for teachers of airborne equipment specialty. Science and Education Guide,36 (2019).
3. Yang Mingxu, Li Lin, Han Zhaofu. Research on the new mode of science refitting of aviation equipment for teachers in military academies. China Modern Education Equipment,(2019).
4. Li Fei, Zhang Kai, Shao Li. Problem Analysis and Countermeasure Research on Electronic Countermeasure Equipment Teaching in Military Academy. Journal of Higher Education Research, ,3 (2019).
5. Wang Guoshi, Yang Shiying, Zhang Chaowei, Han Wei. Reform and practice of radar equipment principle and maintenance course. Journal of Air Force Early Warning Academy, 2(2020).
6. Sui Jiangbo, Ma Tianyi. Thinking and Practice of Instructional Design for New Equipment Training. Educational Science Forum, 9(2020).
7. Zhao Mei, Yue Shuai, Shi Xianming. Strengthening the Three Basic Qualities of Military Academy Teachers with Efficiency as the Core.Science and Education Collection,2(2020).
8. Liu Guangsheng, Li Chengbing, Jia Changzhi, Han Chao. Research on Teaching Reform of Foreign Training of Equipment Curriculum. Modernization of Education, (2019).

Appropriate Solutions for Traditional Media and New Media to Establish in Contemporary Society

Yitian Gong*

Nanjing high school of Jiangsu Province, Jiangsu China, 214441

Abstract. New media penetrates people's lives in thorough aspects, and the majority of people are inclined to receive information through the Internet instead of purchasing newspapers or other approaches to get to know the events around the world. Some experts think it is undisputed that the promising prospect of new media will harm traditional media. As a result, the status of the traditional media arouse the public concern and at the same time provoke a heated discussion among individuals. The total printing volume has declined for nine consecutive years since 2012, and especially since 2020, the COVID-19 impacted the traditional media severely. The article aims at claiming the benefits of the fusion of new media and traditional media and studying the attitude towards each of the media and the fusion of two kinds of media. The research involves the literature reference method to illustrate the two-sides of each type of media. The article analyzes a lot of data from official website and the essay [1]. The impact of new media on traditional media was written by Lavanya Rajendran and Preethi Thesinghraja to identify the tendency of these media. Eventually, conclusions are drawn that it is profitable to combine the new media with the traditional media.

1 Introduction

With the development of contemporary new media, many conveniences and unique possibilities are created. It is time-saving and efficient to type in keywords on the Internet and search for the information they want instantaneously. Google Scholar contains 4750000 research papers that are associated with the media revolution, and 0.08 seconds of spending on loaded time showing on the screen. On the contrary, traditional media is implausible to search for information immediately which can deal with urgent demands because traditional media like newspapers only supply once a day[2]. Otherwise, other attributes like high-frequency interaction, a variety of manifestations, more choices for audiences, etc.

As a result, a contentious topic concerning the fate of traditional media and new media ignites a long-simmering debate about whether there is a tendency to replace traditional media with new media and what the consequences are if the new and traditional media merge [3].

Our paper concentrate on whether the fusion of two kinds of media is able to solve the embarrassing situations within the new media and traditional media. There is no denying that the status of traditional media is deteriorating over time; therefore, a strategy that combines the two media is advantageous for future development.

*Corresponding author: Gongtt0202@163.com

2 The discrepancies between the new media and traditional media

There are a lot of differences between the traditional media and new media, in the following statement, some significant differences like timeliness, interactivity, authority and authenticity are proposed to verify the opinions.

2.1 The different timeliness of the new media and traditional media.

The processes of producing traditional media such as newspapers are complicated; many rigorous and strict steps must be taken and followed. To put it briefly, three main steps are involved, the first step is consist of collecting and editing, examining and approving manuscripts, soliciting contributions, composing, proofreading and final judgment, the following steps are plate making and printing, transporting and publishing are the ultimate processes.

To sum up, the overall procedures are authoritative and convictive, however, traditional media causes time lag after these complex procedures. People delay for a long time to get to know the information with traditional media while instant messages sent by new media tend to be more appealing to people, because countless media, even individuals can compile alone and publish opinions on their social apps, therefore, substantial information and news are produced every few minutes for people to discuss and receive.

2.2 The different degrees of interactivity between the new media and traditional media.

Traditional media is defined as a unidirectional transmission, it is more likely to be a one-to-many mapping[4], people receive the information and can not give the feedback to the contents of the information instantly. Also, some traditional media are supervised by the government, and some content which harmful to the nation's image is prohibited to report. Thirdly, there are different targeted audiences through the specialized press, to be specific, three kinds of newspapers target three groups of people.

For example, the party-political press is dedicated to the task of activation, information, and organization, these party newspapers lie in the attachment which shares the same allegiance. However, the new media seem to act a part in both governor and producer, people can review and express their unique understandings beneath the articles without restrictions when they read the news on the Internet.

In addition, news on the Internet transcends time and space, information spread beyond counties easily and cause cosmopolitan sensations, news on the Internet gather people with different backgrounds, educational level, and class into interaction and discussion. The cost of information exchange is lower than the cost of purchasing traditional media, so people are more willing to exchange information on the Internet.

As Marshall McLuhan proposed, the world is a global village, he coined the term " global village " to illustrate the phenomenon of advantages and drawbacks at the same time due to the prevailing technological progress which is responsible for the momentary sharing of culture.

2.3 The authority and authenticity of the information between the new media and traditional media.

The newspapers as the first media have a long history and impact the media profoundly, so as the time goes, a lot of perfect journalists with intensive news sensitivity and ample experiences in editing the newspapers, as a result, the abilities of professional dedication and attitude are beyond the network media employees [5].

Compared to the information debris, information uncertainty, and ambiguous information, the majority of people prefer to trust the newspapers.

On the other hand, the normative process of auditing are involved and many assertion and judgment of the trend the news development are contained in the newspapers. So, insightful and precise content in the newspapers is a better opportunity for people to receive the information and recognize the situation. The new media also plays an important role in information dissemination.

However, news released on new media multiply in recent years, and lots of presses compete for the headline news and may take the means to report exaggeratedly. Sometimes, some presses ignore the depth of analysis and further prediction but only focus on the speed of reporting, leading to suspects of the news' authority and authenticity.

Also, the messy convergence of information sways people's ideas.

3 The methods to facilitate the successive collaboration of the traditional media and new media

The fusion of media is a relatively efficient method of relieving traditional media embarrassment and strengthening the reliability of new media. As a result, some methods are hypothesized to promote the implements.

In short, it may work to give play to traditional media's strengths and remove the outdated sections, also, it is important to utilize the compelling skills to broaden the range of influence. In the following statements, some proposes are listed to consult.

Figure 1. Online Media Classification

3.1 Traditional media intensifies their advantages to fancy people which embedded into the age of information explosion

Many fake news or disinformation are produced because they are eager to capture people's attention and compete for rapid reporting speed, as well as to create astonishing points and seize the hot spot.

Eventually, these reports are blind and unreliable for covering the original events. In this condition, the strict standards that traditional media contained are significant to dominate in the whole press.

The newspapers keep constant and provide the Internet with high-quality and high-reliable information to enable the brand of the media maintain the authority because teams subsist on traditional media are exceptional and highly qualified. The integration of the two media give play to self advantages maximally for elevate the news information of social benefits and financial benefits. The principle of the fusion of the media is user-oriented, it is also important to bear in mind that content is king in social media.

Conclusively, superior contents combine with focused media platforms is an appealing opportunity to disseminate.

3.2 Both of the media strive for customers' needs and keep innovating the the pattern of the content

The problems revealed in the traditional media are the shortage of content and the tedious information. Nowadays, the newspaper press usually covers unexpected, far-reaching events among nations, thus leading to the high possibility of resemblance. The quality of the content is the permanent purpose in any media, so, the fusion of the media must insist on innovating the superior content and exploring new behavioral forms.

For the contents, some more popular and meaningful contents in the newspapers can be excavated, and refine these contents for blending into the new media more smoothly. It is not only vital to involve the characteristics within the newspapers like authority and brand effect, but also to combine with some vivid, flexible high-efficiency Internet platforms, for instance, the Wechat, Weibo or Tik Tok are good examples.

And at the same time, maintaining the connections with people and increasing the interaction online and offline are momentous, because more personalized needs are visualized to detect.

3.3 More entertainment merge into the media to catch up the attention

With the contemporary rapid pace lifestyle, people tend to watch videos in limited time, and the variety of choices facing people, so it is important to grab their attention initially. Also, people obtain a lot of options, they select the medias they are inclination to. Therefore, it is becoming vital to attract them for some distinctive reasons.

Many humorous sayings or cool special effects are valid and these attributes are appealing to people. It is wise for traditional media to enroll some relaxing elements instead of some dull and rigid information [6]. There is an example that the CCTV in China register the Tik Tok to broadcast concise news abstract and emphasis, the differences between the normal CCTV is the languages, on that account, some interesting and funny words are used to describe the events and provoke peoples' attention immediately. Even some young people who seldom keep track of the news show great curiosity to these events.

Apparently, we can learn from the example that it is significant to dress the medias more unique and characteristic to attract people, thus more dissemination are generated.

4 Discussion

Many development dilemmas can be solved and some embarrassed conditions are avoided.

There is no denying that each of the media can not displace the other media after the collision of the new media and traditional media, the fusion of the media take advantages of the merits and abandon the drawbacks of the media. The professional and rigorous content quality complement the flaw in the new media, meanwhile, the timeliness and wide-spread assist the traditional media in spreading superior news out.

In addition, the diversity of platforms on the Internet provide infinite chances for information to be notified. Consequently, more rewarding news circulate among people, and more correct information guide them to establish true values.

5 Conclusion

To sum up, it is inevitable for the traditional media to decay along with the development of new media because many attributes of the new media are better than the traditional media undoubtedly. However, several solutions that enable traditional media to survive to depend on exploiting their flash points, thus traditional media will not extinct.

In the course of history, and just the reverse, traditional media make use of features to adapt and blend into the complex media circumstances. Considering all the reasons discussed above, we can safely come to the conclusion that it is necessary to introduce the fusion of the media, which should be attributed to the two main points, one is for traditional media, are paid more attention to by people, they can amplify authority and preciseness to optimize the integral accomplishment. And on the other hand, new media stand out from the crowd under excellent content and innovations. In the process of which, some inadequacies are also not taken into account without hesitation.

As we know, media transformations are not simple, it is tricky to implement, and much more potential troubles are faced to solve in the future. For example, a lot of employees need to get rid of all outdated thinking and ideas and all institutional ailments. It is a long-lasting and arduous assignment, especially for some people who worked for traditional media for a long time it is harsh to alter stubborn thinking and make them adapt the irrelevant mechanisms.

As far as I am concerned, more researches are essential to deal with the possibilities of the fusion of media, and more feasible and detailed plans are advocated to perfect the fusion of the media. Although some examples are implemented successfully now, some traditional media are still in danger. When the fusion of media becomes a normal state and the communication form can be changed freely, people can receive a large amount of high-quality information on the premise of ensuring the quality and authority of the content.

Acknowledge

At the end of the essay, I really want to express my gratitude to my professors during the summer courses sincerely. They are professor Berna from the university of Johns Hopkins University and professor Shen from Xi'an Jiaotong University.

Before attending the courses, I only have some ambiguous knowledge of the media which can not make sense, however a lot of detailed information are taught by these professors.

Thus, I know some specialized knowledge about media and these knowledge support me to finish my first-time essay. In the following, I also want to express heartfelt thanks to my essay teacher. She teaches the frames and exhaustive main points for writing an essay.

References

1. Lavanya Rajendran, Preethi Thesinghraja (2014), The impact of new media on traditional media, Anna University, Chenniai, Illinois Institute of Technology
2. Logan, R. K. (2010). Understanding new media: extending Marshall McLuhan. New York: Peter Lang
3. An Nguyen, Mark Western (2016), The complementary relationship between the Internet and traditional mass media: the case of online news and information, The university of Queensland Social Research Center, The university of Queensland, Brisane, Australia

   sm and Communication, Tsinghua University, Center for new media research, Beijing
4. Wayne Xinzhao, Jing Jiang, Jiangshu Weng, Jing He, Ee-Peng Lim, Hongfei Yan, Xiaoming Li (2011), Compare Twitter and traditional media using topic models, Part of the Lecture Notes in Computer Science book series (LNISA, volume 6611)
5. Xia Mosheng, (2017) Some thoughts on the development of fusion of media https://zhuanlan.zhihu.com/p/31296617
6. Luo libin (2021), Discussion about fusion development between new media and traditional media, Zhengzhou broadcasting and TV station

Research on Training Objectives of Civil Engineering Talents in Local Engineering Colleges Based on OBE Concept

Zhao Yang[a], Liangli Xiao[b]

College of Urban Construction, Wuhan University of Science and Technology, Wuhan, Hubei, China

Abstract: The OBE concept is the concept of results-oriented education and the core concept of engineering education certification. Under the background of rapid industrial development, China has put forward higher requirements for talent training. However, at present, many colleges and universities, especially a large number of local colleges and universities, still have problems such as unclear positioning, insufficient basis and unreasonable standards for the training objectives of professional talents, which need to be studied and improved according to the OBE concept. Taking the civil engineering specialty of Wuhan University of Science and Technology as an example, this paper carries out relevant analysis from four aspects : the formulation basis, specific content, open channel and rationality evaluation of professional training objectives. The research results can provide reference for the research on talent training objectives of local engineering colleges.

1. Introduction

In June 2016, China successfully became a formal member of the *Washington Accord*, marking that the engineering education certification work in China has been fully affirmed internationally and the engineering education in China has taken another major step. The core concepts of engineering education accreditation are " Student-centered " , " Outcomes-oriented " and " Continuous improvement ". Therefore, engineering education certification requires Outcomes-Based Education (OBE).

The research on the concept of foreign OBE education began early, and systematic research was carried out from the aspects of generation, development and specific application [1-3]. However, China's research on the concept of OBE education began relatively late. In recent years, Chinese have gradually realized the role of OBE education concept in higher education, especially in higher engineering education, and made some useful explorations [4-11], we should also clearly realize that in the benchmark analysis with international engineering education certification standards, the orientation of professional talents training objectives in some universities in China is not clear. The basis for the formulation of training objectives is insufficient, and the talent standards set by the training objectives are not reasonable enough. As an important human resource in the strategy of " Made in China 2025 " and " Industry 4.0", the applied engineering talents trained by local engineering colleges largely restrict the speed and scale of industrial modernization.

In this context, it is an example of great theoretical value and practical significance to carry out research on talent training objectives of local engineering colleges based on OBE. Therefore, taking the civil engineering major of Wuhan University of Science and Technology as an example, this paper carries out a research on the objectives of talent training based on the OBE concept, so as to provide reference for the top-level design of relevant professional education in local engineering colleges and universities.

2. Basic situation of civil engineering major of Wuhan University of Science and Technology

Wuhan University of Science and Technology is a provincial key comprehensive university co-constructed by the Ministry of Education. It is one of the national " Central and Western Universities Basic Capacity Building Project " selected universities. The civil engineering specialty originated from the industrial and civil construction engineering specialty of Wuchang Construction Industry School. In 1996, it recruited four-year undergraduates majoring in architectural engineering. In 1998, it was renamed the civil engineering major, focusing on the direction of construction engineering. In 2003, the professional direction of road and bridge engineering was added, and in 2010, the professional direction of Geotechnical and underground engineering was added to civil engineering. The enrollment of master's students began in 2000 and doctoral students began to be recruited in 2014.

[a] Corresponding author: yzwh77@163.com
[b] 605316713@qq.com

3. Basis of training objectives

3.1 College orientation

The college adheres to the socialist direction of running a school, fully implements the party 's education policy, follows the law of development of higher education, adheres to the cultivation of morality, takes talent cultivation, scientific research, social services and cultural innovation and inheritance as its own responsibilities, bases itself on the local area, serves the industry and faces the world, and is committed to building an internationally renowned domestic high-level teaching and research university. Cultivate high-quality talents with all-round development, solid basic theory, reasonable knowledge structure, innovative and entrepreneurial spirit, practical ability and international vision.

3.2 Professional Resource Conditions

Wuhan University of Science and Technology has been running a major in civil engineering for more than 60 years. It has a discipline platform such as a doctoral degree authorization point for second-level disciplines, a master's degree authorization point for first-level disciplines and the right to grant a master's degree in engineering. It has established a provincial key experimental teaching center, a provincial research and development center, as well as a research institute, a design research institute, an internship base and other teaching practice platform. Meanwhile, all professional teachers have engineering practice experience, which provides important support conditions for the training of high-quality applied talents in this major.

3.3 Social demands

At present, the country is implementing the Belt and Road strategy, and there is an urgent need for a large number of high-quality applied talents with an international vision and strong innovation and practical ability. Meanwhile, with the upgrading and transformation of China 's construction industry, the practical application of a large number of new technologies and new methods, as well as more and more complex projects, the industry and enterprises urgently need a large number of high-quality applied talents with solid theoretical foundation, broad professional knowledge, strong practical and innovative ability, a certain international vision, a high sense of social responsibility and good engineering professional ethics, team spirit and comprehensive ability to solve complex civil engineering problems.

3.4 Research and feedback

Through special visits organized by schools and colleges from time to time, investigate the working status and performance of employers ' graduates ; use alumni gatherings, alumni back to school, graduate recruitment conferences, graduate QQ groups and other channels to collect graduate work information, and obtain the evaluation of the training effect of the society and employers on this major; through the organization of talent training seminars involving responsible professor teams, professional teachers and industry experts, the social and industry needs are discussed, the research data of talent training in our school are analyzed, and the effectiveness evaluation of talent training in this major is carried out. The survey and feedback results as the main basis for the formulation or revision of training objectives.

4 Training objectives and opening channels

4.1 Professional training objectives

The professional training objectives is to cultivate high-quality applied talents who can adapt to the needs of modern science and technology and economic development, develop morally, intellectually, physically and aesthetically, master the basic principles and basic knowledge of civil engineering disciplines, and be competent for the technical and management work of various types of projects such as construction engineering, road and bridge engineering, geotechnical and underground engineering through basic training of engineers. It has a solid basic theory, broad professional knowledge, strong practical ability and innovation ability, and has a certain international vision. It can be engaged in technology or management in the investigation, design, construction, management, education, investment and development departments of civil engineering.

It is expected that graduates will achieve the following goals in about 5 years: (1) To have a good dedication, social responsibility and engineering ethics, with quality awareness, environmental awareness and safety awareness, can combine engineering and social sustainable development. (2) To have a solid system of professional knowledge and lifelong learning ability, to adapt to the development of new civil engineering technology, to carry out innovative practice; good communication skills, team spirit and organizational management ability. (3) To be fit for an engineer or have related professional technical ability and conditions; can be engaged in planning, survey, design, construction, management, education and research development work as a high-quality engineering and technical personnel in design, construction enterprises, real estate, investment department, research institutes, government management departments. (4)To have international vision and good foreign language application ability, can carry out cross-cultural communication, competition and cooperation.

4.2 Opening channels and measures for professional training objectives

Teachers can understand the professional training objectives through Wuhan University of Science and Technology undergraduate admissions website, College of Urban Construction website, Wuhan University of Science and Technology educational administration integrated management information system, training program and other channels. In addition, the college helps teachers understand the connotation of training objectives by organizing all professional teachers to learn training objectives, discussing the implementation of training objectives, participating in the revision of training objectives and other measures.

Students can also learn about professional training objectives through Wuhan University of Science and Technology undergraduate admissions website, College of Urban Construction website, Wuhan University of Science and Technology educational administration integrated management information system, training program booklet and other channels. In addition, this major through the enrollment publicity, freshmen education, professional teachers as class teachers, professional introduction, professional courses and lectures and other measures to help students understand the professional training objectives.

The community can understand the professional training objectives through the school undergraduate admissions website, urban construction college website. In addition, this major also widely introduces the training objectives of this major to the society through enrollment publicity, Campus open day, visiting employers, campus recruitment fairs, alumni gatherings, alumni returning to school and other channels.

5 Rationality evaluation of professional training objectives

5.1 The relationship between training objectives and college orientation and social and economic development on talents demand

The school 's talents training objectives is to " cultivate high-quality talents with all-round development of morality, intelligence, physical fitness and beauty, solid basic theory, reasonable knowledge structure, innovative and entrepreneurial spirit, practical ability and international vision ".This major cultivates " high-quality applied talents who can engage in technology or management in departments related to civil engineering survey, design, construction, management, education, investment and development ", which is in line with the school 's positioning.

With the implementation of major strategies such as the country 's Belt and Road, Central China Rise, and the construction of the Yangtze River Economic Belt, as well as the upgrading and transformation of China 's construction industry ; the industry has higher and higher requirements for the theoretical basis, professional knowledge, practice and innovation ability, international vision, social responsibility, professional ethics, team spirit and comprehensive ability to solve complex civil engineering problems. Therefore, the training objectives of this major meet the needs of social and economic development for civil engineering professionals.

5.2 Graduates ' main fields of employment and social competitive advantages

According to the classification employment status of civil engineering students of Wuhan University of Science and Technology in the recent 3 years, as shown in table 1, the graduates of this major are mainly employed in various enterprises of civil engineering, accounting for about 60%, which is consistent with the training objectives and main employment fields of this major. Among them, the proportion of students entering design enterprises is increasing year by year, reflecting the growth of students' professional comprehensive ability. More than 20% of the students study for graduate students, indicating that the graduates of this major have a solid theoretical foundation and professional knowledge, high comprehensive quality, and the ability to continue their studies, which is consistent with the training objectives of this major.

Table 1 Classification employment status of civil engineering students of Wuhan University of Science and Technology in the recent 3 years

Year		2019	2020	2021
Civil construction enterprises	Numbers	138	129	121
	Percentage /%	53.28	63.55	55.25
Civil design enterprises	Numbers	4	2	5
	Percentage /%	1.54	0.99	2.28
Other civil enterprises	Numbers	8	6	6
	Percentage /%	3.09	2.95	2.74
Non-civil enterprises	Numbers	41	16	11
	Percentage /%	15.83	7.88	5.02
Government institutions	Numbers	8	4	21
	Percentage /%	3.09	1.97	9.59
Graduate studies	Numbers	60	46	55
	Percentage /%	23.17	22.66	25.12

From the feedback of graduates and employers, the main social competitive advantages of the graduates are having good self-learning and lifelong learning ability, having good humanities, social science literacy and a sense of social responsibility and mission, having a sound personality, physical and mental health, having good engineering quality and be able to perform professional responsibilities in a complex engineering environment.

5.3 Periodic evaluation and revision system of rationality of training objectives

The evaluation and revision of the professional training objectives are carried out simultaneously with the evaluation and revision of the training program. The school has established a regular evaluation and revision system for the training program, and comprehensively revised the training program every 4 years. Meanwhile, each major can make partial revisions to the training program every year as needed.

When carrying out the evaluation and revision of the training program, the school teaching guidance committee formulates guidance opinions, and the college establishes a working group for the revision of the training program and organizes specific implementation. The main person in charge is the vice president in charge of teaching ; the working group is mainly composed of responsible professor team members, course leaders, professional backbone teachers and industry experts ; the rationality evaluation mainly examine whether the training objectives conform to the orientation of school personnel training, the resource conditions of the major, the questionnaire survey and statistical analysis results of professional graduates and employers, the evaluation and analysis results of the achievement of annual professional graduation requirements, and the evaluation and opinions of industry and enterprise experts in related fields on the training program. After the revised training program was examined and approved by the teaching guidance committee of the college, it was approved and implemented by the school office.

The evaluation and revision process of the professional training program is divided into three stages as follows: (1) Extensively solicit opinions and suggestions from teachers, students, graduates, employers, industry and enterprise experts on the training of professionals. (2) Revision of the training programme organized by the working group; after being reviewed by the College Teaching Steering Committee and proposing amendments, the first draft of the training program was formed. (3) convene the profession expert symposium, carries on the examination to the training plan first draft, revises the latter to report the school teaching steering committee to approve, finalize. In the past four years, this major has revised the training objectives of undergraduates in 2018, 2019, 2020 and 2021.

6 Conclusion

It is of great significance for the whole talent training system to formulate reasonable talent training objectives based on OBE concept. For local engineering colleges, scientific and reasonable determination of talents training objectives should be carried out from the following aspects:

(1) The formulation of training objectives should have sufficient basis, which can be carried out from the aspects of school orientation, professional resources, social needs and research feedback.

(2) Training objectives should be specific, clear, and should be open to teachers, students and society through a variety of channels.

(3) The evaluation of the rationality of the training objectives can be carried out by analyzing the relationship between the training objectives and the school orientation and the demand for talents in social and economic development, counting the main employment fields and social competitive advantages of graduates, and formulating the regular evaluation and revision system of the rationality of the training objectives.

Acknowledgments

The Ministry of Education's Industry-University Cooperative Education Project (202102113046); the key project of undergraduate teaching research in Wuhan University of Science and Technology (2021Z016).

References

1. D., Pradhan, (2021) Effectiveness of outcome based education (obe) toward empowering the students performance in an engineering course. Biology & Philosophy (2).
2. G. Thirumoorthy, (2021) Outcome based education (obe) is need of the hour. International Journal of Research -GRANTHAALAYAH, 9(4), 571-582.
3. M.K. Chan, C. CWang, A.A. Arbai., (2022) Development of dynamic OBE model to quantify student performance. Computer Applications in Engineering Education (5).
4. Y. Zhang et al., (2022) Applied Outcome Based Education to Engineering Courses. Journal of Educational Research and Policies, 4(9).
5. H. Niu, J. Fan, H. Yan. (2022).Research on the Integration of Subject Competition and Professional Course Teaching Mode Based on OBE Concept. Frontiers in Educational Research (16).
6. Y. Zhang, X. Wen, Z. Feng, J. Zhao, K. Wang, (2022). Influence on the Training Mode of Talents in Fire Engineering Colleges Based on OBE. International Journal of Learning and Teaching (2).
7. Q. Ruan, (2022). Research on Teaching Reform of Financial Big Data Analysis Course Based on OBE Concept. Advances in Educational Technology and Psychology (7).
8. J. Luo,S. Xiao, L. Chen, (2022). Research on the evaluation system of practical Teaching Quality based on OBE concept. Computer Informatization and Mechanical System (1).
9. H. Pan, B. Bao, J. Huang., (2019) Exploration and Practice of Flipped Class Mode for "Unit Operation of Mixing" Lecture Under the Outcome Based Education Concept. Education Journal (3).
10. Y. Wang, X. Dong. (2019) Reflections on the Teaching of English Listening Based on Obe (Outcome Based Education)..(eds.)Proceedings of 2019 7th International Education, Economics, Social Science, Arts, Sports and Management Engineering Conference (pp.2601-2604).Clausius Scientific Press, Canada.

X. Shan. (2022).The Teaching Method of Theoretical Mechanics Course Based on OBE Concept. Advances in Educational Technology and Psychology (2).

Analysis of social media's online expression strategy for social hot spots based on the perspective of mass communication-Take "Bilibili" as an example

Shuying Chen*

Hangzhou Normal University, Hangzhou, 310000, China

Abstract. With the rapid development of information technology on the Internet and the establishment of new online media platforms, it has played a great role in the delivery of information and the reception of content. The new media have profoundly influenced people's lives in terms of the way they communicate, the construction of ideas and the execution of processes. This has not only enriched the ways in which news is disseminated, but has also opened up wider avenues and channels for news reporting. Since 2009, Bilibili has grown into a representative domestic secondary platform which is a high-density group of young internet users. The development and dissemination of social events that cater to the moods and emotions of young people can also serve to moderate cultural tensions to some extent, but, if used poorly, can be counterproductive. It is important to pay attention to the issue of culture, both to prevent pan-entertainment and to achieve cultural identity. In the face of online public opinion, government news media should change their previous negative attitude and adopt new and timely methods of response. The correct way to guide public opinion in hot events is a problem that the news media has to face and urgently needs to solve.

1 Introduction

With the popularity and the use of the Internet, the demand for all kinds of information has grown quickly, creating the conditions for people to better understand and participate in society; thus enabling various information platforms to integrate with each other, giving the public more opportunities to express their own views and expand the public perspective.

As the rise of short videos has accelerated people's access to information, hot issues in society have also arisen; the rapid development of the Internet has enabled internet users to better accept news from different sources and of unknown authenticity, thus undermining the health of the online opinion ecology and making the news media less credible [1]. In today's online society where information is flooded and pluralistic, the public tends to prefer information provided by the government media when receiving information, which is the authority and credibility long established by the government media, which has a huge fan base; due to its innate advantage of news dissemination, the government media, in order to give full play to its function of guiding online public opinion, must first grasp the news facts before producing press releases to weaken the effect of their dissemination; at the same time, they should correctly guide the direction of public opinion to keep in line with mainstream public opinion, so that the public can obtain true and authoritative information, and provide sufficient space for communication on the Internet so that public opinion can be fully expressed.

Bilibili is the biggest ACGN entertainment community platform in China. It is a gathering place for young people and its related accounts are suitable for the development of Bilibili. It is also a mature audience group with strong user stickiness [2]. "CCTV News landed on Bilibili on December 8, 2019, expanding the youth audience and spreading mainstream voices in a more relevant and youthful format. "The successful launch of CCTV News on Bilibili is an initiative to further integrate mainstream culture with popular culture. CCTV is the flagship of China's mainstream media, and it is taking the lead in the media convergence trend, trying to build a new communication platform. This paper looks at the marketing strategy of the social hotspot event Bilibili account and analyses the overall content production of the Bilibili account and its various links, providing a useful reference for the development of mainstream media to break the circle. The aim of this paper is to analyse the communication effects of the trending topics that emerged on Bilibili, with a focus on their impact on young users and how to increase them. At the same time, the paper also combines relevant communication theories and provides a reasonable analysis of Bilibili's account, which provides a useful reference for the development of Bilibili.

* Corresponding author: 2727723069@qq.com

2 Literature review

Public opinion refers to the general public perception and opinion of public events and represents the common consciousness of the majority of society. As the concept of public opinion is classified as a social ideology and social psychology, it covers a wide range of issues, making it difficult to develop a unified definition of public opinion. Some common definitions include: "Public opinion represents the common consciousness and opinion of the majority of authoritative groups or individuals in society." [3] "Public opinion is a collection of public opinions on public events, public figures, social phenomena, issues and ideas, etc., which is universal, continuous and consistent, and may influence the development of social events." [4]

Internet opinion guidance refers to making Internet users change their views on Internet opinion with the help of the guidance of the relevant subjects. New media are influencing the public perception of the Internet in all aspects of society. Any major event that may cause strong reactions from users will gradually create a free market for conducting opinion exchange, and the free and open online environment will also generate new issues. However, the free, anonymous, interactive and instantaneous communication characteristics of public opinion inevitably lead to qualitative changes in the dissemination of information. Effective management of public opinion through online opinion guidance is an important issue in modern society, which aims at adjusting netizens' misconceptions about social opinion and turning to actively participate in the construction of online public opinion. In fact, the key to guiding online public opinion lies in establishing scientific means of communication, reasonable agenda setting and an effective feedback and guidance mechanism, so that the public will recognize the positive influence of online public opinion and thus actively cooperate with the construction of online public opinion guidance, so that online public opinion will develop in a more harmonious, rational and direction. [5]

The Internet can cover social aspects that cannot be covered by traditional media and allow for good interaction among netizens. Bilibili is one such platform with wide coverage and strong interaction among netizens, and using Bilibili as an entry point will provide perspectives that cannot be covered by traditional media.

The interactive nature of online information dissemination and netizen participation is a prominent feature of online public opinion. As long as a certain number of public participate in topic discussions on online platforms, there is a chance for a serious public opinion crisis to form within a short period of time, no matter what kind of event it is. When sudden destructive, complex and catastrophic factors appear on a social event at the same time, it is likely to attract the attention of the public on the Internet. At this point, the media, as the party with the most information about the event, needs to sort out information about the event for the public, alleviate the fear and anxiety that the event brings to the public, and guide public opinion positively through platform channels such as microblogs, public numbers and news clients, so as to channel public sentiment and make their comments rational.

3 Method

3.1 Social hot pots about cyber violence

Three new confirmed cases were released in Chengdu on 8 December, and the patient in question, Zhao, is the granddaughter of confirmed the case. The patient, Zhao, was criticised by netizens for her indiscreet private life as she frequented many bars and busy locations prior to her diagnosis, and nine related terms such as "granddaughter of a confirmed patient in Chengdu" have since appeared. Zhao's ID card, address and other private information spread online, and some netizens faked their own friendships and videos, hurling numerous insults at Zhao and even at his staff, calling him a "bad girl". At the same time, many people were affected by the rumours.

Within hours of Zhao being diagnosed with the new coronavirus, someone posted Zhao's name, ID number, home address and other information, and posted his picture on all social media. In just a few hours, public opinion online became increasingly heated, partly because of the content circulating online attacking Zhao, and partly because of the outrage caused by the release of Zhao's personal information. As the dispute escalated, the Chengdu police launched an investigation into the confirmed leak of personal information in the afternoon of December 8 and brought the matter to its first public climax. From December 8 to the evening, several media outlets published condemnatory statements about the leak, with CCTV and other state media publishing articles such as "CCTV comments on the leaked information of the confirmed girl in Chengdu: Anxiety about the epidemic should not be a reason for online violence" "It's not good to point fingers excessively." At the same time, many people were affected by the rumours.

Zhao issued a statement on the incident at 9am on December 9, not only stating that he was unaware of the incident prior to his outing, but also apologising to the people of Chengdu for the incident. This response immediately generated more attention and was the second wave of conversation on the matter [6]. Afterwards, the Chengdu police released a message that the 24-year-old man had been administratively punished for leaking the confirmed girl, and the netizens' enthusiasm for the matter gradually cooled, but were more supportive of Zhao.

Once the circumstances surrounding the online violence against the girl in Chengdu were clarified, the fake public opinion information was gradually declining due to the timely clarification and reporting by the official media. The official media released the news to various platforms within a short period of time and expressed their views in their own voices, thus reversing the course of this hot issue. The official media must give full play to the fast and interactive nature of the news media to guide public opinion correctly and give the public authoritative information in order to stop the spread of such rumours. When confronted with a hot issue, the government media should be mindful of public skepticism and gradually steer it towards stability until the issue is properly resolved [7].

3.2 Elements of a successful communication strategy for social hot pots

On December 13, 2019, CCTV News published a short article titled Medical insurance bureau experts soul cut, enterprise representatives shed tears on the spot, which was uploaded on Bilibili for 40 seconds and sparked the attention of a large number of netizens, with the number of hits exceeding one million and the number of pop-ups exceeding eight thousand. The video was then posted to the official microblogging site of Bilibili, and the content of the video was switched to the entire page so that netizens could get a better idea of event.

The dynamic section focuses on news and news reports, which are published in text and image form. The text form is precisely what complements and supplements the content of the short videos, the key information, and its function is increasingly prominent. young viewers of the Bilibili also have a certain demand for quality news at home and abroad, and paying attention to news and understanding current trends is an important part of young people's development. At the same time, CCTV News provides a platform for Bilibili users to learn about domestic and international issues and to participate in national public affairs.

Fragmented communication has broken the pattern of time and space and communication methods, the rise of the micro power of the audience, the collision of the traditional media speaking power, and the core of the Internet thinking is the user's problem. In the road of social communication, traditional media and our media more from the user needs and use of scenarios, consider audience preferences, and gain market favor. There should be more use of young, popular voices to interpret information and comment on current events. The quality of content and the form of communication should not only match the communication characteristics of the platform itself, but also truly meet the needs of the audience.

With the continuous development and deepening of the Internet, netizens have gradually become the main users with multiple identities, especially in Bilibili which is a place where young people gather. This is the current state of online media literacy of the "Internet indigenous" group [8]. In an environment of information overwhelm, the latest official news is published in a timely manner, and public opinion is proactively shaped to provide immediate information to Bilibili users. After the riots in Hong Kong, the government's official account carried several days of consecutive texts, including those about the killing of elderly people in Hong Kong, the closure of schools in Hong Kong and the release of government statements on the issue. Articles titled The New Coronavirus Pneumonia cases: 136 new cases in Wuhan, 2 confirmed in Beijing, 1 confirmed in Guangdong, Wuhan Health Commission issues precautions for high incidence of viral pneumonia, 217 cases of pneumonia with new coronavirus infection have been confirmed nationwide have also been published. Since then, the official Bilibili account has been announcing the number of confirmed cases of novel coronavirus pneumonia and control measures every day.

3.3 Suggestions about helping Bilibili accounts for spreading social hot pots

The uploaders on the Bilibili should enhance their communication with users by increasing the frequency of their presence in the comments section, by being able to post their views on the home page and by liking them in the comments. Respond to users in the comments section expressing doubts about the content in the contributed videos. Special staff have also been arranged to monitor and reply to messages, to actively focus on changes in user sentiment in hot events, to seize the initiative, to guide public opinion and to correct mistakes in a timely manner; and to avoid wrong tendencies in public opinion.

Official media accounts have strong media resources, professional photography teams, and maintain close cooperation with relevant government departments [9]. Most of the video content submitted comes from other official media, and should increase the proportion of personalised content; so that they can better demonstrate their role. Firstly, expand the team of collaborative other uploaders and cultivate positive, upbeat and connotative uploaders; secondly, the cooperation between CCTV and uploader has increased the number of submissions. The official account should be more aware of the importance of civilianising and keeping the material close to the public. Adding good self-publishers and working with Bilibili's uploaders to make more room for Bilibili's local creators.

Official clients and advanced media users should remain calm and define their own position of objectivity and fairness during the dissemination of social hot pots , rather than blindly co-create hot online events or push the envelope. It is also possible to outline commonalities in social hot pots. Professional analysis and commentary on published works as well as reviews allows the media itself to intervene early in guiding public opinion on online events by improving its foresight of public opinion. The process of guiding public opinion on the Internet is still a process that cannot be achieved through a single report or a single programme, but requiresIt takes a certain amount of sustained coverage over a certain period of time to have a gradual effect. There is a time process, from the occurrence of a hot online event to its resolution, then traditional media need to capture all the links in the process and bring out a detailed, rational analysis to convince the audience of the different

Themes and the ability to set different issues at each stage of the event's development, play a good traditional media agenda-setting ability, comprehensive guidance network hot events

Currently, the public opinion guidance work of China's government news media is mainly through grasping the general trend of social opinion and striving to provide comprehensive and accurate coverage of online public opinion events [10]. When the public focuses on a hot topic and gradually expands its discussion and dissemination, a major social There are many examples of official media and governments responding in a timely manner to curb the deterioration of the situation. For example, in the case of the sinking of the "Eastern Star", the government announced the accident in time and the national media reported the accident in a timely manner. At the same time, foreign journalists were invited to report on the rescue work in Supervalu, Hubei, to ensure that the information was open and accurate, so the matter was handled smoothly online. Another example is the posting on the internet about "sexual abuse of minors in kindergartens and orphanages in Bijie and Kaili, Guizhou", which caused widespread public concern because of the negative social impact. The following day, the Guizhou police investigated the matter; the images circulating on the internet of "rape of underage children in Bijie and Kaili" were obtained by the suspects using internet searches and falsified information. The local government and media came forward within hours, stating their position and actions, and following up on the news to reassure netizens.

4 Conclusion

In the process of dealing with online public opinion, the relevant departments and the news media should follow up and improve it in a timely manner to ensure the correct guidance of public opinion, so that public opinion can be handled in an orderly manner and things can be brought to light [11]. At present, the official account is still in the initial stage of development. It is important to make full use of and explore the features and functions of Bilibili to improve the credibility of the website and its form of communication. They also should make full use of its own quality resources and the characteristics of the Bilibili platform to achieve a perfect combination between quality innovative content, platform attributes and user needs. In this era of ever-changing new media platforms, everyone can post online to express own opinions. The mainstream media should actively guide the correct political direction while calling on the youth community to actively participate in the creation of content; Bilibili is a gathering place for young people who are still at a critical stage of their development and whose positions and perceptions are easily influenced by extreme rhetoric and non-mainstream ideologies. While maintaining their political authority, mainstream media should also play an active role in guiding young people in terms of their values, life and career.

Online self-publishers, claiming to be the deliverers and providers of media communication platforms, should continue to maintain their own sense of leadership, seek to strengthen the establishment of regulatory mechanisms through the development of media codes that can make the media more harmonious, and more strictly regulate their own discipline; secondly, for pornography, infringement of reputation, online The second is to immediately stop information in categories such as pornography, infringement of reputation and online abuse, and to impose appropriate penalties on the publishers and disseminators of undesirable information once they are found. In addition, the various shortcomings of online self-media are inextricably linked to the imperfection of online laws and regulations. Relevant legal departments in China should improve laws and regulations as soon as possible, and in particular should In particular, they should formulate and promulgate special laws and regulations on network information dissemination and network media management as soon as possible, and crack down on all kinds of misconducts in the process of using self-media. In this way, the management of self-media will be more operable and more just and legal.

References

1. Yanpin,L., Mingjie, Z. (2021). Irrational Expressions of Online Public Opinion in the New Media Era: The Case of "Jiang Ge Murder" as an Example. (2018-7):87-90.
2. Duorong, W. (2021) Analysis of the operation mode of UGC video websites--Take B station as an example. Satellite TV and Broadband Multimedia. (11):2.
3. Jianming,L. (1988) Basic Public Opinion Studies. China Renmin University Press, Beijing.
4. Xiaoping, M. (1989) Revealing the Mystery of Public Relations - Public Opinion Studies. China News Press, Beijing.
5. Bingxing,M. (2013) Research on Internet Public Opinion Guidance[D]. Beijing Jiaotong University.
6. Yihan,C. (2021) Information Epidemic Response Strategies in the Context of the New Coronary Pneumonia Epidemic: The Case of the "Chengdu Girl" by the Internet Violence. Journal of News Research., 5:2.
7. Lulu, L. (2022) An Analysis of Ethical Issues in Disaster Reporting in the New Media Era - Taking the Henan Rainstorm Incident as an Example. Western Radio and Television
8. Li, H. (2011) Local government media image communication strategy in the context of online public opinion monitoring [D]. South-Central Minzu University.
9. Changhua,L., Yue, X. (2020) [J]. Emergency Discourse Communication in Mainstream Media in the Era of Short Video--Analysis of Reports on the New Crown Pneumonia Epidemic Based on People's Daily's Shakyin Account. Chinese Journalists, 6:5.
10. Hui, L. (2018) The Propagation Trend of Online Public Opinion in Hot Social Events: "Hangzhou

Nanny Arson Case" as an Example. News Knowledge, 1:3.

11. Ke, Z. (2017) An exploration of media literacy in online opinion dissemination by the media. TV Guide. (18):1.

The Influence Factors of Short Video Marketing on Consumer Purchasing Behavior and the Effective Suggestions

Yaping Zhao*

Tsinghua University, Beijing, 100084, China

ABSTRACT. In recent years, with the rapid development of the Internet, the promotion and application of TikTok and other short video APPs have greatly enriched people's life and ways of leisure. Many enterprises pay attention to the value of short video marketing, and short videos, as a type of commodity, have been used for marketing. All of these have promoted the continuous growth of short video marketing mode. However, at present, some enterprises or businesses are not clear about the impact of short video marketing on consumers' purchasing behavior, which to a certain extent affects the utilization of business resources and the sustainable and stable development of enterprises. This paper mainly analyzes the characteristics of short video marketing, studies the the influence factors of short video marketing on consumer purchasing behavior, and puts forward some effective suggestions. Conclusions can be drawn that short video marketing has the features of original content and a combination of online and offline. One influence factor of the consumer purchasing behavior is the short video marketing platform availability and function availability. Another one is the psychological distance between consumers and short video marketing. Moreover, in order to improve the short video marketing effects, it is essential to make the personalized marketing scheme based on consumer demands and improve the professional quality of short video bloggers. Additionally, there is a need to build a consumer value common trust and strengthen the short video content management. Last but not least, it is also important to focus on the interaction with consumers and organize more diversified promotional activities.

1 Introduction

Short video is a way of Internet communication. With the expansion of mobile terminal coverage and the improvement of network speed, short video platforms represented by TikTok are favored by many people. Major businesses use the short video platform to carry out marketing work, forming a short video marketing mode. Short video platform has the characteristics of multiple users and heavy data traffic. Compared with traditional marketing methods, short video marketing method can interact with consumers and improve consumers' participation and experience, thus producing purchasing behavior. However, in specific practice, the effect of short video marketing activities is different. Some products are popular with the public after short video marketing, but some marketing effects are not satisfactory. This is because some businesses do not have a clear understanding of the role and mechanism of short video marketing. They do not know which factors will have an important impact on consumers' purchasing behavior, and they fail to make scientific use of business resources, thereby affecting the sales volume of goods. In today's era, studying the influencing factors of short video marketing on consumers' purchasing behavior and putting forward scientific and reasonable suggestions can provide an improvement path for enterprises and businesses to carry out marketing activities, improve their competitive advantages, and further promote the development of short video marketing. Therefore, it is necessary to study the influencing factors of short video marketing on consumer purchase behavior, so as to provide reference for related enterprises and businesses.

*Corresponding author: 1344535755@qq.com

2 features of Short video marketing

2.1 The Pace of Fashion Led by the Original Content

At present, many short videos come from the original authors. These short videos are loved by young people and have the characteristics of leading the fashion trend. For example, a user recorded a short video with the caption "spiritual boy". With a unique expression and rhythm, the video was liked by a lot of people. After dissemination, it quickly became popular on the Internet and was imitated by short video users. In addition, "diss him (pán tā)", "grasp (ná niē)", and other expressions are used by multiple users. Due to the interesting facial expressions, they become classic network vocabulary. These fashionable languages have a great impact on the way users express, and can reflect the huge influence of original works [1]. Therefore, enterprises and businesses pay attention to original content, transform web celebrities into the styles popular with the public, and create more humorous content and hot topics, thus laying a good foundation for short video marketing. Then, by making use of the popularity of Internet celebrities to make an influence on products and transforming consumers' attention and love of Internet celebrities to products, enterprises and businesses can thereby enhancing consumers' consumption awareness, and then produce purchase behavior. This marketing mode has obvious advantages over traditional advertising. It is more acceptable to consumers and more suitable for the social development in today's era.

2.2 Online and Offline Combination

Under the short video marketing mode, online and offline products are organically combined, and the products that need marketing will be promoted through online promotion and offline promotion. The integration of short video content and the platform can not only improve the interaction between users and the platform, but also enhance users' sense of participation and stimulate users' desire to buy [2]. At the beginning, most TikTok short videos were directed and performed by the users themselves. They transfer original works to the short video platforms such as TikTok, and attract the majority of users to watch, so that their videos can get more clicks and likes, thus attracting more fans. With the increase of original works, and the improvement of public attention, they will develop into a web celebrity. With the wide application of short video software, many celebrities live broadcast on TikTok and other short video platforms. They carry out various activities, create hot topics at the present stage, and attract the attention of the majority of users, thus producing a good publicity effect. In addition, short video marketing also has an offline promotion mode. Through combining offline promotion activities, such as web celebrity meetings, with online publicity, forming an online and offline interaction, and constantly developing new materials, the short video marketing effect can be improved.

3 The influence factors of short video marketing on consumer purchasing behavior

3.1 Platform Availability and Function Availability of Short Video Marketing

In the process of short video marketing, both the platform availability and the function availability will have an impact on consumers' purchasing behavior. However, the platform availability has a more significant impact, while the impact of functional availability is relatively weak. First is the the impact of platform availability. In the short video marketing, more consumers will be attracted by the new technology and interactive forms. They take a strong interest, and then produce the purchase behavior [3]. These are incomparable to offline shopping. Second is the functional availability. According to the analysis of consumer purchase behavior, the interesting availability of functions has a great impact. Consumers tend to be attracted by the entertainment attribute of short videos. High-quality content and the humorous style can promote consumers to produce purchase behavior. In addition, the means of short video marketing are diverse. Through diversified promotion means, more attention of consumers can be attracted. Besides, marketing bloggers can play more to their own expertise to provide consumers with professional knowledge and practical operation demonstration. This is easier to impress the consumer, and therefore improve product sales. For example, the marketing blogger Li Jiaqi conducts cosmetics marketing through short videos, and uses professional knowledge and personal experience to provide professional explanation for consumers. As a result, the cosmetics he recommends attracts many consumers to buy. Finally is the social availability. The impact of social availability on consumer buying behavior is less obvious [4]. At the present stage, most consumers carry out social activities through Wechat, QQ, etc. In short video marketing, the social settings are not prominent enough, so the social attribute is not strong, and it has little impact on consumers.

3.2 The Psychological Distance Between Consumers and Short Video Marketing

On the one side, availability can have a very significant impact on consumers' purchasing behavior through psychological distance. With the help of platform availability, the psychological distance between consumers and short video marketing can be shortened, thus bringing a positive impact on consumers' purchasing behavior. Through comprehensively displaying the marketing products and sharing the shopping videos, consumers can directly see the product information and the visualization of marketing products can be achieved. Moreover, the short video blogger's interpretation and interaction with consumers help consumers accept this way of marketing compeletly and then buy the product on the short video platform [5]. Compared with the platform availability, the influence of functional availability is relatively weak. In the process of short video shopping, consumers will be attracted by some price discounts, subsidies, and other activities. Professional product introduction and highly interesting video content can also shorten the psychological distance between consumers and short video marketing, and then lead consumers to complete the product purchase on the short video platform. However, the impact of social supply is not significant. Therefore, it can be seen that in short video marketing, consumers' do not have too much social demand. Therefore, social supply plays a less significant role in narrowing the psychological distance between consumers and short video marketing, and is less influential on consumers' purchase behavior.

On the other side, from the perspective of consumers,

a certain psychological distance will be produced when consumers perceive the supply of short video marketing products. If the psychological distance becomes shorter, it means that consumers' satisfaction with short video marketing is greater, and a high corresponding acceptance degree will increase the probability of purchase behavior. If the psychological distance becomes longer, it means that consumers' satisfaction with short video marketing is lower. It is easy for consumers to produce resistance, thus reducing the probability of purchase behavior [6]. Generally, when buying goods through short videos, consumers will perceive the availability and supply from various dimensions, such as the marketing content, commodity price, professionalism, and interactivity. These supply factors will bring different degrees of satisfaction to consumers, resulting in a long or short psychological distance. The longer the psychological distance is, the less the chance of buying behavior will be. In the process of shopping, most consumers will seek the satisfaction of psychological needs. When the psychological needs are maximized, consumers are more likely to produce purchasing behavior. Therefore, when applying the short video marketing, enterprises and businessed need to clarify the factors affecting consumers' purchasing behavior, grasp the different pain points of consumers' needs, and develop scientific and reasonable marketing plans, thereby further improving the effectiveness of short video marketing.

4 Advice for improving short video marketing effects

4.1 Setting up Personalized Marketing Plans According to Consumer Needs

By analyzing the influencing factors of short video marketing on consumers' purchase behavior, it can be learned that the focuses of consumers are different in different stages of the purchase process, and different consumers also have different needs. Therefore, to improve the effects of short video marketing, enterprises and businesses need to set up personalized and diversified marketing programs for different stages, so as to meet the needs of different consumers and improve the pertinence of marketing. In the process of formulating short video marketing programs, the most important thing is to grasp the needs of the target consumer groups, and accurately select the marketing content based on this basis. It is also essential to establish the corresponding incentive mechanism and stimulate consumers to buy. Therefore, it can be seen that the demand of target consumer groups is the core of short video marketing, and the arranged marketing scenarios and marketing forms used should be built based on this core content [7]. In the process of construction, enterprises and businesses can make full use of artificial intelligence technology, big data technology, etc., digging deep into the purchase demand of the target consumer group, designing the scene of marketing, and putting it timely into the short video, so as to attract the attention of consumers, and effectively improve the sales of goods as well as the purchasing power of consumers.

4.2 Improving the Professionalism of Short Video Bloggers

In short video marketing, there will be special bloggers to explain the marketing products, including the product composition, appearance characteristics, using skills, and other content, so that consumers can have a full understanding of the products and their interests in buying can be stimulated, thus producing purchasing behavior. Therefore, the professional level of short video bloggers is directly related to the purchase behavior of consumers. Especially in the introduction of product ingredients and using skills, there is a need to provide consumers with professional explanations and guidance, so that consumers can have a good shopping experience. Therefore, enterprises and businesses should pay attention to the training of short video bloggers, strengthen and improve the specialization degree of bloggers [8]. In addition, short video bloggers should establish the concept of lifelong learning, keep pace with the times, actively improve their professional quality, establish a professional image in front of consumers, shorten the psychological distance between short video marketing and consumers, and then complete product sales, thus improving the effects of short video marketing.

4.3 Building a Consumer Value Common Trust

With the rapid development of the Internet, more and more people go deep into all levels of the Internet and use the Internet to carry out various work. However, the Internet is virtual. When carrying out short video marketing work, enterprises should build consumer value common trust, guide the values of consumers, promote the purification of short video marketing environment, provide consumers with a good short video shopping environment, and promote the benign development of short video marketing. In addition, short video platforms should strictly standardize the account registration and formulate corresponding mechanisms to avoid the emergence of illegal accounts. The underlying algorithm should be used to recommend high-quality short videos to the majority of consumers, and the commodity trading link of the short video platform should be improved, so as to reassure consumers of the security of short video shopping.

4.4 Strengthening the Management of Short Video Content

In the context of short video marketing era, enterprises or businesses must pay attention to short video content by strengthening the content management and creating high-quality short videos with strong fun and useful ability to achieve marketing success. First, the fun of short videos can be shown through the title and content. After investigating and researching, the title and content of the highly popular short videos are mostly vivid and interesting, or with suspense, which can arouse users' curiosity, attract them to click and watch short videos, and increase the number of likes of short videos. Secondly, the usefulness of the short videos is mainly shown through the information in the short videos. By watching the short videos, consumers can understand and judge the product, and then make the purchase decision [9]. Therefore, in the process of short video marketing, it is necessary to stand in the perspective of consumers, and dig and refine the information useful to consumers on the basis of clarifying consumer needs. It is also essential to interpret the characteristics of the products, so that consumers can realize that the product information is useful to themselves, thereby producing the purchase behavior. Finally, in short video marketing, the core that really causes consumers to buy is the content. Only by making useful short video content can the psychological distance between short videos and consumers be shortened. Therefore, it is necessary to strengthen the management of short video content, strictly standardize the production process of short video content, clarify the person in charge, strengthen the audit and supervision of short video content, and avoid the occurrence of false or illegal content.

4.5 Focusing on Interacting with Consumers

In the short video marketing, we should pay attention to the interaction with consumers and enhance their sense of experience. After releasing marketing videos, marketing subjects should pay close attention to the messages in the comments section, and timely answer their doubts, so that consumers can feel that they are valued and their sense of trust can be enhanced. At the same time, through the interaction in the comment area, the marketing subjects can also timely understand the actual needs of consumers, provide constructive opinions and suggestions for consumers in the interaction process, and make consumers to be more trustful and loyal to the short video marketing, so as to guide consumers to make the purchase decision [10]. In addition, the short video bloggers should also give full play to their own guiding role. In the process of commodity marketing, short video bloggers should attract the attention of consumers through various ways of interaction, such as replying their messages and sending them hongbao. They also need to actively guide the topic to encourage more consumers to participate in the whole process and experience the product, thus improving the effectiveness of short video marketing.

4.6 Organizing and Carrying out Diversified Promotional Activities

When carrying out short video marketing, all enterprises and businesses should consider the current situation of their own development, formulate a scientific and reasonable pricing system, and use diversified price means to stimulate consumers' desire to buy. Price has always been an important concern for consumers, and most consumers are sensitive to price. Therefore, when carrying out short video marketing, enterprises should choose reasonable promotional means, and organize various forms of promotional activities. Promotional activities can mobilize the initiative and enthusiasm of consumers to make them buy and consume. In addition, enterprises should reasonably determine the commodity prices according to the economic conditions of the target consumer groups, so as to reduce the economic pressure of consumers, improve consumers' desire to buy, and enable them to buy cost-effective goods.

5 Conclusion

To sum up, short video marketing is currently a popular marketing mode, which is favored by many consumers, and it has also become one of the strategic marketing methods for most enterprises. The platform availability and function availability of short video marketing and consumer psychological distance will have an important impact on consumers' purchasing behavior. When enterprises and businesses use short videos for marketing, they need to set up personalized marketing programs according to consumer needs, build mutual trust of consumer value, pay attention to interaction with consumers, and strengthen short video content management to improve consumers' trust and loyalty, so that the marketing effects can be further improved. However, it should be noted that the methods of short video marketing for different regions and age groups are different, and the marketing effects are also different. In future research, these factors can be used as variable factors to find the best way of marketing for related enterprises.

References

1. L. Cheng, Research on the Influence of Short Video Content Marketing Based on Two-way Perspective [D], Jilin University, 2022.
2. F. Chai, The Impact of Short Video on Tourism Marketing—takes TikTok Short Video as an Example [J], Travel Overview, 2022 (07): 191-193.
3. J.J. Yang, Impact of Short Video Marketing on Consumer Purchase Decisions [J], Modern Marketing (Business Edition), 2021(11):58-60. DOI:10.19921/j.cnki.1009-2994.2021-11-0058-020
4. L.P. Huang, Z.G. Huang, The Current Situation and Countermeasures of Agricultural Products Marketing under the Prevailing Influence of Short Video—Take Shatian Pomelo in Rong County as an Example [J], Southern Agricultural Machinery, 2021, 52(18): 112-114.
5. Ream Amaraliphear. Impact of Short Video Marketing in Phnom Penh, Cambodia on Consumer Brand Attitudes [D], Beijing Foreign Studies University, 2021. DOI:10.26962/d.cnki.gbjwu.2021.000581.
6. X.P. Sheng, Effect of Short Video Marketing Content on Consumers' Willingness to Buy [D], Donghua University, 2021. DOI:10.27012/d.cnki.gdhuu.2021.001371.
7. Q. Yin, Study on the Effect of Short Video Scene Marketing on Consumer Purchase Will [D], Chang'an

University, 2021. DOI:10.26976/d.cnki.gchau. 2021.000918.

8. Y.H. Wu, Impact of Short Video Marketing on Tourism in Hangzhou [J], Marketing Community, 2021, (08): 3-4.
9. Y. Wang, G. Liu, Research on the Influence of Short Video Social Platforms on Marketing Methods [J], Technology Tank, 2020, (11):69-74. DOI:10.19881/j.cnki.1006-3676. 2020.11.12.
10. Y.F. Xiao, The Impact of Short Video on E-commerce Marketing [J], Marketing Community, 2020, (42): 197-198.

Reflection on the Communication of Urban Image of Chaoshan Area in the Era of New Media

Huan Fang[1,a], Tian Zhang[2,b], Yanting Zhang[3,*]

[1]Guangzhou College of Technology and Business
[2]Guangdong Fengyi New Energy Technology Co., Ltd.
[3]Guangdong Industry Polytechnic

Abstract. Since the "13th Five-Year Plan", China has made remarkable achievements in the construction of new urbanization, and its development has also shifted from increment and expansion to a new stage dominated by stock renewal. In the tide of diversified construction of city, Chaoshan area with strong regional characteristics and cultural deposit has been frequently frustrated in shaping its urban image. Many untrue public opinions and rumors even appear on new media platforms and are widespread. How to effectively communicate the urban brand image of Chaoshan area and therefore promote regional economic development has become an aspiration of every Chaoshan person. In this paper, by collecting and analyzing questionnaires, we design a "CBIO" model. In order to improve communication effect of urban image of Chaoshan area, communication subjects should paid attention to key factors such as communication tactics, urban brand positioning and urban infrastructure construction.

1 Introduction

The competition between cities has gradually shifted from simple "economic competition" to "comprehensive competition" that incorporates the dimension of urban image. How to form and develop an urban image with regional characteristics has been in the spotlight among people. A favorable urban image can contribute to the improvement of economic level, and be internalized to increase the community's sense of belonging and cohesion, further promote citizens in the locality to build a positive image for the construction and development of urban image and become a strong group motive to propel the dissemination of urban image. In the era of new media, video platforms, social platforms and live streaming platforms increasingly become gathering places for people to learn about the world and express their viewpoints. With the high-speed flow of information and decentralization of platforms, "multiple subjects" in the Internet world have replaced the previous dominant position of mass media. While injecting vitality into the society, the high-speed flow of information has also raised uncertainty, including those realistic-looking rumors.

Chaoshan area enjoys a global fame for preserving and inheriting traditional Chinese culture. But under the tide of new media, the image of Chaoshan area began to be questioned and smeared on the Internet. For example, over the last year, the discussions on the image of Chaoshan area on domestic and foreign mainstream social platforms (Weibo, Douban, Quora, etc.) have formed a "distorted" "pseudo-reality"[1], and Chaoshan area has been pasted negative labels such as "men are superior to women" and "feudal superstition" and partly formed a "stereotype" against Chaoshan area through the wide and rapid diffusion of Internet media. This brings potential hazards to the development of urban image in the region. Therefore, how to get Chaoshan image out of the dilemma of "pseudo-reality" on the Internet platform, let more people understand an objective and real Chaoshan area, and promote urban development has become an issue to be concerned about. In this paper, to address this issue, by proceeding from the public's cognition and expectation about current Chaoshan image, we determine whether generally negative comments on Internet platforms represent true views of most people in reality. Combined with the analysis results of questionnaire, we attempt to sketch factors affecting the communication effect of the urban image of Chaoshan area, with a view to lay the groundwork for urban image of Chaoshan area to step out of "pseudo-reality"[1] and achieve positive development in the future.

[a] Email: huan_fh@163.com
[b] Email: 572416525@qq.com
* Email: zyt621_3@163.com

[1] Pseudo-reality is borrowed from the concept of "pseudo environment" by Walter Lippmann, a U.S. news critic, which refers to a "reality" processed by the communicator. Since this kind of processing is generally performed within the communicator, the audience often regards the processed "reality" as an objective "reality".

2 The Communication System of Urban Image

In the era of globalization, "urban communication" is still an important proposition of urban development. With the acceleration of urbanization in China and reinforced trend of economic integration throughout the world, the urban brand construction in China is faced with unprecedented opportunities and challenges. In general, the urban image communication system is mainly composed of the subject and object of communication, and disseminated through mass media and new media platforms. It should be noted that compared with the era of traditional media, the relationship between the subject and object of communication is no longer a one-way process as in the past. The audience has higher voice, stronger ability to participate, and gains more sense of participation in the transmission of Internet. With the advent of the era of new media, people's subjective initiative has been reinforced like never before. The audience is no longer simply information recipients but become communicators at a new level. For this reason, the communication subject of urban image has become more diversified, and the subject and object of communication no longer have rigorous and clear boundaries.

Generally speaking, the urban image communication in the era of new media mainly involves the three participants.

2.1 Local government

There is a close relationship between the government and urban image, and they are mutually promoting and influencing [1] (Xie, Li and Li, 2000). The benign development of urban image is inseparable from government. Government agencies have abundant material resources, human resources, financial resources and information in terms of management. As the most authentic source, the government has unparalleled advantages over other communication subjects. The key to urban image communication lies in how to screen and extract excellent traditional cultural image and inner spiritual elements in the urban image [2] (Chen and Xie, 2021), on the premise that high-quality communication content is guaranteed, we should also learn about the characteristics of communication audience, grasp the data fed back by audience, constantly adjust and optimize the outdated communication methods, and improve the effect of urban image communication, by associating with new media platforms.

2.2 Local citizens

The traditional one-way communication mode of information has become a thing of the past. As an emerging medium, the new media platform has soon captured the public's attention with its advantages, such as rapid, convenient, informative and interactive. Citizens are not only able to take advantage of new media to keep updated, but also able to become disseminator at any time to export content to the outside world [3] (Gao and Liu, 2016). The shift from one-way to two-way represents the improvement of the public's subjective initiative and social involvement, and also plays a supervisory and guiding role in public opinion. Taken together, under the influence of new media platform, citizens have dual identities and roles, and can serve as active disseminators of local urban image.

2.3 City visitors

City visitors are probably tourists who come to the city for sightseeing, businessmen on business trips, people who visit, study and work. One of the purposes of urban image communication is to enable visitors to gain an insight into the history and traditional culture of the city within a certain period of time, feel the unique charm of the city, and stimulate their willingness to publicize and share the areas that they have visited. The visitors are mainly from different regions of the world, instead of those living inside the city. Their arrival further spreads the image of the whole city to their respective social circles. For example, they may disseminate their traveling experience on the Internet in the form of picture or video, and eventually form a positive or negative word of mouth. Practice has proved that the number of pictures commented and shared will have a dramatic effect on the number of tourists received by scenic spots in the city [4] (Lai, Tang and Zhu, 2011).

As mentioned earlier, although more and more audience plunge themselves into the production process of information, local citizens, city visitors and netizens that have never been to the target city all have the right to express their viewpoints on the Internet. However, due to different backgrounds and experiences of the audience, their degrees of rationality and emotional expression differ. Individuals often show certain subjectivity, which finally alienates the real environment through the wide spread of the Internet. To date, the social image of Chaoshan area on the Internet is being built into "labeled pseudo-reality". Although "netizens who have never been to the target city" will also make comments that are not necessarily true or are ungrounded on Internet platforms, their credibility is not high for the rational public, so they are not included as main participants of communication activities.

3 The Macro Environment in Chaoshan Area

To unscramble the nature of urban image communication activities in Chaoshan area more clearly, this paper holds that we should not stop short at t analyzing the communication process and communication elements, but take into account the effects of the macro-environment, including politics, economy, social culture and technology (i.e. PEST theory[2]).

[2] PEST refers to the analysis of macro environment. The macro environment is also known as general environment. The target issue is analyzed from four aspects: politics, economy, social culture and technology.

3.1 Political environment

The development of urban image branding is inseparable from the vigorous push and dissemination of government policies. In recent years, with the help of profound cultural deposits in Chaoshan area, municipal governments in Chaoshan area show regard for the protection and inheritance of intangible cultural heritage and take the initiative to apply for intangible cultural heritage projects [5] (wodsy.com, 2020). A variety of folk traditional culture festivals are held regularly, and well-known media producers are invited to produce promos or short videos belonging to Chaoshan image (e.g. "I am from Chaoshan", etc.), to intensify publicity and build Chaoshan image, with the aim of telling Chaoshan stories well. At the same time, it actively accesses new media platforms to carry out communication activities, such as the 1st "Intangible Cultural Heritage" Online Shopping Festival in Shantou in 2020 [6] (cnr.cn, 2021).

3.2 Economic environment

The benign development of economy is the necessary prerequisite to stimulate the city to seek greater cultural influence and promote urban image communication. Over the past two or three years, the annual GDP growth rate in Chaoshan area has increased year by year. According to the data of economic operation in 2021, the year-on-year economic growths in Chaozhou City, Shantou City and Jieyang City in 2021 were 9.3%, 2.1% and 6.1% respectively. The sustained improvement of the macroeconomic environment in Chaoshan area is providing sufficient space to push forward the branding of urban image. However, due to the vague and oscillating positioning of three cities in Chaoshan, even Shantou, a special economic zone, can hardly find a solid and powerful economic growth point, while Chaozhou, as a "famous cultural city", particularly needs to adhere to the road of branding on the basis of pinpointing cultural and tourism resources [7] (Cai, 2016).

3.3 Sociocultural environment

The folk culture in Chaoshan area has a long history of and profound deposit, and most Chaoshan people attach great importance to the inheritance of traditional culture [8] (Wang and Cao, 2019). The sharing of solidarity and mutual assistance among Chaoshan people and their high cohesion disable Chaoshan culture from changing fundamentally, no matter how many wars and changes of times it has experienced [9] (Hong, 2018). However, it should be noted that whether for language or folk custom, Chaoshan natives should pay attention to intergenerational inheritance [10] (Hu, 2013), fully mobilize Chaoshan people's passion for local inheritance in a way that young people are more delighted to hear and see, find communication schemes that conform to the values of contemporary young people, are favored by young people, and realize the diversification of value, and foster cultural confidence.

3.4 Technical environment

With the rapid development of information technology and media technology, the new media operation mode (represented by the Internet) changes the way people communicate and spread information in an indirect way. The object of communication is no longer a pure object, but becomes a theme that can actively utilize media to participate in the communication. The number of users of major new media keeps surging, and people's awareness of actively transmitting information has been constantly improved, which has provided unprecedented opportunities and challenges [11] (Song, Zhan and Hu, 2019).

4 Questionnaire Analysis of the Urban Image of Chaoshan Area from the Perspective of Audience

On the whole, macro conditions can lay a foundation for the development of urban image of Chaoshan area. However, the weak communication effect in the era of new media, as well as negative information and rumors on the Internet keeps impacting the urban image of Chaoshan area. Considering the non-authenticity of "pseudo-environment", whether negative labels on the Internet represent the real viewpoints of most people is yet to be verified and mined. On this cognition premise, the author designed questionnaires and distributed them through two platforms, Weibo ("public discussion" media) and WeChat ("acquaintances socializing" media) and tried to avoid the impact of value similarity [12] (Shi et al., 2020). Finally, 369 valid questionnaires were retrieved. The content mainly includes three parts:

Part I- Basic information. Basic information was collected, such as gender, age and native place.

Part II- The attitude towards the construction mode of the urban image of Chaoshan, the cognition of urban image of Chaoshan and channels to understand the urban image of Chaoshan.

Part III- The attitude towards traditional culture, as well as the attitude, cognition and expectation for Chaoshan folk culture.

4.1 Discussion about main findings

The participants of our questionnaire included 131 males and 238 females. The age structure was concentrated in the age group of 18-35 (see Figure 1, 'yrs' means years old, such as 18-35 years old), and the proportions of Chaoshan natives and non-natives were similar (see Figure 2)

X\Y	Under 18	18-35 yrs	35-60 yrs	Above 60	Subtotal
Male	10(7.63%)	114(87.02%)	5(3.82%)	2(1.53%)	131
Female	7(2.94%)	223(93.70%)	6(2.52%)	2(0.84%)	238

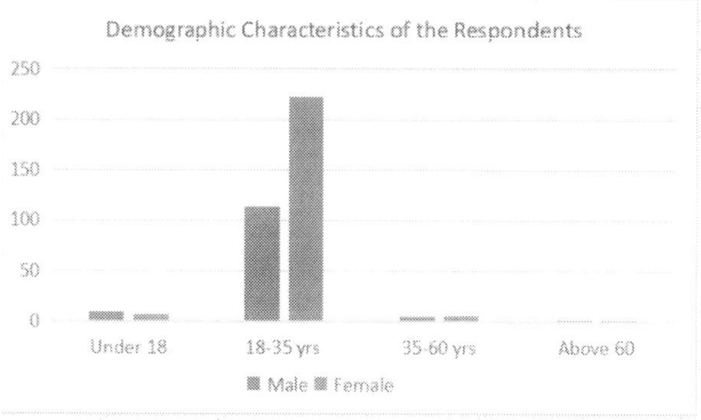

Figure 1 Demographic Characteristics of the Respondent

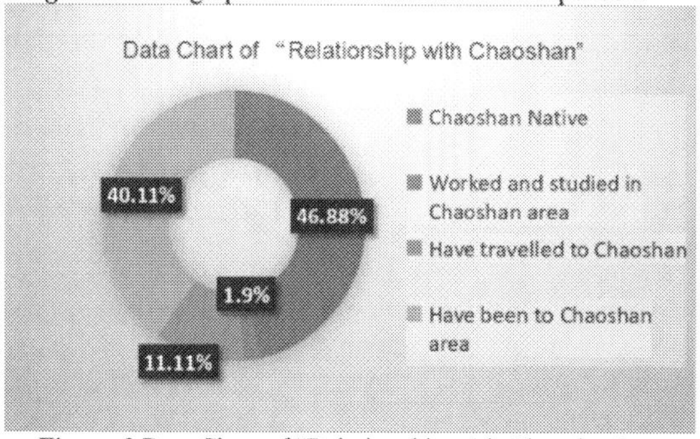

Figure 2 Data Chart of "Relationship with Chaoshan"

In terms of the cognition and attitude towards the spread of Chaoshan traditional culture, most respondents gave positive comments and the keyword included: strong cultural atmosphere, long history, extensive and profound culture, etc. About 25% of the respondents thought that Chaoshan culture was "old and out-fashioned" and "lack vitality" (see Figure 3). Considering that 40.11% of the respondents of the samples had been to Chaoshan area, but only 25% thought that Chaoshan image was negative, this was inconsistent with the generally negative and radical expressions of Chaoshan area on the Internet (e.g., Weibo, Douban and other open platforms). But whether their native place belonged to Chaoshan area or not, their cognition about "Chaoshan people's appreciation of family inheritance" was basically the same (see Figure 4).

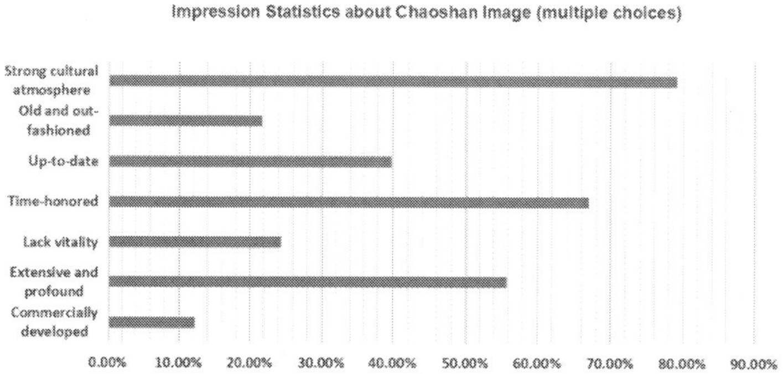

Figure 3 Impression Statistics about Chaoshan Image (multiple choices)

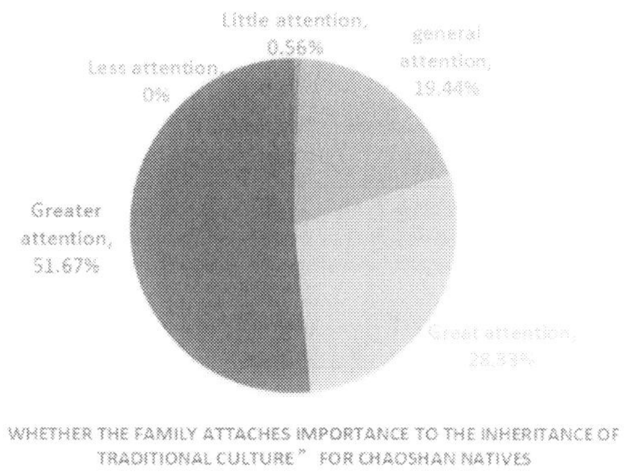

Figure 4 "Whether the family attaches importance to the inheritance of traditional culture" for Chaoshan Natives

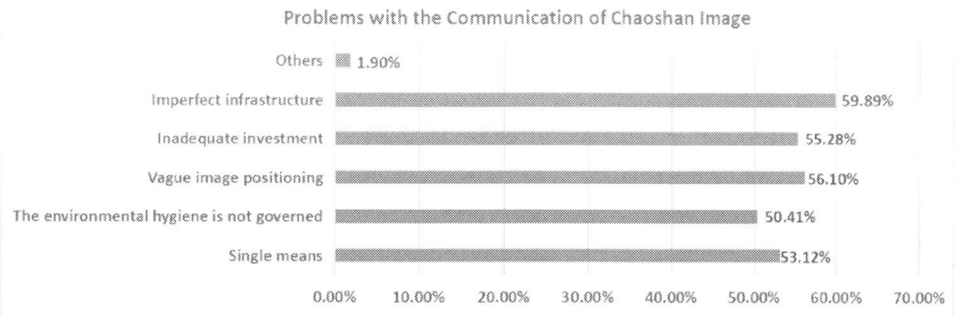

Figure 5 "Problems with the Communication of Chaoshan Image"

In view of the current weak communication of the urban image of Chaoshan area, most respondents ascribed it to the vague positioning of the city and imperfect infrastructure (see Figure 5). Specially, this was manifested in three aspects (see Figure 6 and 7):

① A clear urban image was not built. Even though the municipal governments in Chaoshan area had designed urban image promos, and the audience had learned about Chaoshan culture, most respondents' cognition source of Chaoshan culture was still interpersonal communication. Chaoshan area didn't have an authentic and credible image focus and this probably left room for the diffusion of unreal rumors about Chaoshan image on the Internet.

② There was no unified planning and correct guidance, and various communication subjects in the society (principally governments, organizations, non-organizational groups and individuals) didn't form a synergy on communication. In the era of new media, most of the fragments about Chaoshan were improvisations without professional editing, and the content was of low quality. What's more, in the existing popular film and TV works, Chaoshan elements were obviously scarce. These conditions were not conducive to the formation and development of the real urban image of Chaoshan in the heart of audience.

③ The urban image lacked brand characteristics of its own. The urban image brand strategy has been adopted and proved to be effective by more and more cities. Chen and Xie [13] (2021) argued that the application of city branding to the shaping of urban image required the city to form identification with the audience by means of marketing and communication according to its unique cultural deposits, environment, humanistic style and economic strength, etc., taking the functional and spiritual attributes of the city as the content, so as to win the audience's recognition of urban brand image. Regrettably, in this respect, nearly 50% of the respondents considered that the development of Chaoshan city didn't form its own brand characteristics, which echoed with ① and once again proved that in the development of urban image of Chaoshan area, the communication effect at the level of audience cognition was indistinctive.

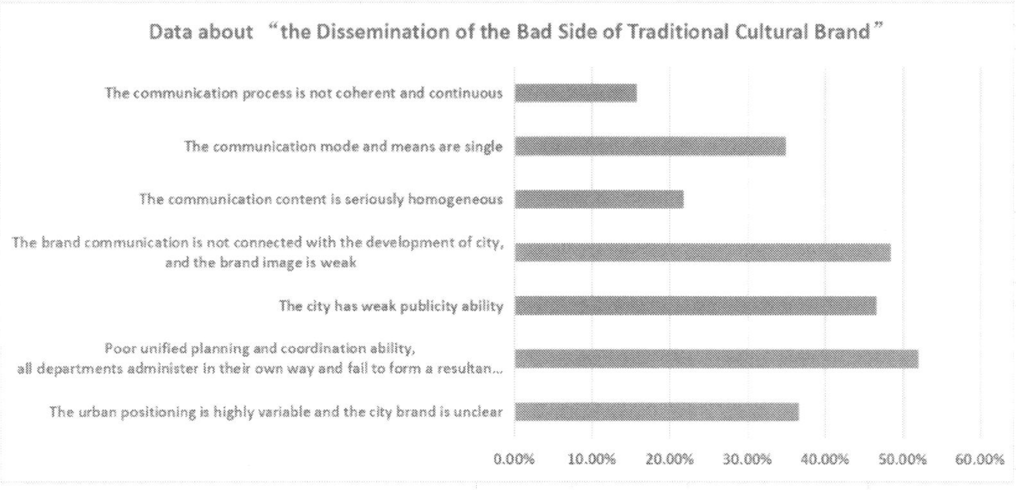

Figure 6 Data about "the Dissemination of the Bad Side of Traditional Cultural Brand"

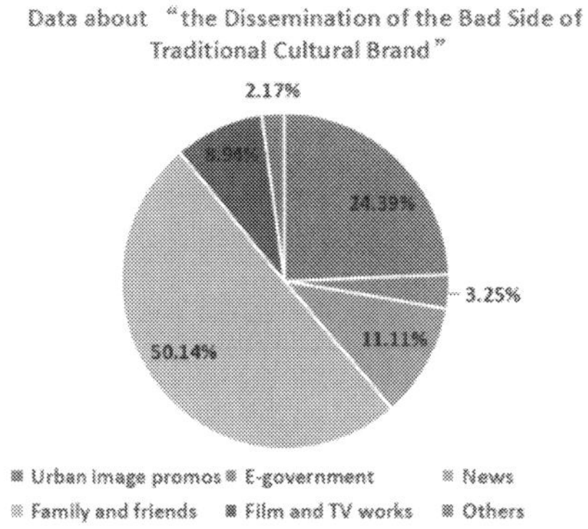

Figure 7 Statistics of "Channels for obtaining Chaoshan Cultural Information"

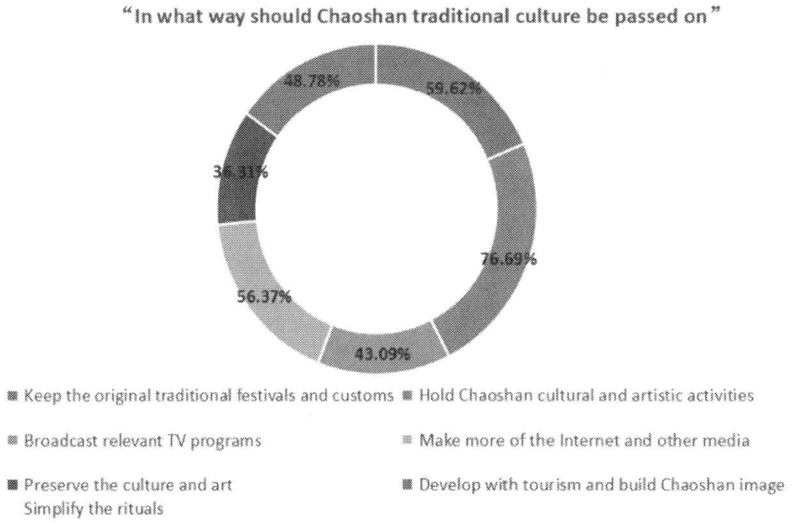

Figure 8 "In what way should Chaoshan traditional culture be passed on"

Another noteworthy result indicated that even if the respondents were generally young and fully interested in Chaoshan culture (see Figure 8), even 59.62% of the respondents wished that Chaoshan culture can maintain its traditional visage. This result also contradicted the label of "feudal superstition" which the Internet platform attached to Chaoshan image. As a matter of fact, in the eyes of most respondents, the positive role played by the spread and inheritance of traditional culture in Chaoshan area was worthy of recognition, and they can make people produce a sense of identity and cohesion, while pleasing their body and mind (see Figure 9).

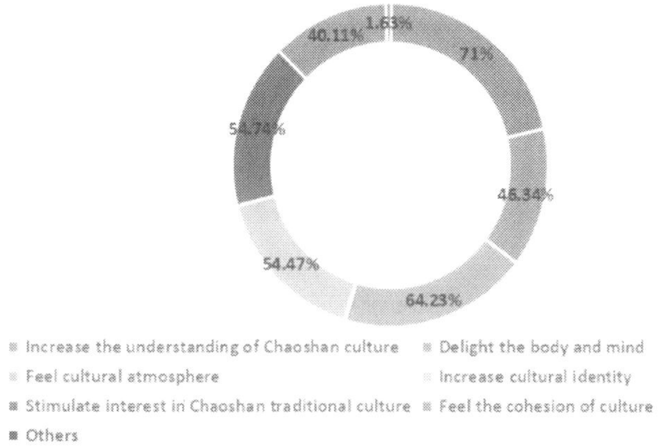

Figure 9 "Harvest from Participating in Chaoshan Traditional Cultural Activities"

5Influence Factors of the Communication Effect of Urban Image of Chaoshan Area

Combined with the results of questionnaire survey and relevant academic literature, "CBIO factors affecting the communication effect of the urban image of Chaoshan area" can be summed up (see Figure 10). That is to say, with the subject and object of communication being clarified, we considered that the key to truly change the urban image of Chaoshan in a positive direction was whether the communication subject (i.e., all municipal government departments, non-governmental organizations or groups in Chaoshan area; Chaoshan natives) paid sufficient attention to "communication tactics", "urban brand positioning", "infrastructure construction" and others.

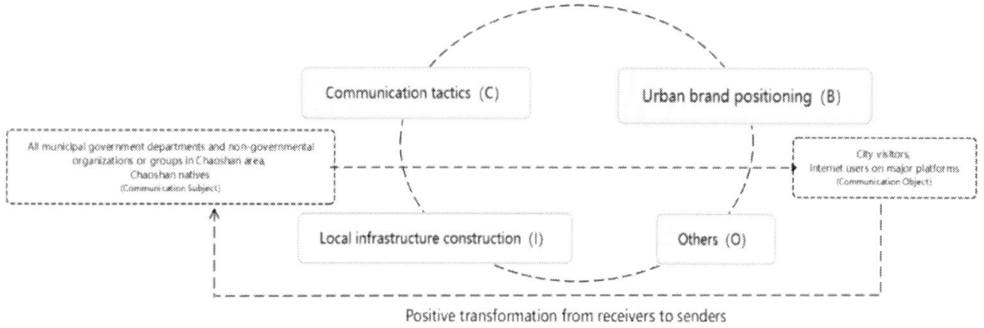

Figure 10 CBIO Factors affecting the Communication Effect of the Urban Image of Chaozhou

Specifically, the communication tactics of urban image of Chaoshan area guided by brand communication should connect the subject and object of communication. Meanwhile, the subject and object of communication were interchangeable in the era of new media. If the communication subject can exert a communication effect and bring positive experience to the communication object, the communication object was likely to become communication subject at a new level and get involved in the communication activities of urban image spontaneously (e.g. shared pictures and videos). However, as of the writing of this article, Chaoshan area didn't perform satisfactorily in four factors affecting CBIO, which was mainly manifested as follows: (1) communication tactics (the communication means were monotonous and the publicity structure was vague); (2) urban brand positioning (vague image positioning and insufficient connection); (3) infrastructure construction (imperfect infrastructure); (4) others (wanton pricing of characteristic cultural products, etc.).

6.Conclusion

Based on the negative information and labeling of Urban image of Chaoshan area on the Internet in recent years, by taking the form of questionnaire, combined with existing literature, this paper puts forward "CBIO factors affecting the communication effect of urban image of Chaoshan area", in an attempt to present a potential solution to the currently relatively weak urban image communication in Chaoshan area. Objectively speaking, more follow-up efforts are required to convey the urban image of Chaoshan area, such as the use of empirical analysis, and the collection of larger sample size to verify the correlation between CBIO factors and the communication effect of urban image of Chaoshan area. In addition, it should be noted that the model proposed in this paper is still established on the traditional concept of communication process. In the era of new media, the boundary between communicator and recipient is no longer clear. Their roles are interchangeable and appear concurrently. It is necessary to deliberate on and examine the traditional communication model more deeply.

References

1. Xie, Li and Li. The role of governmental public relations in the shaping of urban image [J]. *Science & Technology Progress and Policy*, 2000(10):185-186.
2. Chen and Xie. Application of brand character theory in the shaping of urban image [J]. *Beauty & Times* (City Edition), 2021.
3. Gao and Liu. The evolution of hosting communication pattern in the new media environment [J]. *Journal of International Communication*, 2016,38(03):6-19. DOI: 10.13495/j.cnki.cjjc.2016.03.001.
4. Lai, Tang and Zhu. On the influence of online word-of-mouth on tourists' choice of destinations [J]. *Management Review*, 2011,23(06):68-75. DOI: 10.14120/j.cnki.cn11-5057/f.2011.06.012.
5. ihchina.cn. How much do you know about Chaozhou intangible cultural heritage [EB/OL]. [Jun 24, 2022]. https://www.ihchina.cn/Article/Index/detail?id=21784
6. cnr.cn. Respect traditional underpinning research and "thicken" cultural connotations of Chaoshan [EB/OL]. (Jan 1, 2021) [Jun 24, 2022]. https://baijiahao.baidu.com/s?id=1687661959580552063&wfr=spider&for=pc
7. Cai. On the brand image communication system and strategies of chaozhou cultural tourism city under 5W mode [J]. *Journalism & Communication*, 2016(08):7-10.
8. Wang and Cao. Cultural review and modern inheritance of "Yinglaoye" custom in Chaoshan area [J]. *New Western*, 2019(06):50-52.
9. Hong. Analysis of the cultural confidence of chaoshan [J]. *Comparative Study of Cultural Innovation*, 2018, 2(27):33-35.
10. Hu. Hakka and chaoshan people in Nanling [J]. *Life World*, 2013(07):36-39.
11. Song. Zhan and Hu. A PEST study of the living inheritance of Jinxiu (Guangxi) folk sports [J] *Journal of Guiyang University (Social Science)*, 2019.
12. Shi, Gui, Huang and Zheng. Typology, mechanism and influence of "cocoons effect" of online media-intermediary analysis based on a 'Survey of Social Mentality of Chinese College Students' (2020) [J]. *Journalism & Communication*, 2022, 29(05):43-59+126-127.
13. Chen and Xie. Application of brand character theory in the shaping of urban image [J]. *Beauty & Times* (City Edition), 2021

Research on Film Short Video Marketing Strategy from the Perspective of the New Marketing Theory of 4C

Jingyi Xu*

School of Film, Communication University Of Zhejiang, Zhejiang, China, 310018

ABSTRACT. Short video is a new symbol of information dissemination in the era of mobile Internet. The rapid development of mobile Internet is changing people's lives but also changed the traditional marketing model In the process of the development of the mobile Internet, it catered to the present audience fragmentation reading habits, make the participatory culture fans interaction in the process of movie marketing. Nowadays, the short video led by TikTok has become the standard distribution of film marketing in China, and the content and form are constantly innovative. Based on the New 4C Marketing theory, this paper starts with the special attributes and marketing strategies of short video in combination with China's national conditions and specific cases, and analyzes and explores this short video marketing model. It then obtains the reasons and strategies for the good effect of short video in China at present.

1 Introduction

In 1990, Professor Robert Lauterborn, an American scholar, put forward the 4Cs marketing theory corresponding to the 4P of traditional marketing in his special article "new marketing lithium: four PS pass é: c-words take over". This theory aiming at the needs and expectations of consumers. Nevertheless, with the further development of Internet technology, 4C theory has further evolved into "New Marketing Theory of 4C", namely: connection, communication, commerce and co-operation.

As a new way of film marketing in the mobile Internet era, short video marketing has special attributes and advantages that traditional media marketing does not have. For one thing, the situation of rapid development of science and technology promotes the film'S content and form to keep pace with the times. For another, the popularity of short videos also brings new development opportunities for the publicity and distribution of films. Under such circumstances, the author takes " the New Marketing Theory of 4C" as the theoretical basis, and discusses how the film industry can use short videos to carry out effective and accurate marketing around the dissemination characteristics of short videos and the application of film marketing in short videos. Through research, this paper reveals that we can put advertisements more accurately and effectively on short video apps like Tiktok, stabilize the target population and explore potential customers. Achieve better publicity effects with fewer operating expenses.

*Corresponding author: find_jingyixu@163.com

2 The special attributes of short video

2.1 Strong social contact attributes

At present, the number of domestic short-video users is huge and growing. On August 27, 2021, the China Internet Network Information Center released the 48th statistical report on the development of China's Internet. As of June 2021, the number of users of online video (including short videos) in China had reached 944 million, including 888 million short video users, an increase of 14.4 million over December 2020, accounting for 87.8% of the total Internet users. One billion users have access to the Internet, forming the world's largest and dynamic digital society [statistical report on the development of China's Internet [1]. In the mobile Internet era, where traffic plays a decisive role, the growth of user scale means the size and potential of the market, and more traffic means higher revenue. For film publicity and marketing, the wider the crowd covered by short videos, the more potential movie audiences can be explored.

With the rapid development of the Internet, the original model of word of mouth on the quality of goods is difficult to adapt to the new business environment, and under such circumstances, online word of mouth comes into being. It relies on the Internet to spread rapidly, and shows new characteristics in the way of transmission, which are respectively spreading, diversity, interactivity and anonymity [2]. They not only comment on it but forward it to their friends and even share it on their social media. These recommendations from peers have high credibility and low transmission costs. The word-of-mouth effect broadens the scope of expansion, and fission transmission strengthens the influence. Therefore, the strong social attribute enables the short video to obtain the audience's real idea when promoting the film. The producer can timely adjust its marketing strategy and carry out efficient marketing according to the real-time changes of comments, retweets, likes and other data on the short video platform.

2.2 Solid emotional links

In 2017, video clips such as "a lovelorn woman crying and eating mangoes in front of the camera," "the incumbent fights after meeting his predecessor" and "crying alone on the back of a chair" on Tiktok were widely shared, causing heated discussions among users. After seeing these clips, many people went to watch the film "The Ex-file: The Return Of The Exes"", making it an unexpectedly high box office hit. Love is a timeless topic that troubles men and women. These topics and segments with strong emotional conflicts have strong emotional links with the audience, attracting a large number of audiences to the cinema to pay for this film. This was the year that Tiktok became a big hit in China, and the film marketing started in Tiktok.

On the one hand, such fragments can capture the core emotional content of the story. In addition to such clips, shooting gags is also a way to attract the attention of the audience. People can quickly have empathy by releasing the film's shooting gags or editing the most emotional clips of the film on the short video platform, coupled with episodes that conform to the artistic conception. According to the survey data, more than 50% of users have been exposed to video gags or essence clips on the Internet, and these clips are the most likely to arouse users' desire to watch movies [3]. For example, in the content released by the official Tiktok account of "Hi, Mom," the hot video content includes "a mother always leaves the best for her children, regardless of herself", "Jia Ling shoots a farewell play with her mother in the hospital, and she can't shoot with extreme tears," etc. These are the emotional climaxes in the film and highlights. Director Jia Ling's touching story about her and her mother, the movie prototype, has also appeared in short videos many times. Through such clips, the affection between mother and children has aroused public resonance. Many users of Tiktok have told their stories with their mother in the comment area, forming a secondary discussion, and achieving good communication results.

On the other hand, the producer can use short videos to present the audience's reaction and feelings. From the early "The Ex-file, The Return Of the Exes" to various later emotional films, short videos will record the content of the audience's tears when watching the film and the audience's emotional outpouring and expression after watching the film. Real cases are often more moving than fictional stories. Film short video marketing guides the audience to review and discuss the film again from multiple angles in this way, so as to enhance the emotional persuasion of the film and mobilize the audience's desire to watch the film again.

2.3 Lower marketing cost

On the one hand, short video marketing is more cost-effective. Compared with traditional media, Internet platforms can capture user information with the help of big data, accurately push content to users according to their daily habits, and attract target users to the greatest extent [4]. For example, in first-tier cities or heavily trafficked subway stations, it may cost hundreds of thousands of yuan or more to display information about the film on an LED screen. Hundreds of thousands of passers-by will see the advertisement during the promotional period. Half of these indiscriminate passers-by are probably not the target audience of the film, and some people who are busy with work may not pay too much attention to the surrounding advertisements. But if you put in a short video, the cost will be much lower, and the number of people watching it will be more than hundreds of thousands. As a short video platform with the largest user scale in China, Tiktok has an accurate algorithm that can portrait users. It can be said that on the short video platform, those who can see the video are most likely those who are interested in the film. The information the audience sees is not only the starring information, but also other content about the film, which can further deepen the target audience's understanding of the film and increase their probability of watching the film.

On the other hand, the content generated by short video marketing is more convenient and diverse. In addition to press conferences, posters, and theater activities, traditional film marketing will purchase advertising space on buses, subways, radio, television, and other channels. These methods are costly and difficult to replicate. The advertisement in the form of a poster is only a static picture, and the content that can be promoted is very limited. The promotional videos put on TV stations, websites, outdoor electronic screens, etc., once successfully produced, cannot be easily replaced due to the high production cost. The biggest difference between short video and traditional media marketing is that it is easier to produce and cheaper. According to users' preferences, they can push advertisements similar to private customized advertisements that meet their wishes. These advertisements can have different content and different presentation forms. In this way, diversified and accurate advertisements can achieve the maximum range of dissemination.

3 Short video marketing strategy

3.1 Precise targeting and advertising

First, locate the highlights of the film. As an important category of short video platform, film and television have a large audience for film commentary and film highlights. The platform has corresponding detailed data, which can be evaluated from multiple dimensions. According to the preferences of short video users, the short video platform will summarize and analyze their films of the same type, obtain the data such as the explosion point of the film in the short video platform, determine the direction of the main content, and then refine the marketing strategy.

Secondly, target groups should be targeted for precise delivery. By analyzing the user portraits of the targeted historical films, understanding the various indicators of the audience, and adding the feedback from the pre screening and film trial meetings, we can have a more detailed positioning of the film's target population, including gender, urban distribution, age, interest preferences, and so on. In the process of short video push, we should first control the core audience, label the target users of the film, and then launch it accurately to reach the people who are interested in the film, so as to maximize the effect of communication and transformation [5]. Therefore, the short video marketing strategy is to spread outward from the core circle, publicize the interested groups, actively develop potential users, and expand the film increment. For example, the users of the movie " Love you forever " consuming content in Tiktok station are highly coincident with the data of the ticketing platform in terms of urban areas and gender ratio, which also confirms the precise touch ability of short videos in the distribution of film and television content.

Finally, data monitoring and flexible adjustment of marketing plans are also essential. The propagandist should monitor the popularity and word of mouth of each topic at each stage of film marketing, and adjust the marketing plan at any time. After that, we will further plan and release hot topics, and launch more popular content and secondary creation of characters. Finally, it is spread to more audiences through private messages, hot searches, push, etc.

3.2 Combining products and distributeing benefits

The short video platform will cooperate with the ticketing platform in the film's publicity and distribution work, and bring small programs to achieve the accurate transformation of "want to see" and "buy tickets". In the short video transmission, the cat's eye and ticket panning applets will be in the lower left corner. In the process of popular publicity materials, once the audience is interested, they click to enter the applet to generate the conversion of "want to see" and box office. Based on the needs of activities and box office conversion, short videos related to films will also be linked with lottery or ticket activities. Users enter the specially designed activity page and complete the tasks of "paying attention to the official number of the film", "watching the trailer", "marking the cat's eye / tickets to see" and so on. Then they can receive coupons or draw movie tickets, which can encourage the audience to enter the cinema to watch movies.

3.3 Creation of a live streaming platform to interact with fans

As an efficient and direct publicity and distribution transformation strategy, live streaming ticket selling has gradually attracted the attention of header films since the end of 2019. This is the most direct where that film marketing has reached a closer link with the audience through cooperation with the short video platform. As a short video platform with a large audience, live streaming has made the topic popular. The promotion and distribution linkage of the box office trinity is becoming more and more mature. The content generated in the live broadcast process can also be released twice to trigger discussion and increase its popularity. Every live broadcast brings a value increment for the film's publicity and distribution.

Cinema films are mainly commercial films, and the fan economy plays a major role in the box office of commercial films. For example, during the publicity period of the movie "End Game", the popular actor Andy Lau settled in Tiktok and gained more than 50 million fans in a few days. During the live broadcast of Tiktok, the number of people online exceeded 30 million, and the number of people who wanted to see cat's eye increased significantly, breaking the record, which also helped "the crowd surge" buck the trend in the second half of the Spring Festival, with the film arrangement rising and the box office proportion rising.

Short video live broadcast also solves the inconvenience of holding offline press conferences during the epidemic. "Hi, Mon" opened its first Tiktok "cloud road show" before the screening, which combines live experience with remote interaction. The leading actor was broadcast live on the Tiktok platform, and thousands of offline viewers were also broadcast simultaneously in cinemas around the world, attracting many users to interact with the live broadcast room. The creator gained the favor of a large number of onlookers through online interaction with the audience, making more viewers have the desire to watch the movie.

3.4 Attach importance to scene marketing and combine movies with life

The concept of "Scene" itself comes from the film, which refers to the scene in the film. Scenario marketing is a marketing method that provides customers with a better future life scenario to increase the sense of experience and realize consumption. Merowitz, an American scholar, believes that "scenarioism breaks through the limitations of space and gives a new emotional connotation to 'field and scene', that is, the environmental atmosphere of behavior and psychology created by media information. He believes that the most fundamental thing of electronic media is not to affect us through its content, but to affect us through changing the scene [6]."

Scene marketing will be an important marketing mode in the future. Relying on its strong and rich scene building ability, the film has the natural advantages of scene marketing. Many stories are set in ancient movies, such as "The Lost Tomb" and "Once Upon a Time", which will model the iconic scenes in proportion and put them in squares, parks, shopping malls and other places with large traffic. When citizens see these scenes when they pass by, they will not only think of the film immediately, but also have the feeling of being in the movie situation. Although this can also have a good effect, the publicity cost is high, the cycle is short, and the targeted population is relatively limited. After the rise of short video, it is only necessary to set the corresponding scene special effects in the short video app to reproduce the scene in the movie. As long as people hold up their mobile phones, they can shoot short videos in the movie scene. Not only the movie scene, but at present, short videos have also penetrated into many users' life scenes. When the movie uses short videos for scene marketing, by linking the entry point of publicity with the life scene, it can wake up the specific emotions of the target users, so that the audience can feel the connection between the movie and themselves, so as to stimulate the movie viewing desire.

3.5 Deep integration and going deep into the upstream of the industrial chain

Due to the epidemic, some films cannot be shown in cinemas. The 2019 Spring Festival movie "Lost in Russia" was forced to withdraw due to the situation, and then used streaming media as a screening platform to meet the public. Such a strategy is a concrete manifestation of the attempt of the Spring Festival movie to premiere online. What is more worth mentioning is that the company that bought the movie, ByteDance, has its most famous product, the short video app Tiktok. ByteDance's purchase of the exclusive broadcasting rights of the whole network of "Lost in Russia" in a sense achieved the goal of timely curbing the financial losses of the film under the dilemma of being affected by the epidemic. With the measure of letting the whole population watch movies for free, the download volume and traffic of ByteDance's Tiktok, watermelon video, today's headlines and other products have been comprehensively increased. While winning public praise and reputation, it has skillfully used the traffic of the Spring Festival file to improve its popularity.

In addition to being a general cooperation platform for film marketing, short videos will further penetrate the film industry chain. Public data shows that Tiktok users account for more than 90% of Tiktok users under the age of 35, which is highly coincident with mainstream film consumers. In 2021, Tiktok participated in a total of 13 films as a producer and co-producer, three times as many as in 2020. This advanced method of Tiktok is actually the same as the way in which platforms such as cat's eye and taopiao enter the film industry. They both build irreplaceable important value in the first link of the industrial chain, and then enter the upper reaches of the industrial chain through value maximization. Most of the projects Tiktok participated in the joint production are head movies, which in turn confirms that Tiktok's voice in the industrial chain is strengthening. Under this relationship, the resource support and flow support of short videos for film marketing will be further enhanced, and Tiktok will gradually become the most important and effective publicity and distribution channel for film marketing [7].

4 Conclusion

The New Marketing Theory of 4C has fundamentally changed the theoretical basis of marketing. Traditional marketing is centered on the products of enterprises, and customers are passive, while integrated marketing is centered on the needs of customers, and maximizes the profits of enterprises on the premise of meeting the needs of customers. Network marketing connects enterprises and customers with a convenient and fast way of communication, emphasizes the exchange and transmission of information, takes meeting the diversified needs of customers as the center, and "integrates" the marketing objectives with customers' needs [1]. It can be said that the short video platform has created new content and form in film marketing, and achieved a better communication effect, while the 4C rule has played a guiding role in film short video marketing. In China, the integrated marketing of "film + short video" is in its infancy, with rapid development, remarkable effects and considerable benefits. However, the rapid development will also bring a series of deficiencies, such as serious content homogenization, lack of innovation, lack of connotation and so on. Faced with such situations and trends, practitioners need to seize the opportunities of the times and calmly think about the potential problems and deficiencies, so as to contribute to the sustainable development of film short video marketing. We cannot ignore the content and connotation of short videos in order to achieve marketing purposes. With the rapid development of the mobile Internet, short video has increasingly become an effective means of film marketing. For the effective combination of short video and film content, we still needs to continue to explore and innovate.

In addition, although the above research is based on some public data and theoretical information, this paper still has some limitations due to the lack of personal experience. To illustrate, the sample scope of this paper is not broad enough. In the future, research need to be more comprehensively, collect data from more platforms and channels, and discuss a wider range of categories.

REFERENCES

1. China Internet network development state statistic report [EB/OL]. Baidu wikipedia, https://baike.baidu.com/item/ China Internet network development state statistic report / 8278359? Fr = Aladdin
2. Ning Hang. Short video reengineering marketing scene [J]. News research guide,2019,176 (20):155+219
3. Analysis of the digital development of Chinese film marketing 2021 [R] Analysys, 2021-09-29.
4. Research on film marketing strategy in the era of short video [j]. sound screen world 2021 (1): 74-75
5. Chen Xiaonan. Analysis of short video platform marketing strategy in film publicity [j] Shanxi Youth,2020(7):101-102.
6. Wu Xianxia. Research on online word-of-mouth marketing strategy [j] Market research, 2010379 (11): 37-39.
7. Qiao Nan. Research on the marketing strategy of short videos in film publicity and distribution

The Application of Post-Humanism in Digital Media Visual Design——Cyberpunk 2077 as an Example

Xuefei Zhang*

Winchester School of Art University of Southampton

ABSTRACT. This paper answers the meaning of "post-humanism" and explains the characteristics of Cybopunk digital media style under the influence of post-humanism. Based on this, we analyze the application of this art style in games by taking Cybopunk 2077 as an example. Finally, I will explain the value of post-humanism to modern media technology and digital culture. To explain the value of post-humanism in modern media technology and digital culture is to better convey the art philosophy of "post-humanism" and the artistic attitude behind it to the public, promote the development of media technology, and further promote and expand the diversity of digital culture.

1 Introduction

Since the 1960s, with the accelerating development of artificial intelligence, biological technology, information technology, and cognitive science, humankind has been able to modify and expand through technology, both physically and mentally, thereby the nature of the past. Meanwhile, with the rapid development of technologies such as artificial intelligence (AI) and virtual reality (VR), human artificial products such as robots are becoming more and more quasi-human as highly intelligent creatures. Human subjects, both in reality and in games, begin to be created and reconstructed in virtual spaces. The human body is no longer inextricably linked to "nature", but increasingly embraces the design and modification of technology. The idea of man is more and more exposed to relentless questioning, but the definition of "humanism" never emerges. Philosophy, science, political science, economics, sociology, anthropology, medicine, ecology, geography, literary criticism, film studies, gender studies, cultural studies and other departments and fields have been involved in the global scale, setting off a wave of post-anthropological research. Philosophical and scientific concepts often lead to paradigm shifts in the way people think about the relationship between the world itself and the universe. They also change the development of digital culture and modern media technology, and interweave literature and art in the traditional sense in cyberspace, forming a new form of mutual penetration and integration of literature, art and technology. In fact, the posthumanistic tide that began in the 1990s has already found rudimentary signs of blind optimism in transhumanism / warning of excessive intervention and domination in the life of technology. Reflection has become a research topic that posthumanism is paying attention to. One of the central terms of the posthumanist thesis is "cyborg," and "the post-human era dominated by cyborgs has arrived [1]." Therefore, this paper answers the meaning of the term "post-humanism" and then explains the characteristics of the Cybopunk style of digital media under the influence of post-humanism. On this basis, taking Cybopunk 2077 as an example, the paper analyzes the application of post-humanism to modern media technology and digital culture, and finally explains the value of post-humanism to modern media technology and digital culture.

2 Posthumanism

Posthumanism is also called posthumanism. There are two main schools of this theory, including the critical post-humanism and transcendental/hyper-humanism. The latter believes that technology can help humanity go beyond its limits and become perfect and refuse to mix with other organisms. In essence, it still has an illusion of humanism and an optimistic attitude towards the development of a science and technology society. The former settles the anthropocentrism of humanism, builds independence centered on "universal vitality", reflects on the inundation of modern social technology. Technology is not only the extension of human body function, but also human independence. Indeed, the post-humanist wave that began in the 1990s realized that transhumanism/transhumanism had begun to become blind optimism and later warnings and regrets about excessive intervention and domination of technological life. It has become a research topic of humanistic concern. In 1985, American scholar Donna Haraway published her famous "Cybernetic Manifesto." Although the word "post-human" does not appear directly in the text, she proposes the collapse of three boundaries,

*Corresponding author: xz20u21@soton.ac.uk

a mixture of humans who have long dominated the three boundaries. Changed to "robot" and became a hybrid "cyborg" (Haraway, 2006). Haraway said that humanism has relied on the differences between humans and animals, living organisms and machines, and substances and non-materials, but modern science, science fiction novels, technology, capitalism, racial studies, minority groups, animal studies, feminism, etc. He pointed out that the mass development in terms of scholarship has destabilized such a strict difference and has been politically questioned. Mechanical creatures are a mixture of theoretical and manufactured mechanical creatures. In short, we are electronic humans. Electronic people are our body and have given us politics. " [3]

After the widespread dissemination of the Cyborg Manifesto, many posthumanists began to focus on how modern technological culture fundamentally undermines the existence of the human body, reality and fiction, human-computer interaction, and artificial intelligence. Terms such as intelligence and self-creation have begun to enter the debate, and a post-human era dominated by cyborg humans has arrived. An electronic human is a mixture that is a cross-border mixture of an organism and an inorganic (machine). The e-human problem involves philosophy of technology, philosophy of science, culture of science and technology, ethics of science and technology, cultural studies and other multidisciplinary fields. In today's landscape society, where artificial intelligence is ubiquitous, the degree of human interaction is constantly increasing, subverting the classical concept of "human" as defined by the Renaissance humanities. Everyone isobsessed with the production and consumption of images, and we live in a world highly mediated by science and technology. While enjoying the benefits of high tech, the negative impacts associated with it must be digested.

3 Characteristics of Cyberpunk Digital Media Styles Under the Influence of Posthumanism

First, the integration of diverse cultures. The worldview of Cybopunk is often built on the predictions of future scientific and technological developments and reflects the fusion of multi-ethnic and multi-ethnic cultures in its social structure. Take the 1982 film Blade Runner as an example, it offers many typical Cybpunk elements as one of the classic works of Cybpunk visualization, underground slums, skyscrapers, people in folk costumes, colorful neon lights. These elements are all placed in the same space in the style of late modernism. Therefore, many cybopunk works are set against the background of international cities that intersect east and west. For example, in Hong Kong, China, and Tokyo, Japan, prosperity and congestion, wealth and poverty are key words in their image, urban landscape and multidimensional cultural ecology. Sufficiently meets the demand for cybopunk. Second, "there are many futurist science fiction scenes" [4]. There are many futuristic sci-fi scenarios in the electro-punk works. The world of e-punk is highly developed in terms of artificial intelligence and virtual reality. As an important part of the visual representation of Cybpunk, sci-fi characters are generally bionic or modified humans, with body parts transformed into robotic structures and cables everywhere. The scene architecture used to convey information in is usually a huge high-rise building with a futuristic style and has a certain sense of visual pressure. " [5]. We also have high-tech products such as VR eyeglasses, projection mapping and aircraft, which is also a visual element that can reflect the Cybo punk style. In terms of color operation, using high-purity colors such as blue, red, purple, and green, a sense of science and technology such as neon signs that are symbolic in movies and virtual network spaces filled with green electronic information symbols. To emphasize. Third, there are various image representation formats of Cybopunk digital media. Early electronic punk visual art was limited by the computer graphics technology of the time. Except for the actual constructed scene, it is generally presented in two dimensions. Nowadays, more and more video works use 3D image technology to construct shapes and even whole spatial scenes. Creators can express their thoughts more accurately through 3D video, create more virtual video, present a more multidimensional future science fiction environment, lack dynamic expression and immersive experience of 2D video Can be supplemented.

4 Example Analysis of Post-humanistic Art in Cyberpunk 2077

Cybopunk 2077 is a role-playing game based on the novel of the same name by famed science fiction author Mike Goldsmith. It is set in a dark, highly technologically advanced and corrupt future world. In the context of the game, technological advances have undermined the authority of the state, creating an autonomous region centered on corporate land. Although the progress of technology only enriches the lives of the rich, the hardening of the class is serious, the people at the bottom of the class live a difficult life and have to work hard to enjoy the life brought by technology. The game's official CDPR has "entropy" for this state[6]. The 2077 worldview basically adopts the traditional electronic punk setting, high-tech bottom life, corporate colonialism, deep cyber cyberspace, and the most fashionable technology in the world. The worldview of 2077 basically uses the usual cybopunk setting, high-tech bottom life, corporate colonialism, network space in the deep net, and human beings as a combination of body and machine.

4.1 Styling

The world of Cybopunk 2077 is divided into several levels, with companies at the top, residents at the middle, and urban dwellers at the bottom. The three categories correspond to the perspectives of the three players in the game: strong, gang, and homeless. The game officially created the name for each level of clothing style, neo-militarism, sculpturalism, neo-sculpturalism and moisture gradualism. The regimental camp is neo-militarist, with a style of clothing between classic and functional, with the responsible person wearing traditional kimono. Regimental combatants remodel their entire bodies into a functional fighting style, with dark overall tones and cold, hard lines emphasizing ruthlessness and transcendence. Supermarkets need combat clothing to fight for profits. These clothes are not only practical, but also very structuralist. There are two main styles in the power camp, Kinkyism and the new Kinkyism, and "Kinkyism" was born out of the desire to find bright colors to brighten the gray world, which is always stuffy. The main feature is the exaggerated and strange visual style of clothing, operating a large amount of high-tech fabrics, fluorescent colors, translucent materials, pvc, laser fabrics, metals and more. Neo-kitschism attracts the attention of those who wish to brighten up their world with bright colors. The new Kinkyism captures the surface aesthetics of Kinkyism in a more luxurious and rich style. The lower class clothing is "entropy" and is characterized by a style that is more practical than the appearance and coverage. Take Animal Gang, for example, which rejects technology because the body is taller than anything else, focusing more on the display of muscles than the justice of the many elements of the body. There is also a tech-worshiping, hacker-savvy Poison group, whose costumes are a mix of the primitive rough elements of Haiti's original inhabitants and the electronic punk elements of hackers. In addition to Animal Gang and Voodoo Gang, there are American six Street Gang, Japanese Tiger Claw Gang, and Valentino Gang, a fusion of Italian and Mexican styles.

The distortion of fake limbs is also an important part of the game's look. By morphing the body into a weapon or tool in the game world, various prosthetic limbs enrich combat and diving. How to play. In the game, imitations that originally belonged to the army flowed into the market through underground routes, giving people more living space and resources, and the people at the bottom committed crimes to get better imitations. Some people lose the consensus after a lot of network transformation, look down on their own lives, think their bodies are inferior and vulnerable. By 2077 in the game, fakes have become symbols of fashion and class. The mechanical prosthetic bodies of the powerful are the brilliant red, gold, and black colors, the rogue prosthetic bodies are more crude, the simplicity of the inhabitants, the homeless and the homeless at the bottom are patchwork wilderness.

4.2 Scene Design

The building is mainly cold, dark and low. The building materials are mainly hard metal, plastic, steel or cement, brick and other raw materials, showing a simple, rough, concise and decadent effect, which is combined with the external environment. It is a unified whole. It is unified as a whole. Materials such as natural materials and plants with vital characteristics are rarely found. The building is filled with signs that are a source of neon and cultural elements, and some of the alley sign art is flawed and visual. The ground floor is a dark, dank living environment with graffiti-covered underground malls and sophisticated casinos, where crime and violence abound. The upper area has a very high sense of science and technology, and the height rules and deep tones of the color scheme give the player a sense of inviolability from above. The game not only uses the classic electronic punk style of Eastern and Western art combination, but also adopts the folk art style of many countries. Compared to previous cybopunk movies, novels and games, 2077 has enriched every aspect of the internet world, allowing players from many countries to find shadows of their culture. Compared with the stereotypes of other cultures in traditional web-themed films, 2077 focuses more on the integration of cultural elements. Play it on the radio while you're driving, you can listen to the songs of Japanese idol singers, or you can listen to the song "Train Journey" by a Chinese band in the important story of the story. You can hear the song "The train is flying in the sky, the dream is 9 days" by the Chinese band. This understanding of pop culture in other countries is respect for athletes.

4.3 Prop Design

In addition to highly developed information technology, the shadow of modern weapons will be easily seen on vehicles and weapons. The item design of 2077 combines modern tools with the game itself, leaving players with an alternative process, without leaving a modern shadow. These designs also include high-tech weapons that may exist in the future world, which are also concrete manifestations of the cybopunk element.

4.4 Plot Design

The hero dreams of creating a world in a small town at night and becoming a "legendary" person. Surprisingly, he gets caught up in the case of the strongest company, Arasaka, and takes the time to play the whole game, which becomes a story. The story has many turning points, with unpredictable consequences. At the heart of the plot is the personal tragedy of a small role, and he fell into the torrent of the times in a class-differentiated, centralized society. There are many criticisms about how to play the game, but the episodic experience allows the player to experience the powerlessness, emotion, and sadness of the world. At the heart of cybopunk culture may not be the visual effects of the city's weird impostors and dazzling neon, but the tragedy of humans' inability to control their destiny in social floods and science and technology games. In this regard, 2077 not only presents a more realistic world of electronic punk, but also reflects well on the development of human society.

5 Application Value of Post-humanism to Contemporary Media Technology and Digital Culture

With the rapid development of new media technology, postmodernism, an important literary genre, has evolved into the style of cyberpunk. The application of posthumanism in mobile interaction design is characterised by flamboyant kinetic design and the construction of virtual three-dimensional spaces. The tip of the neck, the credible metaphor, the passage, and the completion. For example, the humanistic globalization of the future, the eastern and western cultural elements of the city, the future of the humanistic network, the future of the future, and the reality of the imaginary arts. Once this is a trivial information tool, the representative form is the representative form of the humanities, and it has the potential to be used in practical design. Fusion provides a systematic visual effect, while being an integrated stimulus and integrated reasonable elements. Third, the feeling of immersion in the alternating interaction design. Posthumanistic literary genre. At the same time, a certain degree of 3D alternating scene structure is established. Easy realization depth immersion.

The "post-humanist" literary background of cyberpunk determines its metaphorical function as a design style, and the design works use cyberpunk visual elements to convey anti-utopian and independent thinking cultural demands. The metaphorical function of image information makes the abstract information of the design object more easily understood and accepted by the audience through visual communication. In today's information explosion, the diversity of information leads to the difference of ideology and culture, and gives rise to different cultural fields and social groups. It is hoped that in the future, other cultural content compatible with cyberpunk values can be displayed in the style of cyberpunk to promote cultural diversity.

Cyberpunk is essentially a product of mankind's fantasy of the future, and its creative philosophy is instructive for future design. With the development of the times, Cyberpunk works, whether in literature or in art and design, need a contemporary interpretation of the future world, which is manifested as the renewal of old visual elements and the creation of new visual forms in visual design. Cyberpunk, as a future-oriented design style, is not static. It has no set design rules, which allows it to continuously feed data to the design domain. The development of digital society has made some of the early ideas of cyberpunk a reality, which has gone beyond the surface of words and gradually permeated into modern life. It has been widely used in digital media design, which not only conforms to the trend of The Times, but also establishes a realistic emotional connection with the audience. In future design practice, cyberpunk creators should make full use of technology to explore new ways of interacting and presenting visual graphics, to restore the functional conception of images in the cyberpunk worldview to reality, and to apply more interactive technologies to digital media design to provide a richer experience for the audience.

6 Conclusion

Later humanism was a criticism of humanism as a philosophical tide, advocating "de-humanism". However, when we discuss posthumanism, we are dealing not only with themes of anthropocentrism and evolution, ecology or technological cooperation, but also with how different ideas confront these types of themes. We are discussing what we are doing and the challenges we have to face.[8] The development of digital society makes some ideas of post-humanism theory become a reality, making it less superficial and more universal in modern life, so that it can be widely used in digital media design and adapt to the times. Not only do, but build a realistic emotional connection with the viewer. In the future design practice, electronic punk creators will make full use of technical means to explore new visual graphic interaction and representation, and reconstruct the concept of image function in the world view of electronic punk. More interaction techniques must be used in digital media design to provide viewers with a richer experience. Posthumanism also submits a view of nature in harmony, rational use of high technology, science and technology restores beneficial functions to humanity, sustains the sustainable development of society, loves non-humanity, and makes a difference. He admitted that he embraced and respected others, and advocated the construction of a posthuman society in which various species coexist in harmony. Thus, post-humanism theory not only helps to clarify the intrinsic requirements of emerging technologies, but also provides a new starting point and new research methods for reexamining the relationship between man and technology, man and animal, man and environment.

Acknowledgment

Special thanks to Jennie, who provided valuable help in writing the text.

References

1. Haraway, D., 1985. A cyborg manifesto: Science, technology, and socialist-feminism in the late 20th century. In The international handbook of virtual learning environments (pp. 117-158). Springer, Dordrecht.
2. Nayar, P.K., 2018. Posthumanism. John Wiley & Sons.
3. Wolfe, C., 2010. What is posthumanism? (Vol. 8). U of Minnesota Press.
4. Murphy, G.J. and Vint, S. eds., 2010. Beyond Cyberpunk: New Critical Perspectives (Vol. 3). Routledge
5. Zhang, H. and Zhang, M., 2020, July. Research on Cyberpunk Images in the VisualDigital Media. In 2020 International Conference on Computer Vision, Image and Deep Learning (CVIDL) (pp. 39-43). IEEE.
6. Marchand, J.W., 1975. Towards a Sociology of Linguistics.
7. Athique, A., 2013. Digital media and society: An introduction. John Wiley & Sons.
8. Wolfe, C., 2010. What is posthumanism? (Vol. 8). U of Minnesota Press.

Reasons for Teenagers' Habitual Use of Social Media: A Case Study of TikTok

Yihan Yang*

Oldenburg Academy, Oldenburg, IN, 47006, the United States

Abstract. With the development of Internet and 5G, new social media has been constantly developing and updating. People are getting more and more used to get information from social media. People's lives have been filled with applications, such as TikTok and Instagram. These applications not only bring much fun and convenience to people, but also make it possible for people's fragmented time to be used wisely. However, at the same time, many people, especially teenagers with poor self-control, would easily be overdependent on the social media. As one of the most famous social media at present, with the help of big data, TikTok has successfully made some teenagers seriously depend on its platform by inferring the users' minds and accurately showing them the content they demand. This paper takes TikTok as a case study and teenagers as the research object to analysis the reasons why teenagers use social media habitually, and provide some reasonable solutions to reduce teenagers' media dependency. In short, teenagers get addicted to TikTok primarily because of their self-control is not strong enough so that they fell into the trap of TikTok. TikTok and other social media use big data to predict users' preference, and take advantage of users' psychology to make teenagers get addicted to social media without realizing. In order to get rid of the traps, network supervision departments should strengthen the management, TikTok and other social media should recommend more useful short videos to teenagers, and teenagers themselves can also take advantage of big data.

1 Introduction

We are now living in an era of mobile Internet. Social media have been embedded in all aspects of people's lives. Population of social media is increasing every single day. By 2020, there are around 3.81 billion users being active on social media all around the world [1]. Social media have huge influence on people, especially contemporary teenagers who are under a new media environment. Their connection with new media is much more dense and inseparable. Social medias provide our life a lot of convinence. Because of social medias, people can make better use of fragmented time, the connection between people is closer, and there is more fun in our lives ... However, it is not a good idea to let teenagers get in touch with too many social medias. Many teenagers are still in the process of physical and mental growth. Due to the monotony of school life and their low self-control, many teenagers are currently addicted to the colorful world created by the media, where they can obtain information, socialize, and entertain [2]. TikTok, as one of the most famous new media, has more than 800 million daily active users as of August 2020, and 49% of them are teenagers. This kind of dependence on social media has caused serious problems to teenagers, such as unhealthy lifestyles and declined grades. This paper is going to talk about how social media, such as TikTok, work to make teenagers get media dependency. The paper also puts forward some targeted strategies for these pathological symptoms, striving to provide teenagers with a healthy and effective way of study and life.

*Corresponding author: yihanhanyang@gmail.com

2 Reasons for Teenagers' Overuse of Social Media

2.1 Big Data

Big data is now working well on its value orientation. First of all, the information in the Internet is much more abundant than tradition social media, while big data can help pick up valuable information, then organize it, and, at last, present the information. Secondly, big data can explain the connection between things. By using big data technology, people can analyze the relationship among different kinds of data information. It can also present the relationship more intuitively and comprehensively, which has great significance to enhance the value of data information. Finally, big data has the ability to predict. It can predict the development trend of unknown data according to the existing data information, and provide the possible basis to make decisions [3]. Big data has huge significance on the development of new media. It helps select the topics, enrich the types of news products, and improve the accuracy of news reports. However, TikTok and other similar applications are also using big data to make people, especially teenagers, get addicted to it. Teenagers are considered the future hope of a country. They are the most energetic group. However, contemporary teenagers grow up in an increasingly affluent environment. With greater study pressure and a crowded urban environment, some teenagers show anxiety, depression, and other negative emotions in an early age. Online media provide teenagers a virtual world that they could make up their own names, heights, and even identities. It might bring a time of peace for teenagers to relax from the stressing work. However, there would not be any other good impacts on teenagers' growth if they indulge in the virtual online world too early, including TikTok short videos and other video games. According to Niu, up to 41.00% of teenagers in China said they would watch TikTok whenever and wherever they were free. There are 7.50% of teenagers said they would watch short videos several times a week. Only 3.00% of teenagers said they would watch it several times a month. This data demonstrates that the frequency of most teenagers watching TikTok is pretty high [4]. Teenagers are still too young to distinguish and separate between the real world and the online world. They can not make good use of the intelligence of big data since they are still too young to have self-control. When teenagers like, collect, or leave the comments on the TikTok platform, big data would start to analysis this data information, predict what kind of types of videos they are interested in, and then keep sending them the similar kinds of videos. In this way, as long as they open the application, something they are willing to keep watching would always show up. As time goes by, teenagers would slowly get addicted to TikTok videos.

2.2 Fear Of Missing Out

As fingers continue to slide, new content will continue to emerge. It is quite common to see people watching TikTok on the way to and from work, during work breaks, and even in the bathroom. All those social media have become part of our daily lives. We watch them, talk about them, and participant in them. Research shows that deep participation in social networks is partly due to the "fear of missing out" (fomo). Fomo refers to a sense of concern that one could miss out the information from other social media users. Fomo is a relevant predictor of social media addiction [6]. Teenagers like to talk about popular dance, music, and short videos from TikTok at school. If a teenager did not check TikTok recently, he or she might worry about being laughed at by other teenagers and is scared of being unable to join the conversation of other teenagers due to the missing out information from TikTok. Therefore, he or she could not help but keep checking TikTok.

2.3 Competition Relationship

Human beings not only need to keep in touch, but also like to participate in social competitions. This may also be a factor driving the habitual use of social media [5]. Once there is a "like" button, it means that a person can record the number of likes they receive. Like will have a value, and teenagers use these statistics to improve self-esteem or satisfy vanity. They are at the age of comparing and flaunting. Once TikTok satisfies their vanity, they will spend more time on thinking how to make a better short video, so that they can gain more likes. They may also watch the videos with many likes to study from them. This makes social media users form a habit of checking their social platforms frequently.

2.4 A Way to Build Social Connections

Another reason that social media are so addictive is that they play one of the most basic roles of human beings - the establishment of social connections with others. Social media, such as TikTok, are very easy for users to post. After posting, one will wait for others to like the posts or leave a comment. Seeing someone "like" the posts can directly make the users' brain release dopamine and get pleasure. This can satisfy the desire to be recognized and make the users rely on it, especially teenagers who are at the age of proving that they have grown up. The returns are unpredictable. You would never know how many likes, comments, or followers you would get when you log in. As we all know, intermittent and unpredictable returns are the most addictive. At the same time, it is thought to be a kind of social reciprocity when someone likes or leaves a comment on others' posts. It feels as if people are establishing contact with each other. Schools are the communities for teenagers. Social media provides teenagers a place to build relationships with people online. They share videos with their friends online. In this way, they have common topics to talk about. According to Xue's study shown in Table 1, 58.33% of teenagers get to know TikTok from recommendation of their classmates and friends [8].

Table 1. How teenagers get to know TikTok [8].

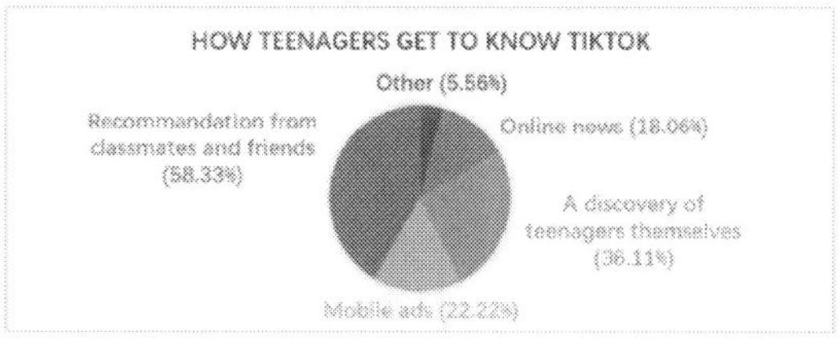

2.5 Pavlovian Response

As time goes by, teenagers may have trouble controlling themselves. In other words, they will check TikTok no matter it is necessary or not as long as they have phones in their hands. Just like Ivan Pavlov's experiments, the salivary reaction happening when putting food into a dog's mouth is a brain reaction. It is a direct connection between sensory and motor nerves in the spine or lower brain center. In contrast, conditioned reflexes, such as the salivation when hearing ringing tones or other previously neutral stimulating sounds, are the result of new reflex channels established in the cerebral cortex during the formation of conditions [10]. As long as teenagers develop a conditioned reflex in their minds that the use of phone is to watch TikTok videos, then they would always watch TikTok when they have the phones even thought that was not what they meant to do in original. This is the worst condition.

3 Solutions

3.1 Strengthening the Management of the Internet

Controlling the Internet is not only a need to promote the healthy development of the Internet, but also an inevitable requirement to protect teenagers from its negative effects. China should continue to strengthen the management and restraint of the Internet. Taking Australia as an example, Australia is relatively strict in Internet management. Australia has established the communications and media authority agency. The agency is specifically responsible for restricting the content of Australian local websites. It will use filtering software to test investigating online gambling. As long as it involves in child pornography, sexual violence, abetting crime, racial hatred, terrorism, and other content, it will be strictly prohibited from spreading [7]. TikTok should learn from this and strengthen the audit. Even though it has been better after TikTok being sued, however, many users who have a large number of fans are not sealed. In order to make more profits, they take advantage of the sensitive words, keep selling things secretly, or spread some bad news information, which should be prohibited immediately.

3.2 Recommending Useful Videos

In another way, it is important for TikTok and other social media to recommend useful videos first. We are now in an era of entertainments. There are too many performances in the network world, and sometimes it is not clear where the boundary between the real and unreal is. Teenagers will find themselves in the information flow and gradually lose themselves. There are actually valuable short videos in TikTok. However, most of them are wasted. If TikTok can recommend short videos, such as cooking, fraud prevention, and life tips, it would improve teenagers' knowledge and help them live a better life [8]. TikTok can also give priority to add a "current events" column, so teenagers can pay more attention on latest national trends.

3.3 Taking Advantage of Big Data

Not only TikTok system itself needs to be advanced, but teenagers, as users, can also take advantage of the fact that "big data knows teenagers too well." As the author mentioned before, big data can analysis the information, and then keep recommending what users like and want to see. It is obviously one of the primary reasons that teenagers get addicted to TikTok. From this logic, if teenagers want to stop watching too many TikTok short videos, they can intentionally watch some videos that they do not actually like. For example, if a teenager were not interested in watching short videos talking about animals at all, in order to stop over-watching TikTok short videos, the teenager needs to force himself or herself to finish watching all of short videos about animals, and then leave comments. At the same time, as long as the teenager sees funny videos he or she likes or are interested in, the videos can be quickly slid away. In this way, big data would think the teenager likes the content about animals, but not funny videos. As time goes by, as long as teenagers open the TikTok, it will always recommend videos that they do not like. Teenagers will then lose the interest in TikTok and other social media because there is nothing attractive. In short, teenagers need to develop their subjective critical consciousness. Heidegger believes that human nature has the ability to think independently. In order not to fall victim to the new media era, we should always realize that media is only a tool for human beings, not a product of dominating our spiritual world [9].

4 Conclusion

Human beings now have entered the era of new social media. The Internet has brought us so much convenience. However, it also means that the educational information technology innovation represented by the Internet will accompany the growing process of teenagers. TikTok and other new media are good at taking advantage of big data to analyse teenagers' preferences, and then keep recommending the contents they like. As long as teenagers open the application, short videos they like will always come up. Therefore, TikTok short videos are always attractive to them. Teenagers' self-control abilities are still not mature enough. Gradually, they would get addicted to the TikTok short videos and neglect their study. Furthermore, everyone, especially teenagers who are still not mature enough, has the desire to be seen, to be responded, and to be confirmed. The setting of "like" in the TikTok satisfies this desire. The "like" button allows teenagers to somehow satisfies their vanities. Furthermore, teenagers do not have the ability to distinguish right from wrong. Most of the time, they follow the crowd. It is a potential factor that teenagers get addicted to TikTok new media. The worst situation is when teenagers have conditioned reflexes that they always start to watch the TikTok videos as long as they have phones in hands. It is important for teenagers to focus on their study instead of TikTok short videos. It is necessary for China to continue to strengthen the management and restraint of the Internet. There is a need to strictly prohibit the dissemination of bad videos. Besides, schools should educate teenagers to improve their subjective critical consciousness. TikTok short videos could be useful for teenagers' studies. It is all depend on how they use it. If they keep watching the short videos that they are actually not interested in, TikTok will gradually lose the attraction to them. In short, this paper talks about how TikTok and other social media take advantage of big data and teenagers' social and behavior psychology to make them get addicted to TikTok gradually.

In order to avoid those traps, network supervision departments should strengthen the management, modify the system to recommend useful short videos to teenagers, and teenagers themselves can also take advantage of big data. Future studies can focus on the big data to further analyse its mechanism mentioned in this paper and provide more statistical evidence.

References

1. Dean, B.: Social Network Usage & Growth Statistics: How Many People Use Social Media in 2022? (2021), https://backlinko.com/social-media-users.
2. Pang, B.: Research on Coping Strategies of Teenagers' Media Dependence. Heilongjiang University (2019).
3. Qin, Y.: On The Deep Integration of Internet Big Data and External Communication. Media Economy, 59–60 (2017).
4. Niu, F.: Analysis of The Impact of Short Video on Teenagers in the Internet Environment and Its Countermeasures. Modern Vocational Education, 100–102 (2021).
5. Griffiths, M., Richard, T.A.W.: Risk Factors in Adolesence: The Case of Gambling, Videogame Playing, and the Internet. Journal of Gambling Studies, 199–225 (2000).
6. Milyavskaya, M., Saffran, M. et al.: Fear of Missing out: Prevalence, dynamics, and consequences of experiencing FOMO. Motivation and Emotion 42(3), 725–737 (2018).
7. Wang W.Y.: How to Control the Negative Impact of the Internet on Teenagers. Special Focus, 119–120 (2014).
8. Xue, L.L., An Analysis of College Students' Use of Short Video on Mobile Terminal and their attitude towards it. Guangdong Canye 52(04), 140–141 (2018).
9. Li, R.Q., Zhou, L.: Research on Short Video Dependence Under Information Cocoon Room: A Case Study of TikTok. Journalism & Media Studies, 75–76, 127 (2021).
10. Wu, Q.E.: Dog's Simultaneous Discrimination Conditioned Reflex to Figure. Journal of psychology, 178–182 (1959).

The Impact of Live Streaming Behavior on Consumers' Consumption Intention

Jiayao Ying*

School of Business and Public Administration, Wenzhou-Kean University, Wen zhou,325060, China

Abstract. The emergence of the phenomenon of live streaming has led to the development of many industries at home and abroad at present, especially in the context of the COVID-19, live streaming has facilitated people's consumer life. In order to explore the current popularity of live streaming in China and to provide reasonable suggestions for the development of the live streaming industry, this paper investigated samples of various age groups around us by means of a questionnaire distribution. Research results show that the popularity of live behavior is not very high, there are still a lot of consumers more accustomed to the form of offline consumption, which also requires continuous innovation in the form of live, applicable to a variety of audience groups, live product quality reputation continues to improve, so that consumers' purchases are guaranteed. This paper conducts in-depth research on live broadcasting behavior, and also provides feasible suggestions for self-improvement of live broadcasting e-commerce, from the perspective of consumer preferences and consumer psychology.

1 Introduction

As a result of the epidemic, approximately 23% of consumers in China have shifted from offline to online purchases. If the epidemic continues, nearly 46% of consumers will shift to online purchases. In the case of such a large-scale consumption form of transformation, live broadcast will become one of the mainstream promotion mode of online consumption. Live broadcast has the characteristics of mass and wide spread, everyone can participate in live broadcast; at the same time, the impact of live broadcast on consumption is very wide, even if not through the form of direct participation in consumption activities, live broadcast can still improve visibility through its traffic, playing an indirect role in the promotion.

Most of the current shopping and gaming-based industries use live streaming to increase the heat, and the audience groups are generally younger, so increasing the popularity of live streaming behavior among the general public is the key to increasing consumer spending.

In the future era of rapid digital development, live streaming as an online fast communication method can provide great convenience for consumers. Consumers can select goods through live streaming and experience face-to-face live shopping. Live streaming opens up a new type of leisure and entertainment as one more channel for the public to get information.

The rapid development of live streaming has also bred many drawbacks. Due to the lower threshold of the live industry, after solving the employment problem also brought the professional level of practitioners is too low, the spread of vulgar content in the live streaming as well as the integrity of the live broadcasters, have made consumers have concerns about the live broadcast. The serious Matthew effect of the live streaming industry makes it impossible for young anchors to accumulate enough traffic without being able to innovate and reform the live broadcast format. Therefore, rectifying the problems of the live streaming industry is also a top priority.

2 Literature Review

2.1 Factors Influencing Consumers' Consumption Intention

There are various factors that affect consumers' consumption intention, and this paper classifies the factors that affect consumers' willingness to consume into direct and indirect factors based on whether each type of factor has a direct or indirect effect on consumers' willingness to consume.

2.1.1 Direct Factors.

Direct factors play a decisive role in consumers' shopping behavior. Among the direct factors affecting consumers' willingness to spend, consumers' disposable income and consumer demand are the two most important direct factors.

According to data from the National Bureau of Statistics, China's per capita disposable income in the first

*Corresponding author: 1851409289@qq.com

quarter of 2022 was 10,345 yuan, a nominal increase of 6.3% over the previous year and a real increase of 5.1% after deducting price factors.[1] At the same time, the national per capita consumption expenditure of 6,393 yuan, a nominal increase of 6.9% over the previous year, net of the impact of price factors, real growth of 5.7%.[2] It follows that with the positive effect of per capita disposable income on per capita consumption, consumers' willingness to consume will also be influenced by consumers' real disposable income, that is, their financial capacity.

Consumers have different needs for different consumption objects. According to the substantive content of consumption needs, they can be divided into material consumption needs and spiritual consumption needs. Material consumption demand refers to people's need for material goods for living. In the act of live streaming, consumers buy basic household goods in the activity of live streaming with goods is the material consumption demand. While spiritual consumption demand is the process by which consumers consume their needs for spiritual activities.[3] The degree of expansion and contraction of material consumption demand is relatively small, while the degree of expansion and contraction of spiritual consumption demand is large, therefore, the consumer's willingness to consume will also have a direct impact when choosing different content of consumption demand.

2.1.2 Indirect Factor.

Indirect factors play a role in aiding consumers' consumption decisions. Among the indirect factors affecting consumers' consumption intention, brand effect and star effect are the two most important factors.

Brand effect refers to the positive effect brought by the company through establishing a good brand image, gaining the recognition of consumers and making the brand known and approved by everyone. A good brand relies on various marketing tools to broaden consumers' perceptions of it, and likewise, the strengths and weaknesses of the brand influence consumers' choices. [4] The brand effect determines the positioning of the goods, and a good brand has the consumer motivation force.

The star effect means that the stars drive the popularity and sales of the products they endorse through their own fame, shaping a good image of the company and bringing positive effects at the same time. In order to establish their brand image and create brand awareness, companies will invite stars to endorse their brands, an emerging marketing tool that is very effective in today's increasingly developed film and television industry. In the case of webcasts, the popularity of the spokesperson or even the anchor of the product being sold will have an indirect effect on the willingness to consume.

2.2 The Vein Of Development Of The Live Streaming Economy

2.2.1 The History Of The Live Streaming Economy.

The development of the live streaming economy is due to the rapid development of technology today and the popularity of big data. Starting from the application of 3G technology and the promotion of smartphone use in 2008, video websites gradually emerged and handheld games also started to step into the Chinese market.[5] In 2008, the "video cabout 2010, China appeared live broadcast platform, mainly live games and other entertainment activities, the network live market is further vertical refinement. According to the statistics of CNNIC, as of June 2016, the scale of domestic webcasting users reached 325 million, accounting for 45.8% of the total number of Internet users. Among them, the scale of the active volume of users in the live video market has exceeded more than 20 million, marking the arrival of the era of universal live streaming.[6]

With the impact of the epidemic, e-commerce is becoming an important channel for people's consumption, according to data provided by the China Internet Network Information Center, by the end of 2021, the scale of China's online shopping users reached 842 million, of which, the scale of live e-commerce users reached 462 million, live has not only existed in entertainment activities, but also in a series of economic activities such as e-commerce shopping guide penetration, live groups will also be oriented to more age groups.[7]

Since 2015, due to the influx of capital, the live industry staged a "thousand broadcast war", not only the emergence of phenomenal traffic anchor, more stars have joined the live broadcast, and even government officials also personally help live with goods. With the development of the live economy, live goods began to live from the "hundred flowers" to the "giant hegemony" pattern over.[8] The flow and capital are gathering to the head platform at the same time, the live e-commerce sales have gained explosive growth.

According to the "2019 Taobao Live Ecological Development Trend Report" data, Taobao live platform in 2018 monthly live users grew 100% year-on-year, with over 100 billion goods, a year-on-year growth rate of nearly 400%, and more than 400 live rooms with more than 1 million goods per month, which has created a 100 billion dollar market.[9]

2.2.2 The Current State Of The Live Streaming Economy.

Live e-commerce is the fastest growing business model innovation in China in recent years.[10] Live streaming as a new situation of consumption attracts consumers to make impulse purchases and drive incremental consumption. Live streaming is constantly evolving to become adapted to a wider range of consumer groups. From the perspective of the elderly population, at the end of 2020, China's elderly population over 60 years of age 264.02 million people, accounting for 18.7% of the total population, in response to the phenomenon of a larger proportion of the elderly population, live streaming is also constantly lowering the threshold, increasing its richness, interactivity and sociality. [11]At the same time, the content of live streaming has been broadened, and the form of live streaming initially gained economic growth by spreading from culture to today's integration with agriculture, sports, and business to meet the diverse needs of society.

3 Questionnaire Design

3.1 Sample Description

This paper investigates and counts the extent to which most consumers participate in the live streaming economic life by means of a questionnaire. This questionnaire is open to all age groups and aims to derive the proportion of different age groups involved in live e-commerce activities and the impact of live e-commerce on people's economic activities. A total of 150 questionnaires were distributed and 132 were returned, with an effective rate of 88%. Considering the efficiency of questionnaire distribution, all the questionnaires were distributed through online form. The demographics of the sample are shown in table 1-3.

Table 1. Demographics of the sample Table—— gender(Table credit: original)

	Gender	Frequency	Percentage	Effective percentage	Cumulative percentage
Effective	Male	28	21.2	21.2	21.2
	Female	104	78.8	78.8	100.0
	Total	132	100.0	100.0	

Table 2. Demographics of the sample Table—— age(Table credit: original)

	Age	Frequency	Percentage	Effective percentage	Cumulative percentage
Effective	Under 20	25	18.9	18.9	18.9
	21-30	29	22.0	22.0	40.9
	31-40	19	14.4	14.4	55.3
	41-50	20	15.2	15.2	70.5
	51-60	11	8.3	8.3	78.8
	60+	28	21.2	21.2	100.0
	Total	132	100.0	100.0	

Table 3. Demographics of the sample Table—— profession(Table credit: original)

	Type	Frequency	Percentage	Effective percentage	Cumulative percentage
Effective	State-owned enterprises	29	22.0	22.0	22.0
	Private enterprisesv	14	10.6	10.6	32.6
	Other (teacher)	1	0.8	0.8	33.3
	Other (retired)	1	0.8	0.8	34.1
	Other (bank)	1	0.8	0.8	34.9
	Other (Clothing)	1	0.8	0.8	35.7
	Business unit	5	3.8	3.8	39.5
	Retired	30	22.7	22.7	62.2
	Foreign companies	2	1.5	1.5	63.7
	Students	37	28.0	28.0	91.7
	Freelancer	11	8.3	8.3	100.0
	Total	132	100.0	100.0	

As can be seen from the figure, 21.2% of the sample were male and 78.8% were female, with significantly more women than men in the sample. The age distribution of the sample is more even, except for the 51-60 years old class, the difference in the percentage of the remaining ages is small.

3.2 Questionnaire Design

The questionnaire was designed with 17 questions, divided into two main sections, including basic information about the respondents (1-3 questions), and information about the respondents' preferences (4-17 questions). The distinction is made by the question "Do you usually have the habit of watching live broadcasts", respondents will answer "yes" or "no" to determine the next questions, which will give a general idea of the popularity of live streaming in people's daily lives. Meanwhile, in order to ensure the reliability and validity of the questionnaire, Likert's 5-level indicator measurement method was used for the measurement of variables in questions 14-17. In other words, "1" means "strongly disagree" and "5" means "strongly agree". The larger the number, the higher the degree of agreement. The questions of the questionnaire were set as shown in Table 4:

Table 4. The questions of the questionnaire (Table credit: original)

Type	Question No.	Question
Basic information of the respondents	1	Your gender
	2	Your age
	3	Your profession
Information on respondents' preferences	4	The most important shopping channel you usually use
	5	Whether you usually have the habit of watching live streaming
	6	Your most watched live streaming platform
	7	The type of live streaming you usually watch
	8	Your weekly shopping frequency
	9	How often you shop in the live room in a month
	10	Your monthly income level
	11	Your monthly living expenses
	12	Your monthly shopping amount (non-student)
	13	Your monthly shopping amount student)
	14	Factors that influence your choice to shop live
	15	Whether you have ever spent money impulsively in a live streaming
	16	Your reasons for not choosing live streaming shopping
	17	Whether you have ever over-hoarded in the live streaming

4. Data Analysis

4.1 Preference Of The Sample For Live Streaming Methods

From the analysis of the most important shopping channels of the samples, there were 71 samples, that is, 53.79% of the consumers would choose to spend through the app shopping platform, and only 1.52% of the consumers on overseas shopping. As seen through Table 3, men are more dependent on app shopping platforms than women, with 75% of men choosing app shopping platforms for shopping and only 48.1% of women. In addition, women prefer to shop through offline shopping malls (42.3%), which may be related to the fact that women are more interested in the quality of goods, and the quality assurance of offline shopping malls is higher than that of general app shopping platforms.

Table 5. Shopping channels for the sample (Table credit: original)

	Male	Female	Total	Percentage
Various software live streaming	0	8	8	6.06%
overseas shopping	0	2	2	1.52%
shopping in the mall	7	44	51	38.64%
App shopping platform	21	50	71	53.79%
Total	28	104	132	100.00%

The preferences of different age groups for shopping methods are shown in Table 6. Over 76% of consumers under the age of 30 shop by using app shopping platforms, which is in line with the general perception that the younger you are, the more you can adapt to more advanced shopping method. In the sample survey, consumers over 50 years old spend on offline shopping malls, many older consumers pay more attention to the quality of goods, and offline shopping malls, although for the app shopping platform consumers will spend higher time costs, but offline shopping malls as the most traditional way of shopping, providing the convenience of operation on the way to shopping.

Table 6. Shopping styles of all ages(Table credit: original)

	Various software live streaming rooms	Percentage	overseas shopping	Percentage	Shopping in the mall	Percentage	App shopping platform	Percentage	Total
Under 20 years	2	8.00%	0	0.00%	4	16.00%	19	76.00%	25
21-30	1	3.45%	0	0.00%	4	13.79%	24	82.76%	29

31-40	3	15.79%	0	0.00%	4	21.05%	12	63.16%	19
41-50	1	5.00%	0	0.00%	7	35%	12	60.00%	20
51-60	1	9.09%	1	9.09%	7	63.64%	2	18.18%	11
Over 60 years	0	0.00%	1	3.57%	25	89.29%	2	7.14%	28

4.2 Live Streaming Penetration Survey Results

Of the 132 samples surveyed, the popularity of live streaming is not very high at present, with a penetration rate of only 39% within the scope of the survey. Of the sample who do not watch live streaming, 81.25% do not choose to shop live because of the price of the product, the anchor, the sales method, the after-sales service and the quality of the selection. The quality of the live streaming selection was the most important concern of the sample, with 42.5% of the sample believing that the quality of the live streaming selection did not meet expectations. Among the samples watching live streaming, the proportion of the population of different age groups watching live streaming basically shows a decreasing trend with the increase of age, which is in line with people's basic perception, except for the group of 21-30 years old, the proportion of the sample population watching live streaming in this group is even lower than the group of 41-50 years old, so it can be inferred that the current audience age group of live streaming is more on the younger side, which is also in line with the characteristics of young people who are more receptive to new things(table 7).

Table 7. Live streaming penetration rate by age group(Table credit: original)

	Not	Yes	Total	Percentage
Under 20	12	13	25	52.00%
21-30	18	11	29	37.93%
31-40	9	10	19	52.63%
41-50	12	8	20	40.00%
51-60	7	4	11	36.36%
Over 60	22	6	28	21.43%

As for the channel to watch live, 46.2% of the total sample chose short video apps such as Tiktok and Kuaishou, short video apps cover a wide range of content and are more convenient to use, so they can meet the various needs of users. The sample of Huya and Douyu platforms that chose to mainly broadcast games was smaller, mostly young users. Tmall and Taobao, which are mainly shopping apps, accounted for 36.5%. Table 8 of the statistical results, 71.2% of the respondents to watch the shopping live mainly, most of the rest for the game live (21.2%), live currently more in the form of entertainment to serve the user. According to statistics, 618 shopping activities of year 2022, Taobao live broadcast account number year-on-year growth of 12.4%, turnover of more than 100 million live 32, turnover of more than 10 million live 434; big anchor transaction year-on-year growth of 63%, store broadcast transaction year-on-year growth of 22%. At the same time, Tiktok e-commerce released the "2022 Tiktok 618 Goodies Festival" data report, showing the consumption trends of the platform during the event. The report shows that from June 1 to 18, Tiktok e-commerce live broadcast totaled 40.45 million hours, and the short video of hanging shopping carts was played 115.1 billion times. This shows that the short video and shopping platform live broadcast also has a great role in promoting consumer consumption.

Table 8. Sample preferences for different live streaming platforms and categories (Table credit: original)

Platform	No. of people	Types	No. of people
Short video apps such as Tiktok and Kuaishou	24	Shopping live	37
Huya and Douyu platforms	2	Game live	11
Shopping App such as Tmall and Taobao	19	Fitness live	1
Bilbili	4	Financial Management live	0
Red	2	Other	3
Other	1		

Among those surveyed, 42.31% (more than 4) believe they often engage in impulse spending in live streaming, 28.85% (3) believe they occasionally engage in impulse spending in live streaming, and 28.84% (less than 3) believe they have not engaged in impulse spending in live streaming(table 9). From the survey results, it can be seen that live-streaming is a sales practice that can, to a certain extent, stimulate consumers' potential desire to consume and stimulate them to purchase goods that they themselves have not planned to buy.

Table 9. Impulse consumption level(Table credit: original)

		Sample No.
Level of impulsiveness	1	11
	2	2
	3	15
	4	8
	5	14

Of those surveyed, 36.54% believe they often over-hoard items on live streaming rather than buying on demand. 26.92% believe they occasionally hoard items on live streaming, and 36.54% believe they have not hoarded items on live streaming. From the survey results(table 10), it is clear that more consumers hoard goods in live streaming shopping than do not. Excessive hoarding of goods not only causes financial losses, but also creates waste. Live streaming shopping as a consumer behavior can have an impact on causing consumers to consume irrationally.

Table 10. Sample hoarding level (Table credit: original)

		Sample No.
Hoarding level	1	11
	2	8
	3	14
	4	10
	5	9

4.3 Consumption Habit Survey Results

The surveyed sample generally consumed less than 5 times a week, with 93 samples accounting for 70.4% of the total. In order to facilitate the exploration of the relationship between consumers' participation in live-streaming activities and their shopping frequency in the consumption process, the other 80 samples who do not watch live-streaming were removed. From the results in Table 11, it can be seen that 34.6% of consumers who participated in live events purchased the majority of their purchases for the week.

Table 11. Sample shopping frequency (Table credit: original)

	Frequency of shopping in the live streaming room in a week			
Frequency of shopping in a week	Under 2 times	2-5times	5-8times	Above 8 times
Under 5 times	25	8	0	1
5-10 times	5	5	2	1
10-15times	3	0	1	0
Above 15 times	0	0	0	0

4.4 Income And Consumption Level Of The Sample

For the working people sample (table 12), the income level is concentrated in 3,000-6,000, accounting for almost half, and 6,000-9,000 and 9,000 or more account for almost a quarter each. Combined with the consumption level, the sample of each income level, the amount of consumption is basically smaller than the amount of income, in a more rational consumption level. Among them, the sample at the income level of 6000-9000 has a significantly lower consumption amount than the income amount. Combined with the age of the sample, most of the samples at this income level are 30-50 years old and are at the age when they need to raise the next generation and support the previous generation. The income of this age group is mainly used to meet the basic consumption needs such as car loan, mortgage and education, and there is not much money for enjoyment spending.

The survey shows that online shopping has surpassed offline shopping and become the main shopping channel for consumers. And live streaming shopping is not as popular as imagined, only 6.06% of the sample choose live streaming shopping, the development of live economy is only in the primary stage. In the sample, only 39.39% of the sample usually have the habit of watching live streaming, less than half of the total sample, indicating that watching live streaming has not yet become a common way of life, and the popularity of live streaming still needs to be improved.

Table 12. Income and consumption levels of active employees (Table credit: original)

Income level	Consumption level	
	Absolute value	Percentage
Under 3000 (4.21%)	Under 1000	100%
3000-6000 (44.21%)	Under 1000	78.78
	1000-3000	18.19
	3000-5000	3.03
6000-9000 (22.11)	Under 1000	50
	1000-3000	50
Over 9000 (29.47%)	Under 1000	20
	1000-3000	30
	3000-5000	20
	5000	30

5. Conclusion

The results of the study show that live streaming behavior has a positive guiding effect on consumers' consumption intentions, and consumers' intentions can have a counter effect on live streaming behavior. According to research data found that most consumers for live streaming selection based on different, more attention or in the quality of live products, as offline shopping can allow consumers to make direct purchases, consumers can intuitively see the good and bad products, and even some offline physical stores will let provide samples to consumers experience. But live streaming is an online format for selling products, using convenience and affordable prices as a gimmick to attract consumers. The product quality is an important factor for consumers to consider, and this is one of the bases for consumers to choose whether to watch the live streaming and whether to shop through the live streaming.

Live operators still have to improve the quality of their products if they want to make their live broadcasts bigger. First, they can provide a trial period for consumers, consumers can choose whether to buy according to the results of their own experience, which also requires mutual trust between live operators and consumers. Secondly, live operators do not need to keep the price low, although the price is inversely proportional to the demand, but in the case of increasing levels of economic development, it is more important to consider the consumer's consumer psychology, and the price difference between the direct broadcast price and the price of the product in the mall is too large to make consumers doubt the quality of the product. Third, Operators need to innovate live streaming content according to the needs of different age groups. Now the live streaming on the market is still dominated by games and shopping, according to the results of the study, most older consumers do not watch the live streaming, and the live streaming of games and shopping is also uneven, less hot live broadcasters can consider changing the form and content of the live streaming, starting from a different age market may be able to get more business opportunities.

References

1. Dai Yue. Analysis of the development status and trend of live e-commerce industry in the new media era[J]. Business Exhibition Economy, 2022(3):3.
2. Shen Zepei. Discussion on live webcasting economy[J]. New Media Research, 2017, 3(21):2.
3. Wang Yingqi, Zhang Niannian, Mei Gaoya, et al. The problems and development trend of "live economy"[J]. New Business Weekly, 2018, 000(004):30-31.
4. Yan Fang. Research on the problems and countermeasures of live broadcasting economy in the context of "Internet+"[J]. China Market, 2021.
5. Qin Jie. Talking about the problems and development trend of "live economy" [J]. Market Weekly - Theory Edition, 2021(63):3.
6. Wen Haiyan. Analysis of the origin and development of live broadcasting economy[J]. Modern Marketing: Next, 2022(1):3.
7. Yang Menglu. On the current situation and prospects of the development of live economy[J]. Shaanxi transportation science and education research, 2020(3):4.
8. Deng Haijian. What makes the live broadcast economy so hot[J]. Times-Picture, 2020(9):1.
9. Li Zhicheng, Liu, Meilian. Research on consumer behavior in e-commerce environment[J]. China Management Science, 2002, V(6):88-91.
10. Dong, Dahai, Jin Yufang. A study of antecedents of consumer behavior tendencies [J]. Nankai Management Review, 2003, 6(6):6.
11. Yang Yi, Dong Dahai. A review of antecedents of consumer behavior in the Internet environment[J]. Prediction, 2007, 26(4):9.

Predicting Online Shopping Intention: the Theory of Planned Behavior and Live E-Commerce

Ruobing Zhang[1,*], Mengxin Chen[2,b]

[1]Xiamen University, VisualCommunication Design, Xiamen, China
[2]Heilongjiang University of Science and Technology, Communication Engineering, Heilongjiang, China

Abstract. With the development of new media technology, some new changes have been brought to traditional shopping. In this environment, live streaming e-commerce has been loved by online consumers. We adopted the theory of planned behavior (TPB), perceived interactivity and perceived utilitarian value as the theoretical basis for research. In the part of data collection, we surveyed 600 Chinese online consumers, the results showed that attitude, subjective norms, perceived behavior control, perceived interactivity and perceived utilitarian value significantly affected consumer behavior intention. The article also discussed the practice and theoretical contributions for live streaming e-commerce.

1 Introduction

As live streaming increased in popularity, a new way of shopping gradually emerged called live streaming e-commerce. Amazon Live, Facebook Live, TikTok, and Taobao Live are prominent examples of live streaming e-commerce where customers can simultaneously shop and be entertained. At the time of this publication, there are two e-commerce streaming types: 1) shopping sites, such as Amazon, live style code, while Taobao.com and JD.com add a live streaming function to e-commerce; and 2) Live.me and Livby add e-commerce to live streaming [8].

In China, almost all e-commerce platforms possess live channels, such as Taobao.com, Douyin, JD.com, etc. On October 20, 2021, Taobao's anchor Jiaqi Li, saw a sales volume of single live streaming that reached 10.653 billion CNY, while Taobao's anchor Weiya, had a sales volume of single live streaming that reached 8.252 billion CNY. In 2021, "double eleven" will be broadcast live in the comprehensive e-commerce GMV of 952.3 billion CNY [11]. This data shows that live streaming e-commerce has occupied a large part of the market share in the retail industry and has become an important sales method.

Given the rapid development of science and technology, live streaming has become an increasingly popular form of media, accompanied by a large number of studies on this topic. Some of the research is about the technology of live streaming system [15, 27, 41]; there is also research on specific platforms, such as YouTube Live [20, 36], Twich [7, 36], and Periscope [15, 39]. With the development of live broadcasting, the media form of live broadcasting has been added to online shopping, forming a new form of e-commerce, namely, live e-commerce. Live e-commerce is a new way of shopping that combines live broadcasting with e-commerce, defined as a subset of e-commerce with real-time social interactivity and media [7].

Buying goods through live broadcasting has greatly changed the way people shop, which is reflected in the annual "Double Eleven Shopping Carnival" and the emergence of key opinion leaders, such as Li Jiaqi. The author wants to understand why live shopping can gradually occupy the market share, which is why consumers like to participate in live streaming e-commerce shopping. This paper will use TPB to study people's behavioral intentions to participate in live streaming e-commerce shopping, and also extends the traditional TPB model by considering two factors: Perceived interactivity; and Perceived utilitarian value. The created measurement scale was used to evaluate the influence of Perceived interactivity and Perceived utilitarian value on TPB variables, as well as the influence of attitude, perceived behavioral control, and subjective norms on behavioral intentions, so as to predict which factors affect people's behavior to participate in live streaming e-commerce shopping.

2 Literature review

2.1. The theory of planned behavior

Azjen [2] developed the theory of planned behavior (TPB) to predict people's willingness to act. Based on TPB, attitude, subjective norm, and Perceived behavioral control (PBC) are the key antecedents of three behavioral intentions. Ajzen and Fishbein [3] prove that behavioral intention can represent actual behavior well, and the behavioral intention is determined by the individual's attitude towards the behavior, subjective

* Corresponding author: 627360474@qq.com
b328771751@qq.com

norms and PBC (Figure 1). Azjen's [1] research shows that an individual's attitude towards the behavior is based on the individual's overall evaluation of the behavior. Individuals with a more positive attitude toward a behavior are more likely to perform that behavior. Subjective norms are individuals' perceptions of their expectations of important reference groups around them, and their expectations of these reference groups. PBC refers to perceived behavioral control; actual behavioral control is very important. PBC is the perceived ease or difficulty of an individual in performing a behavior of interest.

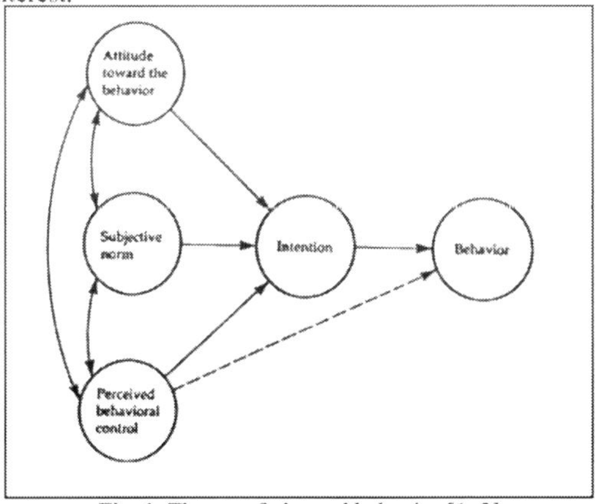

Fig. 1. Theory of planned behavior [1, 2]

The use of TPB is well supported in a wide range of areas. For example, it is widely used to predict behavior, such as health [12], the environment [18] and environmental behavioral intentions [23] TPB has been widely used to predict consumer behavior and is well supported in predicting behavior. Several studies of Internet buying behavior are also based on TPB [5, 16, 25, 27, 32]. George [17] used the TPB to study how much consumers trust Internet credibility and found that consumers were more likely to shop online than those who did not have this belief. Battacherjee [5] found TPB to be useful in explaining that acceptance of e-commerce services, with live e-commerce also having had the characteristics of e-commerce; hence, the author believes that TPB is also effective in the field of live e-commerce. However, in the environment of live e-commerce, compared with traditional e-commerce, live e-commerce has more media and real-time interaction, and in this case, consumers' shopping motivation may change. Therefore, it is necessary to use TPB to study consumer behavioral intentions in the live streaming e-commerce.

Therefore, we propose the following hypothesis:

H1: Consumers' attitudes towards live e-commerce are positively correlated with their behavioral intentions to participate in live e-commerce shopping.

H2: Subjective norms are positively correlated with the behavioral intention of participating in live e-commerce shopping.

H3: Perceived behavioral control is positively correlated with the behavioral intention of participating in live e-commerce shopping.

2.2 Extended TPB model: including antecedent variables

In addition to the three core constructs of attitude, subjective norm, and perceived behavioral control, the TPB also considers antecedents that affect the core factors [18]. These factors, as antecedent variables of the TPB model, affect the core factors, thus affecting the behavioral intention. Some previous studies have combined external variables with TPB to expand the TPB model and found that it can effectively predict consumer behavior. For example, Chen and Tung [9] established an extended TPB research model that combines ethics and the consequences of recycling to explain consumers' recycling intentions and found that this extended TPB research model can explain consumers' recycling intentions well. Hoeksma et al. [24] used the TPB model extended by personal norms to predict consumers' willingness to buy MSU meat, proving that the extended model can predict consumers' purchasing behavior more comprehensively than the TPB model alone. In Yadav and Pathak's study [43], environmental concern and environmental knowledge were added to the traditional TPB model as antecedents of green purchase intention to study consumers' green purchase intention, and the results showed that the explanation of purchase intention was better than that of the TPB model when environmental concern and environmental knowledge were added to TPB. This also shows that the extended the TPB model is effective in predicting consumer behavior intention. Extending the TPB model has a relatively high ability to explain the intention of purchasing behavior, the author boldly introduces two antecedent variables to expand the TPB theoretical model.

In the selection of antecedent variables, the author refers to some previous literature studies and finds that perceived utilitarian value is often used to predict behavioral intentions in some past studies related to live e-commerce [7, 35], therefore, the author adds the perceived utilitarian value antecedent variables to the TPB model to predict consumers' behavioral intentions.

Compared with traditional e-commerce, live e-commerce has more real-time social interaction [7], which is an important feature of live e-commerce. Cai and Wohn [8] found that enjoying interaction and leading the trend are important motivations for consumers to participate in live streaming e-commerce. Therefore, the author adds perceived interactivity to the TPB model as another antecedent variable.

2.3 Perceived interactivity

Leo's [30] research shows that the perceived interactivity directly and indirectly affects the user's attitude towards the website, and the live streaming e-commerce also has the characteristics of network interaction. We therefore believe that perceived interactivity affects consumers' attitudes towards live e-commerce. Lin's [31] research shows that the more interactivity consumers perceive, the higher their purchase intention, and the higher their positive attitude towards the website. From the previous literature, it was found that perceived interactivity was closely related to consumers' attitudes and behavioral intentions. Furthermore, live e-commerce has more real-time social interactivity. Therefore, in this paper, perceived interactivity is put into the TPB model as an antecedent variable to explore whether perceived interactivity in the TPB model also has a positive impact on attitudes and behavioral intentions.

Therefore, we propose the following hypothesis:

H4: Perceived interactivity is positively correlated with consumer attitudes.

H5: Perceived interactivity is positively related to consumer behavioral intention.

In TPB, perceived behavioral control and behavioral intention jointly control the generation of behavior [2]. Perceived behavioral control refers to people's perception of ease or difficulty in performing the behavior of interest. Compared with information systems-related work, the effectiveness and efficiency are similar to the perceived usefulness and perceived ease of use in the technology acceptance model (TAM) [13], and other studies have noted the strong similarities between the two groups of structures [6]. In Teo's [40] research, interactivity is related to effectiveness and efficiency. In other words, if the website can be considered more interactive, it can be regarded as more effective and more efficient. Combined with Benbasat and Barki's previous research, we believe that interactivity has similarities with perceived usefulness and perceived ease of use, and the stronger the perceived usefulness and perceived ease of use of the website. The Hansen et al. [21] study shows that, in the context of social media transactions, "Perceived ease of use" moderates the effect of "perceived behavioral control." The empirical results support the hypothesis that perceived ease of use (from TAM) significantly amplifies the effect of perceived behavioral control (from TPB) about the intention to use social networks for transactions. Therefore, we believe that in the network trading environment, perceived behavior control will be affected by consumers' perceived other factors. Combined with previous studies in the literature, we know that perceived interaction has a positive correlation with perceived ease of use, which has also been shown to positively influence PBC in past studies. In this paper, we add perceptual interaction to the TPB, and combined with these studies above, we propose the idea of whether perceived interaction has a relationship to the PBC in the TPB model, and whether this relationship is positive.

Therefore, we propose the following hypothesis:

H6: Perceived interaction affects the perceived behavioral control, and it is positively correlated.

2.4 Perceived utilitarian value

Studies have shown that consumer attitudes can significantly predict purchase intentions [37]. In the past, most of the research on consumer behavior intention focused on the influence of hedonic value and utilitarian value on consumer behavior intention [7, 8, 22]. Utilitarianism means work capability, instrumentality and practicability. Utilitarian outcomes are the result of the "conscious pursuit of desired outcomes" [4], and utilitarian motivations include convenience and cost reduction (i.e., money, time, and effort) [29]. The convenience and cost reduction that individuals perceive in the process of participating in live shopping is perceived utilitarianism.

Perceived value during shopping positively influences consumers' attitudes and behavioral intentions [34]. Khare's [29] research also showed that perceived value is positively related to consumers' attitudes, including the utilitarian value perceived by consumers, as demonstrated by many studies [20, 36]. Utilitarian values play a strong predictive role in predicting consumers' attitudes and future intentions toward online retailers, and it can be seen that online consumers' shopping preferences are mainly due to utilitarian reasons [37], such as price savings and greater convenience. Cai et al. [7] found that consumers prefer to participate in live e-commerce rather than traditional online shopping, and there are many motivations related to convenience, discounts, and products, such as more time-saving, more favorable prices, and access to the latest product-related information and demonstrations. These are the utilitarian motives of live shopping. It can be said that the utilitarian value perceived by consumers from live shopping greatly affects their purchase intention (behavioral intention). However, in these past studies, there is no literature adding perceived utilitarian value into the TPB to study its influence on attitude and behavioral intention. The authors want to understand whether perceived utilitarian value also positively affects attitude and behavioral intention in the TPB model.

Therefore, we propose the following hypothesis:

H7: Perceived utilitarian value positively affects consumer attitudes.

H8: Perceived utilitarian value positively affects consumer behavior intention.

In previous content, we have mentioned that perceived ease of use positively affects perceived behavior control [22], while in the definition of perceived utilitarian value, utilitarian interests can be explained as the result of ease of use and satisfaction [7], so we believe that perceived utilitarian value has the characteristics of perceived ease and can also affect perceived behavior control. This paper aims to study how perceived utilitarian value affecting perceived behavior control is positive, as shown in past literature.

Therefore, we propose the following hypothesis:

H9: Perceived utilitarian value was positively correlated with perceived behavioral control.

In TPB, subjective norms interact with attitude and PBC, and subjective norms refer to perceived social pressure to perform or not to perform behavior [2], derived from the reference group that individuals consider important. Consumers perceive the utilitarian value in the process of participating in live-streaming e-commerce, such as more favorable prices and time saving. Whether these characteristics will also affect the subjective norms, that is, the important reference group that consumers think that they will do this, for example, to participate in live-streaming e-commerce shopping because of a lower price.

The author did not find the relevant literature shows the influence of perceived utilitarian value and subjective norms, but in the perceived utilitarian value as an antecedent variable in TPB, studies how important the perceived utilitarian value is on affecting the TPB variable. Hence, this paper aims to understand whether perceived utilitarian value has an impact on subjective norms, and we put forward the following research questions:

RQ1: Does perceived utilitarian value affect subjective norms? Is it a positive correlation or a negative correlation?

The hypotheses included in the model are listed below (Figure 2):

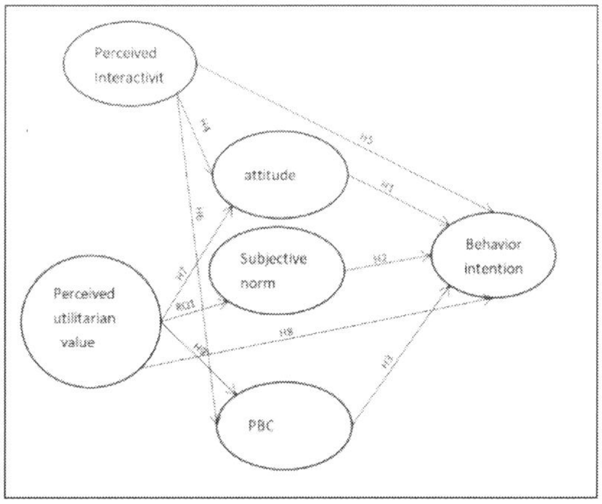

Fig. 2. hypotheses included in the model

3 Methods

3.1 Data collection

From July 15 to July 31, 2022, we conducted a national online survey. Six hundred questionnaires were collected from China. Before conducting the survey, we obtained the informed consent at the beginning of the questionnaire. Only the participants who read and agreed to the informed consent form could continue to answer the questions.

Participants came from 27 provinces in China, with Fujian (14.5%), Jiangsu (14.33%), Liaoning (14%), Anhui (10.67%), and Zhejiang (10.5%) accounting for 64 percent. In total, 245 males (40.83) and 355 females participated in the questionnaire. More than half of them had a bachelor's degree or higher (53%), and 73.66% of people had a monthly income of more than 5000 CNY. Of the participants, 75.33% had the experience of participating in live streaming e-c shopping.

3.2 Measures

The author developed the questionnaire by adjusting the existing reliable and valid questionnaire scales.

3.2.1 Attitude.

Attitudes are measured according to the three dimensions employed by Sadiq et al. [38], and George [17]. Participants measured their attitudes towards participating in live streaming e-c shopping through the following dimensions: (a) I like to participate in live streaming e-c; (b) Participating in live streaming e-c will make me feel happy; and (c) I think live streaming e-c is very useful, not only to save money, but also to let me better understand the products. The Likert 5-point scale was used, from strongly disagree to strongly agree, and these items were averaged to form a composite index for positive coding, in which the higher the score, the better the attitude towards live streaming e-c shopping behavior. (M=3.630, SD=1.000, Cronbach's alpha=0.900).

3.2.2 Subjective norm.

Subjective norms are measured on a four-point scale that combines the scales of George [17] and Sadiq et al. [38], in which participants rated the following statements on a scale of 1 (strongly disagree) to 5 (strongly agree): (a) My friends or family support me to participate in live shopping e-c shoping; (b) My friends or family members think it is valuable to participate in live shopping e-c shoping; (c) My friends or family are very willing to do live shopping e-c shopping; (d) I can often hear information about live shopping e-c shoping from my family or friends, such as 618 discount activities, double 11 discount activities, etc. The items are averaged to create a composite index, the higher the score, the higher the group referred to by individual recognition of live streaming e-c shopping. (M=3.590, SD=1.010, Cronbach's alpha=0.901)

3.2.3 Perceived behavioral control.

Perceived behavioral control was adapted from George's [17] scale, measured in four dimensions, in which participants rated the following statements on a scale of 1 (strongly disagree) to 5 (strongly agree): (a) I have time to participate in live shopping e-c shopping; (b) I have the financial ability to participate in live shopping e-c shopping; (c) I can participate in live shopping whenever I want; and (d) Whether I participate in live shopping e-c shopping is entirely up to me. The items are averaged to create a composite index; the score indicates the degree of ease and difficulty that consumers perceive when participating in live shopping. The higher the score, the easier consumers feel when participating in live streaming e-c shopping. (M=3.61, SD=1.01, Cronbach's alpha=0.901)

3.2.4 Perceived Interactivity.

Perceived interactivity was was adapted from Xiang and Chae [42], in which participants rated the following statements on a scale of 1 (strongly disagree) to 5 (strongly agree):"Do you agree that the following behaviors will make you like to shop by participating in live streaming e-c shopping in live streaming e-c shopping?": (a) Send a bullet screen to interact with others; (b) Other users respond positively to my bullet screen; (c) My bullet screen (or question) can get a quick reply from the host.; (d) the host will be very patient to reply to my bullet screen or give me purchase suggestions. The items are averaged to create a composite index, scores indicate the effect of perceived interaction on attitudes, subjective norms, PBC, and consumer behavioral intentions, with the higher the score, the greater the effect (M=3.650, SD=1.010, Cronbach's alpha=0.896).

3.2.5 Perceived utilitarian value.

Perceived utilitarian value is measured in three dimensions by combining the scales of Chiu et al. [11], and Cai and Wohn [8], in which participants rated the following statements on a scale of 1 (strongly disagree) to 5 (strongly agree): "Do you agree that the following characteristics will make you like to shop by participating in live streaming e-c shopping in live streaming e-c shopping?": (a) There are often preferential activities in the live broadcast room, which are cheaper than usual (such as 618 activities, or exclusive preferential activities for Internet celebrity anchors, etc.; (b) Live broadcast enables me to see the details of the product more intuitively; (c) In the live broadcast room, I can see what other buyers have bought, so that I can keep up with the trend. The items are averaged to create a composite index, scores indicate the effect of perceived utilitarian value on attitudes, subjective norms, PBC, and consumer behavioral intentions, with the higher the score, the greater the effect (M=3.680, SD=1.020, Cronbach's alpha=0.894).

3.2.6 Behavioral intention to participate in live streaming e-c shopping.

Behavioral intention is adapted from the scale of George [17], which has three dimensions to measure, in which participants rated the following statements on a scale of 1 (strongly disagree) to 5 (strongly agree): (a)I am willing to spend money and time to participate in live shopping e-c shopping; (b) I want to participate in live shopping e-c shopping later; (c) I have plans to participate in live shopping e-c shopping in the future. The items are averaged to create a composite index. The higher the score, the stronger the consumer's behavioral intention, and the higher the possibility of participating in live shopping in the future (M=3.570, SD=1.060, Cronbach's alpha=0.897).

4 Results

First of all, we import the survey data into SPSS, and analyze the reliability and scale statistics of each dimension of the variables. After the reliability of the dimensions is determined to be up to the standard, the data of each dimension is integrated into one variable by calculating the variables in SPSS.

Next, we used AMOS to construct a model graph and import data to run a structural equation model to test the relationship between model fit and study variables on the basis of sufficient sample size, no outliers, and normal distribution of all measurement items, a goodness-of-fit statistic measuring the model indicated a good fit to the data (CFI=0.99; TLI=0.99; $PCMIN/DF$=1.298 and $RMSEA$=0.02).

Table 1. Reliability and correlations

	CR	ATT	SN	PBC	PI	PUI
ATT	0.900	1				
SN	0.901	0.395	1			
PBC	0.901	0.43	0.440	1		
PI	0.896	0.424	0.408	0.393	1	
PUV	0.894	0.432	0.425	0.431	0.402	1

Note(s): ATT: Attitude; SN: subjective norms; PBC: perceived behavioral control: PI: Perceived Interactivity; PUV: Perceived utilitarian value; CR: composite reliability

Table 2. Results of hypotheses testing

Hypothesis	Path	β	P Label
H1	ATT→BI	0.177	***
H2	SN→BI	0.181	***
H3	PBC→BI	0.195	***
H4	PI→ATT	0.298	***
H5	PI→BI	0.183	***
H6	PI→PBC	0.255	***
H7	PUV→ATT	0.311	***
H8	PUV→BI	0.202	***
H9	PUV→PBC	0.327	***
RQ1	PUV→SN	0.900	***

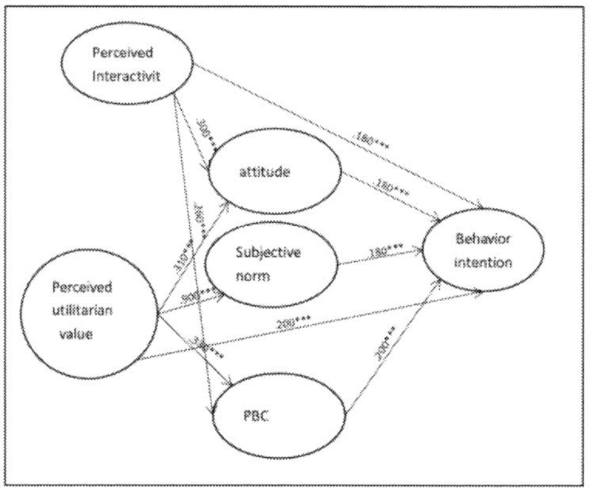

Fig. 3. Evaluated model

The reliability of all items is higher than 0.6, indicating that the measurement validity is acceptable. In the present study, at least in this sample, attitude and behavioral intention showed a significant positive correlation (β=0.177, p <0.05), which is consistent with our hypothesis, and thus judging that H1 is supported. Subjective norms showed a significant positive correlation with the behavioral intention (β=0.181, p <0.05), as supported by H2. The PBC showed a significant positive correlation with the behavioral intention (β=0.195, p <0.05), and the H3 was supported. As the three original variables in the TPB model, the relationship with the behavioral intention was once again supported in this study. It can be concluded that the TPB model is also very effective in the research of live streaming e-commerce.

According to Table II: new added antecedent variables were also supported in the study, perception interaction was positively correlated with attitude presentation (β=0.298, p<0.05); the H4's are supported; perceived interaction positively influences the behavioral intention (β=0.183, p<0.05) that supported the H5, as in past studies; a positive correlation was presented between perceived interactivity and attitude and behavioral willingness. Meanwhile, the difference was that, verifying perceived interactivity equally positively influences attitudes and behavioral intentions in the TPB model; and in the environment of live streaming e-commerce, in other words, the interactivity perceived by consumers in live streaming e-commerce can positively affect consumers' attitude and behavior intentions.

Perceived interactivity positively affects PBC (β= 0.255, p < 0.05), supporting H6. In previous studies, we know that perceived ease of use can positively affect PBC, so we believe that PBC will be affected by antecedents in TPB. In this paper, we boldly assume that perceived interactivity can also affect PBC as antecedent variables, and this effect is positive. The results of the data prove that, at least in this sample, this hypothesis is true, and perceived interactivity positively affects PBC.

Perceived utilitarian value positively affects attitude (β= 0.311, p< 0.05), which supports H7. Perceived utilitarian value positively affects behavioral intention (β= 0.202, p< 0.05), which supports H8. Different from previous studies, it is verified that the perceived utilitarian value in TPB also has a positive impact on attitudes and behavioral intentions, and it can be said that the utilitarian value of live e-commerce is an important motivation of consumer behavioral intentions. Perceived utilitarian value has a positive effect on PBC (β= 0.327, p< 0.05), which supports H9.

In the above, a research question is raised, whether perceived utilitarian value affects SN, and whether the effect is positive or negative. In the research data, it is shown that perceived utilitarian value significantly affects SN, and the effect is positive. (β=0.900, p < 0.05. This paper aims to study how the perceived utilitarian value affects each factor in the TPB model.

5 Discussion

The purpose of this paper is to understand what factors affect people's behavioral intention to participate in live streaming e-commerce, and the TPB theory is used as the theoretical basis of the research. The empirical results show that attitude, subjective norms, and perceived behavior control can predict consumers' intention to participate in live streaming e-commerce. Then, we explore the impact of perceived interaction and perceived utilitarian value on consumers' willingness to participate in live-streaming e-commerce, and we add perceived interaction and perceived utilitarian value into the TPB model as preexisting variables to separately discuss their relationship with attitudes, subjective norms, perceived behavioral control, and behavioral intention. As expected, we found that perceived interactivity and perceived utilitarian value significantly and positively affected consumers' intention, attitude, and subjective norms to participate in live-streaming e-commerce. Perceived behavioral control significantly and positively affects consumers' intention to participate in live e-commerce. Moreover, the two pre-variables we added also affect the three original variables of TPB as assumed. The behavior intention of consumers to participate in live streaming e-commerce depends on the comprehensive role of the five variables: perceived interactivity, perceived utilitarian value, attitude, subjective norms and perceived behavior control.

This study has made several contributions to the current literature. First, this study discusses the influencing factors of consumers' participation in live e-commerce, and further enriches the participation behavior of live e-commerce. Second, this study expanded the TPB theory by adding two variables. Our findings confirm that there is a significant positive correlation between perceived interactivity and perceived utility value on the attitude, perceived behavior control and behavior intention of participating in live streaming e-commerce, and a positive association between perceived utility value and subjective norms. Finally, this study confirms the effectiveness of the expanded TPB model. The expanded TPB theory increases the R side interpretation power by 8% more than the original TPB model and has a good prediction ability (the original TPB model R side is 0.406, and the expanded TPB model R side is 0.486).

The findings suggest that TPB can predict consumer behavioral intentions, consistent with previous findings [5, 16, 25, 27, 32]. The same applies in the live streaming e-commerce environment. Attitudes, subjective norms, and PBC positively influence consumers' behavior intention to participate in live streaming e-commerce.

As we expected, perceived interactivity was positively associated with consumer behavior intention, which shared the findings of Lin [31], the stronger consumers' perceived interactivity, the higher their intention to buy and the higher their positive attitude towards the website. The essence of live streaming e-commerce is a way of sales. Consumers purchase intention is closely related to consumers' experience in the purchase process. The research results show that higher interactions can create a good experience for consumers. Moreover, perceived interaction also positively affects attitudes, similar to the findings of Leo [30]. This means that providing higher interaction to consumers is beneficial to facilitate consumer purchase, which is a sum effect on attitude and intention, affecting intention by influencing attitude, but also directly affecting intention.

In this paper, the impact of perceived interaction on PBC is a highly exploratory topic, and the results are the same as our expectations, and the perceived interaction, which can positively affect PBC. As mentioned in Teo [40], the stronger and more interactive the website is, the more effective and efficient. In the context of the live streaming e-commerce environment, the stronger the interaction, the stronger the perceived ease of use (efficiency) and perceived usefulness (effectiveness) for consumers. In our results, perceived interaction is positively correlated with PBC, while Hansen et al. [21] showed that perceived ease of use regulates the effect of perceived behavioral control. It can be seen that this study also confirms the relationship between interaction and effectiveness and efficiency in the Teo [40] study. We guess that the interactivity of live streaming e-commerce improves the efficiency of consumers in the shopping process. Through the interactive ability of live streaming, consumers can buy goods faster and more accurately, thus improving consumers' intention of participation. Moreover, generally speaking, the more favorable attitudes and subjective the norms are, the stronger the perceived behavioral control, and the stronger the consumer behavior intentions [1]. However, in this study, it can be seen that perceived interaction significantly positively affects attitudes, and we believe that this is also an important reason for the positive correlation of perceived interaction with PBC. We also share the same view in explaining the relevance of perceived utilitarian value and PBC.

Perceived utilitarian value has been used as variables in past studies of consumer behavior [7, 8, 22], as perceived utilitarian value has a positive effect on the behavioral intention.

Perceptual utilitarian value has been widely used as a variable in past research on consumer behavior [7, 8, 22], and the results also demonstrate that perceived utilitarian value positively affects behavioral intention and perceived utilitarian value and attitude. According to the research results, the perceived utility value still plays a strong role in predicting consumer behavior. From the perspective of content analysis, perceived utilitarian value has a very positive impact on SN. We find that preferential prices and convenience of shopping are important factors affecting SN. The real-time interactivity of live e-commerce is also an important motivation for behavioral intention, such as the positive response to the information released in the live broadcasting room (57% agreed), and the host's reply and answer to consumer information.

From the perspective of content analysis, perceived utilitarian value has a very positive impact on SN. We find that preferential prices and convenience of shopping are important factors affecting SN. The real-time interactivity of live e-commerce is also an important motivation for behavioral intention, such as the positive response to the information released in the live broadcasting room (57% agreed), and the host's reply and answer to consumer information.

6 Impacts and limitations

This study has both theoretical and practical implications. In terms of theoretical significance, the results of this study once again prove the effectiveness of the theory of planned behavior in explaining consumers' online shopping behavior, and many studies have successfully used TPB as a theoretical framework to explain the intention of online shopping or other e-commerce activities [5, 25, 17]. In this study, TPB is added to perceived interactivity and perceived utilitarian value, which enriches the antecedents of TPB in the field of online purchase behavior research. Increasing studies on the influencing factors of online purchase behavior use the TPB theoretical framework, and we can find out which factors are more important. This would be conducive to more research on online purchasing behavior in the future.

This article may highlight some tips for retailers and enterprises that plan to enter live e-commerce in the future. For example, consumers' perceived utilitarian value is an important motivation for them to participate in shopping, including the latest and most detailed product displays and preferential activities. Businesses can carry out some live, exclusive, preferential activities on specific dates or festivals, such as Singles' Day. At the same time, through the publicity of activities, potential consumers will forward and share with people around them. Research data shows that SN can positively influence consumers' behavioral intentions. More than half of the subjects (58.17%) can often hear preferential news about live shopping from friends or family. That is to say, when consumers are shared or invited by friends about activities, they will be very willing to participate in shopping. In terms of real-time interaction, live e-commerce has advantages over traditional e-commerce. The interaction between consumers is also very important, as most consumers (61.83%) who are willing to participate in live e-commerce send a mesages to interact with other consumers, and consumers can get feedback from other users about the goods in the interaction. In addition, businesses can place staff in the audience to mobilize the overall atmosphere of the live broadcasting room. At the same time, the host and consumers cooperate and complete a better interaction.

From a practical point of view, with the rise of live e-commerce, this new consumption mode has gradually occupied the retail market. Our study has implications for these practices: first, our research can better provide suggestions for traditional retailers to change their operation mode, implying that traditional retailers can better understand the shortcomings of their own sales mode, so as to find ways to change or adapt to the market environment. At the same time, for live e-commerce businesses or platforms, our research can better advise them on how to improve their competitiveness, so as to share the market share.

In addition, we find that perceived utilitarian value has a significant positive impact on subjective norms. Simply put, the expectations of live e-commerce conveyed by consumers' important reference groups can effectively influence consumers' purchase intentions. The implication is that businesses can increase consumers' purchasing intentions by increasing the publicity of some preferential activities, such as 618, Double Eleventh and other live broadcasting activities, and at the same time, they can achieve better results through mutual publicity among consumers.

As with any study, the study conducted in this paper has limitations that should be addressed in future research. First of all, we investigate the intention of live shopping behavior as a whole in our study, rather than the intention of live shopping behavior of individual platforms or certain categories of goods (such as cosmetics or clothing). Different platforms or different categories of goods may have different associations. Second, at present, there are two streaming e-commerce types: The first is that shopping sites such as Amazon live style code, while Taobao.com, and JD.com add the live streaming function to e-commerce; the others add e-commerce to live streaming like Live.me and Livby [7]. This paper studies the overall live e-commerce shopping behavior, yet for these two different types of live e-commerce, there is no distinction within the study; they may have different factors, which would also be an interesting direction for future research.

7 Conclusion

In conclusion, the results show that the extended TPB can effectively predict consumer behavior, and The validity of the extended TPB theory is confirmed. The data show that the extended TPB model has better prediction ability than the original TPB model. The two newly added variables: perceived interactivity and perceived utilitarian value will affect the three original variables of TPB, the behavioral intention of consumers to participate in real-time e-commerce depends on the comprehensive effect of five variables: perceived interaction, perceived utilitarian value, attitude, perceived behavioral control and subjective norms.

This study has theoretical and practical significance. From a theoretical point of view, it is helpful for other scholars to do more research on online purchasing behavior in the future. From a practical point of view, it can provide some effective suggestions for retailers and enterprises who plan to enter or have entered today's real-time e-commerce in the future.

References

1. Ajzen, I. (1985). From Intentions to Actions: A Theory of Planned Behavior. In Action Control. In:Julius, K., Jurgen, B.(Eds.), SSSP Springer Series in Social Psychology, Springer, Berlin, Heidelberg. pp. 11–39.
2. Ajzen, I. (1991). Theory of planned behavior. Organizational Behavior and Human Decision Processes, 50, 179-211.
3. Ajzen, I., & Fishbein, M. (2005). The influence of attitudes on behavior. In D. Albarracín, B. T. Johnson, & M. P. Zanna (Eds.), The handbook of attitudes. Mahwah, NJ: Erlbaum Publishers. pp. 173–221.
4. Babin, Barry J., DardenWilliam R., and Mitch Griffin. (1994). Work and/or Fun: Measuring Hedonic and Utilitarian Shopping Value. Journal of 86 Consumer Research 20(4), 644.
5. Battacherjee, A. (2000), Acceptance of e-commerce services: the case of electronic brokerages, IEEE Transactions on Systems, Man, and Cybernetics – Part A: Systems and Humans, Vol. 30 No. 4, pp. 411-20.
6. Benbasat, I., Barki, H., (2007). Quo vadis, TAM? Journal of the Association for Information Systems 8 (4), 211–218.
7. Cai, J., Wohn, D. Y., Mittal, A., & Sureshbabu, D. (2018). Utilitarian and Hedonic Motivations for Live Streaming Shopping. Proceedings of the 2018 ACM International Conference on Interactive Experiences for TV and Online Video - TVX '18.
8. Cai, J., Wohn, D. Y. (2019). Live Streaming Commerce: Uses and Gratifications Approach to Understanding Consumers' Motivations. Proceedings of the 52nd Hawaii International Conference on System Sciences | 2019, pp.2549-2557.
9. Chen, M. F., Tung P. J., (2014). Developing an extended Theory of Planned Behavior model to predict consumers' intention to visit green hotels, International Journal of Hospitality Management, Volume 36,2014,Pages 221-230,ISSN 0278-4319.
10. Chinanews (2021). Nine billion yuan of wealth of the wei Ya couple, the business map collapsed. Retrieved September 5 , 2022 , from https://www.zhihu.com/question/507558964/answer/2280403190
11. Chiu, C.-M., Wang, E.T.G., Fang, Y.-H. and Huang, H.-Y. (2014), Understanding customers' repeat purchase intentions in B2C e-commerce: the roles of utilitarian value, hedonic value and perceived risk[J] . Information systems journal,2014,24(1), 85-114.
12. Conner, M., & Sparks, P. (1996). The theory of planned behavior and health behaviors. In M. Conner, & P. Sparks (Eds.), Predicting health behavior: Research and practice with social cognition models. Buckingham: Open University Press. pp. 121-162.
13. Davis, F., (1989). Perceived usefulness, perceived ease of use and user acceptance of information technology. MIS Quarterly 13 (3), 319–339.
14. E. Veloso, V. Almeida, W. Meira, A. Bestavros, and S.D. Jin. (2017). A hierarchical characterization of a live streaming media workload. IEEE/ACM Transactions on Networking.
15. Friedländer, M.B. (2017) "JISTaP Streamer Motives and User Generated Content on Social Live-Streaming Services", J Inf Sci Theory Pract JISTaP 55(11), 65–84.

16. George, J.F. (2002), "Influences on the intent to make Internet purchases", Internet Research, 12(2), 165-80.
17. George, J. F. (2004). The theory of planned behavior and Internet purchasing. Internet Research, 14(3), 198–212.
18. Greaves, M., Lara D. Z., Chris S. (2013). Using the theory of planned behavior to explore environmental behavioral intentions in the workplace, Journal of Environmental Psychology, 34, 109-120,ISSN 0272-4944.
19. Ha, J, and Jang, S.C. (Shawn). (2009). Perceived values, satisfaction, and behavioral intentions: The role of familiarity in Korean restaurants, International Journal of Hospitality Management, 29(1), 2-13, ISSN 0278-4319
20. Haimson, O.L., and J.C. Tang. (2017), What Makes Live Events Engaging on Facebook Live, Periscope, and Snapchat, Proceedings of the 2017 CHI Conference on Human Factors in Computing Systems, CHI'17, 48–60.
21. Hansen, J.M., Saridakis, G., Benson, V., Risk. (2018). Trust, and the interaction of perceived ease of use and behavioral control in predicting consumers' use of social media for transactions,Computers in Human Behavior, Volume 80, 2018, Pages 197-206,ISSN 0747-5632.
22. Hanzaee, K.H. , & Rezaeyeh, S.P. (2013). Investigation of the effects of hedonic value and utilitarian value on customer satisfaction and behavioural intentions, African Journal of Business Management, 7(11), 818-825.
23. Ho, S.S., Liao, Y., & Rosenthal, S. (2014). Applying the Theory of Planned Behavior and Media Dependency Theory: Predictors of Public Pro-environmental Behavioral Intentions in Singapore. Environmental Communication, 9(1), 77–99.
24. Hoeksma, D.L., Gerritzen, M.A., Lokhorst, A. M., Poortvliet, P.M. (2017). An extended theory of planned behavior to predict consumers' willingness to buy mobile slaughter unit meat, Meat Science, 128, 15-23, ISSN 0309-1740.
25. Jarvenpaa, S.L. and Todd, P.A. (1997), "Is there a future for retailing on the Internet?", in Peterson, R.A. (Eds.), Electronic Marketing and the Consumer, Sage, Thousand Oaks, CA, pp. 139-54.
26. K. Pires, G. Simon, (2015). YouTube live and Twitch: a tour of user-generated live streaming systems. Proceedings of the 6th ACM Multimedia Systems Conference.
27. Khalifa, M. and Limayem, M. (2003), "Drivers of Internet shopping", Communications of the ACM, 46(12), 233-9.
28. Khare, A. (2011), Influence of hedonic and utilitarian values in determining attitude towards malls: A case of Indian small city consumers. J Retail Leisure Property, 9, 429–442.
29. Kim, Y. K, and Kang, J. K.(1997). Consumer perception of shopping costs and its relationship with retail trends. Journal of Shopping Center Research 4, 2: 27–62. Retrieved January 28, 2018. From http://173.254.37.135/JSCR/IndArticles/Kim_N297.pdf
30. Vijayasarathy, L.R. (2004). Predicting consumer intentions to use on-line shopping: the case for an augmented technology acceptance model. Information & Management, 41(6), 747–762.
31. Lin, p.w. (2006). The effects of consumers' online shopping goals and their characteristics on perceived interactivity and shopping behaviors. Retrieved August 24, 2022, from Missouri-Columbia University, Institute for MOspace Home Web site: https://mospace.umsystem.edu/xmlui/handle/10355/4577
32. Limayem, M., Khalifa, M. and Frini, A. (2000), What makes consumers buy from Internet? A longitudinal study of online shopping, IEEE Transactions on Systems, Man, and Cybernetics – Part A: Systems and Humans, 30(4), 421-32.
33. ME, Ruiz-Molina, and I, Gil-Saura. (2008). Perceived value, customer attitude and loyalty in retailing. J Retail Leisure Property, 7, 305–314.
34. Moon, M.A., Khalid, M.J., Awan, H.M., Attiq, S., Rasool, H and Kiran, M. (2017). Consumer's perceptions of website's utilitarian and hedonic attributes and online purchase intentions: A cognitive–affective attitude approach, Spanish Journal of Marketing - ESIC, 21(2), 73-88, ISSN 2444-9695.
35. Overby, J.W., Lee E.J. (2006). The effects of utilitarian and hedonic online shopping value on consumer preference and intentions,Journal of Business Research, 59(10–11), 1160-1166, ISSN 0148-2963.
36. Pires, K., Simon, G. (2015)."YouTube Live and Twitch: A Tour of User-Generated Live Streaming Systems", Proceedings of the 6th ACM Multimedia Systems Conference on - MMSys ', 15, 225–230.
37. Paul, J., Ashwin, M., Jayesh, P. (2016). Predicting green product consumption using theory of planned behavior and reasoned action, Journal of Retailing and Consumer Services, 29, 123-134.
38. Sadiq, M.A., Rajeswari, B., Ansari, L., Kirmani, M.D. (2021).The role of food eating values and exploratory behaviour traits in predicting intention to consume organic foods: An extended planned behaviour approach, Journal of Retailing and Consumer Services, 59, 102352, ISSN 0969-6989.
39. Tang, J.C., G. Venolia, and K.M. Inkpen. (2016). "Meerkat and Periscope: I Stream, You Stream, Apps Stream for Live Streams", Proceedings of the 2016 CHI Conference on Human Factors in Computing Systems - CHI '16, ACM Press, 4770–4780

40. Teo, H., Oh, L., Liu, C., Wei, K. (2003). An empirical study of the effects of interactivity on web user attitude. International Journal of Human–Computer Studies, 58, 281–305.
41. Vieira, A. B., da Silva, A. P. C., Henrique, F., Goncalves, G., & de Carvalho Gomes, P. (2013). SopCast P2P live streaming. Proceedings of the 4th ACM Multimedia Systems Conference on - MMSys '13.
42. Xiang, Y., & Chae, S. W. (2021). Influence of Perceived Interactivity on Continuous Use Intentions on the Danmaku Video Sharing Platform: Belongingness Perspective. International Journal of Human–Computer Interaction, 1–21.
43. Yadav, R., & Pathak, G. S. (2016). Young consumers' intention towards buying green products in a developing nation: Extending the theory of planned behavior. Journal of Cleaner Production, 135, 732–739.

On Strengthening the Dissemination of Positive Chinese Traditional Culture in Colleges and Universities in the New Media Era

Kexin Li*

Shandong Institute of Commerce and Technology

Abstract. General Secretary Xi Jinping once profoundly pointed out that "Positive Chinese traditional culture is the root and soul of the everlasting Chinese nation".[1] However, the traditional communication mode of positive Chinese traditional culture is facing challenges and limitations in colleges and universities in the new media era. The development of a comprehensive new media platform is an inevitable trend to strengthen the dissemination of positive Chinese traditional culture in colleges and universities. The coexistence and collision of traditional media and new media has brought opportunities and challenges to the spread of positive Chinese traditional culture. To attract young audiences, mainly college students, it is necessary to expand the effective communication paths of positive Chinese traditional culture in colleges and universities in the new media era by means of content transformation, as well as the online and offline combination from the aspects of communication mode, communication form and communication content.
[Key words]: New media; Dissemination; Colleges and universities; Positive Chinese traditional culture

1 Introduction

Nowadays, people have the most convenient way to get and exchange information. With the rapid development of science and technology and the wide application of new media, the popular media form has transited from traditional media to new media relying on the social network. The Internet is highly sought after by college students with its instant communication mode and rich information resources.

In this context, it is crucial for positive Chinese traditional culture to break through the limitations, embrace the Internet, and to take advantage of the Internet. More specifically, the Internet allows educators to integrate the ideological and political education in colleges and universities with positive Chinese traditional culture and give a full play to the unique value and function of traditional culture in "building morality and cultivating people" in colleges and universities.

This paper will focus on the significance of strengthening the dissemination of positive Chinese traditional culture in colleges in new media era, explaining the opportunities and challenges, and trying to find the communication strategies for positive Chinese traditional culture to attract college students.

2 The significance of strengthening the dissemination of positive Chinese traditional culture in colleges and universities in the new media era

Inheriting the positive Chinese traditional culture, carrying forward and cultivating national spirit, and improving the comprehensive quality of contemporary college students are the best starting point for college students' cultural quality education. Meanwhile, the inheritance of traditional cultural education and strengthening the cultural quality-oriented education of college students is the call of the times. In new media era, college students spend time on internet and social media more than traditional media. Therefore, colleges and universities should pay more attention on expanding the ways of inheriting positive Chinese traditional culture.

The reasons that colleges and universities need to strengthen the dissemination of positive Chinese traditional culture are the need to cultivate the cultural self-confidence of contemporary college students, the need for effective inheritance of positive Chinese traditional culture in colleges, and the need of cultivating qualified college students by utilizing positive Chinese traditional culture.

* Corresponding author: kli10@kent.edu

https://doi.org/10.1051/shsconf/202315502009

2.1 The need to cultivate the cultural self-confidence of contemporary college students

Establishing cultural self-confidence is a very important part of the construction of socialism with Chinese characteristics in the new era. Positive Chinese traditional culture can provide rich spiritual nourishment and theoretical basis for strengthening cultural self-confidence. General Secretary Xi Jinping pointed out: "Cultural self-confidence is a more basic, broader, and deeper self-confidence. The positive Chinese traditional culture bred in the development of civilization for more than 5,000 years, the revolutionary culture bred in the great struggle between the Party and the people as well as the advanced socialist culture accumulate the deepest spiritual pursuit of the Chinese nation and represent the spiritual symbol of the Chinese nation."[2]

Culture rejuvenates the country, and culture strengthens the nation.[3] Positive Chinese traditional culture is an excellent and empowering culture created by the Chinese nation that was inherited and precipitated for 5,000 years. It abandons the dross in Chinese traditional culture and retains the part that has a positive effect on the development of contemporary society. Contemporary college students are the main force in building a socialist cultural power and will also become the backbone of national development. The cultural self-confidence of college students in the new era inevitably needs the spiritual support of Positive Chinese traditional culture. College students carry the historical responsibility of carrying forward Chinese culture in the new era. It is also an inevitable requirement for strengthening China's cultural soft power and building a socialist cultural power to comprehensively enhance college students' positive Chinese traditional cultural literacy and cultivate their cultural self-confidence.

2.2 The need for effective inheritance of positive Chinese traditional culture in colleges and universities

The inheritance, development, and revival of positive Chinese traditional culture in the new era cannot be separated from the network communication. Compared with traditional media, new media is an evolvement of the traditional media, with a multi-dimensional approach of media integration, which not only conforms the principles of cultural communication but can also adapt to the development trend of new media. Media products use a variety of ways and means to present diverse images with strong platform adaptability and flexible content dissemination methods. It can not only give full play to the original functions of positive Chinese traditional culture and retain the advantages of traditional communication methods, but also better convey the mainstream values of society through a variety of platforms and communication methods.[4]

At the same time, the use of new media as the media of positive Chinese traditional culture can also better meet the needs of young people in colleges and universities, so that the positive Chinese traditional culture can hold a dominant position in terms of communication, influence, and competitiveness. The positive inheritance and development of positive Chinese traditional culture in colleges and universities can enhance the cultural consciousness of college students and help to establish "four self-confidences". General Secretary Xi Jinping pointed out, "Where would Chinese characteristics be if not for the 5000-year history? Without those Chinese characteristics, how can we have such a successful socialist road with Chinese characteristics today?"[5] The profound cultural accumulation of China's positive traditional culture is the unique cultural soft power of the Chinese nation and a cultural symbol that differs us from other civilizations. It has become an important mission of ideological and political education in colleges and universities in the new era to help college students establish cultural self-confidence. It is self-evident that it is important to strengthen the communication of new media so that college teachers and students can have channels and platforms to better understand positive traditional Chinese culture.

2.3 The need of cultivating qualified college students by utilizing positive Chinese traditional culture

In order to cultivate qualified college students, in addition to teaching professional skills to meet the needs of socialist construction, colleges and universities should help college students establish a broad outlook of the world, life, and values, as well as cultivate their patriotism and noble moral sentiments, so as to guide their behaviours. These are inseparable from the study of positive Chinese culture.

"The way of Great Learning lies in the enlightenment of brilliant virtues, the remoulding of people, and the pursuit of ultimate goodness." This famous saying comes from the Confucian classic Great Learning, which is listed as the "top of the four books". It was quoted by General Secretary Xi Jinping in his speech at a symposium for teachers and students at Peking University. General Secretary Xi Jinping also said, "The core value, in fact, is a kind of virtue, which is not only personal virtue, but also a great virtue, that is, the virtue of the country and society. Without virtue, a country cannot thrive, and people cannot stand in society. If a nation and a country do not have common core values, its people will be indifferent and have no sense of belonging, then the nation and the country will not move forward. Such a situation is common in our history and in the world today."[6]

Thus, positive Chinese traditional culture can lead the way for the growth and development of college students. At present, some college students pursue comfort, enjoy what they already have, and tend to go with the flow. The most fundamental reason is the lack of lofty ideals. Confucianism takes "benevolence" as the core and puts morality as the priority of life, while "ritual" is the external code of behavior. By studying positive traditional culture, contemporary college students can continuously improve their cultural taste and accomplishment, cultivate the excellent quality of modesty, prudence, respect for others and pay attention to teamwork spirit, which is crucial to the improvement of their personal ideological realm that allows them to become qualified college students in the process of socialist modernization.

3 The challenges of spreading positive Chinese traditional culture in the communication of colleges and universities in the new media era

In the new media era, colleges and universities will inevitably encounter problems such as lagging technical means, monotonous form, and boring content in the process of spreading positive Chinese traditional culture, which makes it difficult to attract the active participation of young audience. Therefore, it is necessary to conduct an in-depth analysis on the current media platform, communication content, communication methods and audience status.

3.1 The problem of communication platform

For college students, they have a large demand for new media information as they use new media platforms more frequently. According to the 48th Statistical Report on the Development of China's Internet, as of June 2021, the number of online education users in China has reached 325 million. Students are one of the larger groups of Internet users in China, accounting for 32.1%. At the same time, the Internet penetration rate among people with high school degrees and above has reached a high level.

However, the current spread of positive Chinese traditional culture relies too much on traditional media. The traditional cultural programs with high popularity in recent years, such as the Chinese Poetry Conference and its derivative programs, or other programs related to positive Chinese traditional culture, such as The Reader, Words Like Face, and National Treasures, are almost all based on television platforms. Meanwhile, it should also be noted that although traditional media such as TV and newspapers have done a lot of publicity for these programs, it is often not traditional media but new media platforms that make these programs popular. For such well-known traditional Chinese cultural programs as China Poetry Conference, the Reader, and National Treasure, it is not the complete episodes of the program itself that has been widely spread and highly popular among the audiences, but the clips. For example, the "Flying Flower Order" link from the "China Poetry Conference" was made into a short video and quickly spread on the new media platform. The Weibo forwarding volume of relevant videos can reach tens of thousands, and the data of likes, collections, and comments on short video platforms such as Tiktok and Kwai are also high. Thanks to the publicity effect of the new media platform-based fine-cut video, these positive Chinese traditional culture-related programs based on traditional media have gained more and more attention, realizing the flow of audience groups from the new media platform to the traditional media, and injecting new vitality into the traditional media audiences that are gradually losing in the new media era.

It can be seen from the examples of TV programs that if we want to spread positive Chinese traditional culture in colleges and universities in the new media era, we should first make the content amiable, trustworthy, and usable. However, under the circumstance that college students use new media platforms more than traditional media platforms, what should be broken is the limitation of the platform. Only by allowing college students to have access to the positive traditional Chinese culture on a familiar platform can these cultures appear "friendly" and "trustworthy" enough to learn and "use".

3.2 The problem of communication content

In the new media era, the creation form, communication mode and even discourse system of publicity are gradually changing. However, many colleges and universities do not have a correct understanding of the differences between new media platforms and traditional media. Most colleges and universities are continuing the propaganda pattern centred on traditional media, and the inherent thinking of "making propaganda by propaganda", which is too written to adapt to the new media platform, has not changed significantly. For college students in the new era, the communication content of Positive Chinese traditional culture should be adjusted accordingly.

At the same time, with the popularization of Chinese traditional culture in the society, a variety of mixed traditional values are presented in front of college students. The amount of information in the network environment is huge and very complex so it is easy for the audience to get lost. Especially for young people who are not mature enough to distinguish whether the information is true or not in a short time, they are easily confused. The diversified presentation of network information leads to unpredictable communication effect, which then results in misunderstanding in positive Chinese traditional culture. Therefore, we should think about how to highlight the correct value orientation in the complex network information and guide the audience to correctly interpret the excellent traditional Chinese culture.

3.3 The problem of communication audiences

The communication base of positive Chinese traditional culture has always been the traditional media, mainly television, radio, and newspapers. Among the three, television has the most extensive contact with young people in the new era. However, we should also be aware that in China, among the most influential TV media audiences, there is an inevitable age gap.

According to the latest analysis of smart TV users in China in 2021, the group aged 30 to 39 is the main force of TV users, accounting for 66%, while users aged 20 to 29 account for 25% and users aged 19 and below account for less than 5%. [Latest Analysis Report on China's Smart TV Users in 2021] First of all, it should be noted that primary school and preschool children in general families watch TV much more frequently than junior high school and high school students. Under the pressure of entering a higher school, parents start limiting their children's entertainment time, and the first one to be affected is the TV platform with animation as the main contact in childhood. At university stage, most students choose to live on campus with no cable TV access. When they are in close contact with the TV media again, they come to the adulthood who need to work. Therefore, in the youth stage, most student groups have limited access to traditional media, and positive Chinese traditional cultural programs rooted in traditional media do not have a good communication effect on them.

As a result, some college students have a weak foundation of traditional cultural knowledge. Their cultural heritage is slightly insufficient, and they cannot effectively understand the essence of traditional culture. Thus, they are indifferent to traditional cultural knowledge. Setting up courses related to traditional culture in colleges and universities is certainly a solution, but students sometimes have a strong sense of utilitarianism. Traditional culture courses are easy to become "easy as" that "improve the average score" because of their, thus failing to achieve the desired educational effect.

In addition, college students are still forming their outlook to world, life, and value. However, with the development of science and technology, the popularity of the Internet and the wide use of new media platforms around the world, multiculturalism from all over the world has been brought to college students. In the new media era, the diversity of communication methods, the complexity of information content, and the freedom of the audience make the information on the network mixed. On the Internet, college students are very vulnerable to the influence of Western pluralistic ideas. Extreme Western values such as hedonism, refined egoism, and extreme individualism also deeply endanger the physical and mental health of some college students.

4 Effective communication strategies of positive Chinese traditional culture in colleges and universities in the new media era

To expand the effective communication path of positive Chinese traditional culture in colleges and universities in the new media era, we can think from three aspects: the content transformation of positive Chinese traditional culture in the new media platform, the communication form of positive Chinese traditional culture in colleges and universities, and the combination of online and offline communication of new media in colleges and universities.

4.1 Targeted and creative transformation of positive Chinese traditional cultural content based on new media

First of all, the connotation of positive Chinese traditional culture should be deeply explored. In the new media, if a column or topic wants to stand out from the mass information and attract the attention of the audience, especially the young college students, it must have its personalized innovative content. Therefore, it is far from enough to confine the content to the widely known shallow culture. Instead, it is necessary to excavate deep level content, either go into the blind area of the audience's knowledge to make the audience feel refreshed or show familiar content in a new form to attract the interests of audience.[8]

Secondly, adaptive changes should be made. For college students in the new era, the content of the communication of Positive Chinese traditional culture should be adjusted accordingly. On the one hand, we should examine its contemporary value with modern ideas, discard the negative factors and negative energy in traditional culture, and express and spread it in a language close to the reality of modern life. On the other hand, we should take its essence, present it in a language habit close to young people, combine it with the trend of the times, echo with the current hotspots, and make homeopathic science popularization in the aspects that are interesting to and thirst for college students.

4.2 Divergent thinking on communication platform

The development of the Internet and the emergence of new media put forward urgent requirements for the innovation of traditional media and provided a broader platform and more diverse ways for the dissemination of positive Chinese traditional culture. Compared with traditional media such as newspapers and TV, new media do not have the limitation of space and time, so they can disseminate a large amount of information regarding positive Chinese traditional culture, and delete, modify, and add to it at any time. New media communication based on the Internet has broken through the time and space limitations of traditional media and enhanced the interactivity. The media integration of network communication is to combine the collection, editing and production of newspapers, television stations, radio stations and Internet websites, and make use of their similarity, difference, and complementary characteristics to share resources of information, centralize processing, and then disseminate to the audience through different platforms. There are three main forms of communication of new media platforms for college students as follows.

• Text and picture communication: Currently, the text and picture communication by WeChat and microblog is the most widely accepted communication mode for college students and teachers, and it is also the communication method closest to traditional media. The positive Chinese traditional culture is spread by means of words and pictures, which can provide the audience with fragmented and vivid communication effects.

• Audio communication: Audio platforms such as Himalaya and Netease Cloud Radio make the communication of positive Chinese traditional culture more convenient. The audiences of audio platforms are not the mainstream, but they have gathered a large number of young people. Through their favorite audio forms, such as audio books and radio dramas, positive Chinese traditional culture can gradually penetrate young audiences' daily lives.

• Video communication: Such platforms include, Tiktok, Kwai, Xiaohongshu and other short video platforms. Short and concise videos are increasingly sought after by college students. Although a single video is not long, the short video platform has a kind of "addiction" attraction, which makes it difficult for the audience to close out the app after opening it, so the amount of information consumed on such platforms is very large. Such a way of communication can make the communication of positive Chinese traditional culture more vivid, and can attract young audiences, mainly college students.

4.3 The Diversified Communication Strategy of positive Chinese Traditional Culture for College Students as Audiences

4.3.1 Strengthen the construction of campus culture with the theme of promoting positive traditional culture.

By building a new media practice platform, we could hold campus cultural activities and new media online activities with the theme of promoting the positive traditional Chinese culture, and take the content produced by students themselves as the main body of communication to spread to a wider and deeper audience of students. Campus cultural activities are an important carrier of "educating people with culture". The "Ancient Poetry Conference", "Traditional Calligraphy and Painting Competition", "Han Costume Exhibition" and other activities held in campus activities of colleges and universities can stimulate enthusiasm of students to carry forward positive traditional Chinese culture through online and offline participation, new media content compilation, offline display, online selection, and other methods. Student clubs and extra-curricular activities can also involve students in performing traditional Chinese cultural stories through short plays, musicals, and other performances. On the one hand, these performances can be offline practical activities; on the other hand, they can also generate new media materials such as texts, pictures, videos, which can be uploaded to relevant new media platforms in the school after editing, such as the WeChat official account of school, official Weibo account or other short video platforms, so that students can feel the inheritance of traditional culture firsthand. When this mechanism of "audience produced content" is gradually established on campus, it actually integrates the identity of the sender and receiver of the communication contents. When users of the new media platform are also the main content providers of the platform, it can not only greatly improve the audience's interest in information, but also help the audience expand the breadth and depth of relevant information, thus strengthening the communication effect of positive Chinese traditional culture in colleges and universities.

4.3.2 Big data can be used to effectively and accurately understand and research college students.

On the one hand, understanding the needs of new media users of college students allows us to produce content of positive Chinese traditional cultural communication that is more in line with user preferences; on the other hand, understanding college students' new media reading habits and reading preferences, and pushing content according to their preferences can achieve point-to-point arrival of new media content, and complete personalized and decentralized dissemination of content. However, at the same time, to avoid the information cocoon effect, quality contents should be widely disseminated and pushed, so that potential audiences can have the opportunity to access relevant information and turn into interested audiences.

5 Conclusion

Positive Chinese traditional culture contains rich human civilization and national wisdom, and its strong vitality makes it persistent and continuous, while possessing great contemporary value and advancing with the times. Learning positive Chinese traditional culture and promoting its dissemination in colleges and universities help establish the cultural self-confidence of college students and have far-reaching significance for cultivating qualified college students in the process of socialist modernization.

Nowadays, the world is experiencing the fourth communication revolution in human history brought about by the development of computer network technology, and the way of information communication in colleges and universities is also facing innovation. If positive traditional Chinese culture wants to keep pace with the information age and go deep into the campus of colleges and universities, it is necessary to rely on the new media platform with a high utilization rate of young media audience, improve the inherent thinking of "making propaganda by propaganda", and innovate the communication contents according to the audience needs of college students and the communication characteristics of new media. We should innovate the content and create vivid and novel works related to positive Chinese traditional culture according to different types of new media platforms, such as graphic science and popularization, short video mini-theater, radio drama audiobooks and so on. At the same time, we should make full use of the advantages of online media and offline activities in colleges and universities to fully interact, strengthen the construction of campus culture, use big data to make relevant information more accurate, and finally make the positive Chinese traditional culture enter the daily activities of college students in and out of class, so that the spread of positive Chinese traditional culture can be landed on the online and offline platforms of colleges and universities.

References

1. Xi J.P: Speech at the 15th Anniversary of Macao's Return to the Motherland and the Inauguration Ceremony of the Fourth Government of the Macao Special Administrative Region (December 20, 2014) People's Daily (December 21, 2014)
2. Xi J.P: Speech at the Celebration of the 95th Anniversary of the Founding of the Communist Party of China (July 1, 2016) People's Daily (July 2, 2016)
3. Xi J.P: Report at the 19th National Congress of the Communist Party of China (October 18, 2017)
4. Xu Q.Z., Li K.X.: On the Spread of Positive Chinese Traditional Culture in New Mainstream Media Journal of Shandong Administration College (October 2020)
5. Xi J.P: Xi Jinping pointed out during his inspection in Wuyishan City, Fujian Province (March 22, 2021)
6. Xi J.P: Speech at the Symposium of Teachers and Students of Peking University (May 2, 2018) People's Daily (May 3, 2018)
7. The Latest Analysis Report of China's Smart TV Users in 2021
8. Li J.Y., Li F., Xu H., Li F.X.: Creative Transformation of Positive Chinese Traditional Culture in the View of New Media -- Taking National Treasure as an Example, Frontier Economy and Culture (January 2022)

A Study on the Success of Short Videos Themed on Beijing Cuisines and Its Influence on the City's Image

Yixuan Li*, a

Agricultural University of Hebei, 071001, Baoding, Hebei province, China

Abstract. With the boom of the short video industry in recent years, cities with tremendous internet influences have witnessed rapid growth attributable to the rise of short video platforms where cuisine-themed short videos have become a new channel for the communication of a city's image. On one of the platforms, TikTok, there exists a huge gap in terms of communication performance among cuisine-themed short videos of different kinds. Taking content analysis as the tool, this paper is engaged with the quantitative analysis of the content characteristics, food types, city image, video length and language characteristics of the short videos themed on Beijing cuisines that go viral on the Internet with the coded information extracted from the top 30 related short videos with most likes published on TikTok, so as to investigate their commonalities and shared features. An in-depth analysis is provided as the result with specific cases taken into account to propose for the production and communication of similar videos and to provide certain references for the construction and communication of the image of corresponding cities.

1 Introduction

1.1. Research Background

According to the latest statistics released by China Internet Information Center, the user volume of short video products in China has reached 937 million, which can be treated as a sign that mobile short video apps have become people's common access to entertainment.

Driven by their low barrier of entry, strong universality, diversified user profile and innovation means, and a good traffic-making capacity, cuisine-themed short videos have become the first choice of massive practitioners of self-media, as the theme-specific total play of the term "cuisine" has amounted to 639.97 billion when searched with themed hashtags on TikTok.

With the boom of cuisine-themed short videos, there emerged a variety of subcategories such as cooking tutorial, restaurant hunting, food hunting, food live-streaming, rural cuisines, etc, which all attach great importance to the demonstration of cuisines' charm to satisfy people's covet for delicacies with wonderful visual experience and to the combination of food cooking, entertainment, creativity, and many other elements to pioneer a new horizon for cuisine-themed short videos.

1.2. Literature Review

Firstly, regarding the study of city image, in the article authored by Li Guangbin, Wang Yong and Yuan Zhongjin [1] published in 2006, the connotation of city image was mentioned as such, "as an important business field of city government, the objective of city image is to promote the city to the market and attract more external resources through a well-shaped city image, so as to boost the city's economic development and recruitment, expand the city's financial resources, and sharpen the supply capacity of "public goods" of the city".

City image is the external artistic expression of the inner characteristics of a city and a concentrated reflection of the quality, taste and culture of the city as a whole. Later in 2009, Chen Ying proposed a comprehensive conceptualization of city as a complex entity by taking into account both domestic and international research on the concept of city image, that is, "city image refers to the specific perception, overall view and comprehensive evaluation of the general public on the implicit integrated strength, explicit vitality and future prospect of a city, which is a manifestation of the general characteristics and features of the city." [2]

In He Guoping's article, there provides a more concise definition of city image, that is, "city image is people's subjective view and perception of the city as well as the visualized image or mirror image formed thereby, which consists of three hierarchies: mental image (beliefs, ideas, etc.), behavioral image and visual representation (image and identification system, etc.)." [3]

Secondly, in terms of food-themed short videos, a search was conducted in the database of CNKI with "food-themed short videos" as the keyword, and judging from the current state of art, there remains a scarcity of research findings on food-themed short videos in China,

* Corresponding author: 2436508569@qq.com
a Corresponding author: a19931256177@163.com

and the most of which are contributed by journal papers but not yet academic monographs, with the relevant research fields mainly focusing on short video research and food communication research separately.

Since 2015 when "food live-streaming" was introduced into China and flourished rapidly, many domestic scholars have conducted studies on this subject. According to Cao Xiran (2017), there are several reasons for the popularity of " food live-streaming": the first is its low barrier of entry which provides the general public easy access to it, the second is the diversity of content that can meet the varied needs of different audiences, and finally it is the capability of satisfying the audience's appetite in an indirect manner and serving as a companion when they feel lonely36.

And with the passage of time, cuisine-themed short videos have also been expanded and innovated into a variety of sub-categories such as restaurant hunting, popular amusement and so on.

2 METHOD

2.1. Research Contents

By searching with the hashtags "Beijing" and "cuisine" on TikTok, this paper gathers the top 30 cuisine-themed short videos with the most likes as the research objects. And taking the research results as an entry, solutions are proposed for the possible problems that may emerge in the development course of cuisine-themed short videos, so as to provide certain references for the future growth of cuisine-themed short videos and the short video industry as a whole. Finally, based on the research data, relevant suggestions and strategies are proposed for the construction and communication of Beijing's city image from the perspective of cuisine-themed short videos.

2.2. Research Methodology

Content analysis method is intended to be adopted in this thesis. A quantitative analysis is proposed to be made to provide a quantitative and accurate description of the communication content of short videos themed on Beijing cuisines published on TikTok, which will be sampled and taken into the construction of a category system based on the elements presented in the videos (content characteristics, food types, city image, video length and language characteristics), where they will be further coded and exported into valid data, with which the communication performance of these videos will be analyzed and summarized.

2.3. Research Design

2.3.1. Research Methods and Sampling Methods

This paper adopts the content analysis method and selects 30 short videos themed on Beijing cuisines published on the short video platform "TikTok" as the research object through "simple random sampling". In the same vein, content analysis method will be adopted in this study to screen out 30 cuisine-themed short videos out of all the videos collected from TikTok with different posting times, number of likes, individuals and video lengths, which will be further coded for subsequent analysis. A simple random sampling will be conducted with the random number table by searching with the keyword "Beijing cuisine" to collect related short videos published on TikTok between January 2022 and July 2022, with 30 sampled videos selected as the final research objects. A series of multi-dimensional content indicators covering all forms of video information will be introduced in this study, and video information will contain content characteristics, food types, city image, video length and language characteristics.

2.3.2. Coding and Data Presentation

2.3.2.1. Specification of Multi-dimensional Indicators

a. content characteristics: relaxing, funny, popularization

b. food types: traditional Beijing snacks, Beijing specialties, foreign cuisines, other regional cuisines

c. city image: with rich cultural heritage, modernity, diversity

d. video length: within 1 min, above 1 min

e. language characteristics: Beijing dialect, Mandarin

Table 1. Multidimensional Information Coding Table

	Relationship indicators (code serial number)
content characteristics	relaxing0, funny1, popularization2
food types	traditional Beijing snacks0, Beijing specialties1, foreign cuisines2, other regional cuisines3
city image	with rich cultural heritage0, modernity1, diversity2
video length	within 1 min0, above 1 min1
language characteristics	Beijing dialect0, Mandarin1

Form Source: Self Drawn

Table 2. Table of Coded Video Information

	Relationship indicators (code serial number)	Number of videos (N=30)	Effective percentage
content characteristics	relaxing0, funny1, popularization2	relaxing14, funny2, popularization 14	relaxing 46.67%, funny 6.66%, popularization 46.67%
food types	traditional Beijing snacks0, Beijing specialties1, foreign cuisines2, other regional	traditional Beijing snacks16, Beijing specialties8, foreign cuisines4, other regional cuisines2	traditional Beijing snacks53.33%, Beijing specialties26.67 %, foreign cuisines13.33%, other regional cuisines6.67%

	cuisines3		
city image	with rich cultural heritage0, modernity1, diversity2	with rich cultural heritage11, modernity4, diversity15	with rich cultural heritage36.67%, modernity13.33 %, diversity50%
video length	within 1 min0, above 1 min1	within 1 min 12, above 1 min 18	within 1 min 40%, above 1 min 60%
language characteristics	Beijing dialect0, Mandarin1	Beijing dialect5, Mandarin25	Beijing dialect16.67%, Mandarin86.33 %

Form Source: Self Drawn

3 ANALYSIS AND DISCUSSION

3.1. Data Analysis

The above coding statistics indicate that the content of short videos themed on Beijing cuisines is characterized by a light-hearted atmosphere and a focus on the popularization of Beijing cuisine. The city's image embodied in these short videos is mainly linked to the cultural diversity and rich cultural heritage of Beijing. The descriptions of food in these videos mainly concentrate on the introduction of traditional Beijing snacks and Beijing's specialties. In terms of video length, 40% of the short food videos are limited within one minute, while 60% others are longer than one minute. Meanwhile, some videos are dubbed in Beijing dialect to enhance the attractiveness of the videos and help viewers understand Beijing culture better.

3.2. Characteristics of Successful Short Videos Themed on Beijing Cuisines

3.2.1. With a Short and Flexible Length

According to the analysis of the data from the 30 selected videos, the average length is around 1.5 minutes, which can well retain the attractiveness to users until the end of the video, excluding poor experience from inappropriate length. A length less than 5 minutes but with strong flexibility is the first characteristics of successful short videos themed on Beijing cuisines.

3.2.2. With Diversified Content and Unique Characteristics of Beijing

All 30 videos are fully representative in promoting Beijing's local characteristics and international diversity. The percentage of "short videos themed on finding special cuisines" (traditional Beijing cuisines) exceeds 50%, which is a phenomenon unique to this kind of short videos. It presents the distinctive Beijing characteristics in full Beijing style.

3.2.3. Synergy of Multiple Elements in Enhancing the Visual Experience

Among the selected videos, there is a re-occurrence of Beijing dialect in sound and Beijing's hutongs and courtyards in sight, which is relaxing and refreshing, thus enhancing the viewing experience of the videos, facilitating the expression of emotions and reflecting the image of Beijing as the ancient cultural capital of China.

3.3. Communication Problems and Governance of Beijing's City Image in Cuisine-themed Short Videos

3.3.1. Problems in Shaping and Communicating Beijing's City Image as a Center of Cuisines

The video content presents a serious homogenization and a one-sided image of the city. Through the analysis of the 30 short videos themed on Beijing cuisines, a high degree of homogeneity is found between the Beijing specialties introduced and the commercial spots featured in the videos, and there is a tendency of content imitating, resulting in a large body of short videos with similar video styles, similar content themes, and even similar soundtrack and copy.

3.3.2. Inadequate Government Involvement

Among the 30 highly acclaimed short videos themed on Beijing cuisines extracted from TikTok, the production and publishing subjects only involve personal accounts and TikTok KOLs, with no government accounts or local media accounts included. Among the 30 highly acclaimed short videos themed on Beijing cuisines extracted from TikTok, the production and publishing subjects only involve personal accounts and TikTok KOLs, with no government accounts or local media accounts included. The official government accounts failed to give full play to their role in communication given their magnificent says.

3.4. Governance of the Shaping and Communication of Beijing's City Image as a Center of Cuisines

3.4.1. Enhance the quality of content creation and explore the cultural heritage of the city

Video creators should give full play to Beijing's unique resource advantages as a center of cuisines by tapping into the city's culture, improving the quality of creation, innovating the expression and content, and uncovering more cultural symbols of food to shape and communicate the image of the city. Meanwhile, TikTok officials should enhance their supervision efforts, limit the incentive to create homogeneous contents, and support and encourage creations with cultural profoundness to guide the city image short videos along the right direction.

3.4.2. Strengthen government involvement and clarify development orientation

Government video accounts should give full play to their great influence in shaping and communicating the image of the city, transform the outdated traditional concepts, and actively cooperate with major platforms to deliver quality short video content. When with sound planning of new media communication, the interaction with the audience should also be strengthened to fully heed the opinions of the people, respect the communication rules of the new media and clarify the orientation of the development of the city.

4 CONCLUSION

Through the study of short videos themed on Beijing cuisines published on TikTok, this paper analyzed the reasons behind their success as well as their deficiencies in the communication and construction of Beijing's city image, thereby proposing corresponding governance solutions. The following conclusions are drawn from the study: First, in terms of content presentation, successful short videos themed on Beijing cuisines are short in length but with adequate flexibility and unique Beijing characteristics, as well as a strong coordination of all these elements.

Secondly, in terms of the communication of city image, it manifests a series of problems that still need to be addressed, such as serious homogenization of video content, one-sided city image and inadequate government involvement.

Finally, solutions to the existing problems are proposed such as enhancing the quality of content creation, exploring the cultural heritage of the city, strengthening government involvement and clarifying development orientation.

References

1. He Guoping.(2010) City Image Communication:Framework and Strategy. J. Modern Communication(Journal of Communication University of China), 08: 13-17.
2. Chen Ying. (2009) Media Construction of City Image: Conceptual Analysis and Theoretical Framework. J. Journalism, 05:103-104+118.
3. Zhang Yi, Zhang Rui. (2006) and Theory. J. Management,04:468-476.
4. Li Guangbin, Wang Yong, Yuan Zhongjin.City Characteristics and Urban Image Building[J].Urban Planning,2006,(02):79-82.
5. Cao Xiran: Loneliness and Carnival: An Analysis of the Phenomenon of China's 'Eating and Broadcasting' from the Perspective of Communication Studies [J], Communication and Copyright,2017, (02),102-103.
6. Huang Lili. The construction of the symbolic meaning of short food videos in the context of consumer culture: Taking "Li Ziqi" as an example [J]. New Media Research,2019,5(07):117-119.
7. Yang Liming, Zhou Wenyi. Study on Guangzhou City Image Construction and Communication Strategy in Short Food Videos [J]. City Observation,2022,(01):116-128+163.
8. Zhou Xuyang. Research on the shaping and dissemination of the image of Northeast urban agglomeration by Douyin short video [D]. Mentor: Chi Qiang. Liaoning University,2021.
9. Liu Qianjun. Research on the Shaping and Dissemination of Xi'an City Image in Douyin Short Videos [D]. Mentor: Tang Jiageng. Yunnan University of Finance and Economics,2021.
10. Bi Ran. Research on the production characteristics and audience effect of short food video content at station B [D]. Southwest University of Political Science and Law,2020.

A Study of the Language of Bullet Comments in Bilibili Auto-tune Remix-themed Content Videos

Pengyun Li*

School of Cultural Industries & Management, Community University of Shanxi, Jinzhong, Shanxi, 030619, China

Abstract. As a new and popular medium, bullet comments are popular among young people, and the presence of bullet comments is particularly indispensable in bilibili auto-tune remix-themed content video. The pop-ups themselves have not been a one-day success, but there are many political, economic and social and cultural influences, and the culture of the bullet comments, which originated in Japan, has fused with the culture of Chinese subtitling groups in China, making our bullet comment culture more locally. In fact, many of the internet buzzwords come from online pop-ups, and the auto-tune remix-themed content area, as a gathering place for very talented uploaders, is naturally not short of inspired viewers, and the classic bullet comments from them are gradually becoming visible in many mainstream cultures, and are also used by many influential celebrities or a significant portion of the public, causing a wider spread.From these cultural phenomena, it is easy to see that the producers of auto-tune remix-themed bullet comments have a strong secondary production capacity for specific cultures and a vigorous vitality for cultural products. Therefore, the study of the language of auto-tune remix-themed bullet comments has great significance and reference value for the diversification of cultural industries in the era of online culture and new media.This paper uses literature research, participant observation and inductive analysis to analyse the characteristics and significance of the language of bullet comments in bilibili auto-tune remix-themed content video.

1 INTRODUCTION

First of all, let's talk about bullet comment video sites and the birth of pop-ups. niconico Video began offering an experimental video service on December 12, 2006, after a period of private operation that was not open to the public. At the time, niconico video was the first video site in the world to have a "bullet comment screen" feature, and it became the first video site in the world to have a "bullet comment screen" feature. niconico video became the world's first video site with a "bullet comment" function and has grown to this day. Although the niconico video business has expanded with the commercialisation of its content, the bullet comment video business has always been its core business.

In 2007, a group of otaku culture enthusiasts in China, inspired by the interesting form of video commentary on niconico videos, began experimenting with it on their own site, "Fat Pig", and called this form of commentary "bullet comments". "In June 2007, AcFun was established, which can be said to be the originator of the Chinese bullet comment video website in China, and also one of the earliest bullet comment video websites established in the world. In March 2008, AcFun also began to draw on this form of video commentary, almost as a "transplant" or "Chineseization" of a large number of submissions from niconico video sites. AcFun has also begun to borrow this form of video commentary, almost by "transplanting" or "Chineseizing" a large number of submissions, phrases and other cultural artefacts from the niconico video site to AcFun, under the explicit and direct banner of a bullet comment video site. After Fat Pig shut down its station due to unspecified problems, AcFun's trend was a roaring success, becoming an influential local video site within China's otaku culture circle, and developing to this day, AcFun is arguably the earliest surviving bullet comment video site in China.

Again, the development. As a personal website, AcFun was often unstable in its early days due to problems such as server attacks, and in serious cases was even forced to shut down temporarily for repairs. After AcFun resumed operations, mikufans did not disappear completely. On 24 January 2010, mikufans was renamed bilibili, diverting some of AcFun's core users and other regular users who had become accustomed to bilibili as the environment on AcFun deteriorated. After the deterioration of AcFun's internal environment and the diversion of some of AcFun's core users and others who were already used to bilibili, bilibili developed as an independent bullet comment video site. During this period, several similar sites such as 2dland and Phantom World emerged, while bilibili gradually gained a foothold in China's otaku culture circle with its good user experience, rich in-site resources, and a number of popular UP owners2, splitting the world with AcFun. At the same time, other types of otaku-oriented websites have also begun to borrow this concept to launch "bullet

* Corresponding author: lpy050906@163.com

comment" interactive features, such as the trolling function of the online original comic site "Youbi".

Finally, there is expansion. In recent years, bilbili has overtaken AcFun as the number one otaku-oriented bullet comment video site in China, both in terms of visits and user activity. 126th. In the last six months, bilibili has been far ahead of AcFun in terms of average daily visits per million people and daily page views per million people. bilibili was founded in 2009 and has gradually developed into a fully-fledged bullet comment video site that is not to be underestimated today in every respect.

Bilibili was chosen for this study for several reasons: First, Bilibili is currently the largest bullet comment video site in China, and has even become the industry leader in bullet comment video in China. Secondly, Bilibili has a relatively comprehensive and mature profit model, including the purchase of licensed new releases and game co-op. Thirdly, Bilibii is recognised as one of the more successful bullet comment video platforms in terms of popularisation and third-genisation, which of course has also been a source of criticism and controversy. Overall, Bilibili seems to be basically the best choice for our research on bullet comment video platforms and spaces.

In the current context, the term "bilibili auto-tune remix-themed content" is not a descriptive term for extremely derogatory behaviour, but more of a brainwashing treatment of some curiosities and niche cultures. A bilibili auto-tune remix-themed content video is a new collection of video footage and audio material that is edited to create a specific rhythm. Bullet comments are textual content that appears on the video screen when watching a film, independent of the text of the video itself, and are a form of text often with commentary or trolling.

Table 1. Comparison among niconico, AcFun and bilibili

	Name	Founded	Trait
Japan	niconico	2006	In terms of format, it is the originator of the pop-up video; in terms of content, it promotes the ACG culture that is part of the Japanese Honshu culture
China	AcFun	2008	Relatively few resources, heat, good content and users users are gradually becoming less and less
	bilibili	2009	A lot of resources, each division has quality video content output; the website establishes a good image and culture

2 CLASSIFICATION OF AUTO-TUNE REMIX-THEMED CONTENT VEDIO BULLET COMMENTS

According to the broad categories, they can be divided into form and content. In terms of form, they can be divided into scrolling curtains, fixed curtains and advanced curtains. Scrolling bullet comments are the most common cluster of bullet comments, often white and kept at a certain word count, floating across the screen at a fixed speed. Fixed bullet comments are fixed in the centre of the video, either at the top or bottom or anywhere in the middle, and can be used for emphasis, with the colour chosen according to the nature of the content being highlighted. Advanced bullet comments are a special form of bilibili, which can be freely set in terms of position, size, colour, tilt angle and so on, often with a more "screen dominating" effect.

In terms of content, they are broadly divided into the following categories: 1. deconstruction and construction of the video itself; 2. deconstruction (e.g. misheard lyrics bullet comments) and construction of the video's textual content (e.g. upscaled bullet comments); 3. explanation and popularisation of current events and slots; 4. trolling; and 5. screen-painting (no real meaning,For example, some of the posters will act as "counters" in the bullet comments section, so to speak, "not listening to what's going on outside the window, but only counting the sages", which is sometimes very interesting).

3 ANALYSIS OF THE BULLET COMMENTS VEDIO UNDER THE THEORY OF "MEDIA QUADRUPLE LAW"

In McLuhan's book "The Laws of Media: The New Science", the "quadruple law of media" has been published and systematically explained, the four key points of which are "1. What does the media make amplified, enhanced or expanded? 4. what changes or reverses into what form when it is squeezed?" These four points were later distilled into "elevation, obsolescence, reversal, reproduction".[1]

3.1 Elevation: bilibili auto-tune remix-themed content bullet comments videos go from hot to cold, with a strong sense of audience engagement

There are two important criteria for distinguishing between hot and cold media: the clarity of the information presented and the level of audience engagement. Cold media has low clarity of information and high audience engagement; hot media has low clarity of information and low audience engagement.[2]

For example, on May 18, 2020, the "Strongest Fahai" released by the Pushing Back Bunny uploader had been watched by more than 20,000 people online at the same time, and up to now the total number of bullet comments posted on the video reached 144,829, while the bullet comments had to let the later ones cover the earlier ones due to the limited capacity.

The bilibili auto-tune remix-themed content video itself conveys what the creator expresses, while the emergence of bullet comments makes the relationship between the video content and the audience closer and closer, and the audience's degree of thinking and desire to evaluate the video is more profound and strong. Thus, it can be seen that bullet comments have improved the efficiency of video information dissemination and changed the audience's viewing behavior and mentality.

3.2 Obsolescence: videos are overturned and reset by bullet comments again and again, each time as something new

Bullet comments give audiences the right to speak, allowing them to turn the original video into their own product at any time and in their own interpretation, as is the case with misheard lyrics bullet comments, which refer to the deliberate hearing of one language as being the same as another (or vaguely the same language). Wordplay of similar words and phrases, rewriting the content in the harmonics of another language for the purpose of spoofing or double entendre.

For example, the original source material of "Ode to Boo Boo" uploaded by the gunshot track uploader on 8 January 2021 was a video of a six-year-old girl performing a Korean children's song on stage, which was widely second-created and second-altered because of its benevolent expressions and movements, and many empty-eared bullet comments were created for the magical Korean language, which drew the audience in.

3.3 Reversal: bullet comments embedded, attention drawn

When the original bullet comments did not appear in the video, people were more focused on the work itself. When they saw content that was "groovy", and when people did not want to interrupt the video to look at the comment section, the bullet comments would hold their attention. As people become more adaptable, they will read more bullet comments and find a new balance and pleasure in watching them.

For example, "Out of the Country" released by AB Fresh uploader on 11 June 2021, using three different countries' explosive music to change one of the hottest music at that time

Honey Snow Ice City, triggered countless bullet comments in the bullet comment section for eager discussion on song titles, song genres and exotic styles.

3.4 Reproduction: the collective viewing model creates a new type of social venue

The combination of the bullet comments and the video has created an impulse for the audience to become familiar with it and to spread it, gradually joining the camp of the bullet comments and turning them from recipients to spreaders. This is a time when watching videos is no longer a single purpose, but rather a world where there are few like-minded people in the real world, but where there are voices everywhere that share the same position.

For example, in the video "[Trump] No one knows how to imitate better than I do" uploaded by the uploader of Gunshot Tracks on May 15, 2020, the child in the video imitates Trump's classic movements and gestures, while making sure to add the most classic "China" to his mouth, which inevitably triggered the collective hilarity of netizens.

4 THE SIGNIFICANCE OF THE BILIBILI AUTO-TUNE REMIX-THEMED CONTENT VEDIO BULLET COMMENTS AS A KIND OF ARTISTIC

4.1 A new way of expressing popular culture

The Internet environment is so diverse that a culture that captures the attention of the youth of today can be considered a "success" in the sense that it easily expresses itself while gaining countless attention, and the deconstruction and construction of videos by bullet comments has become a "favorite" of the contemporary audience. [3]

4.2 The catharsis and mingling of emotions

The audience expresses their own opinions in the bullet comments, and different emotions are intertwined, not only expanding the flow of information, but also forming a kind of emotional carnival mode. The various types of trolling and swiping are a reflection of the emotions they place in the bullet comments grid.When emotions gather to a certain extent, a huge emotional camp is formed.

4.3 Resistance to the hegemonic culture

Not all mainstream cultures today are accepted by niche youths. They try to guard their favourite circles, or the unique "black humour" of bilibili auto-tune remix-themed content, or the popular hot spots and music that are often used as material for bilibili auto-tune remix-themed content, and they are not willing to lose such subcultures.

5 FANS OF AUTO-TUNE REMIX-THEMED CONTENT VIDEO BULLET COMMENTS

Most of the material or immaterial culture is made by "people", so "people", as the terminal of the bullet comments culture, naturally plays a significant role.

5.1 What is meant by "fan groups"?

The term "fandom" is essentially a combination of "fandom" and "community". The term "fan" is derived from what we commonly refer to as "fans", but the term 'community' is close to 'community', except that the term 'community' is often thought of in a geo-administrative sense, while its sociological meaning is rarely thought of. It refers to "a group of people who share the same values and help each other in a humane way". Nowadays, with the huge number of Internet users, virtual communities are often the object of multiple references, and the concept of 'fan groups' - in most cases 'fan' - has emerged. The activities of the role-players within a certain area satisfy the notion of community, and thus the 'fandom' is formed.[4]

5.2 The way in which fanzines operate

De Sédou suggests that, in general, "poachers" plunder texts and interpret them for themselves, recreating texts that are useful to them in the process of the consumption stage of cultural products. Jenkins points out that the second identity of the 'poacher' is in fact that of a 'nomad', who wanders between various 'fan' cultures and re-engages with different cultures, holding different texts hostage. These newly created discourses have a great potential and a vibrant and dynamic appearance.

5.3 The psychological activity of the fans

Identity is divided into self-identity and social identity. The fandom in the bullet comments includes both self-identification and social identification. In virtual community-based online social networking, audiences form a strong sense of psychological identity and empathy by watching others' performances, and thus communicate through bullet comment conversations, which happen to deepen the audience's sense of identity indefinitely.

6 CONCLUSION

This paper uses various methods such as literature research, participant observation and inductive analysis to study and explore the classification methods, posting methods, language forms, expression meanings, as well as the behavioral rules and psychological activities of pop-up posters on the bilibili bullet comments video website.

The paper argues that auto-tune remix-themed content videos, as an important strand of youth subculture, have survived in the relatively well-established and well-regulated video media system of bilibili and continue to exert a deafening influence, creating a postmodern cultural landscape. In this paper, we borrow McLuhan's "quadratic law" theory to analyse the release mechanism of auto-tune remix-themed content video pop-ups, which enrich the texts of auto-tune remix-themed content videos, spread auto-tune remix-themed content culture, and link up with auto-tune remix-themed content videos to build a carnival ritual in the square.

The cultural connotations of auto-tune remix-themed content videos have changed in response to the changing discourse patterns. Auto-tune remix-themed content videos have developed out of the circle and have appeared to be politically and commercially adopted. The fame of good auto-tune remix-themed content videos is bound to mean a rise in the level of bullet comments. A good subculture should be mutually fulfilling in every aspect, and how to inspire more quality bullet comments is not only something that bullet comments publishers should consider, but also something that video uploaders should ponder.

Reference

1. Li L. (2016). McLuhan's "quadruple law of the media": an interpretation and discernment. Northwest Humanities Review (00), 247-250.
2. Shi, R.R. & Zhang, N. (2015). Analysis of pop-up videos under the theory of "quadratic law". Media (07), 75-77.
3. Chen, S.S. (2015). The linguistic characteristics and expressive meaning of pop-up videos. News World (07), 105-106.
4. Yin, F.Y. (2020). Ghost videos as interactive rituals: The carnival spirit of "two-dimensional rooms". Journal of Youth (01), 41-48.
5. Yan, Mei-ting. (2021). The social causes and existing problems of the pop-up subculture: The example of Bilibili pop-up network. Media Forum (06), 129-130.
6. Wang, Hui. (2021). A study on the interaction mechanism of pop-up communities - taking the pop-up video site Bilibili as an example. Audiovisual (07),153-154.doi:10.19395/j.cnki.1674-246x.2021.07.066.
7. Yang Xiaohe. (2018). Analysis of pop-up culture in the subcultural perspective - taking bilibili.com as an example. Popular Literature and Art (10), 163.
8. Yang M.G., Wang L.Y. & Wan H.L.. (2017). An exploration of new forms of audience interaction in the live webcast environment - taking bilibili video pop-up network as an example. Design(21),24-25.
9. Su, Biying. (2016). Analysis and utilization of information dissemination mode of pop-up websites - taking BILIBILI website as an example. Lantai World (14), 83-84. doi:10.16565/j.cnki.1006-7744.2016.14.35.
10. Yan Sisi. (2020).Research on the interactive communication of Bilibili users from the perspective of ACG subculture (Master's thesis, Jiangxi University of Finance and Economics). https://kns.cnki.net/KCMS/detail/detail.aspx?dbname=CMFD202101&filename= 1020976878.nh

A Review on Strategies and Development of the Live Broadcast for Agricultural Products under COVID-19

Kejia Li*

South China Normal University, Guangzhou, Guangdong Province, 510631, China

Abstract. Due to the outbreak of the COVID-19 epidemic in 2019, a large number of agricultural products can not be sold. In this situation, live webcasting is used to help farmers sell agricultural products. Most consumers who cannot go out to buy agricultural products because of the epidemic can experience the same shopping pleasure as offline shopping through live broadcasting. Through the method of literature review, this paper summarizes four modes of live broadcast for agricultural products, including the live broadcast initiated by mainstream media, Internet celebrities, officials, and farmers themselves. At the same time, it summarizes the current existing problems, such as the lack of professional knowledge of live-streamers, high logistics prices, and a low rate of agricultural product purchase. On this basis, the future development prospects of live broadcasting in assisting agriculture are discussed.

1 Introduction

In recent years, with the development of the Internet industry, live broadcasting has gradually become a new way of shopping. By June 2020, the number of Internet users in China had reached 940 million, of which 285 million were in rural areas. Accounting for 30.4% of the total number of Internet users. the number of Internet users in rural areas achieves an increase of 36.25 million compared with March 2020 [1]. At the same time, the agricultural products in many poor areas of China are facing the dilemma of poor sales due to the impact of the COVID-19 epidemic in 2019. In this context, the agricultural products industry and the Internet cooperated to form a new form of selling for agricultural products, that is, live broadcasting for farmers. This measure not only caters to the poverty alleviation policy strongly supported by the Chinese government in recent years, but also innovates the sales mode of agricultural products. In view of this emerging phenomenon, the communication academia has launched a series of studies. It can be roughly divided into three research directions [2]. The first category is the study on the form of agricultural live broadcasting, that is, to explore the existing mode of agricultural live broadcasting from the perspective of tools. The second category is mainly about the research on effects, that is, what new impacts the agricultural live broadcasting will bring to the agricultural product selling industry. The third category mainly studies the problems existing in the agricultural live broadcasting from a critical perspective. This paper reviews the modes, existing problems, and development prospects of agricultural live broadcasting. These summaries will help future scholars to conduct more in-depth research and develop more perspectives on this basis.

2 Current Modes of Agricultural Live Broadcasting

According to Lasswell's 5W communication mode, the type of webcast can be subdivided from different dimensions, such as the identity of the live-streamers, the purpose of selling, and the platform for delivering goods. This paper simply divides the forms of agricultural live broadcasting in terms of the identity of the live-streamers. Since the agricultural live broadcasting can be easily started anytime and anywhere with only a mobile phone and a bracket [3], it attracts a large number of live-streamers with different identity backgrounds to enter this field.

2.1 Live Broadcast for Agricultural Products Initiated by Mainstream Media

The advantage of the mainstream media in the form of agricultural live broadcasting is its strong professional ability and credibility. The combination of professional anchors and Internet celebrities is proved to be popular. For instance, the combination of Zhu Guangquan and Li Jiaqi became a trending topic and the top search on Weibo in real time on the day of their live broadcast [4]. Additionally, the mainstream media have a relatively mature live broadcasting team, which can quickly build a high-quality live broadcasting room. Professional live-streamers show excellent live broadcast skills in introducing products and interacting with audiences. By playing a full part in the live broadcasting platform, they

*Corresponding author: 1402916612@qq.com

can give consumers a strong sense of the scene [5]. At the same time, the strong credibility of the mainstream media can enhance consumers' trust in the agricultural products they buy. Consumers tend to believe that the agricultural products selected in the live broadcast room have relatively strong quality assurance, which also helps improve their purchase intention. However, it is undeniable that the mainstream media themselves do not know about the agricultural knowledge as much as farmers. The introduction of products will also stay firmly in the agricultural products themselves.

2.2 Live Broadcast for Agricultural Products Initiated by Internet Celebrities

The advantage of famous live-streamers lies in their huge fan base and strong interactivity. Agricultural products do not have strong product characteristics, and they have been faced with problems such as few sales channels, insufficient popularity, and insufficient publicity [6]. However, the distinctive live broadcasting style and exaggerated expression of live-streamers can quickly attract the audience's attention, thus improving the product's popularity. For example, Dong Yuhui, a popular anchor, becomes famous because of his distinctive teaching style during the live broadcast. When selling agricultural products, he will connect the products with various literary works, so that the audience can learn knowledge while purchasing agricultural products. What is more, a large number of professional anchors who help farmers do not charge extra fees, and this nature of commonweal can quickly gain the audience's favor and stimulate their desire to buy.

2.3 Live Broadcast for Agricultural Products Initiated by Government Officials

The live broadcast that helps farmers and initiated by government officials can quickly attract audiences thanks to a sense of contrast. On the one side, officials generally appear in formal occasions with a serious image. However, in the live broadcast, in order to make their image more suitable for a live-streamer, they use network language and actions to form a sense of contrast and gain more user recognition. On the other side, officials can use the credibility and authority of the government to promote products and win the trust of consumers [6]. This kind of equal interaction with the public can strengthen the emotional connection between consumers and the live broadcast, which is easier to produce purchase behavior. However, it is worth noting that whether this sense of contrast and excessive exposure will affect the authority of the government and whether officials will accept bribes through live broadcasting.

2.4 Live Broadcast for Agricultural Products Initiated by Farmers

In the traditional cultural power structure, the Chinese farmers have a low level of education and generally lack the right to speak in the media [7]. But the development of new media and the Internet allows them to speak for themselves. Generally, farmers have a simple image, which brings a sense of intimacy to the audience. Most of their live broadcasts are done in the farmland. These farmers will introduce their own agricultural products in dialects, thereby shortening the distance with the audience and attracting the audience to buy their agricultural products. Some farmers will also sell agricultural products through the combination of short video and live broadcast. A typical example is Li Ziqi. These farmers will show the natural planting process of products in short videos. The visible planting process allows the audience to buy products with more confidence.

Different forms of live broadcasting for agriculture have their own advantages. To a certain extent, they have helped farmers in poor areas to obtain more income. However, this field is still in its infancy and there are still many imperfections.

3 Existing Problems in Agricultural Live Broadcast

3.1 A Lack of Professionalism for Live-streamers

Under the impact of the current epidemic situation, the agricultural live broadcast that helps farmers still needs to exist for a long time. However, the training from mainstream media can not support the professionalism of the entire industry, nor can it support the sustainable development of the entire industry. The current live broadcast that helps farmers is too dependent on the data traffic brought by the Internet celebrity. Most of the non-professional live-streamers introduce the products in a single and unattractive way, which is easy to cause audience aesthetic fatigue. It should also be added that some live-streamers exaggerate the advantages of the product in order to attract more attention. This behavior will not only affect the consumer experience, but also lose consumers' recognition on the nature of commonweal of agricultural live broadcasting. Another problem worth noting is that, without professional training and a solid fan base, farmers as live-streamers generally attract less attention. The content they make have a strong homogeneity, and the product quality cannot be guaranteed.

3.2 Expensive Transportation Expenses and Slow Delivery Speed

Most agricultural products are fresh products. The shelf life of these products determines that they are not suitable for long-term logistics transportation [8]. It is also not applicable to the seven-day no reason return for most current e-commerce products. However, the transportation prices of the efficient logistics express in China, such as SF and EMS, are relatively high. Most farmers cannot afford this logistics cost. At the same time, the logistics industry in rural areas is underdeveloped and there are only a few express stations. The quantity of agricultural products is huge, but the transportation efficiency is low. This situation will prolong the logistics time and lead to the decay of agricultural products during transportation.

3.3 The Low Repurchase Rate of Agricultural Products

There are two reasons for the low repurchase rate of agricultural products. The first is the lack of branding. Most farmers who sell agricultural products online do not have brand awareness. There are a large number of similar agricultural products in the market. Therefore, it is easy to lose consumers and users' viscosity is low. The second reason is the poor after-sales service. Most of the agricultural products will be damaged during transportation, so the merchants need to resend new agricultural products. However, the existing dilemma is that most of the farmers who need help cannot support the high after-sales costs. As a result, the consumer experience is poor and the repurchase rate is not high.

4 Sustainable Development for Agricultural Live Broadcast

Wang Xinbing and Liu Xianghui mentioned in their paper that the characteristics of the live-streamers, the display of product characteristics, the content quality of agricultural live broadcasts, and the interactive communication in the live broadcast directly affect consumers' willingness to buy agricultural products [9]. Therefore, in order to keep the commercial attribute of agricultural live broadcast as well as taking the public welfare into account, there is a need to optimize the agricultural live broadcast from the following two aspects.

4.1 Encouraging Young People to Start Businesses in Rural Areas

Emerging industries need young people to join. They will bring fresh blood and promote the innovative development of the industry. Most farmers lack professional live broadcasting knowledge and sales knowledge. But young people are more likely to accept new knowledge, understand their hometown culture, and lead villagers to become rich. For example, in order to actively attract young talents to participate, China Youth Daily has formulated a detailed support plan according to the actual situation [10], hoping to bring vitality to the countryside through young people. At the same time, the combination of short video and live broadcast should be paid attention to, and agricultural products should be combined with local culture. For example, Li Ziqi uses a short video to show the local beauty and food in Chengdu, giving the audience a sense of freshness and also stimulating their desire to buy.

4.2 Reducing Logistics Prices and Improving After-sales Service Quality

The reduction of logistics price needs the help of the government to carry out some special plans. The government could lower the transportation cost of the rural products through cooperating with the logistics company. Money subsidies can also be given to farmers to reduce the pressure on their transportation costs. In terms of after-sales service, the government needs to supervise the agricultural live broadcasting, close the live broadcasting rooms that exaggerate the advantages of agricultural products and have poor after-sales service.

The sustainable development of agricultural live broadcasting is a topic worthy of continuous discussion. With the development of the epidemic situation and the rapid changes in the Internet industry, the existing discussion is still at a relatively shallow level. However, the live broadcast that helps farmers has received high attention. From January to February 2020, the online retail sales of physical goods nationwide increased by 3.0% year on year, realizing a contrarian growth [11]. From this point of view, the sales of agriculture were mainly done through off line sales. Therefore, live broadcasting that helps farmers needs to adapt to the changing new situation and continue to develop the online market after solving the existing problems.

5 Conclusion

This paper summarizes and reviews the research results of new agricultural sales modes in the context of epidemic situation. First, from the perspective of the identity of live-streamers, four modes of live broadcast for agriculture are described, namely the live broadcast initiated by mainstream media, Internet celebrities, officials, and farmers themselves. Then from a more macro perspective, this paper discusses the common problems of the agricultural live broadcast and puts forward solutions. Existing problems include the lack of professional knowledge of live-streamers, high logistics prices, and a low rate of agricultural product purchase. Finally, in terms of the future development for agricultural live broadcast, young people can be encouraged to start businesses in rural areas, and the government can help reduce logistics costs and improve the after-sales service through platform supervision.

Agricultural development is fundamental to a country. The new form of agricultural live broadcast combines traditional industries with new media, which not only brings new economic forms, but also makes outstanding contributions to poverty eradication and rural revitalization. In addition to exploring new modes, more live broadcast styles can also be developed in the future. At the same time, the agricultural live broadcasting may also face more industrial problems, such as the industrialization of online sales and offline sales. Besides, the academic research in the field of live broadcast for agriculture products will also be combined with more disciplines and continue to be expanded and deepened. Meanwhile, there are a lot of fields involved in the agricultural live broadcasting, and it is difficult to fully explain all the existing research results in one paper. Therefore, the author hopes that the new development of live broadcasting can be expounded from a more micro perspective in the follow-up research.

References

1. China Internet Information Center. The 46th statistical report on the development of Internet in China [eb/ol] http://www.cac.gov.cn/2020-09/29/c_1602939918747816.htm. 2020/09/29, retrieved from 2022/08/26.
2. Chen, X.C., Tang, R.M.: The Narrative Significance of public welfare live broadcasting in the context of Rural Revitalization. Journal of Southwest University for Nationalities (Humanities and Social Sciences) (11), 156-161 (2021).
3. Yu, R.: The current situation, motivation and optimization path of the "short video + live" poverty alleviation model under the background of Rural Revitalization. Business Economics (09), 125-127 (2021). Doi:10.19905/j.cnkiSyjj19822021.09.042.
4. Liang, W., Ma, M.: From the perspective of rural revitalization, the practical form and deep meaning of webcast with goods. News Knowledge (05), 17-26 (2021).
5. Yang, Y.Y.: Research on the strategy of media "live broadcast with goods" to help rural revitalization. News World (05), 60-62 (2021). Doi:10.19497/j.cnki1005-5932.2021.05.020.
6. Li, J., Cheng, J.L. The sustainable development path of "help agriculture live broadcasting" under the Rural Revitalization Strategy. Western Radio and Television (06), 20-22 (2022).
7. Kuang, W.B. The presentation of local culture in short video and the development of the situation. People's Forum (27), 137-139 (2020).
8. Tang, S.S., Yao, M.F., Hu, J.D.: Research on the application of agricultural products "live broadcasting with goods" in rural e-commerce. Inner Mongolia Science and Technology and Economy (16), 51-52 (2021).
9. Wang, X.B., Liu, X.H.: Research on the influencing factors of consumers' willingness to buy agricultural products under the mode of "e-commerce live + helping farmers". Strait Science (01), 82-86 (2022).
10. Wang, S.L., Liu, J.J.: Research on the innovation of the mode of Rural Revitalization assisted by news media -- a case study of the "Youth Project" of China. Youth Daily Media (17), 38-40 (2021).
11. China Internet Information Center. The 45th statistical report on the development of Internet in China [eb/ol] http://www.cac.gov.cn/2020-04/27/c_1589535470378587.htm. 2020/04/28, retrieved from 2022/08/26.

The Impact of Positive and Negative News about Celebrity Endorsers on the Consumer Behavior of Generation Z — Take the Chinese Market as an Example

Weiran Wang*

City University of Hong Kong, Tat Chee Avenue, Kowloon, Hong Kong SAR, China

Abstract. This study examines whether positive and negative news about celebrity endorsers affects the consumer behavior of Chinese Gen Z consumers. A quantitative research was conducted to measure Changes in Chinese Gen Z consumers' attitudes, favorability, and purchase intentions before and after the time of positive and negative celebrity endorsement events. The results of the study show that whe a celebrity endorser has negative news, Gen Z consumers' favorability towards the endorser and the brand as well as their purchase intention will be significantly reduced. On the contrary, when a celebrity endorser has positive news, there is a slight, but not significant increase in Gen Z's favorability towards endorser and the brand, and their purchase intention.

1 Introduction

Generation Z - Broadly speaking, people born between 1995 and 2010 [1]. In the United States alone, Generation Z spends around $143 billion per year, and their spending is increasing every year [2]. Moreover, Gen Z is a huge age group, not only they have occupied a high proportion of the U.S. population, but also in China, they are the age group that needs to be paid more attention to. Gen Z, which accounts for about 15% of China's population, represents the next engine of domestic consumption growth [3]. Thus, it is important to put more attention to Chinese Gen Z consumers. Notably, Gen Z is considered to be a generation that is closely connected to digital, they tend to be more sensitive to information and they have a very different way of thinking compared to previous generations. Therefore, once there is positive or negative news about a celebrity endorser, most Gen Z will be aware of the information. On the other hand, according to the data from U.S. International Trade Administration, China is still the second largest economy in the world with a gross domestic product (GDP) of $14.7 trillion, following the $20.9 trillion GDP of the United States in 2020 [4]. Meanwhile, China is a country with a typical collectivist culture that places great importance on the exemplary role of positive role models and the negative influence of negative examples on others [5]. Therefore, it is necessary to study the Chinese market, which is not only a huge market but also a market where consumers are more sensitive to positive and negative news about celebrity endorsers. However, even though China is a huge market, the research on the consumer behavior of Chinese Generation Z consumers is limited. Thus, the purpose of this study is to investigate how Chinese Gen Z consumers, who are also mainstream consumers soon, how their consumer behavior will be influenced by the positive and negative news of celebrity endorsers.

2 Literature review

2.1 Negative News & Consumer Behavior

While the potential benefits of utilizing celebrity endorsers are great, so are the costs and risks. The benefits of celebrity use can be reversibly reversed if celebrities suddenly change their profile, decline in popularity, fall into a state of moral turpitude, lose credibility with over-endorsements, or overshadow the products they endorse. Negative information about celebrity endorsers affects not only consumer perceptions of celebrities but also the products being endorsed [6].

In addition, negative consumer perceptions of brand endorser misconduct can spill over to the brand. The endorser's words and actions represent the brand's image to a large extent, and the impact of a negative incident will undoubtedly destroy the good impression of the brand in consumers' minds. This will not only reduce consumers' willingness to buy but also lead to consumer complaints, brand switching, and negative word-of-mouth spreading [7]. However, not all studies confirm that negative news about celebrity endorsers has a negative impact. Mohanty [8] argued that the market does not pay any attention to negative news about celebrities. Bailey [9] argued that negative celebrity news has little or no impact on the brands they endorse or the companies they work with. Carrigan and Attalla [10] also indicated that negative news about the ethical

* Corresponding author: wangwr2018@126.com

nature of endorsers had no negative impact on consumers' propensity to purchase. On the contrary, Berger et al. [11] indicated that we believe that negative publicity may sometimes increase the likelihood of purchase and sales. While the above findings are somewhat surprising, more studies have shown that negative press from celebrity endorsers often has a negative impact. A quantitative study on how negative news about brand endorsers affects young consumers' purchase intentions shows that consumers are significantly less likely to purchase a brand after negative news about a brand endorser compared to before negative news about a brand endorser. The experimental data and in-depth interview results show that contemporary consumers are more sensitive to the perception of negative news about a brand endorser, and after learning of negative news about an endorser, they quickly scroll through major websites and make new purchase decisions about the brands they endorse [12]. In like manner, people react more to the negative events of the celebrity compared to the positive events of the celebrity. Moreover, when a celebrity has a negative event, people will closely associate the celebrity with the company and question the selection and employment of the company, thus the positive event of the celebrity employee does not have as much impact as the negative event [13]. Moreover, consumers may process the negative events of brand spokespersons psychologically, which may lead to different levels of consumer crisis emotions, and eventually affect consumers' avoidance behavior toward brands [14]. Although the literature has examined the impact of negative news about celebrity endorsers on consumers' attitudes and consumption behaviors, their studies have been conducted on previous generations (Gen X &Gen Y, etc.). The literature examining how the consumption behavior of Generation Z in China would be influenced by negative news about celebrity endorsers is still limited. Thus, the following hypotheses are proposed:

H1a: Negative news about celebrities will reduce the reputation of brands in Chinese Gen Z consumers' minds.

H1b: Negative news about celebrity endorsers will reduce Chinese Gen Z consumers' favorable perceptions of the brand and thus reduce or avoid purchasing its products.

2.2 Positive News & Consumer Behavior

Generally, we call events that can bring positive effects to the subject positive events, and similarly, events that can bring negative effects to the subject negative events. In this paper, a positive event for a celebrity is defined as an event that generates positive public opinion and influence on the celebrity, such as winning an award or doing public service [15]. The current study revealed that positive images and congruence between the image of celebrity and brand are important to adolescents [16]. Besides, McIntyre and Gibson [17] argued that few studies have investigated the impact of positive news. Yet, a recent study discovered that individuals seem to be more attracted to positive news than negative news. New York Times readers share positive news faster and in greater numbers on social media than they do negative news. Positive reporting is the main body of news reporting, playing an important role in leading public opinion. It plays an important role in guiding public opinion, establishing the power of role models, and building a harmonious society [18]. Positive news about celebrities increases the market value of the companies they endorse and erodes the market value of the companies they compete with. The difference-in-difference analysis also confirms this view. It also proves that positive news favorably affects the sales of these products [19]. Another literature that uses a quantitative approach to study the impact of celebrity events on firm value shows that people are quick to react to positive celebrity events. But for the degree of impact on the company received, the impact of positive celebrity events on the market value of the company is not as great as the degree of impact brought by negative events [20]. Just like what McIntyre and Gibson [21] have argued above, there do have little research about the impact of positive news about celebrities, and the study in this field is relatively limited today. To date, there is very little literature that has examined how positive news about celebrity endorsers affects consumer behavior. Most have examined how positive news about celebrity endorsers affects the value of a company, and their findings all vary. Some studies show that positive news about endorsers has little effect on company value. Others show that positive news about the spokesperson can increase the value of the company. Given factors such as the very different consumption perceptions of Generation Z from previous generations, and to fill the gap in the field of how positive news about celebrity endorsers affects the consumption behavior of Chinese Gen Z, the following hypothesis is proposed:

H2: Positive news about celebrity endorsers will increase Chinese Gen Z consumers' favorable perception of the brand and thus purchase its products when in demand.

3 Methodology

A quantitative research was conducted to examine how positive and negative news about celebrity endorsers affects the consumer behavior of Chinese Gen Z. The participants of the study were undergraduate students from a university and students from its affiliated high school in mainland China. Meanwhile, in order to ensure that the study has a greater realistic guiding significance, choosing stars that most of generation Z know as examples will have a certain universality. Therefore, news about two top stars in China was chosen, "Yi Yangqianxi's irregular admission to the National Theatre" and "Wang YiBo's support for flood relief in Henan Province" as the negative and positive news of stars respectively. The respondents were randomly selected using the cluster sampling method, and a total of 180 questionnaires were collected. Thirty-two invalid questionnaires were excluded, and 154 valid questionnaires were collected, with an efficiency rate of

85.6%. From Table 1, it can be seen that 43% of the respondents were male and 57% were female, 45% were high school students, and 55% were undergraduates.

Table 1. Frequency analysis of demographic variables

Variables	Cat.	FQ.	Prop.	M	SD
Age	Under 18	68	44%	1.56	0.498
	18-27	86	56%		
Gender	Male	66	43%	1.57	0.496
	Female	88	57%		
Education	High School	69	45%	2.55	0.499
	Bachelor	85	55%		
Occupation	Student	154	100%	1	0

Cat .=Categories, FQ=Frequency, Prop.=Proportion

Note: Own work

4 Results and discussions

The questionnaire was measured using a 5-point Likert scale, with choices (1) and (5) indicating strong disagreement and strong agreement, respectively. with a reliability of 0.857 for the seven questions on how positive news about celebrities affects Generation Z's consumer behavior and a reliability of 0.79 for the seven questions on how negative news about celebrities affects Generation Z's consumer behavior. The reliability of each variable measured in spss 24.0 was at an acceptable level of 0.70 [22].

From Tables 2 and 3, it can be seen that the event of "Yi Yangqianxi's participation in the National Theatre examination" was considered as negative news by 85.06% of the respondents. After this event was reported, the mean value of Generation Z's favorability towards him plummeted from the original 3.66 to 2.36, and the reputation of the brand he endorses also dropped significantly, from the original 3.56 to 2.52. Meanwhile, the mean value of Generation Z consumers' purchase intention to the brands endorsed by Yi also dropped dramatically, from the original 3.92 to 2.57, with H1a and H1b being supported.

Table 2. How Gen Z determine the two news

	WSH	YNT
Positive	97.4%	85.06%
Unsure	2.6%	7.14%
Negative	0%	7.79%

WSH=Wang supports the front line of flood relief in Henan, YNT=Yi was illegally admitted to the National Theatre

Note: Own work

Table 3. Changes before and after the news

	Fv of W	Fv of Y	BR of W	BR of Y	W's PI	Y's PI
Before	3.53	3.66	3.69	3.56	3.87	3.92
After	3.88	2.36	3.79	2.52	4.01	2.57

Fv=Favoribility, BR=Brand Reputation, PI=Purchase Intention, W=WANG Yibo,Y=YI Yangqianxi

Note: Own work

In addition, the news that "Wang Yibo supported disaster relief in Henan Province" was regarded as positive news by 97.4% of the respondents. Before the news was reported, the mean value of Generation Z's favorability towards him was 3.53, while after the news was reported, the mean value had a slight increase to 3.88. Meanwhile, the mean value of the reputation of the brands endorsed by Wang slightly increased from 3.69 to 3.79. Notably, the mean value of Generation Z consumers' purchase intention to the brands endorsed by Wang has risen from 3.87 to 4.01. Although the mean value has increased but to a lesser extent, therefore H2 is not supported.

Table 4. Gen Z's purchasing power and attention to news

BPEFC	ACN
80.52%	86.36%

BPEFC=Buy products endorsed by your favorite celebrities, ACN=Attention to celebrity news

Note: Own work

According to Table 4, 86.36% of Gen Z respondents indicated that they often follow celebrity news, and 80.52% of Gen Z respondents expressed that they have purchased products endorsed by their favorite celebrities. This shows that Gen Z is more attentive, sensitive, and responsive to celebrity news and that the purchasing power of Gen Z in China is substantial. Brands and celebrity endorsers need to pay more attention to maintaining relationships with Gen Z consumers in the future and make strategic adjustments to address their preferences, which are very different from those of previous generations.

As shown in Fig.1, 56.49% of Gen Z respondents believe that the image of the brand endorser is closely related to the brand image, 30.52% of Gen Z respondents believe that the image of the brand endorser is related to the brand image. Therefore, it is necessary for brands to do a clear background check when selecting endorsers, and it is also advisable for brand endorsers to be careful with their words and actions.

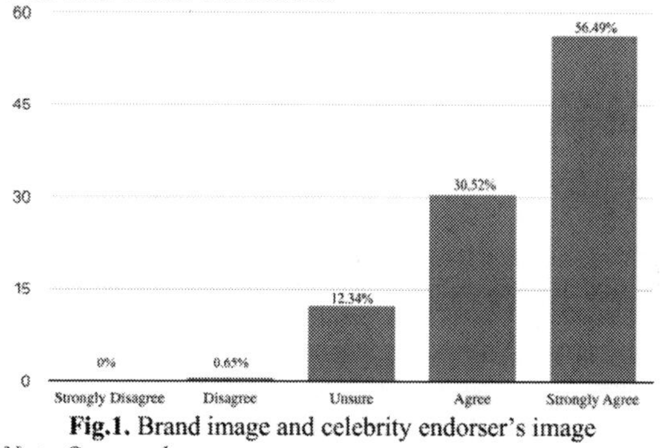

Fig.1. Brand image and celebrity endorser's image
Note: Own work

5 Conclusion

Based on the results of the data analysis, the following three main conclusions and insights were drawn:

(1) Negative news about a brand endorser has different definitions in different people's minds. When

negative news about a celebrity appears, Gen Z will make their own judgments about whether this news is negative news. Once they decide that it is negative news, they will quickly react in terms of their favorability to this celebrity, brand image, brand reputation, and their purchasing behavior, etc. This data shows that 87.01% of the respondents believe that the image of the endorser is related to the brand image. Therefore, the majority of Gen Z will be less favorable to celebrities with negative news, and thus the reputation of the brand will be implicated and reduced among Gen Z. Based on the current situation in China, the incident of Yi made the public feel the extreme unfairness and tilt of social resources, which is one of the reasons why most of the Gen Z considered this incident as negative news, and also indirectly led to the negative impact of Gen Z on the favorability of Yi, the reputation of the brands he endorsed, and the consumer behavior.

(2) Similarly, positive news about a brand endorser also has different definitions in the minds of different people. In this survey, more than 97% of the Gen Z respondents considered "Wang supports floods in Henan province" to be positive news. Although the vast majority of the subjects thought it was positive, there was only a slight increase in their favorability towards Wang and their impression of his brand image and consumer behavior. This shows that the positive effect caused by positive news is not significant compared to negative news, but on the contrary, negative news can cause a greater negative effect.

(3) Most of the Chinese Gen Z has a high level of attention and sensitivity to news about celebrities. At the same time, more than 80% of Gen Z believe that the image of the endorser is closely related to the brand image. For the brand, it needs to do a more detailed background investigation when choosing an endorser. Only when the celebrity endorser and the brand are acting within the scope of reasonableness and legality can the effectiveness of celebrity endorsement be truly improved. Moreover, there is a fill-in-the-blank question in the questionnaire, "What kind of qualities do you think a celebrity endorser should have", and several words appear very frequently. For example, "high credibility", "positive image", "responsible", "professional", "dedicated", etc. It is evident that in Gen Z's mind, a qualified and suitable endorser should have at least these qualities. Therefore, brands should follow the trend and be more cautious in choosing endorsers, and choose celebrity endorsers with both "virtues and talents" for better meaning and effectiveness

References

1. Francis , T., & Hoefel, F. (2018). Tracy Francis and Fernanda Hoefel - Dr. Thomas Wu. McKinsey&Company. Retrieved July 1, 2022, from http://drthomaswu.com/uicmpaccsmac/Gen%20Z.pdf
2. Dabija, D.-C., & Lung, L. (2019). Millennials versus gen Z: Online shopping behaviour in an emerging market. SpringerLink. Retrieved July 1, 2022, from https://link.springer.com/chapter/10.1007/978-3-030-17215-2_1
3. Zhou, J., Poh, F., Zhang, C., & Zipser, D. (2020, December 8). China's gen Z are coming of age: Here's what marketers need to know: Greater china. McKinsey & Company. Retrieved June 13, 2022, from https://www.mckinsey.com/cn/our-insights/our-insights/chinas-gen-z-are- coming-of-age-heres-what-marketers-need-to-know
4. ITA. (2022). China - market overview. International Trade Administration | Trade.gov. Retrieved July 6, 2022, from https://www.trade.gov/knowledge-product/exporting-china-market-overview
5. Li, N. (2017). Study on the impact of negative news about spokespersons on consumers' risk perceptions. cnki. Retrieved July 10, 2022, from https://kns.cnki.net/kcms/detail/detail.aspx?dbcode=CMFD&dbname=CMFD201902&filename=1017266208.nh&uniplatform=NZKPT&v=jljKXFDMOZYg4W7c6OacjPaEkRIxbD3QEZlBKg1DcXBusiEHoqyAuOQljMrcOsfo
6. Erdogan, B. Z. (1999). Celebrity endorsement: A literature review. Journal of Marketing Management, 15(4), 291–314. https://doi.org/10.1362/026725799784870379
7. Mao, J. (2022). A study of consumer boycott behavior triggered by negative events of brand endorsers. cnki. Retrieved July 8, 2022, from https://kns.cnki.net/kcms/detail/detail.aspx?dbcode=CJFD&dbname=CJFDLAST2022&filename=SGLK202202014&uniplatform=NZKPT&v=T5pkoM8mq34knqbkV4kj745gD7mlWq4PyfsmbX6o6qh_T4YzRer6V7QAYnBzk00O
8. Mohanty, M. (2008). The financial information content of celebrity endorser: An Event Study Analysis. Paradigm, 12(1), 89–97. https://doi.org/10.1177/0971890720080111
9. Bailey, A. A. (2007). Public information and consumer skepticism effects on celebrity endorsements: Studies among young consumers. Journal of Marketing Communications, 13(2), 85–107. https://doi.org/10.1080/13527260601058248
10. Carrigan, M. and Attalla, A. (2001), "The myth of the ethical consumer – do ethics matter in purchase behaviour?", Journal of Consumer Marketing, Vol. 18 No. 7, pp. 560-578. https://doi.org/10.1108/07363760110410263
11. Berger, J., Sorensen, A. T., & Rasmussen, S. J. (2010). Positive effects of negative publicity: When negative reviews increase sales. Marketing Science, 29(5), 815–827. https://doi.org/10.1287/mksc.1090.0557
12. Hong, X. (2019). An empirical study of the impact of negative brand spokesperson news on young consumers' purchase intention. cnki. Retrieved July 15, 2022, from https://kns.cnki.net/kcms/detail/detail.aspx?dbcode=CMFD&dbname=CMFD202101&filename=102070

5650.nh&uniplatform=NZKPT&v=Lymxcc1uNBhFGmwmTMl8l1AhyXgERDhHLSbHjc0D10VcyskaGHVBJMeavAgTqbc5

13. Duan, Y. (2020). A study of the impact of celebrity events on the market value of companies. cnki. Retrieved July 13, 2022, from https://kns.cnki.net/kcms/detail/detail.aspx?dbcode=CMFD&dbname=CMFD202002&filename=1020035505.nh&uniplatform=NZKPT&v=b8-rS277XXqfBslNzEx9VEdjmKm6yYfoYozsSCuhXwvOqPo-3jA4GBH9p9qmvBdS
14. Mao, J. (2022). A study of consumer boycott behavior triggered by negative events of brand endorsers. cnki. Retrieved July 8, 2022, from https://kns.cnki.net/kcms/detail/detail.aspx?dbcode=CJFD&dbname=CJFDLAST2022&filename=SGLK202202014&uniplatform=NZKPT&v=T5pkoM8mq34knqbkV4kj745gD7mlWq4PyfsmbX6o6qh_T4YzRer6V7QAYnBzk00O
15. Duan, Y. (2020). A study of the impact of celebrity events on the market value of companies. cnki. Retrieved July 13, 2022, from https://kns.cnki.net/kcms/detail/detail.aspx?dbcode=CMFD&dbname=CMFD202002&filename=1020035505.nh&uniplatform=NZKPT&v=b8-rS277XXqfBslNzEx9VEdjmKm6yYfoYozsSCuhXwvOqPo-3jA4GBH9p9qmvBdS
16. Chan, K., Leung Ng, Y., & Luk, E. K. (2013). Impact of celebrity endorsement in advertising on Brand Image among Chinese adolescents. Young Consumers, 14(2), 167–179. https://doi.org/10.1108/17473611311325564
17. McIntyre, K. E., & Gibson, R. (2016). Positive news makes readers feel good: A "silver-lining" approach to negative news can attract audiences. Southern Communication Journal, 81(5), 304–315. https://doi.org/10.1080/1041794x.2016.1171892
18. Chen, H., & Lyu, H. (2013). An analysis of the audience's negative interpretation of positive news reports. cnki. Retrieved July 18, 2022, from https://kns.cnki.net/kcms/detail/detail.aspx?dbcode=CJFD&dbname=CJFD2013&filename=JJYS201303033&uniplatform=NZKPT&v=nXpmMeD8u8MwwhLkax6gnoBDYVaRlc2u9tp0h4_jGupmXXcSobFsWks-MaJjlsiY
19. Mohanty, M. (2008). The financial information content of celebrity endorser: An Event Study Analysis. Paradigm, 12(1), 89–97. https://doi.org/10.1177/0971890720080111
20. Duan, Y. (2020). A study of the impact of celebrity events on the market value of companies. cnki. Retrieved July 13, 2022, from https://kns.cnki.net/kcms/detail/detail.aspx?dbcode=CMFD&dbname=CMFD202002&filename=1020035505.nh&uniplatform=NZKPT&v=b8-rS277XXqfBslNzEx9VEdjmKm6yYfoYozsSCuhXwvOqPo-3jA4GBH9p9qmvBdS
21. McIntyre, K. E., & Gibson, R. (2016). Positive news makes readers feel good: A "silver-lining" approach to negative news can attract audiences. Southern Communication Journal, 81(5), 304–315. https://doi.org/10.1080/1041794x.2016.1171892
22. Churchill, G. A. (1979). A paradigm for developing better measures of marketing constructs. Journal of Marketing Research, 16(1), 64–73. https://doi.org/10.1177/002224377901600110

The Development of Contemporary Women's Issues

Song Jialu[1,2]*

[1] University of Hong Kong, 999077 Pokfulam, Hong Kong, China
[2] Song Jialu, University of Hong Kong, Pokfulam, Hong Kong, China

Abstract: With the rapid development of the world economy, human thinking and ideas are also progressing. Today's society should break the gender dichotomy and popularize women's right to education. Compared with the feudal society, women's issues in today's society have made great progress, but there are still some serious social problems. This paper discusses women's issues in education, reproduction and gender, and how contemporary women's issues have evolved. In order to solve the problem of women, we should first eliminate the influence of the patriarchal culture. Although gender culture has a certain lag, its development needs the joint efforts of the whole society. After the whole society generally pays attention to women's issues, there will be more outstanding women, which is more conducive to the development of the society.

1. Introduction

In ancient times, women often became victims of politics. The feudal states became political alliances through marriage to consolidate the imperial power, which was related to political stability. In a male-dominated society, only women married far away from the union. Men rarely married far away into the family, so women became the victims of this political deal. Moreover, it is ironic that in a polygamous society, a wonderful woman who marries far away may be married to an older man who is too old for his age. One of the most famous political marriages in Chinese history was the marriage of the Wencheng Princess of the Tang dynasty to the 73-year-old Songzan Ganbu, whose positive impact on the cultural exchange cannot be overlooked, but how sad for a woman. Women had little voice in feudal society, and even in modern times, women are still subject to the influence of a male-dominated society with many restrictions. As women's reform continues, more women's issues have surfaced. Examples include female employment difficulties, political participation crisis, female layoffs, and girls missing out on school. Corrupt gender stereotypes have exacerbated women's tragedy under the market economy's influence. Therefore, today's society should break the gender dichotomy and universalize women's right to education instead of shouting slogans under the banner of feminism and gender equality. It must be admitted that women's issues in today's society have made significant progress compared to feudal society, but they are still severe social problems. A truly equal society would not want to pursue equality, just as a free man would not desire freedom. People will only desire something when they lack it. Even if men and women are equal in law, there is still a need to please the stereotypes in human governance. This paper will discuss feminine issues in education, reproduction, and gender and how contemporary feminine issues are evolving.

Tel:15253556196, E-mail: ardensong@qq.com

2. Results & Discussion

The fight against gender discrimination is the essential practical goal of women's issues. However, the concept of difference exists in gender itself, and gender difference is the logical basis of gender discrimination. In the view of feminists, masculinity overemphasizes the differences between men and women, and they assume some natural superiority of men based on these differences. Feminists see this expression of superiority as a form of sexism. Gender difference is a neutral theoretical concept. In nature, such differences are like the differences between species but do not have a particular social significance. Masculinity is as blindly confident as the existence of anthropocentrism. In the context of this theory of sociality, the meaning of difference is then influenced and driven by value orientations and political purposes. Thus, strictly speaking, there is a difference between sexuality and gender, with "sexuality" being descriptive and "gender" inevitably having a particular value orientation. Although Darwin revealed the superiority of men in his theory of biological evolution, the expression of this superiority is only an expression of natural differences. It does not carry the value of sexism and gender oppression in the sense of political purposes. Men are more powerful than women, but this is a biological difference. Physiological superiority is not a logical basis for gender discrimination. In Le Deuxième Sexe, Simone Beauvoir mentions that the word "female" is derogatory, not because it roots women in nature, but because it confines. It is because it confines the woman to her sex. Wittig cites the idea of Colette Guillaumin as an example. Studies on race indicate that the concept of race as perceived in modern society did not exist prior to the formation of the black economic system, or at least was not constituted in a modern understanding. Following this theoretical analysis, it is ironic that black rights were preached in white supremacist America. Feminism should think outside the box to consider women's issues, rather than making unnecessary arguments under the influence of a patriarchal society.

Women worldwide are in a situation that is not comparable to that of men simply because of their gender. For minority women, older women, women with disabilities, and female immigrants and refugees, the situation is even direr. While women's rights have come a long way in recent decades, with discriminatory laws being repealed and the number of girls in schools increasing, we now face strong resistance. Some countries are weakening legal protections against rape and domestic abuse, while others are pursuing policies that punish women, such as austerity and forced childbirth. Women's sexual and reproductive rights are under threat from all sides. Women are still excluded from the top, from the government to state-owned enterprises to the private sector. Women leaders and public figures are harassed, threatened, and abused online and offline. This power gender gap manifests itself through the gender gap in pay. Some of society's data is also often based on "default male" data. Men are the norm. Women are the exception. Women also struggle with centuries of misogyny and the erasure of female achievement. Women are ridiculed for their sensuality and emotional volatility. Society needs to use women's sensuality wisely. It is no coincidence that women-led governments are incorporating well-being and sustainable development into new definitions of economic success.

We should stop trying to change women and change the systems that prevent women from reaching their potential. Women face gender discrimination and deception all the time and are blamed as victims. Various factors affect women profoundly and have become obstacles to overcoming today's many challenges and threats. To achieve a fair globalization that works for everyone, we must base our policies on genuine consideration of women's issues. Gender imbalances continue to exist in every sector of the world today. There is a straight line between violence against women, domestic oppression, and conflict. How a society treats women, who are half its population, is an essential indicator of how that society treats other groups. Even in peaceful societies, many women do not escape mortal danger in their homes. Because of biological differences, women are less powerful than men, and domestic violence becomes one of the significant issues that put women at risk. It is sad that in times of peace, no women die in war, but many die in domestic violence.

Schooling is often a prerequisite for women to stand up against inequality and be integrated into the market economy. In many Asian and Latin African countries, religious and cultural practices and poverty have resulted in women being unable to get an education like men. Because a family is usually unable to provide education for all their children, they leave priority to their sons, even if their son is a fool. There we see sexism, sexual violence, misogynistic values, domestic violence, virginity, and other phenomena that leave women as if they were still in the purgatory of the 18th and 19th centuries. WuDunn in Half the Sky mentions that more girls have been killed worldwide in the last 50 years than men in all the wars of the entire 20th century. Simply because of their gender. The truth is that when girls drop out of school early, they are faced with working for their families and getting married early. It is a vicious cycle for them; only studying can change their fate. In India, for example, some rural girls who have not studied are cheated or trafficked to brothels and forced into prostitution. If they were educated, it would reduce their chances of being cheated. If the education level of the whole society is increased, it will also increase society's concern for women's safety. It is inconceivable that in the 21st century, there will be women kidnapped, trafficked, raped, and abused just like the "black slaves" were 200 years ago. The technical means and political will to crush modern enslaved people are not lacking; what is lacking is the mass awareness to protect women. Influenced by thousands of years of traditional thinking about the superiority of men over women, women seem to have been in the position of victims. Women never belonged to themselves in Asian and African countries and even in ancient Europe. Only the widespread education of women and the rise of feminism changed women's status in Europe.

Although modern Europe and the United States have unanimously promoted freedom and equality, the 2022 U.S. abortion law still makes much noise. Is it free and equal for women to be unable to make their own decisions about reproductive rights? Although Christianity views the embryo as an independent life, shouldn't the mother have the right to terminate this pregnancy when the fetus is not well developed? When women are raped involuntarily and become pregnant, and they are legally forced to give birth to this child, is this not a double whammy for the victim? Even if one is trying to encourage childbirth, such behavior cannot help but feel disrespectful to women. Women have never been instruments of procreation. The study found that support for abortion depended on the person's age, religious beliefs, Etc., and that they would be opposed to abortion if they supported conservatism and sexism. In the afternoon of November 21, 2007, Li Liyun, who was 41 weeks pregnant, was admitted to the hospital for a difficult delivery. Her husband, Xiao Zhijun, did not agree to a C-section and refused to sign the consent form for the operation. After three hours of routine resuscitation, the doctor pronounced Li Liyun dead. Whatever the reason for Xiao Zhijun's reluctance to sign his wife's consent form for surgery, the fact remains that his lack of cooperation resulted in the death of both his wife and child on the operating table. Pregnancy and childbirth are seen as a rebirth for women. It is impossible to imagine how women could have seven or eight children in a lifetime once upon a time when medical facilities were not perfect. Family planning has not just controlled our population. It has been nothing short of a boon for women. This argument precludes many rural areas from having many children at the risk of being fined for having sons to "carry on the family tradition." This issue comes back to the topic of education. Often people with higher education will change their mindset about this tradition, and only in rural areas where education is not widespread is there a preference for sons over daughters. Since it is illegal for doctors to tell the sex of an embryo in advance, many female babies are abandoned after birth. In this case, the law cannot eliminate this situation, and the most important thing is to change people's perceptions.

Simone De Beauvoir believes that " one is not born a woman, but becomes one" "Women" and "men" are not natural gender being but social class, a political and economic category, rather than eternal. Wittig's perspective breaks down the gender binary. It is extended from her perspective to discuss how contemporary female sexuality is addressed. The gender division of labor has been a universal phenomenon throughout human history. After the Industrial Revolution, with the development of science and technology, the gender advantage of men gradually diminished. Women can perform the same work as men with the help of tools. Although the traditional social division of labor has been transformed, it has remained intact. It has reappeared in a complex and diverse form, and the hierarchical social division of labor has severely constrained women's development. The influence of a patriarchal that has existed for thousands of years has turned into a barrier to the women's development. For example, in negative news coverage, the female gender is frequently and purposefully highlighted whereas the male gender is more frequently invisible. In the framing of stories, women's age, appearance and physical characteristics are emphasized more as landscapes to be gazed at than men's. Women become the central focus of the news, even in cases where women are victims. While there is no shortage of female characters and viewpoints in the news, all of these examples show that the emphasis is always on the wrong perspective.

According to the "Survey Report on the Status of Chinese Women's Workplace in 2021" released by Smartlink Recruitment, 55.8% of women were asked about their marital and maternity status during the job search process. Employers restricted 29.6% of women during the job search process due to gender issues. 18.2% of women were forced to give up their career development to care for their families. From the data, employment is not friendly to women. The first thing that companies consider when hiring women is still women because of family care brought about by the decline in production capacity. The first law of survival of enterprises is profit. A decline in production capacity will inevitably result in reduced profits, so companies are very scrupulous about married women. When they first enter the workplace, there is not much difference between the income levels of men and women. However, their income will decline significantly after women become pregnant and have children. From the day women get pregnant, they have to face the fixed process of pregnancy, menstruation and taking care of the child, and when the child is older and can be taken by parents, they have missed two years of promotion in the workplace. Most women in contemporary China will seek stability after marriage. However, the more they indulge in instability, the more crisis-ridden they are because the enterprise will eliminate them at any time. In the past, the division of labor was "male dominant, female dominant," while contemporary women need to be "both inside and outside." According to a survey conducted by the U.S. National Bureau of Statistics on the total working population and the labor force participation rate of each country, Chinese men's labor force participation rate is as high as 80%. Chinese men's labor force participation rate is as high as 80%. Chinese women's labor force participation rate is 70%, which is higher than that of men in some countries. More and more women are entering the workplace and paying quietly for the family economy. However, their household chores time has remained the same.

Women should not be treated differently. The equalization of the social division of labor is a prerequisite for women's development. Society should give more opportunities to women and not affect their performance in the workplace just because of their reproductive rights. Housewives should be paid a "wage" for their domestic work because they pay the same social value. The dissemination of gender culture is essential to addressing contemporary women's issues, and gender equality education programs should be developed. Students should be educated early on about progressive ideas rather than the influence of traditional ideas passed down from generation to generation. This shows that women's existence is not as sunny as they think. Furthermore, the rise of feminism is also a backlash against the survival situation, but both men's and women's rights benefit one side. We should all clearly understand that affirmative action is the only way. To pursue gender equality, we should make gender equality a common pursuit for all people. Both men and women should put themselves in the correct position and put aside all unequal ideas about gender before pursuing common equality.

It is easy to see the benefits of gender equality for women, more leadership positions, better treatment, and more support at work and home. Men may worry that as women become more established, they will be less established. However, surprisingly, in truth, gender equality is suitable for men, too. Gender equality is not just a women's issue. It requires the efforts of not only women but men as well. Boys will benefit if their mothers play an essential role in the work. It is not just about the impact in the present but about the issues that will affect the next generation. Because an equal family will raise a child with a more cheerful personality than a child raised in an unequal family, to make gender equality a reality, we must change how we advocate for it. Often the focus is on equity. In order to achieve justice, we must provide equal opportunities for women. Men who support gender equality must also stand up and speak out for women.

3. Conclusions

In today's society, although the feudal idea of male power is no longer a hard and fast rule to bind women, its values have been deeply implanted in people's hearts, constituting a deep-rooted value, emotion, and ideological orientation. This invisible and untouchable invisible force binds both men and women. Under market economy conditions, the traditional gender concept still influences social opinion. Therefore, to solve women's problems, we must first eliminate the influence of male-dominated culture. Whether feminist or patriarchal, an excessive emphasis on the opposite side of the development is extreme. A free and equal society should be gender-neutral. The suitability of a job should depend on a person's ability, not on that person's gender. The formation of culture itself is a long historical process that needs to be precipitated and solidified. The economy and technology are the first to develop in the process of social development. In comparison, gender culture has a certain lag. Its development requires the joint efforts of the whole society. Although women have been maltreated throughout history, one cannot ignore the role women have played in the development of history. Both women and men are an integral part of human development. The rise of women is not an end in itself but a process. This process is full of hope. Even in ancient China, where the status of women was low, there were still many outstanding women, such as Fuhao, Wu Zetian, Li Qingzhao, Wencheng Princess, Etc. So after paying attention to women's issues, more excellent women will emerge. This is beneficial to the development of society.

References

1. Wittig, M. (1993). One is not Born a Woman. Feminist Issues, 245-251.
2. Moi, T. (1994) Simone de Beauvoir: the making of an intellectual woman. Blackwell.
3. Kristof, N. D., Kristof, N. D., TB/Politikwissenschaft. (2010). Half the sky. Random House Inc, 124(8), 398-399.
4. Ota, T., Masuda, S., Kido, M., et al. (2009). Half the sky: turning oppression into opportunity for women worldwide, nicholas d. kristof and sheryl wudunn (new york: alfred a. knopf, 2009), 320pp. $28 cloth. Ethics & International Affairs, 23(11), 76-92.
5. Hadi, M. (2019). Feminism: a tool to analyze literature. Transylvanian Review, XXVII, 1012-1018.
6. Henderson, K. A. (1991). The contribution of feminism to an understanding of leisure constraints. Journal of Leisure Research, 23(4), 363-377.
7. Jelena, I., Emanuele, N., Maurizio M. A., et al. (2009). Transplacental transfer of antiretroviral drugs and newborn birth weight in HIV-infected pregnant women. Current Hiv Research, 7(6), 620-625.
8. Tyler, M. (2005). Women in change management: Simone De Beauvoir and the co-optation of women's Otherness. Journal of Organizational Change Management, 18(6), 561-577.
9. Hengehold, L., Bauer, N. (2017). A companion to Simone de Beauvoir and the race. Gender Analogy in The Second Sex Revisited. 47-58.
10. Direk, Z. (2011). Immanence and abjection in Simone de beauvoir. Southern Journal of Philosophy, 49(1), 49–72.

https://doi.org/10.1051/shsconf/202315502015

The Comeback of the Medium: The History and Contemporary Revival of the Vinyl Record Industry

Yiqian Guo*

Jinan University, Guangzhou, China, 510632

Abstract. As an important vehicle in the development of music and recording art, vinyl records have gone through a series of journeys from birth, and development to decline over the past hundred years. However, in recent years, the industry has seen a resurgence and has been called a "renaissance" in the digital music era by some scholars and media practitioners. This paper explores the historical changes and market evolution of the vinyl record industry since its birth, and focuses on the industrial situation, reasons for the revival, characteristics, and future development expectations of the industry during the contemporary "revival" period, in an attempt to explore the artistic value positioning of vinyl records in different periods, and the relationship with the technology and audience psychology of the times, in order to gain useful insights for art communication and its industrial marketization.

1 Introduction

Vinyl records originated from Edison's invention of the cylinder recorder in the late 19th century, and then through the invention of German immigrant Emil Berliner and several inventors' improvements, its early format was formed and gradually became popular in the music market, contributing to the creation and evolution of popular music at that time [1]. In the mid-20th century, Columbia Records improved its original recording and record-engraving technology by inventing the LP record, which solved the problem of recording and playing narrow grooves, thereby increasing the recording capacity of vinyl records and developing a cheaper and more reliable mode of playback, together with the subsequent invention of another portable EP record [2], the vinyl industry reached its peak during this period. However, after the 1980s, with the invention of the CD and the development of Internet technology, the vinyl industry gradually declined and even went into a state of decline where production lines were stagnant for a time, until recent years, when vinyl records returned to some people in a more niche, independent form and its industry revived.

Most of the previous studies on the vinyl record industry have focused on the vinyl record technology itself or the historical and cultural aspects behind the vinyl record, while some studies have focused on the marketing of vinyl records, but relatively little has been done on the relationship between the history of the vinyl record industry itself and the changes in its artistic value, audience consumption and psychology. Therefore, this paper explores the historical changes and market evolution of the vinyl record industry from its birth to the present, and analyzes the industrial situation, reasons for the revival, characteristics, and future development expectations of the vinyl record industry during the contemporary "revival" period, in order to explore the artistic value positioning of vinyl records in different periods, and the relationship with the technology and audience psychology of the times, and then gain useful insights into the marketization of art communication and its industry.

2 The Historical Lineage and Period Division of Vinyl Record Industry

The history of vinyl records and its industrial market can be roughly divided into four periods: the growth period, the development and maturity period, the decline period and the revival period.

The growth period is the period after the invention of the SP vinyl record, from the end of the 19th century and the beginning of the 20th century, when it gradually became known and used by musicians and gave rise to a series of related industries such as turntables, recording equipment and related professions such as recording engineers [3]. Vinyl records were undoubtedly a new thing for people at that time. Compared with the visual, music was a more temporal art, but this "recordable music" was a breakthrough in the space and time boundaries of the art of live performance, and its ability to preserve music meant a new way of producing and enjoying art. In addition, the relative portability of vinyl records meant that "music could be moved". Especially later in this period, after the vinyl industry had taken shape, people were able to use vinyl records to hear music from all over the country and even the world. In a way, vinyl records became a window to the world for the people of that time.

*Corresponding author: 18996192863@163.com

As a result, there was a huge potential market demand for vinyl records catered to both the artistic and cognitive needs of the people at the time, and the number of vinyl records released and the market share of vinyl records showed a rapid increase. For example, the Victor Talking Machine Company had assets of $2,724,016 which grew to $33,235,378 in 1917 [4].

The development and maturity period of the vinyl record industry emerged after the Great Depression in the 1930s. While the relatively complete vinyl record industrial system was severely damaged by the Great Depression [3], the need to revive the industry after the crisis pushed the vinyl industry to seek innovative breakthroughs in both technology and content, such as the invention of LP vinyl records and EP vinyl records in the late 1940s [2]. On the other hand, after World War II, the rapid economic growth of the United States and the public psychological trauma of the war combined to create a huge market for pop music as a spiritual consumer product and also for vinyl records as the main vehicle for pop music. Besides, the emergence of pop stars increased the appeal of some vinyl records to fans as well. This period reached its peak in the 1950s and continued to grow relatively steadily thereafter until the 1980s. Arguably, the vinyl industry dominated the recording and pop music markets for most of the 20th century.

In the 1980s, CDs emerged, bringing disruptive changes to the audio-visual industry, and the vinyl industry entered the decline period. For both listeners and record manufacturers, CDs were a new and potential field. For listeners, CDs were generally cheaper to buy and play; the sound quality was clearer and easier to carry and keep; the design was more fashionable and seemed more relevant to modern life; and for record manufacturers, the production cost of CDs was also lower than that of vinyl records [3], and their storage and transportation were less difficult, which was more conducive to larger-scale distribution. Later, the Internet and digital music were another subversions of the musical art carriers. Their features of free choice, portability, cloud storage, and unlimited capacity have realized the maximum "practical value" of music. As shown in Figure 1, at this time, the demand for vinyl records reached a low point in the market. According to statistics from the Recording Industry Institute of America (RIIA), which has reported music revenues since 1979, the vinyl industry in the United States has entered a period of rapidly declining revenues since the 1980s. In 1981, the total annual revenue of Vinyl Single and EP, SP in the U.S. reached about 2,600 million dollars, but by 2007, the industry's annual revenue had shrunk to about 27 million dollars [5]. And according to a 1990 survey in the New York Times, in 1977, at the height of vinyl's popularity, 344 million vinyl records were sold in the United States. According to a 1990 survey in the New York Times, in 1977, at the height of vinyl's popularity, 344 million vinyl records were sold in the United States. But in 1989, total annual vinyl sales were only 44 million, a figure that dropped by about 90 percent. By 1990, vinyl records accounted for only 4% of total U.S. record industry sales [6]. This period lasted until after the 21st century.

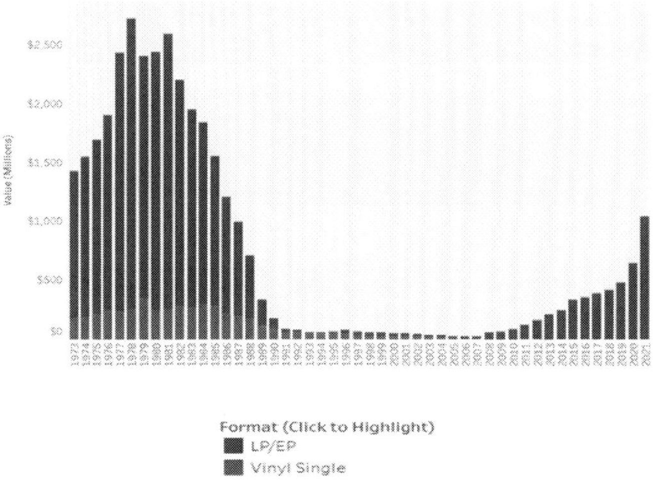

Figure 1. 1979-2019 U.S.Recorded Music Revenues (LP/EP and Vinyl Single) by Format [5]

Source: RIAA

Since around 2006, the vinyl market has shown some recovery and warming up, which can be seen as a new period of "revival" for the vinyl industry. This is the period in which the contemporary vinyl industry has been living, continuing to this day. During this period, the artistic value of vinyl records in people's minds, their market distribution characteristics, and the development prospects of the industry have changed significantly compared to vinyl's heyday, which will be further elaborated in a later article.

3 Contemporary Vinyl Record's Market Situation and Artistic Value Positioning

3.1 General Vinyl Market Situation

As shown in Figure 1, the vinyl industry has regained

a relatively steady growth in revenue since around 2006, and its revenue growth has continued to expand, especially after 2018. According to the Association's Music Revenue Report, in 2020, vinyl records generated more revenue than CDs for the first time in nearly 35 years, and in 2021, Vinyl Single and EP, SP total annual revenue regained more than 1 billion [5], which can be also shown in Figure 2. This renaissance continues today, indicating that the vinyl industry is still showing some life and vitality in today's U.S. music market. IFPI's global music report in recent years shows that the trend in the global market is largely in line with the trend in the U.S [7].

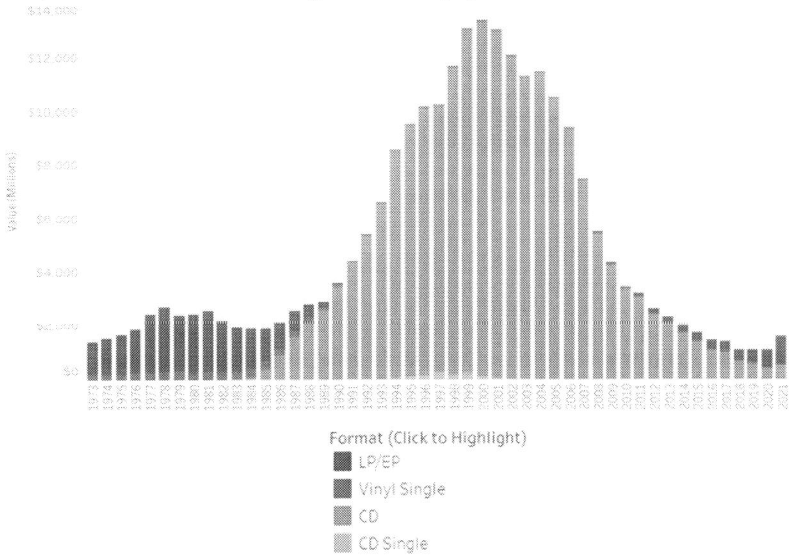

Figure 2. 1979-2019 U.S.Recorded Music Revenues (LP/EP, Vinyl Single) by Format [5]

Source:RIAA

In terms of the age of the consumer audience, the revival vinyl audience is focused on older and younger demographics. According to Marie Charlotte Götting's survey data published in Digital Market Lookout, "In the U.S., consumers in the 25 to 34 age group are just as likely to purchase a vinyl record as consumers 55 and older, with 21% of consumers in both age groups having at least purchased a record. Comparatively speaking, vinyl records were less popular among the 45 to 54-year-old group" [8].

In terms of vinyl consumption, the contemporary vinyl market can be divided into two categories: first, old vinyl records that record "precious sounds". Due to the decline in the number of releases and the destruction and loss of old vinyl records, this type of vinyl has become increasingly rare, and its extended form, such as the reissue of old vinyl records, has also emerged; the second is the vinyl version of contemporary pop music. This type of music generally appears as a physical version of digital music first released on the Internet streaming media, and its sales volume is highly correlated with the popularity that the music and the artist have gained.

3.2 The Reason of Vinyl Record Industry's "Revival"

The New Generation's Return to the Rituals and Physical Art Spaces of the Past. According to the above, vinyl record consumers are concentrated in the older and younger age groups. 55 and older were born during the vinyl industry's maturity and prosperity, and thus are likely to retain nostalgia and a long-term, sustained buying habit for vinyl records. In contrast, the growing interest and willingness to consume vinyl records by a new generation of young people is likely to be a major driver of the increase in revenue from the vinyl industry in recent years.

The numbness and disorientation brought about by the monopoly of media technology development have given rise to the younger generation's return to the ritualistic pursuit of their past lives. From the beginning of the development of the television medium to the rise of the Internet, media technology gradually showed an explosion of monopoly - websites and social media empowered more and more people to become communicators, information filtering systems "failed", and information "exploded", cutting people's lives into pieces of time and space. But unfortunately, as Jameson warned in his postmodernist theory about the "elimination of literary depth", behind the mass and fragmentation of information is the increasing superficiality and shallowness of culture. People are wrapped up in such a culture every day, and can only gain empty emotional experiences [9]. The young generation of today grows up in such an environment. As a result, they have some feelings of inner emptiness and loss, and then begin to seek this sense of gain and life experience from media born in the past, with some cultural characteristics and memories of the past. Vinyl records have captured this need very well: to play vinyl records, people need to prepare a record player, select records, choose a time and place to enjoy music, and, even with a large collection of vinyl records, each person can only play one record at a time. This is a return to "listening", a process in which people spend so much energy that music is no longer "at their fingertips", thus giving themselves a sense of ritual and a sense of concentration, immersion, and tangibility [10].

The emptiness of increasingly hollow data consumption has given rise to the younger generation's return to the pursuit of "owning music". Behind digital music is an open, "on-demand" access structure: data connects users to users, all of whom select and listen to music from the same open online library; music in digital format can be copied infinitely, meaning that the exact same song can be selected countless times, and there is no limit to the music library. There is no upper limit. In this case, music cannot be "owned". The music that each person hears and downloads is more like a scoop of water from this vast torrent of data. In fact, not only music, but all forms of Internet-based information today give audiences this empty feeling of insignificance. However, what vinyl records represent is a more personal, "possessive" structure of access. The process of buying a vinyl record is actually a process of taking it from the public to the private sphere. The vinyl in each listener's hands has a realistic quality; it is not only complete, but the nuances of the production process make it unique. The ritualistic process of listening to vinyl records, as described above, gives the vinyl record a certain degree of "authenticity" and "aura" in the words of Walter Benjamin [11], thus giving the listener a real sense of art appreciation and even art ownership.

The loneliness brought about by the privatization of music appreciation has given rise to the younger generation's return to the pursuit of shared intimate spaces. Contrary to the process of music acquisition, the author argues that the sharability of the current music appreciation process has diminished compared to that in the "vinyl era". The phonograph, the device used to play vinyl records, is a public device located in the private sphere. This characteristic allows for limited music sharing, or "intimate music sharing." People in the mid-20th century might play a jazz record over dinner, or dance the night away with their lovers to the sound of music. However, one of the main features of MP3s, MP4s, headphones and other products born with digital music is the further personalization of the music appreciation process. It is not new to see people sitting in the same living room or at the same dining room table but listening to different music. In this case, music has already become a barrier separating people from each other and from the environment. People could still unplug their headphones for sharing, but compared to the vinyl era, there has already been much less time for people to share music and much more time to shield each other.

The Tendency of the Middle Class to "Elitist" Fun. As mentioned above, compared with digital music, vinyl records are more expensive to listen to (in terms of time and money), have the scarcity that comes with a niche audience, and some older vinyl records even have a certain listening threshold (in terms of musical skill and historical context). The combination of all these reasons gives vinyl records almost all the characteristics of what is generally considered to be a refined cultural interest, "elite culture". Its further effect echoes Bourdieu's theory of "distinction" [12]: while the upper classes, who possess a great deal of cultural capital and the ability to decode art, may be happy to construct their own aesthetic superiority through vinyl, the middle classes are more likely to appreciate and enjoy it in order to escape the limitations of their own class interests. Thus, it can be said that vinyl records cater to the "elitist" tastes of the two classes that have the consumption ability today, and among the two classes, the middle class may show a more enthusiastic pursuit of vinyl records, especially classical vinyl records and old vinyl records.

The Niche Community Gathering is based on the Internet and Social Media. In the Internet era, the distinction brought by the difference of interest can be further diversified from class distinction to cultural circle distinction. Unlike clan communities built on the logic of geography and blood relatives and class groups determined by political and economic status, the construction and consolidation of circles are often directly based on common interests, i.e., a cultural consensus. People within the circle identify, communicate with each other, and even culturally reproduce around their consensus, eventually forming a community with both internal centripetal force and some external exclusionary force. The rise of the Internet and social media has given the growth of such communities a more fertile ground. No matter how niche a culture's "fans" are, through the Internet connection, they may find a group of enthusiasts who can share their interests. This is a transformation from "solitary appreciation" to "group admiration". The social connection and aesthetic value recognition gained in the community will eventually lead to a sense of belonging and stickiness to the circle, which in turn will stimulate the formation of related consumption habits. In fact, as of September 2022, Reddit's "/r/vinly: keep on spinning!" group has had more than 1.1 million members, many of whom bear the distinctive "vinyl fan" label [13].

The Fusion of Artistic Value and Economic Appreciation Space Brought about by "Scarcity". The economist Clare Andrew said, "the reality of art is produced, bought and sold by individuals and institutions working within an economic framework inescapable from material and market constraints" [14]. Therefore, as the market for the vinyl industry recovers, the expansion of the vinyl record's economic value deserves attention. Historically, the advantages of vinyl records in terms of mass production were once dissipated by new forms of music, but today, vinyl records are making a comeback with another characteristic that is the exact opposite of the above - scarcity. Therefore, as the market for the vinyl industry recovers, the expansion of the vinyl record's economic value deserves attention. Historically, the advantages of vinyl records in terms of mass production were once dissipated by new forms of music, but today, vinyl records are making a comeback with another characteristic that is the exact opposite of the above - scarcity. The loss aversion leads to a preference for scarce products, which has led to a surge in demand and prices for scarce products, a trend that has been reinforced by the "speculative economy" triggered by the expectation of their appreciation. A list from collectors' record auction 307 shows that the auction, which will run until July 2022, will sell 1,702 music items, including 1,006 vinyl records

[15]. The potential economic benefits of vinyl records have attracted a large number of collectors and investors, who have paid to give the vinyl industry a new impetus for growth.

Pop Stars' Spin-off Releases Help to Revive the Vinyl Record Market. If vinyl records existed as the primary vehicle for pop stars' releases during the golden age of vinyl's maturity, in contemporary times they have become an incidental part of some pop stars' digital releases. In fact, regardless of when and why the contemporary boom of pop stars releasing vinyl records and triggering a fan frenzy has occurred, this phenomenon has provided a very positive stimulus to the resurgence of vinyl sales. According to billboard charts, the best-selling vinyl record of 2021 was Adele's "30", which sold 234000 copies, followed by Olivia Rodrigo's "Sour" and Taylor Swift's "Red (Taylor's Version)" [16]. For fans, consuming vinyl releases by their favorite stars is a positive attempt to get psychologically close to their idols (as mentioned above, physical vinyl gives listeners a sense of intimacy and ownership that digital music cannot), but also an act of support - declaring their willingness to pay for their idol's work, whether or not they actually open the package and listen to the record on a gramophone in the end.

4 Vinyl Records Future Dissemination and Industry Development Expectations

The vitality over a certain period of time. At least for the time being, the spiritual scarcity of the new generation in the contemporary media and social environment, the quest for "elite" and scarce products, and the industry boosted by the Internet and pop stars are still going on, which means that the factors mentioned above are not going to change in a certain period of time. Moreover, according to the previously mentioned vinyl record sales data, the sales of this art product have been growing steadily in recent years, with no signs of decline. It is clear that the vinyl industry will continue to recover over a period of time.

The development trend of stable internalization. From a deeper perspective, in an era still dominated by mass production, unlimited reproduction and convenient consumption of artworks, vinyl records are actually exploring a path of internalization. The economic and time costs of listening to vinyl records and its "small but sophisticated" market character, which is the basis of consumer appeal and fan community cohesion, are fundamentally contradictory to the goal of "mass appeal". Besides, vinyl consumers brought in by pop stars also lack consumption stickiness (they only followed their idols and have nothing to do with vinyl in essence). As a result, the community of vinyl consumers and enthusiasts may be consolidated and moderately expanded, but the vinyl record industry's relatively marginalized position in the music market will remain unchanged.

The possibility of economic "breaking the circle" is brought about by auctions. In summary, it can be said that in contemporary times, the value of vinyl records as a symbol has far exceeded its practical value. While some consumers seek to live behind the symbol of vinyl, some consumers want to express their support through the act of buying vinyl records, for some vinyl record collectors and old vinyl record dealers, what matters is the economic value behind the symbol. In a logic of capital power operation, the symbolic attachments of old vinyl records are constantly exaggerated and their prices rise, which is what vinyl record auctioneers and even speculators are chasing. It can be seen that the operation of vinyl records in the auction market shows signs of a niche culture being "adopted" commercially, which, even though there may be chaos such as high prices, piracy, and consumer traps, can still serve as a way to popularize vinyl records in the economic field to some extent, and giving vinyl records the possibility of "breaking the circle" in this field.

5 Conclusion

As Marshall McLuhan said, "Media is the extension of man" [17]. Behind the process of media change is a change in the way of people's life and a change in society and culture. The "revival" of a medium may mean both the tendency of contemporary people to return to certain characteristics of the past era in terms of psychology and reality, and the derivation of certain value functions of the medium that can meet the new era and new needs. Vinyl records, however, have embraced and adapted to these circumstances and needs, thus ushering in another "mini-spring" of their lives.

Acknowledgement

The period of writing this article has been somewhat long, and its progress has even been stalled for a while. However, when I look back at the present time when this article is about to be completed, besides reflecting on my mentality and ability during the writing process, I am also grateful to many people and even spaces and scenes, and I would like to express some of them here.

First of all, I would like to thank my advisor in this paper, who kept an eye on me and supervised me during this somewhat long and stuck process of writing the paper, and also discussed with me very kindly and patiently when I suggested to meet with her to discuss the content of the paper. The advice she gave, especially in the area of article writing fundamentals and literature library, helped me more in my paper.

I would also like to thank my friends. They have been very important in the process of writing this paper. They not only encourage me, help me, influence me with their positive energy, but also discuss and even debate with me on topics related to my paper based on their own knowledge. This "brainstorming" process has given me a lot of inspiration.

In addition, I have a strong social support network built by my family and lover. They never put extra pressure on me, but their gentleness and comfort also gave me enough security to regulate myself during all the difficulties and low points.

Finally, I also feel grateful for the spaces and scenes in my life that I love. They include every quiet bookstore I've ever been in, the aromatic cafes, the school library, all the delightful food, and every lonely, warm night.

References

1. Library of Congress, The Gramophone Early Sound Recording Devices, https://www.loc.gov/collections/emile-berliner/articles-and-essays/gramophone/
2. The Vintage Record, The History of Vinyl, by american history now- exploration in digital Curation. Jan27, 2014.http://americanhistorynow.org/2014/01/27/the-history-of-vinyl/.
3. Patmore D, Selling sounds: Recordings and the record business [J] The Cambridge Companion to Recorded Music, 2019, pp.120-134.
4. Kenney W. H., Recorded music in American life: The phonograph and popular memory, 1890-1945 [M] Oxford University Press on Demand, 1999, pp.11.
5. RIAA. 1979-2019 U.S.Recorded Music Revenues by Format, 2022. https://public.tableau.com/views/U_S_RecordedMusicRevenuesbyFormat_0/RevenuesbyFormat?%3AshowVizHome=no.
6. Andrew Pollack, Recording Enters a New Era, And You Can't Find It on LP, by the new york times. Apr1, 1990. https://www.nytimes.com/1990/04/01/us/recording-enters-a-new-era-and-you-can-t-find-it-on-lp.html.
7. IFPI, Global Music Report 2021[J] 2021, pp.9-12.
8. Marie Charlotte Götting, Vinyl record buyers in the U.S. 2018-2019, by age group, by statista. Jun2, 2021. https://www.statista.com/statistics/1008779/vinyl-record-buyers-us-by-age/
9. Kaka Guo, On Jameson's Postmodernist Critical Theory of Culture[J], Journal of Kashi University, vol.42, no.2, 2021, pp.61-62.
10. Xuejia Liu, Vinyl Consumption in the Age of Digital Music [J] Market Forum, 2020, pp.68.
11. Walter Benjamin, A short history of photography [J]. Screen, 1972, vol.13, no.1, pp.19-25.
12. Pierre Bourdieu, Distinction [M] Social Theory Re-Wired, Routledge, 2016, pp.198-215.
13. Reddit, /r/vinyl: keep on spinning! Jan 25, 2008, https://www.reddit.com/r/vinyl/.
14. Dave Beech, Art and Value: Art S Economic Exceptionalism in Classical, Neoclassical and Marxist Economics [M] Historical Materialism Book vol.94, 2015, pp.1.
15. Collectors' Vinyl, collectors' record auction 307 [J] 2022, pp.1-7.
16. Keith Caulfield, Adele's "30" Becomes 2021's Best-Selling Vinyl Album, Holds at No.1 on Top Album Sales Chart, by billboard. Dec12, 2021. https://www.billboard.com/music/chart-beat/adeles-30-2021-best-selling-vinyl-album-1235013224/
17. McLuhan M, Understanding media: The extensions of man [M] MIT press, 1994, pp.1-2.

Analysis on the Application of Psychological and Business Strategies of Douyin to Build up Consumer's Addiction

Xiaoyu Zhu*

Shanghai United International School (gubei campus)-Shanghai, China

Abstract. The modern society has been developing rapidly, and the Internet has become one of the most revolutionary development throughout the eras. Nowadays, people's high reliance on the Internet inevitably leads to the development of diverse online products, one of which is the occurrence of the short video applications. A typical representation of a successfully runned short video platform is Douyin, which is the research objective of this paper. This research aims to investigate the strategic mechanisms utilized in the platform Douyin, in terms of both psychological and commercial aspects. This paper focuses specifically on how the use of strategies forms a habit-forming technological product which causes individual's high addiction. To conclude, the hook model and recommendation algorithms can be the main mechanisms that cause user addiction to Douyin platform.

1 Introduction

The phenomenal and rapid growth of online technologies has significantly impacted people's modern life around the globe. Innovation of mobile technologies has successfully shifted people's reliance on wired technologies to mobile ones, which require less space and bring more convenience. With the surge of electronic devices such as mobile phones, tablets, and laptops, the number of users of such has also increased dramatically. According to Statista's survey released in February 2022 [1], China's registered mobile Internet users increased from 302.73 million in 2010 to 1028.74 million in 2021, accounting for one-fifth of the world's total. According to the China Internet Network Information Center (CNNIC), "the share of Chinese users accessing the internet via mobile phone had already surpassed users who accessed the Internet via PC since 2014." Douyin, the Chinese version of an app known as TikTok, established on the 11th of March 2016 in Beijing, and is now playing a national leading role in the industry, turning itself into one of the most welcomed apps for people. Referring to the Insider Intelligence website [2], the data displays that 639.4 million active daily users are on Douyin, and it is predicted to expand up to 835 million in 2025. According to its official site [3], the platform's obligation is to inspire creativity, inspire life, and help people maintain a beautiful life. Therefore, various kinds of videos with positive and original content are promoted on the platform of Douyin , which contributes to a person's great life.

This paper aims to investigate the commercial and psychological strategies behind the platform Douyin and reveal the mechanisms which generate addiction through an analysis of principles and theories.The author studies how Douyin uses different strategies to build up its product and successfully play a leading role in the modern video platform industry. Moreover, its social influence, especially its negative impacts to consumers are analyzed in this paper in order to better understand its product from more dimensions.

*Corresponding author: zhuxiaoyu2023@outlook.com

2 Strategies of Douyin for Stimulating User Interest

2.1 The Hook Model

In the book 'Hooked' written by Nir Eyal [4], four concepts, also referred to as the hook model, of how to build habit-forming products are elaborately analyzed. The four principles are the four psychological stages a consumer proceeds through, and in the end, they are a loop that build a habit-forming product. The 4 stages start with a trigger, then an action that accomplishes the trigger, next with a variable reward given to the action, and eventually with the investment that gives values to the user. Figure 1 from the website ProductPlan visually presents the hook model [5], illustrating 4 principles that act as a loop, resulting in consumers' frequent and voluntary reliance on a product.

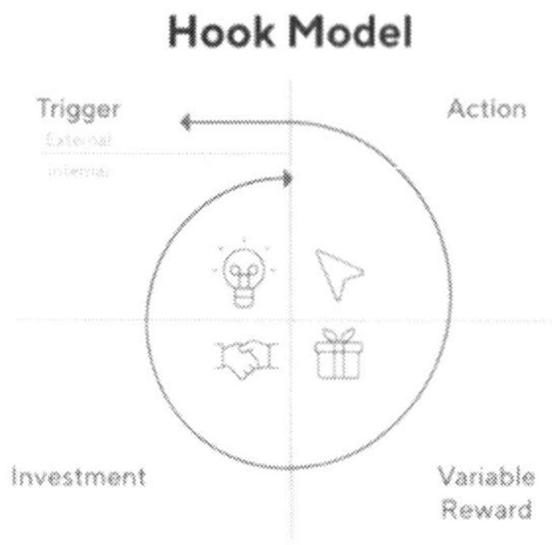

Fig. 1. The Hook Model [5].

Trigger. Trigger is an actuator that stimulates behavior, which initiates people's thoughts of something immediately. There are two types of triggers, external trigger and internal trigger.

External Trigger. External trigger refers to the elements delivered from the outside environment that provokes thought. In the case of Douyin, the design of the app icon, the video notification, and the viral music and dance all contribute to the form of external triggers. These triggers can link with each other. When people think of the music and dance, they automatically relate to the fact that these specific elements occurred repeatedly in Douyin, then the image of Douyin's icon appears in their mind. In addition, Douyin also initiates collaboration with assorted platforms to construct more external triggers. According to the Douyin official site about collaboration [6], platforms across different industries, such as 'Mao Yan Entertainment' from the entertainment industry, 'Blast Boy' from the gaming industry, and 'Capcut' from the video editing industry, all work with Douyin. For instance, a user can directly select a range of music from his or her Douyin saved list when using Capcut, and can also directly share the video edited in Capcut to Douyin without switching the app. According to the official site of Capcut [7], Douyin and Capcut collaborate simultaneously to ensure the creation of high-quality content and its widespread distribution to users, offering great closed-loop marketing for the content creator and advertisement sponsor, due to the successful spread of data and insight. The external triggers of Douyin tempt consumers into not only thinking about the app when related content appears but also raise people's level of interest or attention to the app by collaborating with numerous apps across diversified industries.

Internal Trigger. The internal triggers are triggers that elicit connections to ideas or emotions, which are usually intangible and varied. Negative emotions that last for a short period of time, such as boredom, depression, and loneliness, are some of the typical emotions that link with triggers. Surprisingly, Douyin has the ability to satisfy these emotions together. Boredom can lead consumers to use Douyin to watch videos, which creates a task filled with diversity for people to distract their focus from boredom and to enjoy. Depressed emotions can lead people to use Douyin to search for positive videos that can motivate them. Loneliness leads consumers to connect with followers or friends through commenting, mentioning, or even starting lives and meeting new people on the platform. All of these internal triggers are linked with the desire of using the app.

With both external and internal triggers clearly present in Douyin, the next step in the loop is the action.

Action. With the expectations formed by triggers, consumers demand an action where they take tangible behavior to satisfy the triggers. However, the action should meet three elements according to the Fogg Behavior Model [8], where B (behavior)=MAT (M for motivation, A for ability, and T for trigger). Triggers already exist for Douyin users. The concept of motivation indicates the presence of pleasure, hope, and social acceptance. For one thing, Douyin offers a limitless range of videos, which is a pleasure and hope for the consumers, as they may hope to find joy and positivity by viewing the content. This corresponds with Douyin's core value mentioned before, which is to inspire creativity, inspire life, and help people maintain a beautiful life. For another, people's need for social acceptance can also be satisfied with the interaction with other users on Douyin, such as following each other, giving likes, comments, and reposts. All of these build up one's self-confidence and social acceptance. After motivation and trigger are both secured, Douyin continues to drive the consumers to feel that they have the ability to take an action. Douyin provides short videos that last for only around fifteen seconds to ten minutes [9]. Therefore, users can take a low extra cost to make videos on Douyin since it only requires one mobile device and can be accessed in any external environment as long as the internet is available. It is also very approachable for users to create content quickly, because shooting short videos is time-saving and only requires a few clicks on the screen. Due to the convenience, the number of Douyin users has increased a lot, allowing the platform to be more interactive as more people are willing and able to spend time and energy on it.

Variable Award. The next stage in the loop is the variable award, which is about receiving a reward that is not fixed, therefore it is different each time. The variable award has the same effect as any fixed award, which is encouraging consumers with things that please them, so that they can constantly build up trust, affection, and reliance with the platform. From the perspective of content creators, they must hope their videos to be viewed by more people when the videos are posted publicly. Since videos posted on Douyin are usually short and the platform has a very strong and active user base, one of the most appealing awards for the users is the rapid distribution of their videos. Douyin delivers the video to abundant users, which usually results in the creator receiving likes, comments, reposts, and duets. These positive feedbacks from other users act as an award since positive feedback is a signal of popularity that allows the creator to be more engaged and motivated. It is a variable award but a fixed one due to the diversity of content. Each time the content varies with the music, setting, character, video length, etc, it indicates that each video posted on the platform is unique. Thus, the award varies each time, depending on the quality of content and number of followers. This mechanism inextricably stimulates consumers to create content in the long term, with the goal of producing better content,

https://doi.org/10.1051/shsconf/202315502016

gaining more views, and eventually growing to be an account with a high number of followers.

When considering the point of view of the non-content creator, the mere video-watching consumers, there is also variable awards. All of the videos in Douyin appear on the 'for you page' with full-screen mode according to the system's recommendation algorithms. The curiosity is formed each time when a user scrolls the screen in order to switch to a new video. The consumer cannot predict the content of the next video, so every scroll brings originality and exploration, which encourages them to keep scrolling and immersing with a sense of mysterious and addictive experience.

The presence of variable awards is vital for turning a product into habit forming. Everyone needs new dimensions within a product to feel actively engaged and interested in the long term. If users always know what award they will receive in advance, the expectation for and willingness to use a product can be gradually reduced. The variable award is an important factor contributing to a product's life cycle, focusing on creating an experience beyond what users expect.

Investment. The final step in the hook model is the investment, where consumers invest time, effort, money, and emotions into using a product. The voluntary investment corresponds with the expectation of a gratified output. An increase in investment over time depends on the amount of output a consumer receives. The more investment there is, the more dependence and addiction to Douyin one can form, particularly with the surge of views and followers as variable awards, which causes further investment.

Investment does not mean to force users to spend money, but to shape users' habit-forming behavior toward a product so as to build up trust. Investment also provides a respected and personal interpretation of the product experience, which may help improve the product by reporting any issues or suggesting any advice to the platform. Investment influences the previous three stages of the hook model, leading triggers to be formed more easily, actions to be more feasible to occur, and rewards to be more attractive and desirable.

2.2 Recommendation Algorithms

The recommendation algorithm is a systematic and logical calculation method that deduces accurate and complete recommendation data in the video page of Douyin. It helps the app to obtain efficient output within a limited time and input. Powerful algorithms not only save users' time and cost in finding high-quality content but also enable an efficient operation of the platform. Douyin has rapidly expanded into the Chinese market and accumulated a massive portion of users in a short period of time, and one of the reasons for its success is inevitably the intelligent algorithm system. Douyin adopts an algorithm to determine the dissemination of videos mainly from two dimensions.

Targeting on Users' Personal Information. When a new user registers and logs in to the account, personal information such as name, age, gender, city of residence, and career are all being filled out. It is the first data set the algorithm system is supposed to be in control of. In addition, there is also a choice of logging in using a third-party social account, which leads the algorithm to explore some previously existing personal information and historical social behavior, enabling the system to further investigate the user's behavior, habits, and interest efficiently [10]. The next content that appears on the screen is the selection of labels about interesting fields, including but not limited to comedy, animation, gaming, food, healthy lifestyle, sport, fashion and etc. It is compulsory to complete the selection of some areas of interest before starting the app.

Targeting on Users' Online Interactions. In the digital era, "the users are no longer just passive receivers and consumers of information, but also active senders and producers [11]." Therefore, another method of calculating precise recommendations is based on an analysis of users' social interaction, which can be classified into two categories. The first category is calculated according to the connection of accounts among the user's friends, family, and phone contacts, who are mutual friends on the platform with the user. The algorithmic mechanism is tracking the number of interactions between the user and the friends, where if the more are detected, the closer connection of such social relationship Douyin will assume the user is having. Therefore, it is highly feasible that the algorithm recommends similar or the same content that the friend has liked to the user's page, with the assumption of close friends share a similar field of interest.

The other category is based on analyzing users' interaction of comments, likes, and reposts with unfollowed accounts or non-mutual friends. When the Algorithm detects that a user often likes videos from the same creator, the system infers a high likelihood of the user keep liking new videos made by that creator. As a result, the platform identifies such phenomenon as an attribute of the user's interest, resulting in frequently promoting videos of that creator to the user's page. In other cases, when a user often likes videos of the same type, such as videos using the same hashtags, then the Algorithm will learn to customize such videos into the user's field of interest, thus increasing the chances of users watching such videos.

The recommendation algorithmic mechanism successfully helps Douyin to pinpoint an individual user's domains of interest. This can help raise user stickiness with the product, further enhance users' dependence on the platform, and increase usage time and frequency.

3 Problems of Douyin's Addiction Strategies

3.1 Creating Limited Vision

The utilization of both the hook model and the recommendation algorithmic mechanism systematically meets the needs of the consumer by providing a habit-forming product with customized and personalized content according to the user's preferences. However, the most evident drawbacks are also due to such mechanisms.

Only offering content favored by consumers in a video platform limits the growth of users' vision, narrowing their perception of information and knowledge. Consequently, such mechanisms may result in one's solidification of ideas to an extent, leading the audience get immersed in the virtual world built up in Douyin.

3.2 Losing Self-control

The 4-step loop discussed in the hook model illustrates how consumers can form a habit of using Douyin, which results in an outcome of losing the ability to stop. Once consumers have formed the habit of checking Douyin in their free time, such habit can be worsened to a circumstance where consumers lose self-control and intentionally create spare time to check Douyin videos even during their working time. It is an over-reliance that may cause the lose of self-discipline and make people feel unable to focus on reality.

4 Conclusion

In conclusion, Douyin efficiently utilized the strategies mentioned above to form a highly addictive platform for its users. When shaping its habit-forming product, Douyin adopts the hook model containing four stages, namely the stage of trigger, action, variable award, and investment. The adoption of hook model makes users generate a systematic and complete psychological mindset. Such mindset is the direct trigger that connect Douyin with its users closely, leading the consumers to be highly reliant with the platform. Additionally, the platform's precisely designed recommendation algorithm system calculates the user's field of interest by referring to the past searching record, liked content, interacted account, etc, to form a favorably personalized recommendation page, or known as the "for you page". Application of such a method ensures the information with a high possibility of interest to the users will be disseminated on the page. This has caused users to almost exclusively watch the videos of a similar genre without feeling exhausted since they are interested in those topics. Negative impacts also arise as a consequence of these successful mechanisms. Limited vision due to recommendation algorithms is inevitable, and a loss of self-control due to the addictive system may also interfere with consumer's real life, causing a lack of self-discipline and focus on reality.

In terms of the recommendation algorithms, this paper does not demonstrate the systematic calculation process which involves professional knowledge in mathematics. Therefore, future research can be improved by precisely illustrating the calculation steps of the algorithms, showing a clear and logical process of how exactly data are gathered and analyzed.

References

1. Statista: Number of mobile internet users in China from 2010 to 2021 (in millions), www.statista.com/statistics/273973/number-of-mobile-internet-users-in-china/#:~:text=The%20rapid%20growth%20in%20smartphone%20and%20tablet%20user, last accessed 2022/07/31.
2. Insider Intelligence: Douyin users in China - stats and forecast (2021-2025), www.insiderintelligence.com/charts/douyin-users-china-stats/, last accessed 2022/07/31.
3. Douyin: About us - company value and social responsibility, www.douyin.com/aboutus/, last accessed 2022/07/31.
4. Eyal, N.: Hooked: How to Build Habit-Forming Products. Norwick, Penguin Books, 2016.
5. Newell, S.: Hook Model, www.productplan.com/glossary/hook-model/#:~:text=The%20Hook%20Model%20is%20a%20four-phase%20process%20that. last accessed 2022/07/31.
6. Douyin open platform: open.douyin.com/platform, last accessed 2022/07/31.
7. Lv.ulikecam: Capcut professional version-easily used video editing tool for desktop-easy editing with high quality content, lv.ulikecam.com/adv_coop, last accessed 2022/07/31.
8. Captology_Behavior_Model_for_Persuasive_Design_2009.Pdf. Google Docs, drive.google.com/file/d/19LYba4fuceGM3KhqxTXByV4msmb33t1o/view. Accessed 30 July 2022.
9. Deng, D.S., Seo, S.B., Li, Z.M., Austin, E.W.: What People TikTok (Douyin) about Influencer-Endorsed Short Videos on Wine? An Exploration of Gender and Generational Differences, Journal of Hospitality and Tourism Technology, 2022, 10.1108/jhtt-05-2021-0143.
10. Zhao, Z.W.: Analysis on the "Douyin (Tiktok) Mania" Phenomenon Based on Recommendation Algorithms, E3S Web of Conferences (235), 2021, 10.1051/e3sconf/202123503029.
11. Han, B.C., Butler, E.: In the Swarm: Digital Prospects. Cambridge, Ma, Mit Press, 2017.

Research on the Use of Social Media to Grasp Consumers' Psychology and Create Traffic: A Case Study of Brandy Melville

Zheran Liu*

University of Southampton, Southampton Hampshire, UK

Abstract. In recent years, Brandy Melville (Hereinafter referred to as BM), an Italian clothing brand, has gained popularity in mainland China because of the style of its slim tops and skirts, which echos the "freedom to dress" speech that has been popular in recent years. However, BM only produces one size and only girls who are thin enough can wear it. This makes other girls feel that their bodies are not recognized. Therefore, BM is also known as the "creator of body anxiety". Nevertheless, wearing BM has become a hint to prove one's "good figure", and BM's marketing strategy also cleverly utilizes consumers' vanity to achieve the purpose of marketing itself. So even though there are a lot of critical comments, BM is still gaining popularity from the public. The purpose of this case study is to explore how Brandy Melville grasps consumers' psychology and controls their emotions, and how it takes advantage of hot comments and critical comments to build its own image and drive consumers to buy. Through the case study of Brandy Melville and the interview with its audience and Key Opinion Leaders (KOLs), this paper aims to find out how social media is used to influence and guide the public's psychology and emotions. Results show that social media can be used to build the brand image, lead the fashion trend, echo hot topics, and create public sentiment to grasp consumer's psychology and create traffic.

1 Introduction

Today, people are living in a consumer society, where consumers' desire to buy is constantly stimulated for the market expansion and pursuit of profits [1]. O2O (online to offline) came into being, making the Internet a platform for offline transactions [7]. In this situation, fast fashion brands influence and attract female groups through Internet marketing, and gain benefits by controlling female body desire [1][2]—the desire of consumers to create personal identity through fashion consumption [3]. Fast fashion brands are committed to using the Internet to explore female desire, lead female trends, and indirectly attract women to buy [4]. In East Asia, the aesthetic of "white, thin, and young" occupies a dominant position. In China, white is regarded as beauty since ancient times, and the delicate figure and harmless young face are even more popular [5]. Capital quickly captures the marketing trend of body anxiety, shaping and exacerbating the aesthetic trend of "white, thin, and young", which has become a "perfect" norm and aesthetic standard [5]. Young women are sensitive to the aesthetic standards portrayed on social media. They are led by blind conformity and are easily to be attracted by the figure or dress of celebrities and influencers of social media [4]. Meanwhile, with the support of fan economy, bloggers start to compete for data traffic and hot topics, and make use of their fan effect by adding specific tags [6]. Brandy Melville promotes the high quality content of self-shaping "thin and white" body on the online social platform, and achieves accurate matching through big data, thus harvesting traffic dividends [5]. This paper takes Brandy Melville as an example to explore its media marketing strategies, thereby making reference for similar brands to promote themselves in mainland China.

2 Methodology

2.1 Research Participants

The author conducts interviews from two different groups of interviewee: KOLs (Key Opinion Leaders) and consumers. From KOLs, it can be known whether they have enjoyed the dividends brought by "BM wind", and how to use the BM trend to attract the public's attention. From consumers, their emotional feelings and psychological cognition of BM style, as well as BM's influence on their consumption behavior can be known. All the KOLs and consumers will be connected from the Xiaohongshu, Weibo, and WeChat, because these three platforms currently have the largest number of mainland users [7], and the interviewees from these platforms are more representative. The demographics of participants are shown in Table 1.

*Corresponding author: liuzheransoton@163.com

https://doi.org/10.1051/shsconf/202315502017

Table 1. Demographics of participants.

Participants (only surname or screen name)	Gender	Age	Length of Use Year	Type of Interviewees Consulted (for whom)
Li	F	26	4	consumer
Gao	F	30	3.5	consumer
Huang	F	22	3	consumer
Greenie	F	25	4	KOL
Xu	F	19	2	KOL

2.2 Research Design

This paper uses qualitative research, and the main methods are case study and interview. Because fast fashion brands mainly use consumers' psychology and emotions to carry out social media marketing, a case study (BM as an example) is the most clear and direct way to give readers an object to think and analyze, and can provide detailed information about individuals. However, this paper focuses on the analysis of consumers' emotions and psychology, and there are great subjective differences among individual consumers. However, the conclusion obtained by using the case study only is too theoretical and easy to deviate from the actual situation, and it is difficult to be extended in depth. Therefore, an interview will be analyzed together with the case study in the form of thematic divisions. The interview questions are designed as follows:

To KOLs:

1. What is your general opinion on BM style?
2. Have you ever enjoyed the dividends brought by "BM style"? What way does BM use to attract traffic?
3. What do you think about BM's creating body anxiety? Do you ever feel body anxiety from BM style? If so, will such negative emotions motivate you to buy BM and strive for identity, or will you choose to boycott it directly?

To Consumers:

1. What is your general opinion on BM style?
2. How does the "BM style" notes on social platforms affect your mood? (Does it make you want to follow or does it cause body anxiety and disgust?)
3. What do you think about "BM style" creating body anxiety? Do you ever feel body anxiety from BM style? If so, will such negative emotions motivate you to buy BM and strive for identity, or will you choose to boycott it directly?
4. Will you try BM because of the blogger's promotion (influenced by marketing)?

Above are the preliminary basic interview questions. To better understand the interviewees' emotions and psychology, extra questions may be added based on the communication with the interviewees and their thoughts. The way to gratitude these interviewees will be inviting them to the Suzhou hotel/restaurant for free experience (it is easy to get these experience because of the author's work), so as to motivate and appreciate interviewees to participate in the survey and actively express their feelings.

3 Results and Analysis

From the above case study and interview, the findings will be analyzed through four aspects: the image building, fashion trend leading, hot topic echoing, and public sentiment creation.

3.1 Building Image

BM adheres to the slogan "one size fits all" and gives girls who fit into BM an honor of showing off their body. This has given BM an image of only suitable for thin girls. As Li and Huang (the consumer) said: "It is suitable for girls with a good shape. For the girls who are not thin enough, it is quite unfriendly. Taking my friend who is a bit fat as an example, she cannot find even one suitable dress in BM. Sometimes, wearing a BM makes me feel that I am in a good shape, and that feeling makes me happy."

"This brand gives me a feeling that I am proved to be thin enough if I can wear it or dress it well. Cause we all subconsciously think that BM is only for thin and young girls, and these two elements are the basic of being a beauty for most of the people. So wearing BM makes me feel good."

By insanely implying that "women who can fit into BM are attractive girls", more and more girls join in the practice of wearing BM under this ideology, thus feeling the self-identity they pursue through this purchase behavior [8]. This has also deepened the "white, thin, and young" aesthetic bias.

As Li admitted: "As more friends praised me like 'you are so thin!' when I wear BM, honestly I feel quite happy. This makes me have a feeling of closeness to this brand, which prompts me to buy more clothes in this style...I was not a pursuers of 'white, thin, and young' aesthetic before, but after I tried this BM style and found that I am quite suitable for it, I started to pay close attention to it and gradually fell in love with it. I still think it is not the only aesthetic standard, but for me, it becomes the first standard."

3.2 Fan Culture

BM takes advantage of fan culture and uses stars and Internet celebrities to market itself. The Chinese star Yang Mi, who has 111 million followers on Weibo, makes BM fashion more popular with her BM outfits. At the same time, Internet celebrities pay more attention to user-generated content (UGC), for example, the bloggers of Xiaohongshu use the platform to spontaneously share their daily BM outfits. Most Internet celebrities and consumers are willing to follow fashion trends.

As Li (the consumer) said: *"I do like Yang Mi, and most of my friends love her BM outfits. Her all-black BM look which is famous on Weibo do surprise me and the next day, I went to the BM shop to find the similar clothes. I have been a fan of Yang Mi for about 5 years, and, coincidentally, she is quite suitable for BM style, which makes this brand even more famous. I think she suits the BM brand style quite well and this has attracted more consumers."*

 https://doi.org/10.1051/shsconf/202315502017

Xu (the KOL) said: "I am thin and 'white, thin, and young' is the label and characteristics I attract my fans. BM got famous these years, and I always need to grasp the trend. I remembered the first time I tried it and posted it on Xiaohongshu with a BM tag, I got 3000+ likes. The fans commented it as amazing/good shapes......and I know I successfully grasp the trend and made them feel I am fashionable. Now BM is not a new trend, but still has its traffic. Most of my fans are willing to click on it when I post something related to BM style."

In conclusion, wearing BM can help celebrities prove their own figures and cater to the label of "white, thin, and young". Additionally, the spontaneous sharing of influencers can also in turn promotes BM fashion to become a trend. As Jia and Yue pointed out, producers and advertisers work together to influence mainstream aesthetics and promote the development of the consumer market for ordinary women by taking advantage of the body demonstration effect of celebrities and Internet celebrities [9].

At the same time, as Shi and Li said, young and beautiful bodies are the capital of stars and Internet celebrities [5]. They are double body performers, that is, a capital performer and an idol performer, so they become the biggest beneficiary of the body commercialization and the most powerful promoter of the body consumption myth.

3.3 Echoing Hot Topics

BM has gained popularity by catering to the current hot topic of "freedom to dress". The social dogma and media environment in the mainland make women constantly struggle in the discipline for a long time, and women's self-awareness makes them have corresponding rebellious behaviors [10].

As Gao (the consumer) said: *"I was shamed to wear such kind of clothes before. My mum is likely to say they are too short and she always doesn't like me to wear clothes that are not conservative. I know it is wrong old thought but I am still affected......but until the appear of BM trend, it seems that all the girls are aware that they can wear the short skirts and tops, and prove that they are not wrong for they have this right. So for me, I have changed a lot from a girl who dares not to wear tight dresses to be willing to show my figure. It is a ideological awakening for me."*

Greenie (the KOL) also said: *"The society seems to become more inclusive now. More people would like to see my diverse outfits, and there is less jarring voice about 'You are wearing too revealing clothes'...It seems that the expression of female body has been accepted by most people."*

All in all, BM's bold expression of femininity makes women accept themselves both psychologically and physiologically. Wearing BM makes them show their personality charm and cater to "freedom to dress" speech. It is also a declaration of war to some rigid feudal thoughts. Therefore, for BM pursuers, the popularity of this style is also a reflection of dress freedom and social progress.

3.4 Creating Public Sentiment—Anxiety

On the one hand, women take the initiative to lead the fashion trend and are also shaped by this fashion trend. They enjoy the pleasure brought by fashion and the perfect visual image. On the other hand, women also feel controlled by the aesthetic power created by the media [11], which brings negative influence to the brand. For example, BM builds a loyal community with an exclusive culture to meet the demands of some women to share and show their "perfect figures" on social platforms to gain superiority. After such "figure showing off" is amplified on social platforms, it causes more women's dissatisfaction and anxiety about their figures. Female consumers will find themselves in the dilemma of subjectivity loss when they pursue the fashion trends again and again [12].

As Gao and Huang (the consumer) said: *"I am not too fat and not too thin. When I see the notes about BM shared by others, I will immediately buy them if I thought they are beautiful. Some of the clothes are not looking ideal on me, and at that time, I will think whether I should lose a little weight so that I can look good in it. And that kind of feeling makes me a bit anxiety because losing weight is not so easy for me. Also, I realize that I am not thin enough and I am not recognized by BM completely."*

"I am strict on myself, so I also have some anxiety. Now, I think the society is pretty unfriendly to girls. Although I like BM very much, I can't deny that it helps rationalize the standard, as if fitting into a BM is a proof of a slim and perfect figure. Sometimes I feel tired of viewing the post of marketing, showing how good the body is by wearing BM. I think whether it is the only standard for girl's shape."

Xu (the KOL) also said: *"I do not always use BM to attract fans. There are some critical comments like 'creating body anxiety'. Dramatically, BM helps women out of the old thoughts and get the freedom to dress, but meanwhile, it puts women into another frame. More people realized that and they tend to boycott it if I post too much BM speech. But in such collisions, topics often appear, and that's why BM has so much traffic though the critical comments."*

To conclude, BM presents the "perfect figure" constructed by the media, which becomes the password for bloggers to attract traffic. When they share their notes about the figure, on the one hand, it satisfies the purpose of bloggers to prove themselves and earn traffic and attention, on the other hand, it intensifies the anxiety of the audience. As a result, the female group with the original entertainment mentality will have self-doubt and an increasing amount of anxiety after browsing the notes [13]. As Gu and Kui said, when more and more girls are able to show off by wearing BM clothes, BM will evolve from a brand name to an open-ended style symbol [14]. Girls who are well dressed are automatically divided into the same camp, and through the filter of body size rating, people are judged according to their clothes, which is an invisible irony to women who fail to try BM style, thus triggering the resistance of some consumers. The appear of collisions and the arising of the topics make the BM get more traffic and become more famous.

https://doi.org/10.1051/shsconf/202315502017

4 Conclusion

Brandy Melville's successful marketing can be seen from four aspects: the image building, fashion trend leading, hot topic echoing, and anxiety creating. It provides a way for women to pursue dress freedom from old thoughts and encourages them to show their beautiful shape confidently. However, it also ties women into another frame—the standard of "white, thin, and young". The label becomes a way of marketing for BM and it helps BM get traffic again. The anxiety it created gives it another topical image. As Yin said, under the pressure of staring and being stared at, women's bodies become objects of social media consumption, constantly constructed as the "anxiety" itself. It is not right to praise or criticize BM blindly. However, what other brands should learn from BM is that it captures the audience psychology, the current hot topics, and characteristics of social media to constantly create traffic for itself and keep the popularity. Nonetheless, the interviewees in this research are all from the author's resource, which is hard to represent all the consumers and KOLs. The further study could find other minority consumers for more comprehensive research.

References

1. Baudrillard, J.: On consumer society, Rethinking the Subject. Sociology & Social Policy, 193–203 (2018).
2. Wang, J.N.: The impact of the social media UGC on its users' purchasing behavior. China Collective Economy (05), 59–60 (2022).
3. Armstrong, C.M.J., Kang, J.Y., Lang, C.M.: Clothing style confidence: The development and validation of a multidimensional scale to explore product longevity. Journal of Consumer Behaviour 17(3) (2018). DOI:10.1002/cb.1739.
4. Yin, Y.T.: Study on body anxiety of young women–Take Xiaohongshu as a case study. New Media Research, 78-84 (2022). DOI:10.16604/j.cnki.issn2096-0360.2022.05.004.
5. Shi, D.Y., Li, L.Y.: The anxiety and reflection of women's "body consumption" under visual culture—A case study of "BM trend". Radio & TV Journal, 159–161 (2021). DOI:10.19395/j.cnki.1674-246x.2021.08.071.
6. Liao, Z.: Fan culture under the influence of media development. Open Journal of Social Sciences, 88–93 (2021). DOI: 10.4236/jss.2021.912007.
7. Wang, J.X., Wang, X.L.: Fast fashion clothing brand marketing strategy under the O2O mode. Mass Investment Guide (01), 85–86 (2021).
8. Chan, T.H., Chen, R.P., Tse, C.H.: How consumers in China perceive brands in online and offline encounters: A framework for brand perception. Journal of Advertising Research, 90–110 (2017). DOI: 10.2501/JAR-2017-041.
9. Jia, W.W., Bie, Y.Y.: Web celebrity economic vision under the influence of the marketing: a research review and outlook. Foreign Economy and Management, 24–41 (2021).
10. Dai, Y., Walther, J.B.: Vicariously experiencing parasocial intimacy with public figures through observations of interactions on social media. Human Communication Research, 322–342 (2018). DOI: 10.1093/hcr/hqy003.
11. Li, X.: Micro-cultural aesthetic character and aesthetic paradox under the perspective of media technology. Journal of Jiangxi University of Finance and Economics, 107–115 (2021).
12. Hung, K.: Why celebrity sells: A dual entertainment path model of brand endorsement. Journal of Advertising 43(2), 155–166 (2014).
13. Fan, M.Y.: Young women's body anxiety: BM trend epidemic phenomenon solution. Journal of Shanxi Youth Vocational College, 23–27 (2021).
14. Gu, X., Wei, Y.Z.: Body anxiety. China Newsweek, 11–13 (2019).

Urban Communication Strategy Based on Short Video Platform: A Case Study of TikTok

Xueyi Wang

Literature of Theatre Film & Television, Xi'an University of Architecture and Technology, Xi'an 710055, Shaanxi Province, China

Abstract. The concept of urban communication bred in the new media era takes short video platforms as the vehicle for emerging Internet-famous cities. The internal relationship between the communication strategy and development orientation of urban communication and the image practice and performance space of short videos is worth studying. Taking Chongqing and Xi'an as the research objects, this paper explores the operation mode and communication concept of urban space in the new media era from the perspectives of communication origin, user participation, and in-depth communication. The research shows that the advent of mobile short-video platforms offers a new vehicle and channel for the communication and reconstruction of urban space and enriches the audience's cognition of city images.

1 Introduction

The external image of a city can be spread through media. Zhou et al.[1] believe that the influence of the Internet on urban space is fundamental, and the evolution of the relationship between physical space and cyberspace is an ontological transformation. Short videos spreading in urban space can be regarded as a product of an informatized "space of flow"[2]. Strangers from all over the world make a virtual tour of Internet-famous cities through short videos. The openness of short video platforms often has a certain imagination of "decentralization" and "freedom and openness."[3-4] It allows strangers from different classes and standpoints to present themselves in the same digital space, which promotes the long-distance communication of the culture and characteristics of Internet-famous cities. Urban communication has thus evolved from one-way communication to multi-way communication, and an individual-centered We-media communication network for visiting Instagram-worthy places has been formed.

"Transduction of space" [5] is fully reflected in the short video release. Landmarks displayed and intra-city recommendations allow scattered strangers to share and interact with each other, which promotes personal presentation. Virtual short videos bring a novel experience to the audience, and then an offline visit to Internet-famous cities helps them feel the cultural atmosphere of the cities. Media worker Wei Chuan once said that in the context of new technology, "visiting Instagram-worthy places" is to win the recognition and resonance of the target audience [6]. Such behavior has become a unique recording behavior, showing "a sense of place and space under the constant intertwining and flowing between people and media" [7].

Through the investigation of videos about visiting Instagram-worthy places released by users of major short video platforms, it is found that the purposes of video release are various, ranging from recording mood, sharing food, visiting places of interest, to returning to tradition. The release of short videos is influenced by bloggers' age, gender, knowledge structure, team operation, and other factors, which verifies a "process of self-invention" [8]. Individuals hope to obtain social attention and recognition through short video platforms; as a result, their images presented on the homepage and interpersonal networks tend to be heterogeneous.

The Global Cultural Industry proposed by Scott Lash and Celia Lury [9] points out that media is undergoing a transformation "from representation to things." Short video promotion, which is popular in the digital age, also has new problems in high-speed operation, which affects the communication of Internet-famous cities. Therefore, this paper takes TikTok as an example to analyze how city images are marketed and promoted through short videos and reveal the linkage relationship among short video makers, followers, Instagram-worthy places, and Internet-famous cities.

2 Theoretical Analysis

2.1 The origin of mass media intervention in urban communication

Activity communication and image communication are two effective ways of urban communication. Mass media is constantly involved in urban communication, and mainstream media give information feedback to the urban image. Certainly, the carrier of city image promotion is not

xueyiwang19@outlook.com

limited to mass media. What we emphasize here is that mass media is an effective means of urban communication. In the research on city image promotion, Li [10] studied all links in urban space from the overall, macro, and government levels. The mass media constructs a pseudo-environment with words, sounds, and images, in a unique symbolic, and non-forced way, which enables the public to skip the process of experiencing the city in person and form cognition and imagination about the city in an easier way.

For mass media, the transformation from newspapers to movies is an evolution from texts to images and then to videos; it is also an evolution process from cool media to hot media [11]. Urban memory is generated in physical space, virtual space, and people's spatial practice [12]. Media writing constructs urban memory and shows the unique vitality of a city. With the aid of short videos taken by influencers, Xi'an, Chongqing, Chengdu, and other cities have become Internet-famous, with local scenic spots emerging as new Instagram-worthy landmarks. The cross-border integration of short videos and culture has promoted the development of the economy, especially in the cultural tourism industry. According to the "TikTok Statistics on Labor Day 2021", Chongqing has become the city that attracts the largest number of visitors on TikTok for the first time, surpassing Shanghai, Shenzhen, and Suzhou. In addition, 112 million people made virtual tours to China via TikTok studios, and more than 500,000 people to ancient Rome, Sicily, and Bali.

2.2 Explanation congruence: Personal perception and preference of short video participants

Based on the random sampling results of TikTok, this paper explores the dynamic mechanism of publishers' subjective intention and participants' aesthetic perception experience. According to Szulc, social media not only creates new tastes and desires, but also enables users to produce more data by providing them with new "connections" [13]. The vogue of trending videos about Internet-famous cities is the external manifestation of the coordination between the subjective perception of publishers and participants and the visualization of Internet-famous cities.

The communication and construction of city image show strong media dependency through short video interactive experiences. From the perspective of participants, the emotional orientation of comments and the increase in the actual number of visits indicate that participants can acquire information more in line with their preferences. Renowned artists and bloggers who make short videos break the "spatial barriers" and "uncertainty obstacles" that prevent participants from generating behavioral intentions[14]. Short videos released by these "opinion leaders" often win more likes, followings, comments, sharing, and discussion, which have an obvious demonstration effect. Moreover, popular short videos released by opinion leaders are connected and systematic, which makes it easier to tap fans in certain fields and greatly stimulates participants to visit Internet-famous cities[15].

2.3 Mediatization: Short video promotion excavates the deep connotation and soul of the city

Urban communication is actually the visualization of urban memory. In image production, short videos are short decoding of urban memory, which flows in urban cultural space and civic life, promotes urban cultural production in the upsurge of visiting Instagram-worthy places, and constructs a new urban culture. Therefore, some scholars have pointed out that memory representation is a process of cultural construction [16]; culture, especially cultural value, is the inherent qualitative definition of cities [17].

Sawyer, who engaged in the study of space theory, has raised a question about the archaeological research of "City First:" "Can cities and urban space be viewed as a specific object of theoretical research, or is it just the result and reflection of a broader social process?" [18] As the carrier of urban exploration, short videos cannot serve as an in-depth and comprehensive urban communication report that allows online-and-offline participants to grasp a better understanding of the city. As Sawyer puts it, the city is only the object of theoretical explanation but cannot explain itself [19]. The author of this paper believes that the functional value of short videos should be fully utilized, so that they can become the "media" of modern cities, which can integrate the physical city and the abstract city, and further dig into the urban connotation and promote the charm of the city [20].

2.4 Summary

In the era of traditional media, government authorities hold the say in city image promotion, while in the era of the Internet, short videos can also tell government officials about the city construction in terms of economic development, civic life, and tourism culture. Short video publishers spontaneously use dynamic images as an important carrier to spread city stories, which is more in line with the expressions and aesthetic preferences of youngsters. Compared with professional videos such as films, TV dramas, and news documentaries, short videos have become a visual expression of people's entertainment in daily life, reducing the gap between virtual images and real life. Taking TikTok as an example, this short video platform has low barriers to production and is decentralized; thus, everyone has the chance to become the builder and spreader of city images.

3 Case Study of Short Video Intervention in Urban Communication

Case study is a common research method in marketing communication. This paper takes the case study as the main research method to investigate the data of viral videos about Chongqing and Xi'an and explore the relationship between the trending content and the construction and promotion of city image.

In the case selection process, we compared the number of likes, comments, content, themes, and picture production skills of short videos published by different types of accounts, and examined the positioning and communication capabilities of opinion leaders of different accounts. A total of 20 short video accounts were selected. The number of likes, comment sentiment, number of scenes, number of topics, and number of contents of these accounts in the process of urban communication are shown in Table 1.

Table 1. Statistics on short videos in the urban communication of Chongqing and Xi'an[Image source: drawn by the author]

Account Type	Account Name	Number of likes	Comment Sentiments	Number of Scenes	Theme	Single/Integrated content
Individual	Lyuxingxiaoyouxia	68.2	6.4 Neutral	Many	Travel Tips	Integrated
Individual	Rouroudasousuo	34.3	3.9 Pleasant	Few	Fat Sausage in Hot Pot	Single
Individual	Jigeliaobude	9.6	1.6 Neutral	Few	Guanyinqiao Food Court	Single
Individual	Phoenix Photo Studio	7.3	0.2 Pleasant	Many	Introduction to Chongqing	Integrated
Official	Chongqing Music Broadcasting	8.2	0.7 Neutral	Few	Street Lantern Show	Single
Official	Chongqing Release	34.2	0.8 Pleasant	Few	Wanzhou North Station	Single
Official	Chongqing Daily	4	0.3 Neutral	Few	Chongqingyukuaibao	Single
Official	Chongqing Today	96.7	0.2 Pleasant	Few	Sanitation Workers	Single
Individual	Foggy Capital Chongqing	1.7	0.07 Neutral	Few	Aerial Chongqing	Single
Individual	Hao'er de Lyutu	2	0.1 Pleasant	Many	Wushan, Chongqing	Integrated
Official	Xi'an Broadcasting Corporation	42.9	2 Pleasant	Few	Bless Xi'an	Single
Official	Xi'an News	39.7	1.4 Pleasant	Few	Xi'an Fighting	Single
Official	Shaanxi Broadcasting Corporation	6.9	0.4 Pleasant	Few	Epidemic Situation of Xi'an	Single
Individual	Xi'an Lehua City	160.2	4.8 Pleasant	Few	Trip to Lehua City	Single
Individual	Laoniepaixi'an	3.9	0.1 Pleasant	Few	Datang Everbright City	Single
Individual	Aerial Qinling Mountains	8	0.3 Pleasant	Many	Music Fountain	Integrated
Individual	Aerial China	17.4	0.7 Neutral	Few	Xi'an City Wall	Single
Individual	Dapang Aerial Xi'an	2.2	0.1 Pleasant	Few	Xi'an Bell Tower	Single
Individual	Moran Photography	4.2	0.4 Pleasant	Few	Chinese Tang Garden	Single
Individual	Xi'ankedaibiao	7.8	0.5 Neutral	Many	Travel Tips	Integrated

Note: (1) Sentiment standard: The consistency and difference between the comments and content of short videos, denoted as pleasant if they are generally consistent, otherwise as neutral. (2) The objects of shots: The shots that focus on different objects are denoted as integrated, otherwise as single. (3) Numbers of scenes

From the analysis in Table 1, it can be seen that videos published by personal accounts and public accounts have the same popularity and excellent communication response. Short video creators associate comprehensive creative factors such as rich life practice, personalized subtitle output, and fitting scene creation with the site-specific communication of local cities, which embodies the prevalence of personal accounts and the greater say of ordinary people. Personalized videos released by high-quality personal accounts such as Lyuxingxiaoyouxia, Rouroudasousuo, Xi'an Lehua City, etc., might be completed by a simple mobile phone, though, the rich flavor of life and personal feelings contained in the videos can arouse participants' curiosity and freshness.

On the other hand, from the perspective of the promotion of city image dissemination, short videos with regional characteristics are more likely to stand out in the current market environment. For example, short videos containing dialects, special effects, funny music, etc. tend to make the content more colorful and vivid. Distinctive short videos, such as those that contain dialects, special effects, funny music, etc., often make the content more colorful and vivid, thus standing out in the unwieldy mass of short videos. In addition, short videos can maximize the combined effects of music and images. Urban music has become an important symbol of urban images, which is also an unprecedented change of site-specific expression in the new media era. Besides, food has also become an important aspect of the site-specific communication of Internet-famous cities such as Chongqing and Xi'an. For example, Chongqing's hot pot grilled fish and Xi'an's marinated meat in a baked bun and minced meat noodles have attracted numerous audiences to explore Internet-famous cities on TikTok. With all these factors, the diversity and authenticity of Internet-famous spaces have been better shaped and presented. As the times require, there are also market phenomena such as the city's official publicity short videos, its own traffic, and the sale of city-specific products through live broadcasts. This further enables operators and practitioners of short videos to obtain financial support through the online model.

4 Conclusion

4.1 Reconstruction of participatory urban space

Media development promotes the transformation and reconstruction of urban space; short video platforms play an important role in displaying the regional cultural characteristics while publicizing the image of Chongqing and Xi'an as Internet-famous cities. In the era of short videos, the multi-party interaction between online publishers and participants transforms the subject and object of urban communication. As a result, media users have access to a completely immersive experience, and people-to-people exchanges trigger the formation of scene chains[21]. Therefore, the author of this paper holds that the change of communication role does not imply that short videos have lost their usefulness for the government to promote regional cultural images. On the contrary, in addition to traditional communication ideas and organizational forms, the competition of regional cultural image communication can enhance the influence of a city, and the integration of communication strategy with the media ecological environment of short videos can guide the public.

4.2 Local appeal of Internet-famous space

The sense of place in the study of human geography includes placeness and attachment. The "sense of place" was advanced by Steele, who put it as an experience produced by place and developed by the person based on the interaction between people and places [22]. It is not only the special nature of place, but also people's attachment and feelings induce to the place [23]. In the new media era, the formation of the human-landscape relationship is featured by the use of short video platforms as the carrier to transmit information, communicate emotions, and promote local appeal and city image.

Compared with the previous "sense of place" formed imperceptibly through personal experience, the new one indirectly shaped by the media is influenced by the characteristics of the media itself, and the content conforming to the characteristics of media communication tends to get more views and attract more public attention. The site-specific communication of the sense of place generally increases the thickness and height of city images through cultural and artistic communication. Therefore, the application of digital images to the site-specific communication of Internet-famous urban spaces has not changed the distinctive urban space of Chongqing and Xi'an, but formed "social synchrony without regional proximity" in urban communication [24], which has made Instagram-worthy places in Chongqing and Xi'an surpass the shackles of their original forms. Only by excavating the spatial connotation of urban Instagram-worthy places and constructing an image-medium city, can we build an emotional connection with the audience in the site-specific communication of popular cities through short videos.

4.3 City identity: Reconstruction of local significance

Short video media transforms urban spatial communication from the traditional relationship between man and land to the relationship between man, land, and media. Both place and media participate in the building of a communication mode and a social information communication mode for the mass [25]. The addition of short video media helps the audience to be exposed to extensive text production at the individual level and community interaction at the collective level[26], so as to enhance urban identity and reshape the local significance. There is a two-way connection between the communication and reshaping of urban space. The frequent and continuous communication effect of viral videos can make "Internet-famous cities" full of news value and publicity value and can transform "attention dividend" into "momentum of development" in city branding [27].

To construct local significance, it is suggested to take tourism short videos as the pivot, dig deep into the connotation of scenes, stimulate the interactive mechanism and increase user viscosity. In the entertainment atmosphere created by short videos, history, culture, and art account for a relatively small proportion and lack in-depth exploration. The author assumes that creators should select topics associated with the history, culture, and art and make distinctive short videos that bear both communication value and content depth. What's more, they should integrate online and offline activities in urban Internet-famous space to reflect the commercial value of tourism, and finally realize the reconstruction of local significance and promote the re-communication of city image.

References

1. Zhou K, Zhang H. T., Xia Y. N., Liu C. New characteristics of urban consumption spaces under social media influence: A case study of Changsha's Instagram-worthy places on Red Book [J]. Modern Urban Research, 2021 (09): 20-27.
2. CASTELLS M. The rise of the network society (second edition) [M]. Malden, MA: Wiley [1] Blackwell, 2010
3. Tong M. Urban society and space in the age of information technology [J]. Journal of Urban Planning, 2008 (5): 22-33.
4. Liang Xin, Zhang P. Spatial evolution mechanism and morphological differentiation of rural settlements from the perspective of network society: A case study

of Henan [J]. Urban and Rural Planning, 2020 (2): 23-31, 41

5. KITCHIN R, DODGE M. Code/space: software and everyday life [M]. Cambridge, MA: MIT Press, 2011
6. Wei C: "The culture of visiting Instagram-worthy places is lowering our aesthetic ability," Times Biweekly, No.18, 2019
7. Wu X. M. Walking on the Internet: Facebook users' tactics for visiting Instagram-worthy places and landmark practice [J]. Mass Communication Research, 2016 (126): 93-131.
8. Pan K. L. Social space theory [M]. Beijing: Central Compilation & Translation Press, 2012: 191-121.
9. Scott Rush, Celia Lurie (2007/2010). Global cultural industry: the mediation of things. Trans. by Yao Xinyue. Beijing: Social Sciences Academic Press (CHINA).
10. Li C. C. Multi-dimensional Perspective: research on mass media and city image construction [D]. Suzhou University, 2014.
11. Marshall McLuhan. Understanding media: the extensions of man (Commented Edition) [M]. Trans. by He Daokuan. Nanjing: Yilin Press, 2011.
12. Deng Z. Construction and communication of urban memory from the perspective of space [J]. Modern Communication (Journal of Communication University of China), 2019, 41 (3): 50-55.
13. L Szulc. Profiles, identities, data: making abundant and anchored selves in a platform society [J]. Communication Theory, 2019 (29): 257-276.
14. Yan X. Globalization and localization: research on world cultural heritage and development of Lijiang national cultural industry cluster [J]. Journal of Southwest Minzu University (Humanities and Social Science), 2019, 40 (2): 34-38.
15. Liu L., Cheng D., Su X. An empirical study on the influence of opinion leaders on the participation behavior of virtual tourism community—a case study of college students [J]. Tourism Tribune, 2018, 33 (9): 83-93.
16. Chris Barker: The SAGE dictionary of cultural studies, Sage Publications, 2004, p. 177.
17. Zhao Q: Mumford's view of cities and its enlightenment, Journal of Soochow University Philosophy & Social Science Edition, No.32, 2011, pp. 47-53.
18. Edward W. Su Jia, City first: recombing the origin of urbanization, selected from A companion to the city edited by Gary Bridge and Sophie Watson and trans. by Chen Jianfeng, Lijiang Publishing Limited, 2015, pp. 35, 31, 30, 31, and 30.
19. Edward W. Su Jia, City first: recombing the origin of urbanization, selected from A companion to the city edited by Gary Bridge and Sophie Watson and trans. by Chen Jianfeng, Lijiang Publishing Limited, 2015, pp. 35, 31, 30, 31, and 30.
20. McQ Uire, S. Geomedia: Networked cities and the future of public space, Polity, 2016.
21. Zhang L. Scene classification and application path analysis of mobile audio-supported reading [J]. Publishing Journal, 2020 (3): 103-107.
22. Steele F. The sense of place [M]. Boston: CBI Publishing, 1981.
23. Shao P. R. Temperature of places: social construction and cultural memory of media geography elements [J]. Journal of Jiangsu Normal University (Philosophy and Social Sciences Edition), 2010, 36 (05): 143-148.
24. Yu G. M. and Qu H. Embeddedness and boundary: a study of the influence of social space-time on media use patterns, Contemporary Communication, No.4, 2020, pp. 9-13.
25. Joshua Meyerowitz, No sense of place: the impact of electronic media on social behavior, trans. by Xiao Zhijun, Beijing: Tsinghua University Press, 2002, p. 34.
26. Cai Q. New media empowerment in the scene era from the perspective of low-budget travel subculture [J]. Modern Communication (Journal of Communication University of China), 2016, 38 (12): 124-128.
27. Shan S. B. Rediscover the city through the aura of Internet fame [DB/OL]. http://Opinion.people.com.cn/n1/2018/0605/c1003-30035288. html.2018-06-05.

Research on the brand building of rural e-commerce live streaming under the background of rural revitalization

Jingwen Liu

School of Journalism and Communication, Zhengzhou University, ZhengZhou, China

Abstract. Driven by the rural revitalization strategy, rural e-commerce live streaming has developed rapidly in the past two years, and the differentiated development trend of e-commerce live broadcast brand building has become more and more obvious. This paper takes some cases of rural e-commerce to live streaming brand building on Kuaishou, Douyin and other platforms as the research object, and finds that the lack of professional talents, insecure product quality and service, weak scene construction, and poor marketing and publicity effect are the main problems hindering the construction of rural e-commerce live broadcast brand.

1 Introduction

In 2022, under the background that the country has achieved comprehensive poverty alleviation, in order to consolidate the achievements of poverty alleviation and effectively connect rural revitalization, the CPC Central Committee and the State Council issued the "Opinions of the CPC Central Committee and the State Council on Comprehensively Promoting the Key Work of Rural Revitalization in 2022", which proposed to implement the "digital business to revitalize agriculture" project and promote e-commerce into rural areas. Promote the healthy development of live streaming of agricultural and sideline products. Carry out agricultural variety cultivation, quality improvement, brand building and standardized production improvement actions. [1]. Under the call of the state to comprehensively promote the rural revitalization strategy and agricultural modernization, Taobao, Jingdong and other e-commerce platforms have sunk into the rural market, developed the rural e-commerce market from multiple levels such as talent and technology, and laid out rural e-commerce live broadcasts to drive the development of rural economy and promote the process of rural revitalization.

ZhuZerong, a brand expert and deputy secretary-general of the Standards and Certification Professional Committee of the China Advertisers Association, pointed out that in the future, live streaming will move towards branded live broadcast, which is an inevitable trend[2]. In rural areas, anchor groups represented by "new farmers" have begun to seek to establish personal live broadcast brands, They regard mobile phones as new agricultural tools and live streaming as new agricultural work, and strive to form differentiated competitive advantages and promote the sustainable development of rural e-commerce live streaming under the fierce competition of rural e-commerce live broadcasting.

2 The significance of rural e-commerce live broadcast brand building

Compared with the single legal protection function of trademarks, brands have rich connotations and corresponding promotional functions and business functions [3]. The promotion function and operation function of the brand plays an important auxiliary role in the ultimate purpose of e-commerce live broadcast sales of products to obtain profits, and are also of great significance to farmers to increase income. At the same time, the brand-building process of rural e-commerce live streaming is also a process of continuous improvement of user stickiness, popularity, recognition and reputation, and in the current fierce market competition, brand building is particularly important and critical to sustainable development of e-commerce live broadcast.

2.1 Improve product selection standards

In the initial stage of rural e-commerce live broadcasting, most operators produced and sold agricultural products by themselves, forming a stable industrial closed loop. Such sales methods also have certain drawbacks, such as whether the production conditions are clean and hygienic, whether the quality control is strict, and whether the product quality ultimately meets the market circulation standards, all of which are difficult to supervise. As an important step in the transformation and upgrading of the

* Corresponding author: 974301717@qq.com

live broadcast room, brand building is also a long-term project with a long way to go, and product selection (that is, the process of screening products sold on live broadcast) is a crucial part of e-commerce live broadcasting, which has an important impact on the reputation and reputation of the brand. Once the e-commerce live broadcast brand is established, it is no longer limited to the past single product, self-produced and self-sold stage, there will be more supply of different qualities in the live broadcast room, at this time the live broadcast operation team will take the initiative to improve the corresponding product selection standards, quality assurance of the reputation of the live broadcast brand with product selection, provide consumers with high-quality consumption, and enhance brand recognition.

2.2 Improve product premium capacity

The text of your paper should be formatted as follows:

From the perspective of commercial attributes, the goal of brand building is to obtain lasting and stable commercial profits [4]. In the past, due to the weak awareness of farmers' management, the sales of agricultural products always relied on other terminals, and there were mainly four circulation modes: "acquisition hawker + wholesaler + terminal retail", "leading enterprise + wholesaler + terminal retail", "cooperative + wholesaler + terminal retail" and "farmer + supermarket model" [5]. In these circulation patterns, farmers are always at the bottom, and the profits they receive are severely squeezed. As the endorsement of product quality and standards, branded e-commerce live streaming can create a sense of quality and trust in consumers' minds that is higher than that of other live broadcast rooms. At this stage, with the rapid development of the economy and the improvement of people's living standards, consumers' brand consumption awareness is getting stronger and stronger, and they are more willing to pay for brand products with quality and reputation guarantees. The brand building of rural e-commerce live streaming can attract consumers who pursue quality life to pay for the brand, directly drive the premium capacity of products, and help farmers increase their income.

2.3 Promote industry transformation and upgra ding

Rural e-commerce live streaming has developed rapidly in the past two years, and more and more farmers have joined the camp of e-commerce live broadcasting, and the homogenization competition has become more and more fierce. Many e-commerce live broadcasts attract consumers through gimmicks such as product uniqueness and ultra-low prices. E-commerce live-streaming brands with a clear banner often have a certain degree of authority and leadership in specific fields, are representatives of quality and reputation, and can be quickly recognized by consumers and stand out from similar competitors in the market. From the perspective of cultural attributes, brands often lead fashion, and fashion is initially an avant-garde behavior that a few people try, but once it is widely accepted and imitated, it can quickly become popular [6]. Therefore, because of the current situation of the small-scale and low-level brand building of rural e-commerce live broadcasting, creating a successful rural e-commerce live broadcast brand model case which can break the situation of a large number of rural e-commerce live broadcasts and few brands in the past, and promote the transformation and upgrading of the entire industry.

3 The dilemma of rural e-commerce live broadcast brand building

3.1. Lack of professional personnel

The streamer lacks personal characteristics. At present, the common problem of rural e-commerce anchors is that they lack outstanding personality characteristics and personality charm, and cannot show a three-dimensional image of farmers. The existing rural e-commerce live broadcast, due to the lack of professional knowledge related to living broadcasting, can not grasp the traffic law, but simply move the daily rural life into the live broadcast room, answer consumers' questions about agricultural products, cookie-cutter live broadcast content, lack of unique personal characteristics, over time will bring audience visual fatigue, user churn rate increases.

The awareness of brand building of the operation team is weak. At this stage, because many farmers only follow the trend to join the wave of e-commerce live broadcasting, they do not have sufficient knowledge and understanding of the industry, resulting in them not seeing the future development direction, and obtaining maximum economic benefits in the short term has become the core purpose of some rural e-commerce operators, not to mention considering the long-term development of the brand. On the other hand, due to the small number of successful cases of the brand building of rural e-commerce live streaming emerging in reality, it is impossible to provide them with reference and reference in brand building, and the realistic nutrients for brand building awareness cultivation are insufficient. Most rural e-commerce live streaming is in an unconscious state of brand development, which is now a more common phenomenon in rural e-commerce live broadcasting.

3.2 Lack of guarantee for the quality and service of agricultural products

The quality of the product is not up to par. With the rise of live broadcasting, the safety and quality of agricultural products continue to emerge, and some products enter the market without passing inspection and quarantine, and the quality and safety cannot be fundamentally guaranteed. Agricultural products sold through live broadcasting, due to the direct connection between buyers and sellers, lack of quality review of intermediate links, supervision is difficult, some unscrupulous merchants will be driven by interests, in order to make the finished product look good, illegally add preservatives, shoddy charging, sell agricultural products with excessive fertilization and medicine, there are major food safety risks.

Insufficient awareness of after-sales service. After-sales as a supporting service for product sales, directly affect the shopping experience of consumers. On the one hand, due to the sale of a large number of agricultural products in a short period, e-commerce live streaming will have a large number of problems with orders in a short period, which tests the emergency serviceability of product after-sales. On the other hand, because the work quality of rural delivery workers is difficult to guarantee, the service risk of barbaric loading and unloading and random throwing is higher [7]. Lack of service awareness and lack of manpower are more common in rural e-commerce live broadcast after-sales service, and the situation of putting consumers' feedback and demands aside frequently, and the current situation of imperfect after-sales service increases consumers' risk perception of purchasing agricultural products, which adversely affects the sustainable development of agricultural products and live broadcast brand building.

3.3 The construction of the scene system is weak

The "field" in the scene is a kind of structural monolithic existence, which is no longer limited to the physical meaning of the place [8]. For rural e-commerce live broadcasting, a single live broadcast room is an independent scene, connecting reality and the network, including not only offline live broadcast scenarios, but also online network service scenarios.

Network service scenarios are not interactive. Network service scenarios are an important driver of consumers' willingness to shop [9]. At present, rural e-commerce live streaming is mainly operated by families, rarely hiring professional staff, and there are still many problems at the professional level of live broadcasting, and live streaming is very arbitrary. Due to the lack of understanding of the importance of the scene by many rural e-commerce live broadcast operators, and ignoring the appeal of the live broadcast atmosphere to consumers, the old-fashioned live broadcast content and monotonous form cannot narrow the distance between the anchor and the user, resulting in the loss of live broadcast users.

3.4 The effect of content marketing is not good

Homogenization of short videos and live content. In rural areas, "short video + e-commerce live broadcast" gives full play to the synergistic effect of short video and live broadcast. Before starting the live broadcast, the host shoots short videos of traffic drainage to promote the featured products. Due to their level limitations, many rural self-media people cannot continue to produce high-quality short video content, so they will imitate other short videos for secondary processing, which is not conducive to improving user stickiness and loyalty. Now, the progress of modern technologies such as 5G and AR has brought more possibilities to e-commerce live broadcasting, and new live broadcast forms such as "immersive live broadcast" and "intelligent virtual live broadcast" are constantly emerging, but rural e-commerce live broadcast is still relatively few cases due to financial and technical limitations.

Insufficient marketing campaigns. To build a durable competitive brand and enhance its popularity, rural e-commerce live streaming also needs strong brand marketing and promotion, so that the rural e-commerce brand image is deeply rooted in the hearts of the people. At present, most of the rural e-commerce live broadcasts are released on a single platform for short-term frequency drainage, and the brand publicity channels are narrow and the publicity effect is limited. Although some rural e-commerce operators have consciously promoted their brands through various social media, the overall effect is not good, which is manifested in problems such as rough poster production, unattractive content, and detailed live broadcast introduction, and some publicity mistakes directly weaken the image of live broadcast brands in the minds of consumers.

4 The path of rural e-commerce live broadcast brand building

4.1 Strengthen the construction of anchor teams

The branding of live streaming is first and foremost the branding of the anchor. An important role of anchors in e-commerce live streaming is to rely on distinctive personal characteristics to attract fans and accumulate traffic for the live broadcast room. If the live broadcaster wants to create a distinctive persona, the first thing to do is to make great efforts to explore the characteristics in the construction of the personal image, and the well-known personality traits, good skills, and personal image of the audience can become the basis for creating a persona. On the other hand, given the lack of professional talents in rural e-commerce live broadcasting, it is necessary to start from the aspects of anchor team and professional level improvement. With the development of the live broadcast industry and the continuous improvement of consumer aesthetics, it is imperative to improve the professional level of anchors. Local governments can actively guide the workers and college students outside the country to return to their hometowns for employment and entrepreneurship, and provide them with training in policies and regulations, e-commerce live broadcasting, brand building, etc., to increase the talent reserve of rural e-commerce live broadcast teams, form a leading role, and improve the professional quality and overall level of anchor groups.

4.2 Consolidate the foundation of the industrial chain

As the most basic and core factor in the rural e-commerce live broadcast industry chain, the quality of agricultural products directly determines the survival and development of rural e-commerce live broadcasting. The control of the quality of agricultural products requires the cooperation of the government, industry and business. The government should give full play to its supervisory role in the circulation of agricultural products, identify whether there are violations of laws and regulations in rural e-commerce live broadcasts through big data and algorithms, establish a traceability system, and make every effort to crack down on substandard products and non-compliant live broadcasts, and give play to quality assurance and regulatory efficiency. Leading e-commerce live-streaming enterprises should also initiate industry self-discipline to restrain merchants from selling substandard products. Live streaming merchants should build a solid brand foundation based on excellent product quality, be guided by consumer demand, improve the quality of product selection, find more high-quality sources, and enhance consumers' recognition and reputation of live streaming brands.

4.3 Build characteristic live broadcast scenarios

Specific scene creation helps to strengthen the core appeal of the brand, establish an emotional connection between the brand and the audience, form an effective brand communication relationship, and make the scene become the link of the user's emotional connection [10]. As an important influencing factor affecting consumers' willingness to shop, the improvement of the interactive atmosphere in network service scenarios plays a very important role in promoting the conversion rate of users in live broadcast rooms. In daily live broadcasts, rural e-commerce anchors should attach importance to live broadcast interaction, design more interactive links in the process of product introduction, and create an entertaining and lively atmosphere in the network scene by guiding fans to send interactive links such as barrage, lottery, chat, and microphone connection, so that users can more easily immerse themselves in the sense of scene created by live streaming Make up for the alienation caused by spatial distance and provide a better shopping experience.

4.4 Build a content dissemination matrix

The common combination of "short video + live broadcast" highlights the importance of short video drainage to e-commerce live broadcasting. Capturing users' attention in a short period and presenting interesting and complete content to users is an important requirement for short video production, which requires improving the professional video production level of creators, continuously improving content innovation capabilities, deeply exploring rural cultural characteristics and regional characteristics, and improving video watchability. In the new media era, to allow more consumers to access brand marketing communication content, it is necessary to distribute and spread content on multiple platforms to expand brand communication. The content distribution of many platforms forms an influential unit matrix, which strengthens the user's impression of the brand in the process of expanding communication and continuous interaction [11]. Rural e-commerce live-streaming brand operators can publish matching communication content according to different platform attributes, which suit the platform attributes and the audience characteristics of the platform.

5 Conclusion

After two years of simple replication and large-scale expansion, serious homogenization, uneven quality of anchors, and imperfect logistics and transportation systems have become the main factors hindering the development of rural e-commerce live broadcasting, and transformation and upgrading are imperative. As the only way to transform and upgrade rural e-commerce live broadcasting, the brand is also the core of agricultural modernization, the brand symbolizes the maturity of e-commerce live broadcast development, and its establishment can form a group of stable consumer groups.

References

1. Xinhua News Agency. Opinions of the CPC Central Committee and the State Council on comprehensivel y promoting the key work of rural revitalization in 2 022. [EB/OL]. (2022-01-04) [2022-09-20]. http://w ww.moa.gov.cn/ztzl/jj2021zyyhwj/zxgz_26476/202 102/t20210221_6361865.htm
2. Sina Finance. Zhu Zerong: Live streaming will mov e towards branded live broadcast[EB/OL]. (2021-05 -09) [2022-09-20]. https://bai jiahao.baidu.com/s?id =1699268952567087924&wfr=spider&for=pc.
3. YU Mingyang, SHU Yongping. On "brand communi cation"[J]. International Press,2002(03):63-68.
4. KONG Lingshun, SONG Tongtong. From IP to Bra nd: Full Commercial Development Based on Fan Ec onomy[J]. Modern Communication(Journal of Com munication University of China),2017,39(12):115-1 19.
5. XU Liyan, ZHOU Linjie. Analysis of the existing di stribution channel mode of agricultural products in China[J]. Business Research,2010(08):189-191.
6. KONG Lingshun, SONG Tongtong. From IP to Bra nd: Full Commercial Development Based on Fan Ec onomy[J]. Modern Communication(Journal of Com munication University of China),2017,39(12):115-1 19.
7. DENG Yumin, YANG Shuangqi, WEI Ziwei, et al. How to solve the "last mile" problem of rural expres

s supermarket[J]. China Business Review,2017(24): 9-10.

8. ZENG Yiwu, MA Changjiang, LI Lili, et al. Live E-commerce and Uplink Value Reconstruction of Agri cultural Products: Mechanism and Implementation P ath[J/OL]. Agricultural Economic Issues:1-10[2021-09-29].https://doi.org/10.13246/j.cnki.iae.20210909.002.
9. Wang Xi. Research on the impact of e-commerce liv e streaming network service scenario on consumers' shopping intention[D]. Shandong: Master's Thesis, S handong University,2021.
10. ZANG Lina, LIU Yuying. Construction of brand co mmunication scene based on SIVA theory[J]. Conte mporary Communication,2019(02):97-100.
11. Ma Chuanming. The impact of short video marketin g on brand building: A case study of palm reading D ouyin matrix[J]. Publishing Wide Angle,2020(14):7 7-79.

https://doi.org/10.1051/shsconf/202315502020

The Study on the Generation and Spread of Internet Rumors about the New Crown Epidemic

Yingzhen Ji

English, Jilin International Studies University, Changchun, Jilin, 130117, China

Abstract. Since the 21st century, Internet rumors as a new thing has received wide attention from scholars, and with the help of the Internet Internet rumors have taken on accelerated wings. Due to the two main characteristics of the New Crown Pneumonia epidemic itself, coupled with the current development of Internet technology to provide people with more convenient communication and communication channels, making the Internet rumors during the New Crown Pneumonia epidemic rampant, a thousand disturb people's normal life. This study investigates the management of online rumors in the context of the New Crown pneumonia epidemic, using the relevant knowledge of communication science and sociology of mass media.

1 INTRODUCTION

As of September 2022, Johns Hopkins University released real-time statistics showing that the cumulative number of novel coronaviruses diagnosed worldwide exceeded 600 million, constituting a major public health emergency in recent years in China and globally. The creation of the new coronavirus has undoubtedly provided the "ground" for rumors, and while people continue to be highly concerned about the progress of the epidemic, the asymmetry in information transmission has led to a proliferation of rumors with the epidemic as the central word, which has had a certain impact on the national epidemic prevention and control efforts. The term "rumor" has frequently appeared in public view due to the emergence of coronavirus, and the understanding and interpretation of rumors vary among Internet users and authorities in professional fields. The American psychologist Alpert has proposed that a rumor is a statement that is allegedly true but not proven by definite evidence[1-2]. Although the definitions of rumors vary widely, they emphasize the unknown and uncertainty of rumors. With the development of Internet technology and the constant updating of new media with microblogs and WeChat as carriers, Internet rumors have changed the way rumors used to spread, breaking the limitations of time and space of traditional rumors and speeding up the spread of rumors. Compared with traditional rumors, online rumors have the characteristics of fast spreading, wide range, anonymity of spreading channels and diversified spreading methods[3]. Internet rumors are no longer limited to a specific group of people, time and space, and the scope of dissemination, this means of dissemination that breaks time and space limitations, once a rumor is generated, within just a few minutes through the likes of netizens comments forwarded, making the rumor spread rapidly; when people no longer face a live person, they tend to become less cautious, "freedom of speech The right to "freedom of speech" in turn makes them make arbitrary comments and spread rumors on the Internet[4-5].

*Corresponding author. Email: Gyingyingpearl@163.com

2 Statistical analysis

2.1 Descriptive statistics and basic information on survey respondents

The descriptive statistics mainly show the mean, standard deviation and correlation of each variable. At the gender level, there were 100 women (47.39%) and 111 men (52.61%), with a balanced distribution of men and women; at the age level, the largest proportion of respondents were aged 19-35 (49.29%), followed by aged 36-60, 1-18 and over 60 (28.91%, 12.32% and 9.48% respectively). At the occupational level, students and other occupations accounted for the largest proportion of respondents, both accounting for 18.96%, followed by those engaged in production, those engaged in services and medical workers, retired, and military personnel, accounting for 15.17%, 14.69%, 9.48%, and 8.06%, respectively. At the level of education, the largest proportion of respondents with education of undergraduate/university specialist, accounting for 33.65%, followed by high school/college, junior high school, primary school, and postgraduate/doctoral/postdoctoral, accounting for 29.86%, 16.11%, 10.43%, and 9.95% respectively, the respondents have a high level of education.

Table 1. asic information on survey respondents (The following charts are self-drawn)

name (of a thing)	options	frequency	Percentage (%)	umulative percentage (%)
sex	women	100	47.39	47.39
	male	111	52.61	100.00
age	1-18 years	26	12.32	12.32
	19-35 years	104	49.29	61.61
	36-60 years	61	28.91	90.52
	60+ years	20	9.48	100.00
careers	Engaged in service industry	31	14.69	14.69
	engaged in the production industry	32	15.17	29.86
	other than it	40	18.96	48.82
	military personnel	17	8.06	56.87
	medical worker	31	14.69	71.56
	students	40	18.96	90.52
	retirement	20	9.48	100.00
Educatiol level	junior high school	34	16.11	16.11
	Undergraduate/ University	71	33.65	49.76
	secondary school	22	10.43	60.19
	Postgraduate/PhD/Postdoctoral	21	9.95	70.14
	High school/college	63	29.86	100.00
	add up the total	211	100.0	100.0

The analysis was carried out using the chi-square

2.2 Rumor channel survey

The analysis was carried out using the chi-squaregoodness of fit test for the uniformity of the distribution of the proportion of choices for each option of the multiple choice questions. As can be seen from the table below, the goodness-of-fit test showed significance ($\chi 2$ =41.558, p=0.000<0.05), implying that the proportion of choices for each item is significantly different, and the difference can be specifically compared by response rate or prevalence rate. Specifically, the response rates of a total of 4 items, 3, (Weibo/friends/space), 2, (various hot search lists), 4, (video ebsites (ShakeYin/Quick Hands/bilibili, etc.), and 6, (relatives/friends/classmates), were 20.62%, 19.54%, 19.00%, and 18.33%, respectively, which were all greater than the average response rate of 16.67%, with significantly higher response rates and prevalence rates high, i.e., survey respondents saw epidemic-related rumors mainly from Weibo/friends' circle/space, various hot lists, video websites (ShakeYin/QuickTime/bilibili, etc.), and relatives/friends/classmates.

Table 2. here do you usually see rumors related to the outbreak

item	respond to		Prevalence rate (n=211)
	n	responsiveness	
1. Mainstream authoritative media	75	10.11%	35.55%
2. Various hot lists	145	19.54%	68.72%
3.Twitter/friendship/space	153	20.62%	72.51%
4. Video sites (Jitterbug/ Quickfire/bilibili, etc.)	141	19.00%	66.82%
5.Newspaper	92	12.40%	43.60%
6.Relatives/friends/classmates	136	18.33%	64.45%
aggregation	742	100%	351.66 %

Goodness-of-fit test: $\chi 2$ =41.558 p=0.000

2.3 Survey on first response to rumors

The results are shown in the figure below. 37.44% of the respondents' first reaction to information about the epidemic was "I know but don't forward it", accounting for the largest proportion, followed by 29.38%, 14.22%, 10.90%, 8.06% and 29.38% respectively for determining tource of information, ignoring, other and forwarding. The largest proportion of respondents were identified as the source of information, ignored, other, and forwarded, accounting for 29.38%, 14.22%, 10.90%, and 8.06%, respectively, and the smallest proportion was forwarded when they saw the information about the epidemic.

2.4 Survey of attitudes towards rumors

The results are shown in the figure below. 39.34% of the respondents' attitude towards the news of the epidemic with a large number of readers and retweets is "half believe, don't care, just take a look", accounting for the largest proportion, followed by "I don't believe in it, I only believe in official authoritative news", "I remain skeptical, but will actively seek to confirm whether it is true", "I believe it, many people say so, it should be true", accounting for 28.44%, 22.27%, 22.27% respectively. 28.44%, 22.27%, 9.95%, and the smallest percentage of people believed the news of the epidemic with a large number of readers and retweets.

2.5 Survey on types of rumours

As can be seen from the table below, the goodness of fit test did not show significance ($\chi 2$ =6.803, p=0.339>0.05), implying that the selection ratio for each item was relatively even and not significantly different, i.e., survey respondents had heard confirmed persons entering their area, new confirmed cases in multiple areas, that oral solution of diflucan can prevent neo-crown pneumonia, that drinking Panax notoginseng can prevent the virus, that pets can transmit the virus , injectable/oral disinfectant water can disinfect, and boiling vinegar can kill virus multiple rumors did not differ significantly.

Table 3. What rumors have you heard?

item	respond to		Prevalence rate (n=211)
	n	responsiveness	
1. Confirmation of	141	14.19%	66.82%

entry of persons into the region			
2. New confirmed cases in several areas	148	14.89%	70.14%
3. Shuanghuanglian oral liquid can prevent neoconjunctivitis	158	15.90%	74.88%
4. Drinking Panax notoginseng can prevent viruses	152	15.290%	72.04%
5. Pets can transmit viruses	129	12.98%	61.14%
6. Injectable/oral disinfectant water can be disinfected	122	12.27%	57.82%
7. Boiling vinegar kills the virus	144	14.49%	68.25%
aggregation	994	100%	471.09%

Goodness-of-fit test: $\chi2$ =6.803 p=0.339

2.6 Survey on the extent of the impact of rumours

The results are shown in the table below. In terms of economic loss (e.g. buying drugs by mistake), rumors had the greatest impact on survey respondents with a little impact, accounting for 73.46%, followed by greater impact and no impact, accounting for 19.43% and 7.11% respectively, with a mean value of 2.54. In terms of physical health (e.g. buying drugs by mistake), rumors had the greatest impact on survey respondents with a little impact, accounting for 65.4%, followed by greater impact, respectively, with a mean value of 2.54. In terms of physical health (e.g. believing rumors and eating by mistake), rumors had the greatest impact on survey respondents with a little impact, accounting for 65.4%, followed by a great impact and no impact, accounting for 21.33% and 13.27% respectively, with a mean value of 2.44. In terms of mental health (e.g. rumor-causing anxiety), rumors had the greatest impact on survey respondents with a little impact, accounting for 65.4%, followed by a great impact and no impact, accounting for 21.33% and 13.27% respectively, with a mean value of 2.44. In terms of scientific cognition (e.g., misunderstanding the nature of neoconiosis), rumors had the greatest impact on the respondents with a little impact, accounting for 70.14%, followed by a great impact and no impact, accounting for 21.33% and 8.53%, respectively, with a mean value of 2.49. In terms of scientific cognition (e.g., misunderstanding the nature of neoconiosis), rumors had the greatest impact on the respondents with a little impact, accounting for 64.93%, followed by a great impact and no impact, respectively, with a mean value of 19.91. Overall, rumors had the greatest impact (68.48%) and the least impact (11.02%). In these four areas, rumors had the greatest impact on the economic loss of the respondents (2.54) and the least impact on the physical health of the respondents (2.44).

Table 4. Please select the level of impact the rumor has had on youSurvey of disinformation channels

effects	Greater impact	No impact	There is a slight impact	Average value
Economic loss (e.g., misinformation about drugs)	41	15	155	2.54
	19.43%	7.11%	73.46%	
Physical health (if you believe the rumor and eat it by mistake)	45	28	138	2.44
	21.33%	13.27%	65.4%	
Mental health (e.g. rumor mongering anxiety)	45	18	148	2.49
	21.33%	8.53%	70.14%	
Scientific perception (e.g., misunderstanding the nature of neocoronary pneumonia)	42	32	137	2.45
	19.91%	15.17%	64.93%	
population (statistics)	43.25	23.25	144.5	2.48
	20.50%	11.02%	68.48%	

2.7 Survey of disinformation channels

As can be seen from the table below, the goodness-of-fit test showed significance ($\chi2$ =134.265, p=0.000<0.05), implying that there is a significant variability in the proportion of choices for each item, and the variability can be specifically compared by response rate or prevalence. Specifically, the response rates of a total of 3 (social software), 2 (search website), and 3 (video website) were, respectively, 34.01%, 32.43%, and 32.43%, all of which were greater than the average response rate of 25.00%, and the response rates and popularity rates were significantly higher, i.e., the way survey respondents found disinformation was mainly through social software, search websites, and video websites.

Table 5. How did you find out about disinformation?

item	Respond to		Prevalence rate (n=211)
	n	responsiveness	
1.Search the site	143	32.43%	67.77%
2. Video sites	143	32.43%	67.77%
3. Social software	150	34.01%	71.09%
4. Other	5	1.13%	2.37%
aggregation	441	100%	209.00%

2.8 Investigation of measures to deal with rumours

As can be seen from the table below, the goodness-of-fittest showed significance ($\chi2$ =74.830, p=0.000<0.05), implying that there is a significant difference in the proportion of choices for each item, and the difference can be specifically compared by response rate or prevalence rate. Specifically, the response rates of a total of 3 items, 2 (maintain a skeptical attitude and do not believe easily), 3 (promote the spirit of proof and be brave to dispel rumors) and 4 (learn more about science and improve the ability to discriminate), were 27.83%, 25.00% and 22.92% respectively, which were all greater than the average response rate of 20.00%, and the response rates and prevalence rates were significantly higher, i.e., the survey respondents believed that measures that could be taken to deal with rumors The main ones are to maintain a skeptical attitude and not to believe easily, to promote the spirit of proof, to be brave in dispelling rumors, and to learn more about science and improve the ability to discriminate[6].

Table 6. What do you think can be done to deal with rumours

item	Respond to		Prevalence rate (n=211)
	n	responsiveness	
1. Reduce internet access, don't want to see too many messages	91	13.54%	43.13%
2. Remain skeptical and do not trust easily	187	27.83%	88.63%
3. Promote the spirit of proof and be brave enough to dispel rumours	168	25.00%	79.62%
4. Learn more about science and improve discernment	154	22.92%	72.99%
5. Just looking, hard to manage even if you don't want to	72	10.71%	34.12%
aggregation	672	100%	318.48%

Goodness-of-fit test: $\chi 2$ =74.830 p=0.000

2.9 Investigation into rumour governance

The results are shown in the chart below. 41.23% of the respondents think the most effective measure is "tackling rumors at the source and cracking down on individuals or groups who publish rumors", accounting for the largestproportion, followed by "tackling rumors at the end of dissemination, disbelieving rumors and not The next most effective measures are "to deal with rumors at the source and crack down on individuals or groups who publish rumors"[7], accounting for 32.70% and 26.07% respectively.

3 conclusion

3.1 Research findings

In conclusion, the spread of rumours was more prominent in the major health events of the New Coronavirus outbreak. First of all, the strong need for life safety is the intrinsic reason; in addition, the suddenness, severity and uncontrollability of the event itself and the special measures taken by the government in different situations, the untimely disclosure of information about the virus by the government and official media, and the weakness of scientific protection propaganda have intensified the panic of the public and provided a breeding ground for rumors. The information text evokes excessive concern and panic about life awareness through the call structure , thus spreading related rumors; Internet users surveyed have high demand for life protection and basically get information about rumors from new media channels; people with higher education are more concerned about social injustice; few people choose to forward directly; rumors cause psychological panic among some people as well as unscientific and incorrect understanding of the epidemic The majority of people have an optimistic and supportive attitude towards rumor management[8].

3.2 Shortcomings and outlook

There are three shortcomings in this paper: First, the survey sample is only representative of a certain group of people, and the findings cannot be generalized to the whole population, which limits the external validity of the analysis. Second, the channels provided in the questionnaire are not the only source of information about rumors among the population, and some people may have access to multiple media at the same time. However, this paper only examines the role of a particular type of use. Third, how rumor formation affects the behavior of society, the state, etc. is not discussed in this study and deserves further in-depth exploration in future studies. Fourth, the data used in this paper is cross-sectional, which does not prove the veracity of the causal relationship and should be further explored in the future using longitudinal data or using survey-experimental methods.

References

1. Wang L, Xie Cenggong. Empirical analysis of online rumor propagation in public events: a study based on online rumor information during 2010-2012 [J]. Journal of Shanghai Jiaotong University (Philosophy and Social Science Edition), 2014 (2): 86-99.
2. Chao Nai-Peng, Huang Xian. Research on the phenomenon of "rumors" in Internet communication [J]. Intelligence Theory and Practice, 2004 (6): 586-589, 575.
3. Harold Lasswell. The structure and function of social communication [M]. He Daokuan, Translation. Beijing: Communication University of China Press, 2012 (12): 35.
4. Braddock, R. An extension of the "Lasswell Formula" [J].
5. Jiang Shenghong. The formation, transmission and public opinion guidance mechanism of Internet rumors [J]. Chongqing Social Science, 2012(6):12-20.

6. Dong Lu. Core theories and concepts of communication science [M]. Beijing: Peking University Press, 2016 (4): 28.
7. Chen XH, Gong LQ. A review of WeChat rumor research [J]. Librarianship Research, 2018(18):2-6.
8. Song Xiangqin. Research on the characteristics of the evolution of online public opinion and its guidance strategy in the post-truth era [J]. Journal of Chongqing University of Technology (Social Science), 2018 (8): 139-144.

Investigation and Research on the Use of Tik Tok of College Students in Beijing

Yingyu Liang

Department of Japanese, The Beijing Forestry University, Haidian District, Beijing, China

Abstract This paper sets university students using Tik Tok APP as the research object, combining theoretical basis with empirical research. Based on use and gratification approach, questionnaire and interview ithe core to convey this research. Analysis of characteristics of Tik Tok lay a foundation for the research as well. At the same time, according to the specific use behavior of college students from Beijing area, the reasons and influences of use of Tik Tok by college students will be analyzed. This paper aims to help college students rationally deal with their own use of Tik Tok and provide suggestions for the further development of Tik Tok in the future.

1 Introduction

Along with the development of economic society and the modern science and technology, smart phones are becoming more and more accessible to the public, the spread form of mass media is making a new innovation. Using a mobile phone to watch a short video becomes a new form of communication and diffusion of information, Short video software represented by Tik Tok is gradually thriving. According to announcement of latest beta released by the CEO of the Beijing Byte Dance Limited Company, until the August of 2020, Including the other version of Tik Tok, Tik Tok has more than 600 million daily active users. As Tik Tok stands out in the whole industry, it is becoming more and more widely used among college students nowadays. The user group of college students is growing day by day, with its coverage rate also increasing, which is changing the media and habits of college students to watch videos. According to statistics, 85% of Tik Tok users are under the age of 24, and a high proportion of them are college students. As of Dec 31, 2022, Tik Tok had more than 26 million college students users, accounting for 80 percent of all college students in China. Students, who receive higher education, master knowledge, are curious about new things and have good acceptance ability, are more likely to be attracted by short videos as a form of communication, and they are also one of the groups with the highest stickiness of Tik Tok users. Tik Tok has a profound impact on young people's social interaction, learning, consumption, entertainment and other aspects, and has penetrated into all aspects of young people's lives.

Therefore, short videos on Tik Tok are developing rapidly and exerting a profound influence on college students, which has great research value. Taking college students in Beijing as an example, this study mainly focuses on the following questions: What is the proportion of college students in Beijing who use Tik Tok? What do they usually tend to watch? What kind of impact does this have on their study and life? Thus, the reasons, purposes and the duration of using Tik Tok, the viewing preference, and the impact on their life and study are summarized.

With the continuous development of short videos, relevant experts and scholars have carried out diversified academic research with a wide range of research fields and rich research results. Among them, it mainly focuses on the motivation and the influence of using Tik Tok among college students, and most of them use questionnaires for quantitative research. On this basis, in addition to quantitative research, this study will conduct qualitative research methods such as in-depth interviews and focus group interviews to study the reasons, experiences and influences of use of Tik Tok by college students.

The number of college students is huge and they are in a important position in communication. What's more, the current era of new media is constantly updated. Taking college students as the starting point, seeing small as big, and studying the use of Tik Tok by college students to explore how to correctly use the media can not only provide a reference for college students to build a healthy and positive study life, but also have great significance for the whole society.

2 literature review

As short videos spread more and more widely among college students, researches on this aspect are gradually expanding. Many scholars at home and abroad have studied the use of Tik Tok by college students from many different levels.

In addition to studying the basic situation, such as time, content of attention, age and gender ratio, etc., the paper

B2102831133@163.com

also studies the influence of Tik Tok on college students' study, entertainment, social communication and many other aspects. It can be seen that Tik Tok has attracted the attention of college students because of its video length and concise content. Among its users, young female users are a group of users with a long time and high stickiness under the premise that young users account for a large proportion. [1] Moreover, it has a positive influence on college students, but also has a negative potential risk. For example, in terms of values, Tik Tok helps college students establish correct interpersonal values, political values and happiness values, but the negative induction of values also distorts the cognition of occupation, morality and aesthetics. [2] In addition, the popularity of Tik Tok also has a variety of effects on college students' moods and emotions. While bringing positive incentives and enhancing college students' happiness, it also causes psychological confusion and anxiety. At the same time, Tik Tok addiction will also hinder the personal development of college students. [3] [4] [5] With the deepening of communication research, the research center gradually shifts from communicator to receiver. This is because, with the continuous expansion of the network information space, more media that can meet the needs appear in the public's view, and the public's needs gradually show the characteristics of multi-level, multi-field and diversification. [6] In foreign studies, the proportion of young adults using Tik Tok accounts for more than half. Researchers mainly study the motivation, experience and attitude of young people using Tik Tok through the use and satisfaction theory. According to the results of the study, young people mainly use it to meet their needs of entertainment, socializing, reading news and pursuing fashion trends. It also studied young people who have not used the APP to find out what needs are not being met by Tik Tok, such as feeling that it is too childish, encroachment on personal privacy and cyber violence.[7] At the same time, there are also studies on the impact of Tik Tok on college students' learning. The results showed that there was an inverse relationship between the duration of Tik Tok use and their ability to focus on study. This proves that Tik Tok has some negative effects on college students' learning.[7]

From the perspective of research field, the current research direction mainly focuses on the causes and influences of Tik Tok use by college students, and the form and content are relatively simple. From the perspective of research methods, previous studies on the use of Tik Tok by college students mainly use questionnaires to understand the basic situation of the use of Tik Tok by college students, and then analyze the data, which is difficult to have a deeper understanding. In terms of theory, most analyses are based on the theory of use and satisfaction.In this research, questionnaire survey and in-depth interview are combined, and both qualitative and quantitative research are used to have a deeper understanding of the basic situation of Tik Tok use by college students and to analyze more profound problems behind it. Based on the theory of use and satisfaction, meme theory and media dependence are also used for discussion.

On the basis of these, this study will continue to study the following issues in depth: first, the use of Tik Tok short video software by college students; Second, the purpose of students using Tik Tok; Third, what kind of impact does the short videos that college students prefer to watch have on their lives and how to deal with them? Through the exploration and analysis of the above problems, the evidence that can verify the theory of use and satisfaction in the process of using Tik Tok by college students is found, and the prominent problems in the process of using Tik Tok are discussed.

3 methods and materials

This research adopts a combination of quantitative and qualitative research methods. In addition to distributing questionnaires, in-depth interviews and focus group interviews are also used for research.

Firstly, among college students in Beijing, questionnaires were distributed through several Internet social platforms, such as Wechat, Weibo and Douyin, and samples from colleges and universities in Beijing were obtained. At the beginning of the questionnaire, the researcher has informed the respondents that the questionnaire will be filled in anonymously without disclosing the personal information of the respondents. The questionnaire will be filled in after the respondents agree. After the questionnaire was collected, it was screened, invalid samples were eliminated, valid samples were retained, and the obtained data were analyzed to grasp the general situation of Tik Tok use by college students in Beijing from a macro perspective, so as to obtain a basic understanding.

Then, 8 students from college students in Beijing were selected as the interviewees, and 4 of them were interviewed in depth one to one, and the other 4 were interviewed in focus group. The in-depth interviews were conducted by phone on the afternoon of August 20, 2022. The college students who participated in the in-depth interviews were from 4 different grades majoring in computer science, civil engineering, English and business administration, among whom two were male and two were female. Before conducting in-depth interviews with the four college students, they were informed in advance that the interviews would be recorded and sorted out for this research, and the interviewees' consent was sought before starting the interviews. During the interview, paper was used to make notes and audio recording equipment was used for recording; And ask some divergent questions to the respondents to get more profound answers than those on the questionnaire. At the end of the interview, the recorded interview content was sorted into written form for further analysis.

Focus group discussions were conducted online. The four college students who participated in the focus group interview were from the four majors of information management and Information system, Japanese, public administration and English, including two sophomores and two juniors, all of whom were female. Before the start of the discussion, please inform all the participants that the meeting will be recorded, no personal privacy will be disclosed, and all the content will be used for research only.

At the beginning of the meeting, the moderator began to ask questions, and the respondents answered in turn. After the focus group interview, the content of the meeting was organized into a document form and finally analyzed. After the in-depth interview and focus group interview, not only the basic information about the use of Tik Tok by these students, but also some more in-depth references can be obtained, such as the various influences of Tik Tok use on life and study, and the existing problems of Tik Tok platform at the present stage.

4 Findings

4.1 Descriptive analysis of questionnaire results

4.1.1 Population distribution characteristics of respondents

As the research object of this paper is college students from Beijing who use Tik Tok, the participants were screened at the beginning of the questionnaire. A total of 254 people participated in the questionnaire survey and 210 valid questionnaires were collected. Among them, male samples accounted for 29.1%, and female samples accounted for 71.9%. The number of female samples was relatively large. The sample of the questionnaire covered natural sciences and humanities, with natural sciences accounting for 40.0% and humanities accounting for 60.0%.

4.1.2 Basic usage analysis

First of all, in terms of the starting time of using Tik Tok, 66.6% of students started using Tik Tok in 2017. Secondly, college students who start to use Tik Tok after being influenced by their family and friends and are attracted by the functions of the software itself account for the highest proportion, followed by acquaintances' recommendation and favorite Internet celebrities or stars. It can be seen that the motivation of college students to use Tik Tok is largely influenced by their social relationships. At the same time, they start to use Tik Tok because some of its own functions can meet their needs to a certain extent. Second, the questionnaire shows that the main purpose of using Tik Tok is to relax body and mind, accounting for the highest proportion, reaching 92%. It can be seen that Tik Tok, as a short music video software, its entertainment properties are in line with the needs of college students to relax.

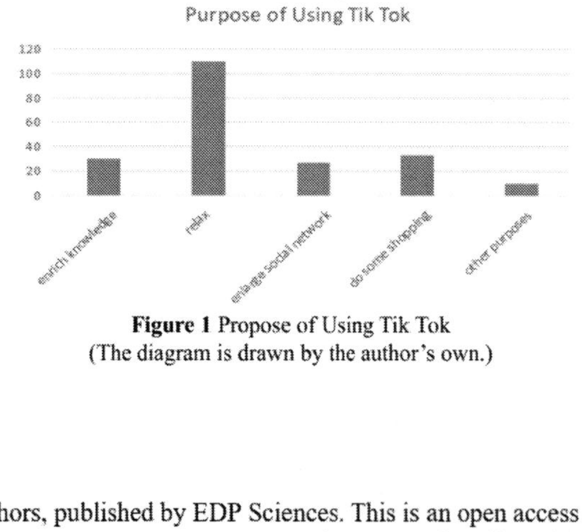

Figure 1 Propose of Using Tik Tok
(The diagram is drawn by the author's own.)

Second, in terms of the average daily use time, the highest proportion is between half an hour and one hour and more than two hours, followed by one hour to two hours and less than half an hour. It can be inferred that among college students who use Tik Tok for more than two hours a day, some of them may be addicted to Tik Tok. Among the evaluation of their addictions to Tik tok, the number of people who are addicted to Tik tok ranks the highest, accounting for 42%, and the rest are evenly distributed. Meanwhile, when asked if they can control how long they use Tik Tok, most people basically agree.

Figure 2 Average Time Spent on Tik Tok Everyday
(The diagram is drawn by the author's own.)

Third, in terms of preferences for watching videos, videos in the field of leisure and entertainment are the most popular among college students, while videos in other fields of beauty, science education and current affairs are evenly distributed. In the questionnaire survey, the most popular types of bloggers watched by college students are food bloggers and funny bloggers. In terms of video content, college students value the innovation of video content most.

Finally, in terms of the influence of Tik tok on college students, most of the samples believe that the positive influence of Tik tok on them is greater than the negative influence, but the negative influence still exists. The positive effects include a relaxed mind and body that can be more productive at work and school, as well as a greater perspective on the world and new knowledge and insights.

4.2 Analysis of the implementation results of in-depth interviews and focus group discussions

4.2.1 Develop an outline for in-depth interviews and focus group discussions

The use of Tik Tok by college students varies from person to person, and it is not possible to fully summarize the use of Tik Tok by college students in Beijing through questionnaire research. In order to explore the use of Tik Tok at a deeper level, the author decided to use in-depth interviews and focus group interviews to conduct in-depth interviews and focus group interviews with college students in Beijing, so as to further understand the deep reasons and influences of the use of Tik Tok by college students from a more extensive dimension.

This interview mainly focuses on the reasons why college students use Tik Tok, and the influence of using Tik Tok is also considered. Therefore, the interview outline is as follows.

1. What is your major?

2. What's your grade?

3. When did you start using Tik Tok APP?

4. Why did you start using it? (See students use themselves so use/network publicity/other reasons)

5. What attracted you to Tik tok?

5. Which bloggers do you watch most on Tik Tok? Why do you like watching them? What are their characteristics?

6. What kind of videos do you watch most often on Tik Tok? Can you describe it in general?

7. How do you think Tik Tok has helped you in your life and study?

8. What negative effects do you think you have experienced when using Tik Tok?

9. What do you think Tik Tok's shortcomings today?

10. Would you like to become a Tik Tok blogger? Please give the reasons and introduce what kind of video do you want to make?

4.2.2 Analysis of results of in-depth interviews and focus group interviews

According to the analysis and induction of the results of questionnaire survey, in-depth interview and focus group discussions, the starting point of using Tik Tok by college students is mostly for the purpose of entertainment, and the fields of the videos they watch also overlap and overlap with the fields they are interested in. In terms of the impact of Tik Tok on college students, it mainly has a significant impact on their life, study, aesthetics and study. While bringing positive effects such as relaxing the body and mind, broadening the vision and showing oneself, there are also negative effects such as addiction to Tik Tok, wasting time and generating negative emotions. At the same time, college students also pointed out that the existing problems of Tik Tok platform, such as repeated advertisements, comment sections and pushed content, will affect the user experience.

The research results of this paper are basically consistent with the previous research conclusions, because their research also proves that the use of Tik Tok by college students mainly meets the cognitive needs, emotional needs and aesthetic needs of college students. Tik tok can not only meet any one of these needs unilaterally, but also meet any two of them at the same time or the integration of all the needs. College students Tik tok users can obtain information on Tik tok platform. The simple process of information interaction makes the process of obtaining information easy to operate, which is in line with the fragmented usage habits of college students. Tik tok can not only entertain college students, but also accompany them and show themselves in a diversified way. In terms of the influence of Tik Tok on college students, both positive and negative effects exist. This paper and previous studies have also proved that Tik Tok can provide emotional and knowledge support to college students, enrich life interest, broaden horizons, realize self-value, and thus gain more self-identity and sense of belonging. But it is also easy to indulge, by the network violence and invasion of privacy and other bad aspects of the influence. At the same time, long-term use of Tik tok will also cause distraction and addiction.

Starting from the use of Tik Tok by college students in Beijing, this study provides a different perspective for the public to understand the situation related to the use of Tik Tok by college students. Tik Tok platform can timely correct questions according to the results of questionnaires and interviews, and college students can also timely adjust the time and way of using Tik Tok. At the same time, the research also verifies the correctness of the theory of use and satisfaction through quantitative research and qualitative methods. For the use of Tik Tok to make deeper research and contribution.

Due to my own academic level is limited, involved in the research of the theory is rich and deep enough, I hope that more and more scholars in the related studies, the community for college students or other representative of a group for further research, expanding the scope of the study, modify the problem of questionnaire, and expand the scope of the questionnaire distributed to explore trill use more questions, To facilitate the continuation of subsequent research.

5 conclusions

According to the analysis and induction of the results of questionnaire survey, in-depth interview and focus group interview, the starting point of using Tik Tok by college students is mostly for the purpose of entertainment, and the fields of the videos they watch also overlap and overlap with the fields they are interested in. In terms of the impact of Tik Tok on college students, it mainly has a significant impact on their life, study, aesthetics and study. While bringing positive effects such as relaxing the body and mind, broadening the vision and showing oneself, there are also negative effects such as addiction to Tik tok, wasting time and generating negative emotions. At the same time, college students also pointed out that the existing problems of Tik Tok platform, such as repeated advertisements, comment sections and pushed content, will affect the user experience.

The research results of this paper are basically consistent with the previous research conclusions, because their research also proves that the use of Tik Tok by college students mainly meets the cognitive needs, emotional needs and aesthetic needs of college students. Tik tok can not only meet any one of these needs unilaterally, but also meet any two of them at the same time or the integration of all the needs. College students Tik tok users can obtain information on Tik tok platform. The simple process of information interaction makes the process of obtaining information easy to operate, which is in line with the fragmented usage habits of college students. Tik tok can not only entertain college students, but also accompany them and show themselves in a diversified way. In terms of the influence of Tik Tok on college students, both positive and negative effects exist. This paper and previous studies have also proved that Tik Tok can provide emotional and knowledge support to college students, enrich life interest, broaden horizons, realize self-value, and thus gain more self-identity and sense of belonging. But it is also easy to indulge, by the network violence and invasion of privacy and other bad aspects of the influence. At the same time, long-term use of Tik tok will also cause distraction and addiction. (Zambrano-Vazquez, L. (2016). The interaction of state and trait worry on response monitoring in those with worry and obsessive-compulsive symptoms [Doctoral dissertation, University of Arizona]. UA Campus Repository. https://repository.arizona.edu/handle/10150/620615; Wen, Y. (2022) Analysis of College students' social media addiction from the perspective of Meme Theory: A case study of Tik Tok. Communication research. 5:119-120)

Starting from the use of Tik Tok by college students in Beijing, this study provides a different perspective for the public to understand the situation related to the use of Tik Tok by college students. Tik Tok platform can timely correct questions according to the results of questionnaires and interviews, and college students can also timely adjust the time and way of using Tik Tok. At the same time, the research also verifies the correctness of the theory of use and satisfaction through quantitative research and qualitative methods. For the use of Tik Tok to make deeper research and contribution.

Due to their own academic level is limited, involved in the research of the theory is rich and deep enough, I hope that more and more scholars in the related studies, the community for college students or other representative of a group for further research, expanding the scope of the study, modify the problem of questionnaire, and expand the scope of the questionnaire distributed to explore trill use more questions, To facilitate the continuation of subsequent research.

REFERENCES

1. Wei, X. (2022) Study on the influence of Tik Tok and other short videos on college students' study life. Electronic Components and Information Technology.2022.6(1):82-84.
2. Yi, L. Zhang, A. (2021) Study on the Influence of Douyin on Contemporary College Students' Values. Appreciation of the classic.92-95
3. Su, Y. Xu W. (2021) A study on the Affective Influence of Tik Tok on College students. Journal of Sino-German Applied University in Tianjin.5:107-110
4. Dong, H. Xu, B. He, S. Liu, W. 2022. https://kns.cnki.net/cms/detail/11.5257.R.20220523.1820.002.html
5. Wen, Y. (2022) Analysis of College students' social media addiction from the perspective of Meme Theory: A case study of Tik Tok. Communication research. 5:119-120
6. Xie, C. Cheng, X. Hu, J. (2020) Analysis of current situation of the use if Tik Tok APP by college students. Journal of Hubei Institute of Technology.37:67-72
7. Vaterlaus, J. M., & Winter, M. (2021). TikTok: an exploratory study of young adults ' uses and gratifications. The Social Science Journal, 1-20.
8. Zambrano-Vazquez, L. (2016). The interaction of state and trait worry on response monitoring in those with worry and obsessive-compulsive symptoms [Doctoral dissertation, University of Arizona]. UA Campus Repository. https://repository.arizona.edu/handle/10150/620615

Research on the Tax Dilemma of Network Broadcast Industry and Countermeasures

Li Yayi[1,*], Liu Yi-ting[2], and Zhang Jing-yi[3]

[1] Jinan University, Faculty of Humanities, Xiangzhou District, Zhuhai, Guangdong, China
[2] Jinan University, Faculty of Humanities, Xiangzhou District, Zhuhai, Guangdong, China
[3] Jinan University, School of Intelligent Systems Science and Engineering, Xiangzhou District, Zhuhai, Guangdong, China

Abstract. With the rapid development of new forms of Internet business, the network broadcast industry has gradually become the tuyer. The huge tax evasion event of network anchors such as Huang Wei and Lin Shanshan broke out, which caused huge social public opinion, but also reflected that there are many special difficulties in the tax work of network broadcast industry at present. However, the existing tax collection and administration system of China can not adapt to the needs of the emerging industry, which induced a series of tax evasion events. This paper will grasp the particularity of the tax work in the network live broadcast industry and analyze the realistic difficulties encountered by the tax authorities in the taxation of network broadcast industry, and put forward the countermeasures for the situation of our country based on the international advanced experience, and propose a new way to deeply integrate the tax work and high-tech.

1 Introduction

In recent years, the development of the Internet has been in full swing, giving rise to many emerging industries. Under this trend, the network broadcast industry emerged and gradually grew stronger, becoming a phenomenon industry in the new form of Internet business. As of June 2021, the number of users has exceeded 638 million, up 75.39 million year on year, accounting for 63.1 percent of all Internet users, according to the 48th release by China Internet Network Information Center (CNNIC). It can be seen that both the capital scale and the user scale of the network broadcast industry are developing rapidly. However, in recent years, practitioners in the network broadcast industry have frequently evaded tax, and the amount is huge, which has brought negative impact to the society, damaged the authority of laws and tax authorities, and triggered adverse social public opinion. For example, in 2021, Huang Wei (a network anchor engaged in e-commerce and live streaming) made false tax declaration by concealing personal income and fictitious the nature of business income conversion, and was prosecuted for tax payment, late fee and fine, totaling 1.341 billion yuan; As soon as the news broke, the Internet denounces, doubts emerge in endlessly. And many network anchors have been evading tax in various ways, which has brought great challenges to the work of tax authorities. In this regard, the author analyzes the unique tax dilemma of the network broadcast industry, in order to give advice and suggestions for the tax collection and management of the network broadcast industry.

2 Characteristics of tax-related matters in the network broadcast industry

According to the Regulations on the Administration of Live Internet Broadcast Services issued by the Cyberspace Administration of China, live network broadcast refers to an activity based on the Internet that continuously releases real-time information to the public in the form of video, audio, graphic and text. It is characterized by convenience, interactivity and real-time performance. Relying on the new form of Internet business, the tax-related matters of network broadcast industry have typical characteristics such as concealment, ambiguity and liquidity.

2.1. Concealment

In the era of digital economy, transaction subjects are gradually dispersed and tax avoidance methods are becoming more and more diverse, which makes the tax related matters in relevant industries more hidden. In terms of the network broadcast industry, users' tipping fees are the main source of income for anchors. With the booming development of online payment platforms, many users will choose to tip privately. Such income is indirect and difficult to be detected, and all parties to the transaction rely on the network, which is highly decentralized and difficult to track. [1]On the other hand,

*2674224550@qq.com

tax avoidance methods in the era of digital economy are also more varied. Due to the characteristics of the Internet industry, which is highly volatile and rapidly updated, tax-related information is also very easy to be hidden and tampered with. Many anchors evade taxes by falsely reporting business and changing the nature of income, which brings challenges to the work of tax authorities. As a result, tax information is not equal, and relevant tax-related matters are difficult to be fully grasped by tax authorities, which reflects the extremely strong concealment of tax-related matters in the network broadcast industry.

2.2 Ambiguity

As a new industry, network broadcast has not formed unified industry norms and practices compared with the traditional industry, and there are still many ambiguities and blank areas about the tax-related matters of anchors. Mainly reflects in: The fuzziness of the definition of income nature and the fuzziness of the division of tax liability. First, the income sources of network anchors are complex and diverse, and the income nature of the current law in China, such as tipping fee and pit fee, is not clear, and there are many opinions in practice; Second, the business models of network anchors and platforms are diverse and flexible, and the tax obligations of anchors and platforms are not clearly divided in law and practice. These ambiguities bring many difficulties to the work of tax authorities.

2.3 Liquidity

Compared with traditional industries, practitioners of network broadcast industry have the characteristics of cross-platform and cross-region. On the one hand, many network anchors often work on several different live streaming platforms and get income. Only through the integration of income from these different sources can the tax payable of anchors be determined. [2] Moreover, the work of network anchors is free, and they often work across regions, and many anchors make use of the tax policy levied by regions to pay less tax. At present, Chinese tax authorities still lack a platform for tax-related information sharing, and cross-regional tax collection and management is still the pain point and difficulty of the work.

3 The dilemma unique to the taxation of network broadcast industry

3.1 Difficulty in visualization of anchors' income

The income sources of network anchors are quite mixed, including but not limited to commissions, pit fees, commissions, rewards and so on. Besides, anchor salary calculation methods of each platform is variety and complex. Some calculate salary according to the broadcasting duration, some adopt commission system, based on the basic salary, plus the commission calculated according to the actual sold goods. There is also a way to calculate the salary, in which the anchor sets the price independently and calculates the difference between the supply price as the basis for the salary calculation. In addition, There is also a hybrid system, which is a proportional combination of various methods of calculating salary. This raises the questions of which revenue sources are included in the various calculation methods, which income is not included in the platform calculation method, and which taxes should be levied, so on. The difference in the amount of income tax paid by different calculation methods is great, which gives opportunities for tax evasion.

3.2 Loopholes at the legislative level

Our current Law of the PRC on the Administration of Tax Collection is formulated on the basis of traditional industries and entity businesses, for the network live industry and other new forms of Internet business industry, the tax regulations still have vacancies. The current laws and regulations do not make clear provisions on the tax registration system, income classification and tax rate of the new Internet business type industries, and there are still loopholes in the tax legislation in this field. For example, in the network broadcast industry, anchors have various business models and income sources, but the existing laws still do not make clear provisions on which income should be taxed and at which tax rate.

In order to adapt to the continuous development of new Internet formats, Law of the PRC on the Administration of Tax Collection is in constant revision and reform, from 1995 to 2015 after four times, but it is still in the traditional economy on the basis of tax regulation emphasizes tax modernization and informatization. [3] But for the digital economy of the emerging industries, such as network broadcast industry, there are no detailed implementation rules; In 2018, China promulgated the Electronic Commerce Law, which to a certain extent has perfected the tax administration system in the field of electronic commerce. However, the definition of the tax liability of each subject of the network broadcast industry and how to implement the tax payment process are still blank at the legal level.

3.3 Information asymmetry in the work of tax authorities

Due to the concealment and paperless of the network broadcast industry, the problem of tax-related information asymmetry is particularly prominent. Current tax collection and management means still can't adapt to the need of network broadcast industry tax management.

To be specific, first, the development of tax collection and administration technology lags behind the development of other new Internet business technologies. With the booming development of the new Internet business type economy, the payment platform and payment method of network broadcast transaction have more choices. Anchors can collect payment by Alipay, wechat and other means, which is difficult for tax authorities to detect and distinguish the nature of each income. At the same time, with the continuous expansion of the scale of the network broadcast industry, the network broadcast platform is associated with a large number of enterprises, business exchanges between enterprises are more complex, and the interest disputes are greater. So it is difficult for tax authorities to track the whole transaction information and ensure the authenticity of tax-related data. Second, the tax authorities have not built an information sharing platform, local tax departments lack communication, and cross-regional tax work is inefficient, which gives cross-regional and mobile network anchors an opportunity to evade taxes. At the same time, tax authorities also lack communication with external platforms. They have not established an information sharing mechanism with payment platforms such as Alipay and wechat, and data sharing involves legal issues of privacy and supervision, so the data provided by third-party payment platforms is very limited. [4]

4 Reference and internalization of international experience

4.1 The introduction of a digital tax

Digital tax was first advocated by the Organization for Economic Cooperation and Development (OECD), which is fully called "digital service tax". It refers to a tax specially levied on the effective profits of the digital service industry. The network broadcast platform mentioned in this paper also belongs to the scope of its levy.

On 19 October 2017, the European Council released a report on digital taxation, in which EU leaders called for "efficient and fair tax systems suitable for the digital age". The EU has been pushing for a digital service tax since May 2015, but had to suspend it due to opposition from some countries. Despite the stumbling blocks in the EU, as of 2021, 43 countries have introduced or are proposing to introduce digital taxes, with national implementation in every state.

Although the full implementation of digital tax is still facing quite a challenge in China, but its concepts and methods are worth learning. The profit and operation modes of the network broadcast industry are quite different from those of the traditional industry. However, our Tax Administration Law is formulated on the basis of traditional industries. If taxation is carried out in accordance with the original provisions, it will be unable to adapt to the economic environment of the new Internet industry, which will leave huge space for tax evasion in the new industry such as network broadcast industry. The establishment of special taxes on these new industries can effectively avoid these problems.

4.2 Information sharing mechanism

On 22 March 2022, the Council of the European Union adopted an amendment to the Tax Administration Cooperation Directive to extend EU tax transparency rules to include sellers of digital economy platforms, as individuals and businesses often do not report the income they receive when using digital platforms, resulting in a loss of tax revenue to member States. Member states are required to automatically exchange information on the income of sellers on platforms to ensure that sellers pay their fair share of taxes. [5] Other parts of the amendment improve the exchange of information and cooperation between the tax authorities of the Member States. By clearly defining the income declaration obligation of platform sellers and the information exchange obligation of member States, it helps member states obtain complete and clear tax-related information of sellers on digital platforms, and promotes the integration and transparency of tax work.

In China, due to the problem of administrative planning, the tax data of various provincial units are not unified or interconnected. The Yangtze River Delta region began to explore fiscal and tax integration in 2020. It can be seen that there are still defects in cross-provincial tax management. Strengthening administrative cooperation and realizing full data sharing among provinces is a big boost to tax regulation.

4.3 Classification tax declaration and management system

Japan adopts the blue and white tax return system in tax declaration. Taxpayers with good tax reputation can fill in the blue statement, reduce their extra tax payment process, and obtain the corresponding tax benefits. At the same time, the tax authorities also carry out irregular spot checks on the blue statement fillers. If the fraud is found, the qualification of the blue statement will be cancelled, and the tax authorities will re-investigate and calculate the tax payable. This system can not only encourage taxpayers to comply with tax payment, promote their active tax payment, effectively regulate the phenomenon of tax evasion, but also improve the work efficiency of tax authorities, making the tax relationship form a virtuous circle. In addition, German tax authorities also adopt a classified management system for network anchors, focusing on monitoring high-income anchors while conducting random spot checks on ordinary income anchors[6], so as to strike a balance between preventing tax evasion and improving government work efficiency.

This classification tax declaration system can be used as reference for our country. However, considering the frequent tax evasion of network anchors in China, we should strengthen the overall supervision of network anchors on the basis of hierarchical management. To be specific, the author suggests that tax authorities should distinguish different types of network anchors and implement hierarchical and classified management according to their income and previous tax credit.

5 The application of emerging technologies in the field of taxation

5.1 "Big Data" + tax collection and management

At present, China has not established a mature third-party information reporting system, and the independent taxpayer reporting system is difficult to implement. It is still a long way from achieving the integrity of tax data. On the other hand, tax declaration lags behind the implementation of tax obligations, and the declaration data examined by tax authorities are static financial information, which lacks the monitoring of taxpayers' financial dynamic changes[7], and the timeliness of tax data is insufficient, leaving a gap for tax evasion. And at present, the data analysis methods of our country's tax departments are mainly based on classification and general description and lack of deep analysis of the tax data. Tax authorities are limited in their ability to analyze and integrate data and cannot detect tax evasion in time. The asymmetry of tax-related information and the backwardness of tax authorities have caused the tax dilemma of network broadcast industry.

Tax big data refers to the multi-dimensional collection of information related to taxpayers on tax information systems and various third-party platforms, including tax-related data generated by all economic behaviors of taxpayers. If these big data are summarized and systematically analyzed, it will be able to effectively screen taxpayers with the risk of tax evasion and effectively break the dilemma of tax work caused by information asymmetry.

At present, our tax authorities use big data tools to a certain extent in their work, but there is still room for improvement in the ways of using and thinking.

First of all, big data can also be used in the top-level design of the tax system, which can be used to predict and analyze the changes in the tax economic relationship scientifically[8], so as to improve the corresponding laws, regulations and system design, and make the tax system more compatible with the economic environment of the new Internet business.

Secondly, at the law enforcement level, tax departments should fundamentally change their working concepts. Using big data, tax authorities can change their working mode from the traditional "problem occuring-problem analysis -reason finding -problem solving" to "data collection -quantitative analysis -relationship finding -process optimization" to better achieve the purpose of preventing tax evasion.

5.2 "Blockchain" + tax collection and administration

Blockchain technology is a new technical means to improve the ability of tax governance. Due to its technical advantages of data imtamability, permanence and transparency, blockchain technology has undoubtedly become an important technical support for tax collection and administration in the digital economy era. [9] In August 2018, Shenzhen Taxation Bureau issued the first blockchain invoice, which was gradually promoted in 2019. By March 2020, Beijing, Guangzhou, Yunnan and other places had successively launched blockchain electronic invoices. Compared with traditional electronic invoices, the biggest feature of blockchain invoices is the use of blockchain's non-tamper-proof, anti-denial, anti-deletion distributed ledger technology. [10] Tax authorities can also use blockchain invoices to trace the source, authenticity and entry information of invoices. The application of blockchain invoice can simplify the tax process, reduce the cost of tax work, and improve the efficiency of tax collection and administration.

However, the process of electronic invoice in our country is still underway, and the popularization of block chain in the field of tax collection and administration is difficult to be achieved. While, in view of the geographical clustering characteristics of the network broadcast industry, the author suggests that we can pilot and promote the blockchain e-invoice in regions with more developed new forms of Internet business. In these regions, we can take the lead in deeply integrating blockchain technology with tax collection and administration, and strengthen the tax collection and administration of the network broadcast industry. This can also facilitate the simplification of the tax process, encourage practitioners to pay taxes independently, and promote the healthy and sustainable development of the industry.

6 Suggestion and countermeasure

6.1 Improve legislation in related areas

Tax collection law is the basis for Chinese citizens to perform tax obligation, but also the fundamental guarantee of our rights and interests. Under the background of the vigorous development of various new Internet industries, we should innovate the tax collection law in keeping with The Times so that it can better serve the economic and social development of our country.

In order to regulate the healthy development of the network live broadcast industry, China has issued the Measures on the Marketing Management of Network Live Broadcast (Trial) and the Opinions on further regulating the profit-making behaviors of network live broadcast to promote the healthy development of the industry. The author believes that the emphasis on the platform's withholding tax obligation is very meaningful. On this basis, the relevant legal documents should further clarify the responsibilities and consequences of the platform's failure to fulfill the withholding tax obligation. In addition, according to the special income nature of the network broadcast industry, a new appropriate tax should be formulated, the tax obligation between the platform and the anchor should be clearly divided, and the information reporting obligation of the platform should be clearly stipulated. At the same time, the obligation of the third-party platform to provide information should be stipulated in legal form, and the scope of information should be divided in detail, so as to protect the privacy of the parties and ensure the transparency of tax-related information.

6.2 Rely on emerging technologies to improve the efficiency of collection and administration

At present, one of the main reasons for the dilemma of tax collection and administration in the network broadcast industry is that the technology of tax collection and administration lags behind that of the new industry. Therefore, tax collection and administration should also keep pace with The Times and actively integrate with high and new technology.

Tax authorities should make good use of "big data", "blockchain" and other emerging technologies, give full play to the function of technology to serve humanity, facilitate the tax work process, improve the efficiency of tax work, and effectively supervise the phenomenon of tax evasion with high and new technology. At present, although Chinese tax authorities also use big data and blockchain technologies, there is still a considerable distance to mature application and complete integration. We should focus on overcoming difficulties, introduce professional and technical personnel, actively develop new technologies, improve our ability to analyze and integrate tax-related information, and explore new ways to combine tax work with big data, blockchain and other technologies.

6.3 Build an information sharing platform

Information asymmetry is a key problem that makes the work of tax authorities difficult, especially in the emerging industries such as network broadcast industry. The digital economy era is a double-edged sword. Taxpayers can easily hide information, and tax authorities can also obtain information.

On the premise of relevant legislative guarantees, we should establish an information sharing mechanism between tax authorities and live-streaming platforms and third-party platforms, so that tax authorities can obtain relevant transaction information of taxpayers in a timely and comprehensive manner, which helps to urge taxpayers to pay taxes in full and prevent tax evasion. At the same time, a unified tax-related information platform should be established within the government to realize the unimpeded flow of information across regions, levels and departments, so as to avoid the phenomenon of double taxation or less tax and no tax due to the mobility of network anchors. In addition, we should also strictly examine the applicable people of the verification collection system to prevent network anchors from abusing local preferential policies to evade taxes.

7 Conclusions

To sum up, the rise of the network broadcast industry and the frequent tax evasion of anchors in recent years have made the unique tax dilemma of the network broadcast industry emerge, including difficulties in visualization of anchors' income, loopholes in legislation, information asymmetry and other problems. From the perspective of learning from international experience, the author proposes that the "digital tax" of the European Union can be used for reference to develop new appropriate taxes, and classification tax management system of Japan and Germany can be copied to conduct tax supervision with different intensity on different types of anchors, and the tax-related information sharing mechanism among EU countries can also be learned. From the perspective of using emerging technologies, the author puts forward two ideas: "big data" + tax collection and management and "blockchain" + tax collection and management.

Finally, the author summarizes the countermeasures at three levels, that is, improving legislation, using new technology, and building an information sharing platform, hoping to offer advice and suggestions for the further development of tax work.

Acknowledgments

Supported by College Students' Innovative Entrepreneurial Training Plan Program of Jinan University.

References

1. Xong, Y., Dong, M., (2022) Analysis of Tax Collection and Management of Individual Income Tax on Network Broadcast Rewards. Journal of Hubei University of Education., 39 (01) : 71-75.
2. Sun, H., (2022) The Dilemma and Breakthrough of Network Anchor Income Tax. Entertainment Method internal Reference., 6: 15.
3. Wang, M., Peng, M., (2022) An Analysis of the impact of Digital Economy Development on the Behavior of Tax Collection Subjects and Policy Recommendations. Economic Review Journal., (08): 93-99.
4. Tan, S., (2020) Institutional Challenges and Policy Recommendations of Tax Collection and Management of Digital Economy. South China Finance., (06): 37-44.
5. Angyong. (2021) The European Union has adopted the Tax Transparency Rules for digital platforms (DAC 7), https://www.investgo.cn/article/hzxxc/anyong/202104/540130.html
6. Guo, J., Wang, C. (2022) Tax Governance of Webcast: Theory Construction and Strategy Selection. Modern Communication (Journal of Communication University of China)., 44(07): 102-111.
7. Liu, S., Sun, J. (2016) Tax Governance: Theoretical Thinking based on Big Data. Review of Economic Research., (09): 3-6.
8. Ma, C., Zhao, D. (2018) The Reform of Taxation Administration under the Background of Internet. Fiscal Science., (09): 18-24.

9. Owens, J., Chen, X. (2020) Blockchain and Tax: Exploring the Next Frontier. International Taxation in China., (02): 10-11.
10. Tang, J. (2020) Blockchain invoice to the middle: Giant has already laid out "new infrastructure", https://finance.sina.com.cn/blockchain/roll/2020-03-30/doc-iimxxsth2433343.shtml

Research on the Ways to Overcome Social Media Addiction

Xiangyu Wu

Kang Chiao International School, Kunshan, Jiangsu Province, China, 215332

ABSTRACT. With the development of technology, social media has been disseminated to the masses, and now basically everyone has their social media accounts. People may not realize that they are addicted to social media, but on the subway, on the bus, while waiting for a meal, when relaxing, or when people are alone, people often use social media to pass their time, they subconsciously open social media to browse, but they can't live without it. The term "low-headed" is also used to describe those people who are always on their phones. The topic of this article is how to overcome social media addiction. Some case studies or personal experiences will be cited. It is divided into "what," "why," and "how" sections so that people understand what social media addiction is, whether they have this symptom, and why they are addicted to it. Finally, it will explain how to choose a better choice, i.e. how to overcome social media addiction. The study is focused on people who are addicted to cell phones. In this era of new media, cell phone addiction has its causes and many external factors, but people can still avoid social media addiction if they are conscious to overcome other aspects such as dependency.

1 Introduction

In today's society, with the support of technology, citizens can easily communicate with others online through social media. Social media is an application or collaborative website that facilitates communication, interaction, sharing, and collaboration. When people are waiting for the bus, riding the subway, or alone, they pick up their phones and open social media to see who has posted messages, what interesting news is out there, etc. This takes up a lot of people's time. But before people know it, they have fallen into an addiction to social media. People tend to talk about addiction to drugs, and addiction to smoking, but rarely do they talk about a person who is addicted to social media. Because of this invisible addiction, people don't know how they are heading to this black hole and they don't realize it. This article is about how to overcome social media addiction and wants to explore the reasons behind social media addiction and how to overcome it. The research method used is a combination of some case studies. The point of this research is to make people aware of the causes of online addiction and their feelings of dependence on social media and their hidden gratification. It is not scary to recognize internet addiction and look at it in a straightforward way, what is scary is that people are unknowingly and deeply addicted to it without doing anything about it or not caring about it. First, this article will clarify what social media addiction is. Then it will explain why people are addicted to social media, both from themselves and their environment. Finally, the paper will suggest some ways to overcome social media addiction.

mandyyyyy543@gmail.com

2 Reasons why people addiction

Three billion people now use online social media, about 40% of the world's population. According to some surveys, people spend an average of two hours a day sharing, liking, tweeting, and updating content on these sites. This means that 500,000 tweets and Snapchat photos are sent every minute around the world. People can see that people have developed a dependence on social media, but this is actually a social media addiction [1]. So, what is social media addiction? It is a behavioral addiction, defined as an excessive focus on social media, an uncontrollable need to log on or use social media, and an investment of so much time and energy into social media that it gets in the way of other key areas of life[2].

2.1 Self

2.1.1 Dependency. First of all, from people's own perspective, people have a deep sense of dependence on social media.

Now, people no longer use newspapers because there are already many social media that will cover news on their platform. For example, Twitter and Facebook, so that people can not have to wait for the newspaper to be delivered at a specific time and worry about whether the content of the day is to their liking. In the past, when citizens read the newspaper, they often feel deeply about a certain content or event in the newspaper and have a lot of their own thoughts and opinions, but people may only be buried in their own hearts or told to their familiar people.

When citizens have some suggestions about the content of the newspaper, they can only find the publisher by writing or calling. This takes up most of people's time. But when the news flows to social media, people can choose what they like to read, and they can comment and express their own opinions and views, which can be seen by strangers first and read by the author first. People can also use the comments section to get different views. This greatly increases the interactivity between us as readers and the author.

Let's look at a statistic. On 2016, Facebook had 1.55 billion active users, with each user spending at least twenty minutes each day reading the newest news in their timeline. This exceeds the population of China or India, making Facebook the world's largest country if it were a country. Furthermore, these populations are expanding: By 2019, the number of active social network users worldwide is expected to be around 2.72 billion, or roughly one-third of the global population[3]. This shows that just the news part of social media makes most people already inseparable from social media.

Social media certainly makes it easier for citizens to communicate and interact with people they know well. People no longer have to worry about whether their different schedules will affect their relationship and communication with each other. It also avoids some of the awkwardness that can arise in chats, where people can think again and again and can respond selectively. The content of the conversation between the two parties is more private for the third party. People can post their recent life or thoughts in their circle of friends. All these features can enhance the emotion between friends and family and reduce the distance.

According to the latest statistics released by We Are Social and Hootsuite, the number of online games worldwide has reached 4.55 billion as of January 2021. There are 4.2 billion social media users, compared to an increase of 490 million people in the past year. And, the number of social media users is equivalent to more than 53% of the total world population now. Some typical social media users now spend 2 hours and 25 minutes a day on social media, in other words, they stay awake about one day a week[4].

2.1.2 Satisfaction. "In atomized social relations, the social life and cultural picture appear fragmented, and inadequate social networks make loneliness a diffuse psychological state, with isolated individuals experiencing unprecedented self-identity confusion and a lack of belonging[5]." People are in a noisy society, in which their initial network of relationships will certainly be damaged due to various reasons, such as the pressure of work and study, or the difference in regions. But with that comes new branches as well. When people feel lonely, they use social media to satisfy their needs, and they use it as a friend to relieve their inner stress and all kinds of insecurities.

In real life, citizens will have many inadequacies that are slowly revealed in the process of being with people, which may be pains or flaws. But these holes will gradually weaken people's self-confidence and make us feel inferior. The emergence of social media allows citizens to enter a virtual world where people can package themselves, they will gain respect, and will get compliments. Because of the protection of privacy, people can talk to strangers and they can export their inner thoughts. People will find their own sense of belonging because they meet like-minded friends and have common topics.

Mass through the virtual fortress built by the new media technology, people can freely enjoy and release their worries. The new media meet people's psychological needs in many ways, penetrating "the key texture of people's psychological structure[5]." They provide people with a spiritual home where they can seek solace, allowing them to rely on and follow them [6].

Once, there was an article called "I lurked in Shanghai "celebrity" group and became a celebrity observer for half a month". The article was about some "celebrities" who are trying to create their own rich persona in the form of collusion. The author said: "Some people may be more vain, like to take pictures, but the money is not enough, so they look for the same needs of people together to put together a single, take pictures and send friends. These people in addition to netizens, there are micro-businesses, some are bloggers, they need good photos to support their operations, but their own money is not a lot. "This is not exactly the satisfaction that social media brings to people, because of their own shortcomings, through social media to befriend some people with a common purpose and goal, they will help each other in the form of mutual help to achieve a win-win situation. When people need some "decoration", social media can provide some very good resources from all over the world, these resources help people to package and beautify themselves.

2.2 Environment

2.2.1 Major trends of the times. In this digital age, the preferred medium for people to access information is social media, which can determine what people are accessing very accurately and quickly. It also helps people to know what is going on in the lives of different people at the same time. TikTok is one of the most popular social media platforms these days, and it is a good way to know what is popular now. When people come across some very funny and useful videos, they will forward them to their friends. But without this social media platform, when other people talk about some popular phrases or hot things, it's hard for those who don't know to participate or join in the discussion of the topic. Over time, there will be a generation gap between people. Why do people often say that there is a generation gap between people of different ages, and where does this generation gap come from? Because nowadays young people know a lot of memes through various social media platforms, while older people don't because they don't read these social media or download them. This time, when they get together, they can rarely resonate with each other, usually about life, which is exactly the topic that some young people now reject, and thus do not want to communicate with older people, which also creates a generation gap.

In addition to TikTok, there is also the output of shopping, audio, social communities, meta-universe, cultural creations, etc[7].

Online shopping is not new to people. China's Taobao has brought great convenience to the Chinese people. Without leaving home, people can simply search for what they want on Taobao, sit at home for a few days, and receive their package very quickly, and many businesses on Taobao accept seven-day, no-questions-asked refunds and returns. Taobao is Asia's largest online retail merchant circle, with 98 million registered members in 2008, which shows its popularity of Taobao[8]. Taobao has become a way of life, completing a seamless connection between the physical and digital experience. Without Taobao, how could the Chinese have ensured their quality of life during the epidemic and under the influence of the environment? Perhaps people would think of offline supermarkets, but what if they lived far from a store? What if citizens don't have easy access to transportation? In this case, people can hardly come to the store, they will definitely buy double to avoid continuing to welcome this inconvenience in a short time. But the shelf life of these things is the same, and people do not control the problem of fresh food. This shows that the advent of Taobao has improved the quality of life for people.

NetEase Cloud Music is also a very popular music platform where people can follow their favorite singers, share songs with good friends, or share songs, and comment on their favorite songs. People no longer use the radio to listen to their favorite music by dialing channels, and they don't have to keep time for fear of missing music. And with one of the latest features that allows people to listen to music in sync with their friends, they can shed the problem of distance and listen to common music directly with a specific person. Imagine if there were no such platforms, how could people know better what music is popular or new now, and how could people quickly and conveniently tell their friends which songs are good.

These social media affect us in the general trend of the times. Citizens may have to, or they may want to, accept these social media. But it is undeniable that without these social media, people would face being eliminated by technology.

2.2.2 App is behind the shenanigans. In the huge market of social media, it is important to compete with each other and it is the goal of a social media company to better attract people's attention and become the most popular social media. A very important point is marketing[9].

In the case of tourism, for example, it is crucial for those working in the tourism industry to understand places, especially towns, and how these should be marketed to and consumed by tourists. By creating and promoting a unique brand destination, new visitors can be attracted to the advertised location. With the rise of social media, travel marketers have more room to explore how they can use their own or other social media sites to communicate with today's technology-connected travelers. They will focus on clearly explaining how to develop branded destinations, with a particular emphasis on product analysis, promoting authenticity, and using social media to create the personalized experience travelers want[10]. So when people open relevant social media about the travel industry or see it on Twitter, Facebook, Instagram, or WeChat, people will click on it. Over time, when people want to go to a particular destination, they will open these social media outlets on their own to browse and read. It's good for both social media and travel companies to help them increase traffic, more people will start creating accounts on social media, travel companies will increase more exposure and readings, thus increasing the cases of successful transactions.

There is another type of marketing called the "celebrity effect". Through celebrity endorsement of a product, it can help a company's products and services stand out in the industry. Traditionally, celebrity advertising has been used to market and communicate by demonstrating their high social influence and authority[11]. They are usually sent to social media platforms, and because celebrities themselves carry a certain amount of traffic and fans, when they are sent to social media platforms, these fans will immediately check the social media, start commenting and liking their favorite celebrities, and then move on to another social media to buy or retweet them for the reason of "support". The content is sent by the star.

The target group of social media is also an important point. Let's take teenagers as an example. With the advent of social media, teenagers may not even watch TV as often as they used to, they will spend most of their time on their smartphones, some social media to watch videos and some other entertainment[12].

Social media content needs to be kept fresh and consistent. It needs to be in some form that brings something tangible to the viewer, whether it's an opinion or a smile[13]. Take TikTok for example, more and more people choose to open TikTok in their free time, there are many videos on TikTok that they are interested in, as long as the platform uses big data to classify their preferences, when people scroll down the video, the next video will be their favorite. As long as they keep pushing relevant videos, people will keep spending time watching them, thus this is why many people like to pass time by using TikTok.

3 Recommendation

Prolonged use of social media can affect people's moods, which can influence their perceptions of something, increase stress, cause anxiety, become irritable and worried because of something, and lead to sleep disturbances and decreased concentration [1]. To get out of the dilemma of addiction to the use of new media on the Internet, people need to awaken the value of the subject, return to the real world of life and art, pay attention to the body's senses, use their mind to feel, go beyond the noise created by consumer culture, and return to spiritual peace and enrichment[14].

First of all, people have to be their masters[14]. As the saying goes, if they want others to change, it is better to let themselves change first. As long as people have determination and self-control, they can do many things they never thought they could do and reach higher goals. In fact, people have a lot of things to accomplish in a day, whether it's big, small, or urgent, people need to go through them all, and many of these things maybe they need to use social media, need to communicate with others. At this point, people can list the things they want to do today in the morning, set goals and priorities for themselves, and list the equipment and time needed for each of these things. This will not only make it clear how long people want to spend on their phone but also prevent them from subconsciously tapping on some social media for irrelevant things when people open their phone. One study showed that the average person lights up their phone screen about 100 times in a day. People can set a small goal for themselves, for example, only allow themselves to light up the screen no more than 50 times in a day. When people reach their goal they can reward themselves with something, which will increase their motivation. Or people can limit the amount of time they spend on social media. People can set a certain period of time to look at social media or to respond.

Furthermore, people need to find ways to keep themselves from getting caught up in the virtual world, when the real world is also very beautiful. People can also keep themselves busy so that their lives are enriched. No more starting to focus on social media as soon as people have time. In their free time, they can read a book, study, exercise, or spend time with their families[15]. If it happens to coincide with a break and a vacation, get out and about too. Go see what the beauty that people swipe on their phone looks like in real life. Nesbitt cautions parents not to let their children get too caught up in the computer, but to guide them away from the computer and the Internet and get outside to experience nature. Because, in the information society, people use more brain power than physical power, they need to use their hands and bodies more in their spare time to balance the overuse of brain power[14].

4 Conclusion

Social media affects people in different ways depending on the prerequisites and the personality of the person[1]. Social media has brought a lot of advantages to people's life and has helped them to improve their quality of life. But as people spend more time using social media, it seems from the reality and some results that it is social media that is controlling them. There are many temptations that people keep encountering in this society, and excessive use of social media can have some negative effects on people. So, how to overcome social media addiction requires willpower and self-control. It is difficult to put down the phone at first, but people have to get used to it and people have to keep saying to themselves that they can do it. People can spend a lot of the time they spend on their phones on other meaningful things such as being close to nature, exercising, and spending time with their families. This paper needs to be improved by adding some surveys on the approximate number of hours of social media use per day by different groups of people, or surveys on the purpose of social media use by different groups of people, which can better support the ideas of this paper.

References

1. Brown,J. (2018), "Is social media harmful? Where is the evidence and the mystery?", https://www.bbc.com/ukchina/simp/vert-fut-42748944
2. Hilliard,J.(2019),"Social Media Addiction" , https://www.addictioncenter.com/drugs/social-media-addiction/
3. Carillo,K. , Scornavacca, E., Za,S. 2017, Datareportal, Volume 54, page 317-335
4. DataReportal, 2022, "Digital 2022 Global Digital Overview,"https://datareportal.com/reports/digital-2022-global-overview-report
5. Fan, K.(2008) Media Cult Theory: The heteronormative relationship between modern man and mass media, 2008, page 3-5
6. Yi,L.P. (2011) "Exploring the causes of audience media dependence in the new media environment", Today Media, Volume 2, page 90-91
7. Ogilvy China, 2022, What's New in Social Media in 2022: Explaining the 7 Trends Impacting Marketing, https://www.ogilvy.com/cn/ideas/2022-nianshejiaomeitixindongxiang-jieduyingxiangyingxiaode-7-daqushi
8. Liu,X.P.(2011), "The Impact of New Media Development on Economic Life Patterns - The Case of Taobao.com.", Today Media, Volume 8, Page 95-96
9. Behera,S., Balaji,P. (2019), "Social Media Marketing: Opportunities And Challenges", International Journal of Management, Technology And Engineering, Volume IX, Page 5639-5641
10. Kolb, B. (2017). Tourism Marketing for Cities and Towns: Using Social Media and Branding to Attract Tourists (2nd ed.). Routledge. London
11. Huang,M.X., Yao,S.Y., Liu,M.H., (2018) Self-improvement or self-deprecation? How celebrity endorsement enhances the marketing effectiveness of social media advertising, Journal of Psychology, Volume 50, Page 907
12. Silvia,S.(2019), The Importance of Social Media and Digital Marketing to Attract Millennials' Behavior as a Consumer, Jounral of International Business Research and Marketing, Volume 4, Page 7-10
13. Today's Source,2019, How to Play Social Media Marketing, https://www.toutiao.com/article/6701909711865774606/?&source=m_redirect
14. Learning Power, 2020, Albert Chow: What about the new media "addiction"? You need a philosophical reflection!,

http://www.xhby.net/index/202010/t20201028_6851929.shtml

15. Xiaobenzi, 2022, How to Overcome Social Media aAddiction, https://life.xiaobenzi.com/3256/%E5%A6%82%E4%BD%95%E5%85%8B%E6%9C%8D%E7%A4%BE%E4%BA%A4%E5%AA%92%E4%BD%93%E4%B8%8A%E7%98%BE.html

Exploring the Effects of Overload and Stress Factors on WeChat Moments Users' Discontinuous Usage Intention: an SSO Perspective

Qianrun Zhang[1,*]

[1] School of Mathematics and Statistics, Beijing Jiaotong University, Beijing, 100044, China

Abstract. This study explores the factors that contribute to social media fatigue and how it affects users' intentions to stop regularly using their WeChat Moments. The findings demonstrate that information overload, social overload, system feature overload, and upward social comparison (stressors) impact social media fatigue (Strain) and ultimately impact the decision to discontinue using WeChat Moments. Social media fatigue acts as a buffer between the stressors of social interaction and the outcome. Users who bear the pressure of the opinion leader role and have low clustering coefficients in the ego network are more likely to express the desire to discontinue using WeChat Moments.

1 Introduction

WeChat social networking was initially known as "acquaintance social networking", but in the development and application process, more strangers are becoming WeChat contacts, and the overload of information and social pressure erodes people's time and energy.

There is no doubt that the negative impact of social media and users' reactions caused by its negative impact has received the attention of the academic field.

Most negative social media usage investigations have focused on Facebook and Twitter, and there are currently fewer studies on discontinuing WeChat Moments. In addition, the antecedents of current research also focus on overload and stress factors, and it remains to be explored whether an individual's position and influence in social networks impact the intention to discontinue the use of their Moments.

2 Research model and hypotheses

The stressor-coping response-outcome models adopted considering the literature mentioned above analysis. We incorporate these variables in the model as controls as the user's gender, age, and prior usage of social media may have an impact on how they make use of such platforms. This is shown in Figure 1.

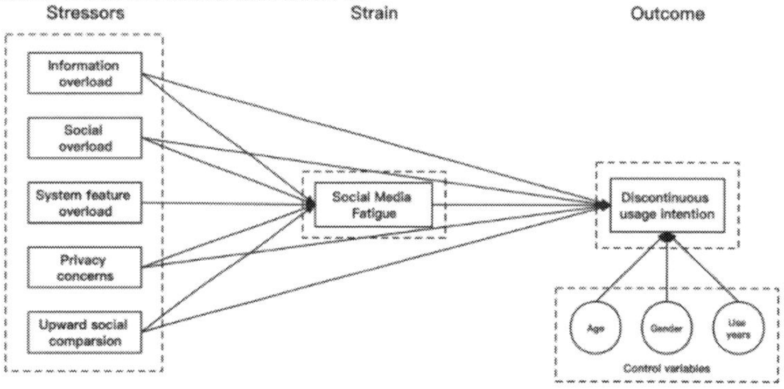

Figure 1 Research model[Owner-draw]

2.1 Effects of overload and stress on social media fatigue

Information, social, and system feature overload are the three latitudes of the overloadability factor. At the same time, Information overload describes invasive information on social media that is more than what a person can handle and comprehend at one time and beyond their cognitive capacity. Users find it more challenging to access crucial

* Corresponding author: qianrunz@163.com

and relevant information due to the mixed information in their WeChat Moments and the frequent photo, text, and video updates.

McCarthy and Saegert first proposed the social psychology notion of "social overload" It primarily refers to scenarios in which individuals respond to the numerous social support demands imposed by internet relationships.(Karr-Wisniewski & Lu, 2010; Maier et al., 2015)[1][2] The effectiveness of human decision-making is impacted by information overload. Social media may be considered virtual online communities where social overload also exists.(Sun et al., 2018)[3] Users of WeChat Moments may appear to be passively browsing their friends even though they are not actively engaging in interactions. They are more prone to experience loneliness and anxiety, leading to social media fatigue. This is due to the rapid growth of weakly related contacts in the Moments. When leading social network service providers introduce additional features to draw subscribers, this is referred to as "system feature overload." While adding new features has the ability to enhance the distinctiveness and productivity of social media programs, having too many features can exhaust users and cause anxiety.(Thompson et al., 2005)[4] Users' information intake and danger of information overload will increase since they will need to put in more time and effort to learn how to use these new capabilities.

H1a: social media fatigue is strongly correlated with information overload

H1b: information overload is positively associated with WeChat Moments discontinuous usage intention

H2a: social media fatigue is strongly correlated with social overload

H2b: social overload is positively associated with WeChat Moments discontinuous usage intention

H3a: social media fatigue is strongly correlated with system function overload

H3b: system feature overload is positively associated with WeChat Moments discontinuous usage intention

Stress factors were derived from privacy concerns and upward social comparison.(Liu Luchuan et al., 2017)[5] Privacy concerns span a wide range of IS-related topics. For instance, people's attitudes and behaviors in instant messaging and electronic health records may be influenced by privacy concerns.(Angst & Agarwal, 2009; Lowry et al., 2011)[6][7] Note that social comparison is more common in social phenomena. The emergence of social media has broadened the ground of social comparison from offline to online, and more and more social comparisons occur on SNSs. Social comparisons are divided into parallel comparisons, upward comparisons, and downward comparisons. Users of SNSs tend to post more positive posts, so comparisons on social networks mainly involve upward social comparisons(Vogel et al., 2015)[8] and upward social comparisons on social networks can negatively affect individuals' mental health.(Jang et al., 2016)[9]

H4a: privacy concerns are positively related to social medial fatigue

H4b: privacy concerns are positively associated with moment discontinuous usage intention

H5a: upward social comparison is positively related to social medial fatigue

H5b: upward social comparison is positively associated with moment discontinuous usage intention

2.2 Effects of social media fatigue on discontinuous usage intention

Psychological research have shown that psychological fatigue lowers performance and engagement and has a detrimental effect on a person's ability to exert continuous effort. People utilize behavioral or emotional coping methods to lessen or avoid these negative experiences when they are stressed. A user's tendency to use or withdraw from social networks may eventually decline as the degree of social network fatigue increases from a transient or mild experience to a deep one.(Liu & Guo, 2017)[10] Social network fatigue and dissatisfaction are brought on by perceived overload, which leads to plans to discontinue using them.(Vanderhaegen et al., 2020)[11]

H6: Social media fatigue is positively related to WeChat moments discontinuous usage intention

Social media fatigue mediates between social overload and social media discontinuance intention by influencing emotional exhaustion. Privacy concerns influence users' intention to use social media persistently by affecting the emotion of worry, which plays a mediating role. Additionally, Niu Jing et al. confirmed the mediating effects of social overload, privacy concerns, information overload, and upward social comparison, and discovered that all these factors—aside from upward social comparison—positively influenced the intention to avoid using social media regularly.(Niu Jing & Chang Mingzhi, 2018)[12]

H7: Social media fatigue mediates the relationship between information overload and WeChat Moments discontinuous usage intention

H8: Social media fatigue mediates the relationship between social overload and WeChat Moments discontinuous usage intention

H9: Social media fatigue mediates the relationship between system feature overload and WeChat Moments discontinuous usage intention

H10: Social media fatigue mediates the relationship between privacy concerns and WeChat Moments discontinuous usage intention

H11: Social media fatigue mediates the relationship between upward social comparison and WeChat Moments discontinuous usage intention

3 Sample and data collection

WeChat users served as the research subjects for this study, while the Questionnaire Star platform was used to distribute the questionnaire. The online survey could only be completed once per respondent.

508 valid responses—a pass percentage of 88.19% — were left after we eliminated 68 ineffective ones. In the sample, there were 360 females (70.87%) and 148 males (29.13%). Most respondents had a bachelor's degree and were between 18 and 25. A total of 201 participants

(39.57%) said they had been in their WeChat Moments for over 6 years. Approximately three-quarters of participants (n = 445; 87.60%) said they used Moments for no more than two hours each day.

4 Data analysis and results

4.1 Measurement model

Structural equation models were utilized to assess the study framework, and SPSS 26.0 was used to conduct reliability studies and confirm factor analyses on the factor loadings in the structural equation models.

Table 1. Reliability and convergent validity analysis[Owner-draw]

Factor	Item	Standardized	Cronbach's alpha	CR	AVE
Information overload	IO1	0.985	0.989	0.990	0.960
	IO2	0.979			
	IO3	0.974			
	IO4	0.981			
Social overload	SO1	0.984	0.992	0.992	0.968
	SO2	0.984			
	SO3	0.987			
	SO4	0.981			
System feature overload	SFO1	0.978	0.993	0.993	0.972
	SFO2	0.986			
	SFO3	0.991			
	SFO4	0.988			
Privacy concerns	PC1	0.990	0.995	0.995	0.978
	PC2	0.991			
	PC3	0.988			
	PC4	0.988			
Upward social comparison	USC1	0.988	0.992	0.992	0.975
	USC2	0.989			
	USC3	0.986			
Social media fatigue	SMF1	0.988	0.992	0.992	0.970
	SMF2	0.987			
	SMF3	0.987			
	SMF4	0.977			
Discontinuous usage intention	DUI1	0.984	0.986	0.986	0.947
	DUI2	0.966			
	DUI3	0.968			
	DUI4	0.974			

Reliability analysis: All the factors satisfied the reliability standards, which suggests that the questionnaire reliability indices were more satisfactory.

Validity analysis: To determine if the model has strong convergent validity, the combined reliability (CR) and average variance extracted (AVE) were computed using the Fornell and Larcker technique. Strong convergent validity was demonstrated by Fornell & Bookstein and Fornell & Larcker, who established standards of 0.7 and 0.5, respectively.(Fornell & Bookstein, 1982; Fornell & Larcker, 1981)[13][14]

The study's measurements are thought to have excellent discriminant validity if the square root of AVE is higher than the correlation coefficient.(Fornell & Larcker, 1981)[14]

Table 2. Discriminant validity analysis[Owner-draw]

	IO	SO	SFO	PC	USC	SMF	DUI
IO	0.9 80						
SO	0.9 77	0.9 84					

SFO	0.977	0.968	0.986				
PC	0.970	0.959	0.976	0.989			
USC	0.966	0.954	0.964	0.960	0.988		
SMF	0.976	0.968	0.975	0.964	0.959	0.985	
DUI	0.978	0.969	0.974	0.965	0.962	0.981	0.973

The bold numbers on the diagonal line are the square root value of AVE.

Overall, the reliability tests for the questionnaire measures were positive.

4.2 Structural model

As Figure 2, social media fatigue was positively impacted by information overload, social media overload, and system feature overload., supporting H1a, H2a, and H3a. The only factor that positively affects the desire to discontinue using Moments is information overload, which is consistent with the H1b hypothesis. Social overload, system function overload, privacy concerns, and upward social comparisons have little to no impact on this intention. H2b, H3b, H4b, and H5b are all false. Privacy concerns and upward social comparison had no significant effect on social media fatigue, supporting H4a、H5a, and social media fatigue was positively associated with discontinuous usage intention, supporting H6.

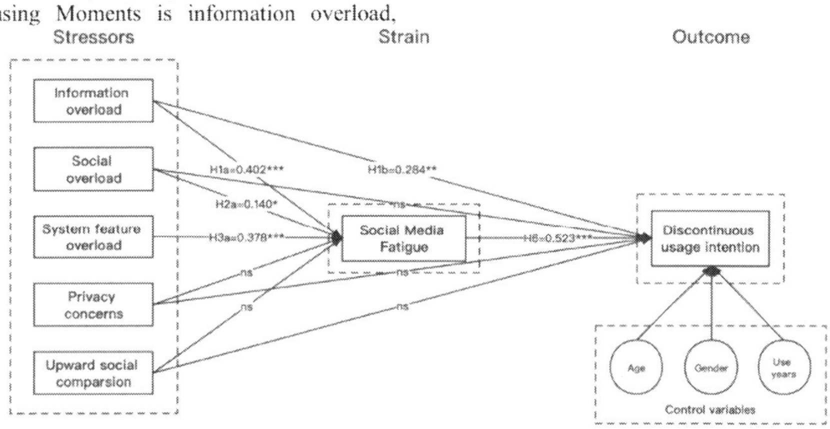

Figure 2 Research model results[Owner-draw]

ns: not statistically significant; *: $p < 0.05$; **: $p < 0.01$; ***: $p < 0.001$.

4.3 Mediation analysis

Regression analysis could be used to assess the mediating effect. Table 3 displays the bootstrap technique mediating effect analysis in SPSSAU, illustrating the mediating influence of SMF on stressors and results.

Information overload and the discontinuous usage intention are both mediated by social media fatigue, supporting research hypothesis H7.

Social overload and the discontinuous usage intention are both mediated by social media fatigue, supporting research hypothesis H8.

System feature overload and the discontinuous usage intention are both mediated by social media fatigue, supporting research hypothesis H9.

Upward social comparison and the discontinuous usage intention are both mediated by social media fatigue, supporting research hypothesis H11.

The study hypothesis H10 was incorrect because social media fatigue did not mediate the association between privacy concern and intention to discontinue usage.

Table 3. Results of the mediation analysis[Owner-draw]

Independent variables	Relationship	Direct Effect	Specific indirect effect	Boot St. error	95% Bias-corrected bootstrap CI	Mediation Effect
IO	IO→SMF→DUI	0.223**	0.353**	0.036	[0.070, 0.213]	Partial mediation
SO	SO→SMF→DUI	0.082*	0.152**	0.025	[0.024, 0.123]	Partial mediation

SFO	SFO→SMF→DUI	0.115**	0.248**	0.035	[0.080, 0.221]	Partial mediation
PC	PC→SMF→DUI	0.035	0.057	0.023	[-0.020, 0.067]	None
USC	USC→SMF→DUI	0.089**	0.127**	0.021	[0.003, 0.085]	Partial mediation

*: $P < 0.05$, **: $P < 0.01$

4.4 Discontinuous usage intention prediction

We further investigate if a person's standing in a social network influences their intention to discontinue and make forecasts from the viewpoint of opinion leaders.

The author chose one random participant from the researchers' WeChat Moments for a thorough investigation to ensure the study's objectivity and rigor. The user's ego network was developed through one-on-one interviews by asking the user's five closest friends and noting how each buddy was socially involved with the user. The three Scales of Connectivity, Persuasiveness, and Maven are suggested by to identify the effect on the user. 14 items in all were generated and measured on a 5-point Likert scale (1- strongly disagree, 2-disagree, 3 - average, 4 - agree, and 5 - strongly agree).

According to the findings, this user's ego network is shown in Figure 3. The clustering coefficient value is 0.2, indicating that only 20% of each person's friends they have spoken to will typically also speak to one another, and triadic closure is not very high. With a mean value of 4 on the scale, this user has a more excellent perception of their opinion leader. This user is more likely to stop using Moments.

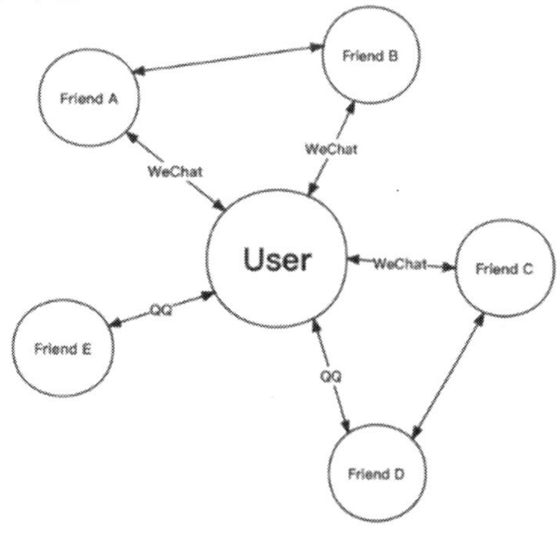

Figure 3 Ego network[Owner-draw]

5 Conclusions

In this study, we found that information overload, social overload, and system function overload all had a favorable impact on social media fatigue. Social media fatigue mediated these overloads and upward social comparisons. Social media fatigue is neither positively influenced by privacy concerns nor does it play a mediating role in privacy concerns and discontinuous usage intentions. Additionally, users' intentions to utilize their WeChat Moments sparingly were influenced to some extent by users' positions in social networks and their perceptions of opinion leaders. Future studies can increase the sample size for further exploration and validation to uncover more influencing factors.

REFERENCES

1. Karr-Wisniewski, P., & Lu, Y. (2010). When more is too much: Operationalizing technology overload and exploring its impact on knowledge worker productivity. Computers in Human Behavior, 26(5), 1061–1072. https://doi.org/10.1016/j.chb.2010.03.008
2. Maier, C., Laumer, S., Eckhardt, A., & Weitzel, T. (2015). Giving too much social support: Social overload on social networking sites. European Journal of Information Systems, 24(5), 447–464. https://doi.org/10.1057/ejis.2014.3
3. Sun, Y., Zhang, Y., Shen, X.-L., Wang, N., Zhang, X., & Wu, Y. (2018). Understanding the trust building mechanisms in social media: Regulatory effectiveness, trust transfer, and gender difference. Aslib Journal of Information Management, 70(5), 498–517. https://doi.org/10.1108/AJIM-03-2018-0072
4. Thompson, D. V., Hamilton, R. W., & Rust, R. T. (2005). Feature Fatigue: When Product Capabilities Become Too Much of a Good Thing. Journal of Marketing Research, 42(4), 431–442. https://doi.org/10.1509/jmkr.2005.42.4.431
5. Liu Luchuan, Li Xu, & Zhang Bingqian. (2017). Research on Social Media Fatigue and Passive Behaviors of Users Based on Grounded Theory. ITA, 40(12), 8.
6. Angst & Agarwal. (2009). Adoption of Electronic Health Records in the Presence of Privacy Concerns: The Elaboration Likelihood Model and Individual Persuasion. MIS Quarterly, 33(2), 339. https://doi.org/10.2307/20650295
7. Lowry, P. B., Cao, J., & Everard, A. (2011). Privacy Concerns Versus Desire for Interpersonal Awareness in Driving the Use of Self-Disclosure Technologies: The Case of Instant Messaging in Two Cultures. Journal of Management Information Systems, 27(4), 163–200. https://doi.org/10.2753/MIS0742-1222270406
8. Vogel, E. A., Rose, J. P., Okdie, B. M., Eckles, K., & Franz, B. (2015). Who compares and despairs? The effect of social comparison orientation on social media use and its outcomes. Personality and Individual Differences, 86, 249–256. https://doi.org/10.1016/j.paid.2015.06.026
9. Jang, K., Park, N., & Song, H. (2016). Social comparison on Facebook: Its antecedents and psychological outcomes. Computers in Human Behavior, 62, 147–154. https://doi.org/10.1016/j.chb.2016.03.082

10. Liu, D., & Guo, X. (2017). Exploring gender differences in acceptance of mobile computing devices among college students. Information Systems and E-Business Management, 15(1), 197–223. https://doi.org/10.1007/s10257-016-0315-x
11. Vanderhaegen, F., Wolff, M., & Mollard, R. (2020). Non-conscious errors in the control of dynamic events synchronized with heartbeats: A new challenge for human reliability study. Safety Science, 129, 104814. https://doi.org/10.1016/j.ssci.2020.104814
12. Niu Jing & Chang Mingzhi. (2018). Research on Social Interaction Stressors and Unsustainable Use Intention in Social Media Use. Journalism & Communication Review, 6, 15.
13. Fornell, C., & Bookstein, F. L. (1982). Two Structural Equation Models: LISREL and PLS Applied to Consumer Exit-Voice Theory. JOURNAL OF MARKETING RESEARCH, 13.
14. Fornell, C., & Larcker, D. F. (1981). Evaluating Structural Equation Models with Unobservable Variables and Measurement Error. Journal of Marketing Research, 18(1), 39. https://doi.org/10.2307/3151312

Research on How Live Streaming Helps Revive the Chinese Economy under COVID-19

Zhiying Wu

Beijing University of Technology, Beijing, 100000, China

Abstract. Since 2019, the epidemic has been the focus of people in all countries. Until now, it continues to frustrate economic development, human health, and social stability. How to help the economy recover under the epidemic has become a daunting task with far-reaching significance in this era. This paper focuses on China, where live streaming is prevalent, and thus explores how it has helped revitalize the Chinese economy. The research method is to cite the data brought by various live streaming platforms and government department reports to illustrate the contribution and importance of live streaming. Finally, this paper demonstrates that live streaming has the potential to make an outstanding contribution to the revitalization of the Chinese economy, including eradicating poverty in rural areas, stimulating consumption and taxation, and promoting the transformation of the economy.

1 Introduction

Since 2019, the epidemic has been one of the most significant issues impeding Chinese economic development, resulting in severe consequences. For example, according to Asian Development Bank (2020), the economic cost to China of extending the outbreak prevention and control period would be as much as $692 billion. Despite the epidemic hampering the economy, China still maintained positive GDP increase rates in the recent three years, differing from many countries that experience sharp declines in GDP. During this period, live streaming can be a powerful platform for the Chinese economic recovery [1], illuminating the country's economy and displaying the Chinese market's strong potential and vitality. Lind defined live marketing as "a form of video delivery increasing in prevalence in the consumer and commercial sectors" [2]. This sales method has rapidly gained popularity and has become a unique consumer scene in China by relying on its advantages of being free of space constraints and interactivity. The Japan Broadcasting Association (NHK) also reported that live streaming is an innovative move by the Chinese in a difficult situation and a sign of their strong vitality. This paper studies how live streaming can help the Chinese economy recover from the epidemic in terms of eradicating poverty, stimulating consumption and taxation, and accelerating the digital transformation of the economy. As the damage caused by the epidemic has been enormous and is likely to remain for years, relying on live streaming to revitalize the economy has far-reaching implications, not only helping restore momentum but also providing a useful guide to future economic woes.

985106804@qq.com

2 The Important Role of Live Streaming in the Development of Rural Areas

2.1 Accelerating the Promotion of Agricultural Products

Poverty eradication is always an essential aspect of Chinese economic progress, and live broadcasts can significantly promote the sales of agricultural products in remote rural areas during the epidemic [3]. The emergence of Internet marketing and e-commerce may play a leading role in the development of modern agriculture, and it can dramatically reduce the distances between agriculture and other industries. The newer the technology, the greater the effect on less developed sectors and regions. Among these modern agriculture technologies, the approach of combining short videos and live streaming is the most popular one, playing a particularly essential role in the exploitation of rural areas [4]. For example, a governor of Zhaosu County in Xinjiang province posted a video of a horse galloping on TikTok. As a result, the video lets more people know about this small county and promotes local rural products to a broader audience.

2.2 Promoting the Sale of Agricultural Products

In addition, the governor and farming enterprises have driven more than 2,000 people to be employed in the local area and helped 10,000 people increase their income. Therefore, this sales method of live broadcasts can break the communication barrier between producers and consumers and change the traditional model of multi-link cascading and profit superimposing sales mode to some extent. It not only facilitates the sale of these agricultural products but also protects the rights and interests of producers and consumers. More importantly, China is implementing the task of comprehensively promoting rural revitalization and accelerating the modernization of agriculture, so live streaming has received substantial financial and technical support from the government [4]. Figure 1 shows the data for online sales of live agricultural products in recent years.

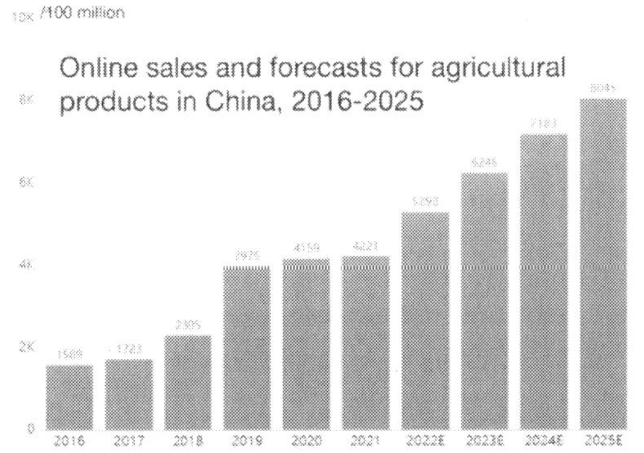

Fig. 1. Online sales and forecasts for agricultural products in China from 2016 to 2025 (iiMedia Research, 2021).

2.3 Improving the Quality of Agricultural Products

However, some people believe that live broadcasts cannot prompt the economic development of remote rural areas due to their weak infrastructures and the massive socio-economic gap between the countryside and cities. The most important feature of Internet programming is the low threshold: with a cell phone, a bracket, and a flash, one can start the journey of the live broadcasts [2]. Thus, webcasts are not so impracticable even in areas with inadequate infrastructure. Furthermore, during the epidemic, people's movement is restricted, and they are more inclined to spend their free time on the Internet [3]. With people having more time to compare similar goods, quality agricultural products are more likely to be found and purchased. Consequently, merchants also strive to improve the quality of their own goods in competition, reducing the number of goods that do not meet their expectations.

In short, it could be claimed that web broadcasts attract a large number of farmers to join the network team, which provides a large number of jobs and promotes economic progress in rural areas.

3 The Important Role of Live Streaming in Increasing Consumption and Tax Revenue

3.1 Stimulating Consumption

In addition to alleviating poverty, live streaming can also contribute to the recovery of the Chinese economy by expanding consumption and increasing tax revenue to some extent. Investment, consumption, and exports are defined as three essential factors of Economics. The two factors, investment and exports, are not so optimistic under the epidemic. Therefore, consumption has become the main driver to bring China's economic development. During the epidemic, one of the starkest portraits of the Chinese web is that everything could be sold live, and everyone could participate in live broadcasts [5]. Internet shows can save the brick-and-mortar retail industry, which is currently experiencing a sharp decline in activity due to the epidemic. Therefore, live streaming is called by foreign media a lifeline for small retailers in China during the outbreak [1]. Specifically, it is possible for numerous businesses to reduce inventory, stimulate consumption, and even expand production through cooperation with anchors and celebrities. For instance, many merchants signed a treaty with the most famous live-streamer Li Jiaqi before the shopping festival, requiring him to introduce their products. According to the official data of Tmall shown in Figure 2, Li Jiaqi reached a total sales volume of 10.6 billion RMB in one day of the Double Eleven shopping spree in 2021. Therefore, the data indicate that live marketing is probably credited for boosting consumption.

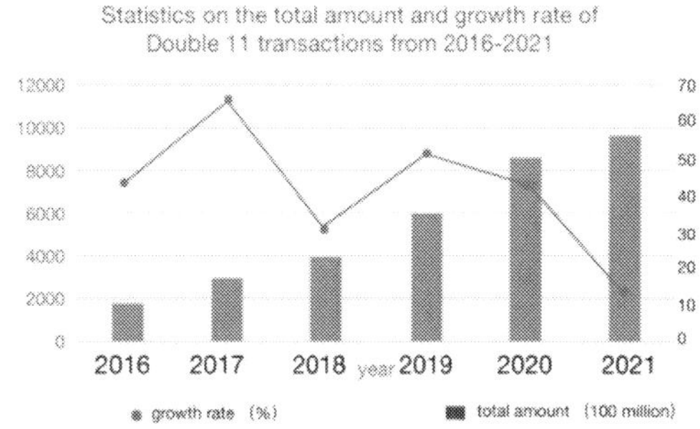

Fig. 2. Statistics on the total amount and growth rate of Double 11 transactions from 2016-2021 (Huaon Research Institute, 2021).

3.2 Increasing the Tax Revenue

Moreover, the Chinese tax revenue from live streaming reaches more than 30%, so the tax revenue generated by webcasts can also contribute significantly to the economic recovery. As China tightens the control over its tax system, tax evasion is gradually decreasing and the tax revenue generated by live streaming contributes significantly to the country's GDP. Regulations and strict laws allow the government to have stronger funds to manage the community and improve the economy in the event of an epidemic. Thus, the increase in the tax revenue and GDP can partly be attributed to the prevalence of live broadcasting.

3.3 Achieving Win-Win Economic results

More importantly, live-streamers moderately earn money themselves, create jobs, and get many people into employment, achieving win-win results. In the case of the live-streamer Li above, Chen states that his outstanding performance in Internet video casts made the Shanghai government give him the bonus of talent settlement [6]. Finally, web broadcasts will be sustainable because merchants create many new ways to guarantee the quality of goods. The most common mode may be offering return-freight insurance, which provides consumers with the cost of returning or exchanging goods. This insurance supervises and urges the merchants to make goods of the same quality as when they were on the Internet. It also provides rights to consumers, thereby reducing the occurrence of unsatisfactory situations [7]. With the improvement of commodity quality, logistics, and regulation, Chinese live industry is getting stronger, and the role of promoting consumption is becoming more obvious.

In brief, the recovery effect of the Internet shows on the Chinese economy during the epidemic can mainly be reflected in the boost in consumption and the raise of tax revenue.

4 The Important Role of Live Streaming in China's Digital Economic Recovery

4.1 Promoting the Digital Economy

Live streaming complies with the trend of networking and digitalization, and it has become a breakthrough in achieving economic transformation and upgrading [8]. The expanding sales boundaries of webcasts during the COVID-19 epidemic can allow more businesses to open new sales channels: opening online stores and combining online diversion and offline consumption. Notably, according to Li, these new ways are driving the further development of the Internet, the improvement of mobile payment platforms, and the networking of the real economy [9]. Secondly, many traditional manufacturing industries facing the double pressure of the Internet transformation dilemma and the epidemic's impact can find a path to crack the difficulties through the live platform. For example, in May 2020, Dong, the president of Gree, attracted more than 16 million net citizens to her live studio and created a record of 310 million yuan in three hours [1]. Thus, it seems that live streaming allows traditional industries to join the digitalization process, increases the proportion of the online digital economy in the national economy, improves the allocation of resources, and finally promotes a high-quality breakthrough in the economy.

4.2 Promoting the Improvement and Upgrading of Economic Structures

Furthermore, live broadcasts accelerate the recovery and reorganization of the Chinese economy by improving the efficiency of e-commerce operations [8]. More specifically, web broadcasts spawn professional warehouse management systems and transportation systems that allow consumers to receive goods in one day. Nevertheless, Qian and Yang believe that live streaming consists mainly of low-end goods, and the lack of technology makes it unsustainable for economic promotion [10]. With the deepening of broadcasting, the merchandise is undergoing an innovation process from low-tech products to emerging products, motivating the improvement of the economic structure [8]. For instance, the TikTok platform has opened up new forms, including live lectures, paid movies, and online concerts. The above consumption is counted as consumption in the tertiary industry, and the increase in the proportion of the tertiary sector is often the improvement and upgrade of the Chinese economic structure. Moreover, as live streaming continues to be modernized, more innovative industries might join in the Internet videocasts, injecting stronger momentum into the transformation of the economy and accelerating foreign trade [9]. Thus, it could be claimed that the involvement of new industries makes live streaming have a bright future and can continue to propel the Chinese economy after the epidemic. In summary, the live broadcasts can promote the restructuring and upgrading of the Chinese economy by facilitating the development of the digital economy, expanding new marketing models, and improving economic efficiency.

5 Conclusion

In conclusion, live streaming can primarily help the Chinese economy recover from the COVID-19 epidemic by raising incomes in rural areas, boosting consumption and tax revenues, and promoting structural improvements and upgrades in the Chinese economy. In addition, web broadcasts have brought employment, improved transportation and other infrastructure, and accelerated the transformation of the digital economy. To draw the above conclusion, this paper mainly cites a large amount of data, but one of the drawbacks is the insufficient data of small businesses, since almost all the data that can be captured belong to successful or head live-streamers and brands. Nevertheless, the value of live streaming cannot be denied. To seize the windfall and dividends of the rapid progress of the live streaming economy, the platform must develop appropriate industry rules, increase efforts to combat fraudulent marketing practices, and actively introduce high-tech and innovative industries. In the longer term, webcasts could not only prevail during the epidemic but also reflect a higher value in the future. Given the new challenges, all industries connected to Internet videocasts should actively explore and further strengthen the role of live broadcasts in promoting the Chinese economy.

References

1. Ni, V., Wang, Y., & Service, B. Coronavirus: Can live-streaming save China's economy? (2020). https://news.sisuer.cn/wp-content/uploads/2020/05/bbc.com-Coronavirus-Can-live-streaming-save-Chinas-economy.pdf.
2. Lind, J.: The Role and Potential of Live Streaming in Marketing for Small and Medium-sized Enterprises (2018).

https://www.theseus.fi/bitstream/handle/10024/149489/Lind_Joonas.pdf?sequence=1

3. Deng, M., Deng, P., Chen, B., Liang, Q., & Deng, G.: Research on Tik Tok platform live streaming e-commerce to help rural revitalization based on SOR model. Academic Journal of Business & Management, 3(6) (2021). https://doi.org/10.25236/AJBM.2021.030614.
4. Peng, L., Lu, G., Pang, K., & Yao, Q.: Optimal farmer's income from farm products sales on live streaming with random rewards: Case from China's rural revitalization strategy. Computers and Electronics in Agriculture 189(8), 106403 (2021). https://doi.org/10.1016/j.compag.2021.106403.
5. Lin, J., & Lu, Z.: The Rise and Proliferation of Live-Streaming in China: Insights and Lessons. Communications in Computer and Information Science, 714, 632–637 (2017). https://doi.org/10.1007/978-3-319-58753-0_89.
6. Chen, Z., Fan, Z.-P., & Zhao, X.: Offering return-freight insurance or not: Strategic analysis of an e-seller's decisions. Omega, 103(6), 102447 (2021). https://doi.org/10.1016/j.omega.2021.102447.
7. Lu, B., & Chen, Z.: Live streaming commerce and consumers' purchase intention: An uncertainty reduction perspective. Information & Management, 58(7), 103509 (2021). https://doi.org/10.1016/j.im.2021.103509.
8. Wang, X.: Research on The Development of Live Streaming Industry of E-business Under Web Celebrity Economy. 2020 the 4th International Conference on E-Business and Internet. 24,16-20 (2020). https://doi.org/10.1145/3436209.3436884.
9. Li, Q.: E-commerce, Free-Trade Zones, and the Linkage Effect to China's Foreign Trade. The Chinese Economy, 54(6) (2021). https://doi.org/10.1080/10971475.2021.1890361.
10. Qian, M., & Yang, H.: Research on the Current Situation and Optimization Countermeasures of China's E-commerce Live Streaming (2020). http://clausiuspress.com/conferences/LNEMSS/FEIM%202020/FEIM2020010.pdf.

Multimedia Advertising and Marketing Methods in the Context of Digital Communication

Shen Zhou

University of Illinois Urbana Champaign, Illinois, the United States

Abstract. This study uses a longitudinal multi-case study design and the objectives of the study are: 1. to explore the application of digital content marketing (DCM) strategies in the context of digital communication; 2. to identify the communication objectives used in their DCM strategies for social media using luxury car brands as a case study; 3. to assess the engagement (reactions, shares, views) of each brand on social media; 4. to use Facebook as a case study to examine the type of media used for his advertising content (text, links, photos, videos, hashtags). The analysis included identifying marketing communication objectives through media content types, social media content marketing strategies and social media indicators. The research results show that brand awareness, brand personality and brand distinctiveness are the most frequently used goals in multimedia advertising and marketing, which makes the focus of digital content advertising marketing used by brands on social media more visible and recognizable, and humanized features and attitudes are reflected on social media.

1 Introduction

Marketing communication is one of the most important components of modern marketing, although it is difficult to develop, and it is facing strong changes. In recent years, driven by the explosive growth of new digital media options, this challenge has become more complex. As one of the digital media platforms, social media marketing is often used to contact customers. Creating digital content marketing (DCM) for each brand is an important ability to develop social media platforms. DCM refers to the management process responsible for identifying relevant digital content, forecasting customer needs and profiting from it. As with advertising, DCM intends to cultivate brand awareness and sales by expanding consumer participation, increasing consumer trust and relationship. However, this is different from advertising, which aims to promote sales. At present, the biggest challenge facing marketing managers is how to develop DCM for brands, so that they can improve social media awareness, participation and sales indicators. The actual results of social media metrics refer to measuring, monitoring, reporting and computing social media content. Nevertheless, as the metrics have not yet reached standardization, it depends on the target of marketing objectives, and the most appropriate indicators for specific measurement standards.

Any DCM strategy in social media must be formulated in the framework of marketing strategy, communication objectives as part of brand integrated marketing communication (IMC) are also included. Although many researchers have explored the concept of DCM as a content strategy, they have theoretically explored the relationship between DCM and marketing communication goals in social media marketing. Although the previous research has put forward the specific goals of social media marketing, it has not been discussed in depth from the perspective of strategic marketing in the IMC programme. It is widely indicated in the relevant marketing literature that the communication strategy must reach the communication goal set in the comprehensive marketing communication plan [1]. In view of the relationship between the content strategy used by the brand on social media and the marketing communication target (part of IMC), it is very meaningful to analyze its content strategy [2]. The objective of study is to analyze the types of marketing communication goals reflected in the DCM strategy of social media marketing and the social media participation indicators of this content.

The marketing communication objectives that can be applied to the traditional and digital marketing platform IMC plan are used in this study. There is no doubt that one of the most influential topics in the future of advertising is how to design, implement and evaluate the comprehensive marketing communication plan. IMC coordinates and integrates multiple communication channels, through which companies can provide consumers with clear, attractive and consistent information about their products and brands [3]. The goal of the IMC project is to integrate all marketing and promotion platforms with the organization's traditional and digital media.

E-mail:Zhous2@illinois.edu

2 Marketing Communication Goals

The eight marketing communication objectives of IMC plan were applied to this study, and these marketing communication objectives are the basis of all communication marketing work in IMC plan.

Creating prominence and awareness. Establishing awareness is the basis of brand and marketing work, which can make the brand get certain attention and thinking in the right way at the right place and time. Therefore, it is easy for them to often recall [4].

Deliver detailed information. To enable consumers to rely on brand performance, they need to agree with products or services and understand why brands can better provide these benefits, because product attributes, features or characteristics can be used as evidence.

Create personality and imagery. Various imagery can be created according to the type of consumers using the brand, when and where to use the brand, etc. Brand personality is a reflection of the humanized characteristics that consumers give to a brand, which can have a certain impact on how consumers view themselves and the brand relationships/connections they form.

Building trust. There are more and more attempts to influence consumers and suspect marketers, and the desire for the authenticity of products and information is also growing. More and more people use social media to communicate, which shows that information from similar and reliable sources, rather than from remote and motivated sources, is becoming increasingly important [5].

Stimulate emotions. Brands can add to these benefits based on the functionality/practicality of their use, thereby increasing their overall perceived value. It is a good example that brand information can increase perceived value-added benefits through the transmission of cultural significance. Different benefits stimulate different types of emotions of consumers, thus affecting their decisions [1].

Motivated actions. Although the brand information received, processed and accepted may form brand preference and choice, there may be inherent disjunction between perception, emotion and behavior, which will not lead to final action. When inducing consumers who have a good understanding of the brand to take actions and behaviors, special information is usually required.

Instilling loyalty. When consumers actually consume, they evaluate their satisfaction and compare their expectations with the explained consumption experience, which may itself be affected by marketing communication. To enhance loyalty and avoid customer churn, building satisfaction is part of the goal[1].

Contact person. Since consumers cannot obtain tangible and credible evidence of product quality and credibility in other ways, brand promotion and word-of-mouth (WOM) are very important for service brands. In order to promote WOM communication and publicity, consumers need to engage and interact with the brand on a regular basis and develop a love for the brand is an important reason for making the brand symbolic and emotional. In addition, brands may need to send brand information to consumers regularly to encourage them to pass on this information to others [5].

Marketers must consider the following when selecting content and communication goals.

(1) The direct impact of the delivered content on consumers, and the types of emotions that may be triggered.

(2) How to complement the selected communication objectives.

(3) How do they contribute to delivering clear content through numerous brand messages [4].

3 Data Analysis

The luxury car brands used as case studies and their information are presented in detail below, along with a separate analysis of each brand. The five brands selected for analysis are shown in Table 1, which includes the Facebook page information of the brand; Including likes, fans and ratings

Table 1. Facebook page information of the brand (Table credit: original)

Facebook	the number of likes	the number of fans	Ratings
Acura de Puerto Rico	93,788	93,113	4.5 out of 5
Alfa Romeo Puerto Rico	2,878	2,907	0
Audi San Juan Puerto Rico	8,296	8,347	4.2 out of 5
Lexus de San Juan	16,448	16,488	0
San Juan Lincoln	10,206	10,242	0

Based on the collected data, the digital content marketing, social media participation indicators (comments, shares, views) and media types (hashtags, links, photos, videos, text) of each brand were analyzed.

During the observation period, Acura de Puerto Rico made 13 posts. According to the result, Acura de Puerto Rico's DCM focused on brand salience, brand awareness, and inspiring emotion through (7 posts) each. In addition, the second highest goal of brand use is to convey detailed information and brand personality, with 5 goals for each. The media types used by Acura de Puerto Rico are, hashtags (12 posts) and photos (7 posts). The other side of the shield, in terms of participation indicators, Acura de Puerto Rico received 150 shares, 1047 comments and 66333 comments on 13 positions in a total of 45 days.

During the observation period, Alfa Romeo Puerto Rico held 23 positions. According to the analysis results, Alfa Romeo Puerto Rico's focus on brand awareness and salience is the focus of DCM. Of the 16 posts reflected by the brand awareness, the brand awareness accounted for 10. Moreover, there are 11 inspirational feelings and 8 brand images. Hashtags (13 posts) and photos (21 posts) are the most commonly used media types of Alfa Romeo Puerto Rico. In addition, there were 451 comments, 369 comments, 67 shares of engagement indicators, and text was used in 23 posts.

During the observation period, Audi San Juan Puerto Rico established 48 positions. According to the analysis, brand personality (22 posts) and brand awareness (18 posts) are the main marketing goals of Audi San Juan Puerto Rico. His most commonly used media types are 47 article texts, 46 article tags and 7 article videos, generating 3939 views in total.

During the observation period, Lexus de San Juan published 15 posts. Brand personality and emotional brand (10 posts per post) are the most commonly used marketing goals in DCM strategy. His most commonly used media types are 15 posts and text with 14 posts.

During the selected observation period, San Juan Lincoln published 10 posts. The results show that brand prominence (7 posts) is their goal, and the most common media types are text (10 articles) and photos (9 posts).

Table 2 summarizes the results of this study. Through research statistics, Table 2 shows the media types selected by each brand and the social media participation indicators generated by each post.

Table 2. Overall analysis of media types and social media engagement indicators (Table credit: original)

Brand names	Media Types					Participation indicators		
	Tags	Links	Photos	Videos	Texts	Shares	Comments	Views
Acura de Puerto Rico	12	2	7	6	13	150	1,047	66,333
Alfa Romeo Puerto Rico	23	10	21	2	23	67	451	369
Audi San Juan Puerto Rico	46	11	41	7	47	479	5,162	3,939
Lexus de San Juan	6	7	15	0	14	1,639	112,695	0
San Juan Lin-coln	0	0	9	1	10	215	3,018	374
Total	83	30	93	15	107	2,550	122,373	71,015

A comprehensive analysis of DCM strategies, media types and social media engagement metrics provides a new perspective for analysing performance of brands. In the marketing communication goals, these brands most commonly use brand awareness (55 articles) and brand personality (49 articles). Text and photos are the preferred media types of the brand from the Table 2. In addition, social media engagement metrics generated a total of 122,373 responses and 71,015 views are generated from the social media engagement metrics.

4 Discussion

Understanding the internal relationship between marketing communication objectives and the types of media used can be carried out through the analysis of social media participation. As displayed in Table 2, 93 posts are attached with photos, making them the most commonly used media type. According to the brands surveyed, Audi San Juan Puerto Rico (41 posts) and Alfa Romeo Puerto Rico (21 posts) use the most photos. In addition, the number of tag repetitions was the largest, with 83 out of 109 posts surveyed. From the above information, it can be determined that the most common media types for brands are text and photos.

The results show that the posts with the highest engagement are those with multiple communication objectives and media types. For Acura de Puerto Rico, the most involved position is position 1. This post has 204 comments, 38 shares and 34000 views. The communication goals consistent with this post are brand personality, building trust, brand awareness, motivating action and instilling loyalty. In addition, the post 7 was the one with the highest participation of Alfa Romeo Puerto Rico, and 163 posts were replied to and 6 shared.

This post uses the following goals: brand awareness, brand salience, brand personality, detailed information, brand personality, building trust and emotional brand building. Post 74 wai Audi San Juan Puerto Rico's highest consumer engagement, and 722 posts were replied to and 24 shared. This post uses the following goals: brand prominence, detailed information, brand personality, and inspiring action. And, post 90 was Lexus of San Juan's highest engagement post, 83,000 posts were replied to and 593 shared. The following communication targets are displayed through the latter: brand prominence, brand awareness, detailed information, brand personality, inspiring action, brand image, and inspiring emotion. Finally, the highest post engagement for San Juan Lincoln had been replied 662 and 46 shared. Brand salience and emotional brand are the communication goals of this post.

The reproducibility of the target and the similarity between media types can be determined by comparison. The result shows that the more integrated the communication goals, the more involved the posts will be. Since photos are the media type with the largest number of repetitions per post, it can be assumed that visual stimulation seems to attract consumers' attention, thus promoting consumers' participation. Furthermore, brand salience (symbol, logo, tagline and brand name) is the most repeated goal, and all communication information should be accompanied by it.

Acura de Puerto Rico is the most popular brand on Facebook, with 93688 likes, followed by Lexus de San Juan, 16448 likes, and finally San Juan Lincoln, 10206 likes. In society, Audi San Juan Puerto Rico has the greatest influence. Among 109 posts analyzed, the brand has released 48 times. However, by comparison, Audi San Juan Puerto Rico did not rank among the top three most popular times on Facebook. Therefore, it can be inferred that having a many likes or followers does not reflect the brand's participation and social influence. The results show that the number of likes on a Facebook page does not equate to the level of engagement with the post, nor does it equate to the amount of content posted versus the quality of the content message delivered.

Brand salience (48 posts) and brand personality (49 posts each) are the most commonly used marketing communication goals

The brand is reflecting the humanized characteristics and attitudes through the dissemination of information, and focuses on making the brand more visible and identifiable; Establish brand association to remind consumers of the brand. According to the research results of media types, text (107 articles), hashtags (83 articles) and photos (93 articles) are the most commonly used media types. Furthermore, it shows that brands limit the number of media types they choose to use in social media. Therefore, we recommend changing the type of media and implementing a different media mix. The combination of communication objectives with different media types will contribute to the positive development of the IMC program. Therefore, it is very meaningful for brands and marketing personnel to consider IMC plan to implement well-designed DCM.

5 Conclusion

From a theoretical perspective, this study contributes to the lack of literature on DCM as a key component of IMC programmes. In addition, there is a lack of data on how brands use DCM in social media marketing and what marketing communication objectives exist in their strategies. These objectives are key to identifying and understanding which aspects of their communication strategy align with desired outcomes based on the performance across digital and traditional media platforms. Previous research has not explored the alignment of DCM with marketing communications objectives. Following the literature review, we found no empirical research on social media marketing strategies linked to marketing communication objectives. The recent literature has focused on fragmented issues such as strategies for effective communication on social media, but has not developed a strategic approach to marketing. Another important theoretical implication is the study of the relationship between communication objectives, DCM in social media marketing, media type and social media engagement metrics used by brands.

From a management perspective, this study suggests a framework that can help marketers better incorporate communication objectives into their social media marketing. When creating communication messages, marketers must have a specific strategy to match the communication objectives. In addition, existing marketing communication objectives must be taken into account and incorporated into the DCM for social media marketing. An important finding of this study is that posts can be grouped into more than one goal, which provides marketers with a variety of communication opportunities. Furthermore, those posts that integrated the most targets seemed to generate more engagement. Furthermore, as photos are the most repeated media type in posts, we can assume that visual representations in posts can generate more engagement. The results show that brand prominence (logos, taglines, symbols, brand names) almost always accompanies the highly engaged content posted by all brands. Marketing managers should use the content analysis criteria developed for this study to analyse which type of DCM strategy in social media marketing is better suited to achieve their brand's marketing communication objectives.

In the field of digital marketing communications, brands have to decide the level of engagement and interaction they wish to have with their brand stakeholders. Through recent years, consumer demand has risen and market tendencies have changed, forcing marketers, academics and brands to adopt and combine old marketing tools with new social media marketing strategies. At this critical time, brands and marketers need to understand that a DCM strategy based on communication objectives will lead to building better consumer relationships and long-term consumer loyalty.

References

1. Sun Feng. Research on the construction of integrated marketing communication model based on customer value [D]. Shandong University, 2006.
2. Shen Guanglong, Liu Yingzhen, He Kemin. Research on integrated marketing communication strategy evaluation index system[J]. Management Science, 2006(01):42-49.
3. Zou Ruitao. Integrated marketing communication strategy and its organizational practice [D]. Wuhan University of Science and Technology, 2008.
4. Yang Nan. On IMC's advertising communication strategy[J]. Contemporary Manager, 2006(11):190-191.
5. Wang Qiwan. Research on the formation mechanism and drive of customer perceived value in integrated marketing communication[D]. China University of Mining and Technology, 2009.

Exploration of Government Collaboration in the Co-Location of Guangzhou and Foshan

MuChuan He*

Guangzhou Foreign Language School, 511466

Abstract: The basic feature of China's economic development in the new era is the shift from the stage of high-speed growth to the stage of high-quality development, and the concept of crosstownization is developed in the context of economic globalization and regional economic integration. Co-location is a requirement of the times, and the essence is close cooperation between cities and towns to achieve good and fast development. This paper explores the governmental collaboration aspect of the Guangzhou-Foshan co-location, discussing three aspects: the current situation, problems and suggestions, analyzing the current situation of the development of the Guangzhou-Foshan co-location, the main problems and putting forward suggestions for future development.

1 Introduction

With the development of society, co-location has become an important form to achieve regional integration and promote regional synergistic development in China. The concept of co-location in China was first proposed by Shenzhen in the Shenzhen 2030 Urban Development Strategy: to strengthen cooperation with Hong Kong in high-end manufacturing, modern services and other areas, and to form a "co-location" development with Hong Kong.[1] Nearly 20 years have passed, and more than 20 pairs of cities nationwide have proposed to promote the development strategy of co-location, which has become the gas pedal of regional high-quality development.[2] As the policy advances, regional co-location faces a series of problems, especially in government collaboration, mainly in inefficient communication between governments, etc. This paper mainly explores the issue of governmental collaboration in Guangzhou-Foshan co-location, with the intention of discussing the problems of governmental collaboration in the context of Guangzhou-Foshan co-location and proposing solutions to provide reference and reference for other regions, thus providing a small contribution to the high-quality development of the national economy.

2 The meaning of Guangzhou-Foshan co-location

Co-location refers to the joint development of two or several neighboring cities while cooperating in various aspects such as economic, social, and natural ecological environment.[3] The co-located cities will echo with other cities in decision-making to achieve common development, complement each other's strengths and take advantage of each other's shortcomings; and integrate and learn from each other in terms of urban characteristics. In a limited resource environment, co-location can drive incremental development and make co-located cities more competitive. It is important to note that co-location is not simply a matter of controlling various data between cities to keep them the same, nor is it a matter of expanding outward to include all surrounding cities in the co-location plan, but rather combining cities with different strengths to produce a more competitive economy.

On March 19, 2009, Guangzhou Mayor Zhang Guangning and Foshan Mayor Chen Yunxian signed the "Guangzhou Foshan City Co-city Construction Cooperation Agreement" and four docking agreements on urban planning, transportation infrastructure, industrial collaboration and environmental protection between the two cities in Nanhai District, Foshan City, which marked the official launch of Guangzhou-Foshan Co-city. After this, the "Guangzhou City Master Plan (2011 - 2020)" and the "Foshan City Master Plan (2011 - 2020)" approved by the State Council also proposed to further promote the co-location of Guangzhou and Foshan.[4] On November 23, 2016, Guangzhou and Foshan held a working exchange forum in Guangzhou, where the two sides planned to promote higher-level co-location and join hands to build a "super city". Guangzhou and Foshan are the first and third largest cities in Guangdong Province, respectively, with bordering borders about 200 km long. In terms of transportation construction, the transportation and railroad network system between the two cities is relatively complete, which helps the flow of people and resources between the cities; the new PRD hub airport is located in Foshan Gaoming, which will form an effective linkage with Baiyun International Airport; Foshan West Station is included in Guangzhou's "five main and four auxiliary" passenger hub pattern, Nansha Port Railway is completed

* xxhefei@hotmail.com

and operational, Guigang High Speed Railway South Guangzhou high-speed railway speed upgrade, Guangzhou-Zhan high-speed railway and other high-speed railway channel to speed up the construction, Guangzhou Station to Guangzhou South Station liaison line and other hub projects steadily advance the preliminary work, Guangzhou-Foshan pole railroad hub pattern has been rapidly optimized and upgraded. In terms of science and technology innovation projects, the two governments and the rest of the cities in the Greater Bay Area are working together in depth to build two national advanced manufacturing clusters, namely Guangzhou-Shenzhen-Foshan-Dongguan intelligent equipment, Guangzhou-Foshan-Huizhou ultra-high-definition video and intelligent home appliances, as well as a national fuel-cell vehicle demonstration city cluster, focusing on making breakthroughs in cooperation in the fields of automobiles, new-generation information technology, biomedicine and health, and forming a pattern of collaborative industrial development based on the division of labor in the industrial chain.[5] In preschool education, Liwan District of Guangzhou City and Chancheng District of Foshan City have started in-depth cooperation on preschool education projects. In terms of environmental protection, the two municipal governments will focus on increasing the protection of the cross-border river surges between Guangzhou and Foshan. The enactment of these policies will enable the two cities to consolidate and enhance government resources and improve efficiency in environmental management projects. The two municipal governments will assess the section water quality to achieve full compliance; systematically carry out Guangzhou&Foshan cross-border and shared drinking water source protection zone construction project clean-up compensation; complete the Pingzhou waterway, Guangzhou&Foshan River, Southwest Chung and other cross-city Bi Road construction and other four cooperation projects, so that the water is clearer and the shore is greener. In terms of co-location of government services, 2,721 government service matters can be handled across cities, and the scope of matters will be further expanded this year; in terms of medical and health cooperation, the synergistic mechanism for prevention and treatment of major infectious diseases and pre-hospital emergency care in the junction area will be improved; in terms of social security cooperation, resource sharing of employment information and skills training will be carried out, and deeper cooperation in social assistance and elderly care services will be promoted, etc.

3 The current situation of Guangzhou-Foshan co-location

3.1 Economic

In 2021, Guangzhou's primary industry grew by 5.5% over the previous year (the same below). Foshan's primary industry grew by 9.5%. The national primary industry increased by 7.1%. The growth level of primary industry in Guangzhou is lower than the national average, while the growth level of Foshan is higher than the national average. Guangzhou secondary industry growth of 8.5%, Foshan secondary industry growth of 9.3%, the national secondary industry growth of 8.2%, Guangzhou and Foshan growth levels are higher than the national level. Guangzhou's tertiary industry increased by 8%, Foshan's tertiary industry grew by 7%, the national tertiary industry grew by 8.2%, Guangzhou and Foshan are below the national growth level. The total annual consumer price level in Guangzhou rose by 1.1%, and the total annual consumer price level in Foshan rose by 1.1%, basically remaining the same. In the external economy, the total value of Guangzhou's annual merchandise import and export increased by 13.5% over the previous year, the total value of Foshan's annual merchandise import and export increased by 21.7% over the previous year, and the national annual goods import and export increased by 21.4% over the previous year. Guangzhou's total import and export growth level is much lower than the national growth level while Foshan's total export value is slightly higher than the national average.

3.2 Social aspects

At the social level, Guangzhou City 2021 (the same below) had a resident population of 18,810,600 at the end of the year, with a natural population growth rate of 6.26‰ and an urbanization rate of 86.46%; Foshan City had a resident population of 9,612,600 at the end of the year, with a natural population growth rate of 6.28‰ and an urbanization rate of 95.21%. The urbanization rate of Guangzhou is lower than the urbanization rate of Foshan. In terms of natural population growth rate, the national natural population growth rate is 0.34‰, and the natural population growth rate of Guangzhou and Foshan are both higher than the national average. The high natural population growth rates in Guangzhou and Foshan are due to the large influx of foreign population. The urban registered unemployment rate in Guangzhou was 2.22% and in Foshan was 2.4%; the national urban survey unemployment rate at the end of the year was 5.1%. Guangzhou City, urban residents per capita disposable income increased by 8.9%, rural residents per capita disposable income increased by 10.4%; Foshan City, urban residents per capita disposable income increased by 8.4% over the previous year, rural residents per capita disposable income increased by 9.6% over the previous year; the national urban residents per capita disposable income increased by 7.1% over the previous year. The per capita disposable income of rural residents increased by 9.7% over the previous year. In contrast, the growth rate of urban and rural per capita disposable income in Guangzhou is higher than the national average, while the growth rate of per capita disposable income in Foshan is only higher than the national average in urban areas, while the rural per capita disposable income is slightly lower than the national average.

3.3. Education and Environment

In terms of compulsory education, elementary school

enrollment in Guangzhou increased by 6.84% over the previous year; junior high school enrollment increased by 4.49% over the previous year. Foshan elementary school enrollment increased by 6.53% over the previous year; junior high school enrollment increased by 7.4% over the previous year. In contrast, the percentage increase for junior high school students in Guangzhou is relatively low, while the percentage increase for elementary school students is almost the same. In terms of higher education, Guangzhou has 11 general higher education schools and 2 adult higher education schools, with 44,600 students enrolled in various forms of higher education and 141,900 students; Foshan has 13 general higher education schools, with 53,500 students enrolled and 153,200 students enrolled. In comparison, (poor summary) the average number of haze days in the city of Guangzhou in 2021 is 4 days, 0.6 days less than the previous year. The number of days with air quality at or better than Level 2 in Foshan was 312. At the same time, both Guangzhou and Foshan have reached 100% of urban household waste disposal rate.

4 Analysis of governmental collaboration in Guangzhou-Foshan co-location

4.1 Inefficient communication leads to competitive internal conflict

The government of Guangzhou&Foshan is not clear about the boundaries of the respective functional areas, and there is serious industrial homogeneity between the two places. The division of labor among regions on professional boundaries is also unclear, leading to waste of resources and ineffective competition. The main reason for the industrial homogeneity and the unclear professional division of labor boundaries in sub-functional areas is the lack of clarity in the planning of the industrial system and the layout of the division of labor between the governments of Guangdong and Foshan for the misalignment of industrial development. At the same time, the lack of communication between the governments also leads to the lack of unified and perfect planning between the two regions. There is no actual geographical distinction between Guangzhou and Foshan, leading to a particularly serious problem of homogenization in the border area between the two places.[6] In terms of urban project development and construction, the government of Guangzhou&Foshan also has the same problem of inefficient communication. Among them, the governments of Guangzhou and Foshan are divided and local protectionism prevails, which is not conducive to the development of Guangzhou-Foshan co-location, nor is it conducive to information sharing and resource complementation between governments. Looking at the results of development and reform in the past decade or so, although the two cities have made great strides in the field of cooperation and cooperation methods, but the degree of cooperation in key areas and major projects still needs to be deepened. From the development of 2011 to the present, not see Guangzhou&Foshan breakthrough class such as in the West River diversion flow and other major infrastructure projects on the big action, the two cities cooperation is still lack of such key planning projects make cooperation and development difficult to break through the bottleneck, to achieve the development of cooperation at a higher level docking.[7] At present, there are strong economic ties between the cities and towns in the Guangzhou-Foshan region, and this connection is manifested in the form of passenger flow, logistics, information flow and other forms of flow clustering and diffusion, but due to the administrative system and management problems, the road traffic network between Guangzhou and Foshan has been difficult to achieve full access, the junction of the two cities still have a number of cut-off roads hindering the travel of the public. In terms of infrastructure, there are still problems with the construction of the natural gas network, the electricity grid and the information network, resulting in a large gap between the natural gas and water tariffs, which causes inconvenience to the public, even though the two areas are clearly located next to each other. Many key projects for the benefit of the people are not reasonably coordinated, lacking in articulation, and unable to fully play the role of regional services. If the lack of communication between Guangzhou and Foshan governments continues, it will become the primary factor limiting the development of Guangzhou-Foshan co-location. (Find an example) If the lack of communication between the Guangzhou and Foshan governments continues, it will become the primary factor limiting the development of Guangzhou-Foshan co-location. At the same time, the emergence of the "isomorphic phenomenon" will cause great waste of development resources and duplication of investment in construction, or worse, will lead to vicious competition in the industry, resulting in the emergence of local protection forces to form a vicious circle. Guangzhou's development in the automotive industry has formed an industrial overlap with Foshan. Guangzhou's new energy vehicle production grew rapidly from 2017-2021, with production reaching 150,000 units in 2021, accounting for 4.2% of the country's new energy vehicle production. In 2021, Guangzhou's total vehicle production was as high as 2,966,400 units, ranking first in the country for the third consecutive year. In 2021, the number of new energy vehicle enterprises in Guangzhou increased by 5210, an increase of 47.88% over the previous year. Among them, Guangzhou Automobile Group as the backbone of the development of Guangzhou's automotive industry has provided a lot of support for the development of the new energy industry. On August 25, 2022, Guangzhou Automobile Group announced to carry out the industrialization of independent battery construction, with a total project investment of 10.9 billion yuan. In December 2018, "Foshan City to accelerate the development of new energy vehicle industry and the promotion and application of a number of policy measures" was announced. This policy also encourages the development of Foshan's new energy vehicle industry by rewarding projects with a total investment of more than 1 billion yuan and put into operation within three years. In the last five years of development, Guangzhou and Foshan have created unnecessary competition in new energy vehicles. Such competition will not only consume government funds, but also the two cities will not reach the expected development in new energy.

4.2 Benefit compensation mechanism is not yet clear, and overlapping policy projects lead to industrial isomorphism

From the interests of consultation mechanisms and cooperation framework agreements signed, although the two cities have signed such as "1 + 4" and "Guangzhou-Foshan co-city cooperation framework agreement" and other multifaceted, broad-level coordination and cooperation in the development of intention documents, but the implementation of the policy is still lack of binding force, the rigidity of norms is still slightly inadequate. Cooperation norms are not legal documents, and do not have effective legal constraints and strong enforcement, coordination is difficult to carry out, law enforcement and related departments do not have the legal effect of the provisions to cooperate with the implementation of the policy, the two cities to cooperate in a number of projects is only on paper. A typical example is the serious conflict of interests between the two cities in environmental protection, such as in the management of the Pearl River water basin, and the fact that the synergistic management mechanism between the two cities in environmental protection has not been fully established, which will have a great resistance to the co-location of the two cities and the realization of mutually beneficial economic cooperation and development.[8] At the same time, the unclear benefit compensation mechanism would allow the governments of Guangdong and Foshan to establish government programs that are extremely similar. Due to the similarity of environmental resources and conditions between the two regions, government subsidy funds are disbursed to similar industries. These industries usually open and develop simultaneously in both locations, thus creating a homogeneous phenomenon. The emergence of isomorphic phenomenon will cause great waste of development resources and duplication of investment in construction, and even more will lead to vicious competition in the industry, resulting in the formation of a vicious circle of local protection forces. In terms of talent resource interchange, Guangzhou has not been able to attract enough high-tech talent to Foshan through its status as the provincial capital. The development of small and medium-sized downstream enterprises is constrained by the inability to improve themselves to let the market gradually solidify, unable to form a healthy competition.

4.3 Excessive exploitation of environmental resources by co-located development

Ecological costs have always been the main concomitant of the urban construction and development process. We can not take the old road of "pollution first and then treatment" in western countries, using the environment as the sacrifice of urban development, and only seeking rapid increase of economic development speed, disregarding the maximum carrying capacity of environmental development. According to relevant information and data, in Guangzhou, the development intensity of Yuexiu, Liwan, Haizhu and other areas are close to 80%, while in the periphery of the city, such as Panyu, Huangpu, Baiyun and other areas, as well as Foshan's Shunde, Nanhai, its development intensity has also reached 40%-60%, only the periphery of Conghua, Zengcheng, Nansha, Gaoming, Sanshui part of the regional development intensity does not exceed The serious lack of back-up land resources will lead to a major reduction in urban development potential.[9] The emergence of concomitant environmental problems such as acid rain, water pollution and eutrophication of water bodies has posed new challenges to the living environment of human beings. For example, in the junction area of Guangzhou-Foshan Fengxiu Gang industrial zone, the industrial zone adjacent to the Pearl River, which contains plastic mold manufacturing companies, metal manufacturing industries and garment manufacturing plants and other highly polluting enterprises. Although the environmental protection supervision department issued a strict policy to eliminate the random discharge of pollutants, but many of these industrial areas located at the junction of Guangzhou and Foshan have become the culprit of pollution of the Pearl River and other waters. After more than a decade of co-location of Guangzhou-Foshan, many junctions of territorial affiliation, administrative areas have been updated. However, many times the law enforcement and administrative departments are not fully unified, which leads to the junction of the "three regardless" zone. The administrative departments and law enforcement agencies in both places have been shirking their responsibilities for a long time after receiving complaints from the public, leading to an increasingly serious pollution situation.

5 Optimization measures of governmental collaboration in Guangzhou-Foshan co-location

5.1 Breaking through the single administrative cage and focusing on cross-regional administrative efficiency are potential requirements for the connotation of co-location

The regional administrative center, as the administrative subject of co-location development, puts it at another level above the local administrative center in terms of development planning layout, i.e. carrying out the establishment of provincial administrative centers and regional administrative subjects. As we can see from the development and construction experience of Western countries, there are two ways to upgrade cross-regional administrative centers: the first is to establish the organizational form of the League of Cities, and the second is to establish a regional council. The regional alliance mechanism is a form of organization similar to a confederation of cities, in which members of various alliances cede public power to each other, redistribute management rights and jurisdictions, and clarify their main responsibilities in urban development. As a combination of urban economic and political alliance and a kind of high-level administrative organization, it will play an important force in regional integration to achieve integrated development, but its emphasis has a high prerequisite for political integration. The requirements of the regional committee are relatively simple, and it only needs to achieve coordinated development among cities in the region in terms of industrial structure layout development, information infrastructure co-build and share, comprehensive land use, environmental co-build and co-protection, etc. to establish a cross-regional management administrative organization category. Guangzhou&Foshan can currently adopt the governance model of cross-regional management committee system. The establishment of a permanent daily leading organ as the main administrative carrier of the regional committee, the implementation of the joint meeting system, the main responsible person is responsible for the relevant organizational coordination work. Joint intra-city and inter-city meetings are held regularly. To set up specific policy implementation agencies in the area of urban development planning and infrastructure industry alignment, to establish a mechanism for mutual trust and consultation between the two sides, and to accomplish the goal of coordinated industrial development in the region. [1].

5.2 Take the overall development situation as a priority, break through the local protectionist ideology, and continuously eliminate policy implementation barriers

As a national central city, Guangzhou should play the role of "big brother" in the reform of the administrative management system of urban governance, take the lead in demonstrating how to break the administrative management model dominated by local interests, improve the formulation of policies and plans for coordinated regional development, play the fundamental role of the market in resource allocation, improve the coordination mechanism for the distribution of interests, and The establishment of a standardized model of cooperation in government behavior to prevent the waste of production resources caused by vicious competition between the two cities. The two municipal governments should achieve complementary advantages in the development of modern industrial systems based on a thorough examination of their respective advantageous industries. To eliminate the homogeneity of industrial structure in the region as the primary principle of regional development, realize the rapid flow of production factors between the region and prevent the occurrence of internal vicious competitive behavior. Specifically, that is, the Guangzhou Municipal Government should focus its major construction efforts on the development of high-end service industries such as international business, logistics park construction, financial services management platform, logistics integration, etc., and focus on the development of highly sophisticated product supply services such as fine chemicals and petroleum material reserves. Foshan government should give full play to the advantages of Foshan and Guangzhou in the geographical location of the immediate vicinity of Guangzhou, Guangzhou as the basis for strengthening its traditional processing and manufacturing industries, such as ceramic building materials, household goods and other industries related to the construction of complementary mode in achieving increased value added products and Guangzhou, the formation of a modern industrial structure system of staggered development.

5.3 Focus on ecological green space and ecological corridor construction and establish effective communication mechanism

Work together to build an ecological protection forest system along the main river trunk, build a structural ecological green corridor, increase the monitoring of ecological damage, focus on restoring the main public green areas in the main urban areas of the city; revitalize the construction of water conservancy infrastructure, strengthen the key maintenance of farmland protection zones in the region; strengthen the integration of water source protection, comprehensive improvement of cross-regional water pollution management problems, clear upstream and downstream pollution Responsibility subjects, and improve the construction of legal regulation system related to water pollution and river protection and prevention. The two regions have already achieved a high level of cooperation in the areas of transportation infrastructure sharing, communication co-location, and health care co-location, issues that received widespread attention when co-location was first proposed, and which have been elevated to the level of public governance theoretical concerns. As a professional group of urbanization construction, the demands of the two municipalities in terms of urban development interests should be constantly met. The two municipal governments should take the initiative to accommodate the interests of these groups in urban development and industrial planning, and guide them in planning for the establishment of a modern enterprise system and guiding the benign development of industries. Furthermore, the government should take the initiative to hold more activities similar to the urban development industry forum, so that researchers and experts and scholars can speak and social opinions can be widely adopted, and an effective communication mechanism between the two sides can be established, so that regular intercommunication can be achieved in the process of crosstown construction; an open information system can be adopted, so that the general public can enjoy a high degree of the right to know in the construction of crosstown.

6 Conclusions

Strengthening regional or city cooperation and promoting regional integration is an important trend in today's world economic development and an important action to adapt to economic globalization and regional integration. Closely related to regional integration, co-location is actually a concept to enhance the competitiveness of cities. The development of Guangzhou-Foshan co-location is conducive to the integration of resources and the formation of a mutually beneficial development pattern between the two cities, which is an important move to achieve economic growth from a high-speed stage to a high-quality development stage. The development of Guangzhou-Foshan co-location still has problems such as inefficient communication and unclear compensation mechanism for benefits. The over-exploitation of environmental resources must also be given greater attention. To promote the integration of Guangzhou and Foshan, we should insist on breaking through the administrative cage and strengthening administrative efficiency; speed up the establishment of a compensation mechanism to integrate the interests of the two cities to achieve the complementary advantages of the two places, coordinated development, while strengthening the common protection of the ecological environment. For this reason, it is necessary to pay attention to the construction of ecological green areas and ecological corridors, and to establish an effective communication mechanism. The promotion of the above optimization measures will be more conducive to the advancement of Guangzhou-Foshan co-city, providing important support for the high-quality economic development of the Guangdong-Hong Kong-Macao Greater Bay Area.

References

1. Liao Kai-Huai, Fu Lan, Breitung Werner, Cai Yun-Nan, Tan Xiao-Hong. A study of co-location from the perspective of border integration theory--Guangfo as an example[J]. Urban Development Research, 2021, 28(07):115-123.
2. Huang Jiangfan. A study on the countermeasures of Guangzhou-Foshan whole-area co-location in the context of Guangdong-Hong Kong-Macao Greater Bay Area[J]. Quality and Market, 2020(04): 28-29+34.
3. Li Hao. Exploration of the planning and implementation mechanism of the Guangzhou-Foshan co-location area[J]. Housing and Real Estate, 2018(13): 108.
4. Yin Laishang. Study on the effect evaluation and enhancement policy of co-location - taking Guangfo co-location as an example[J]. Urban Observation, 2016(03): 53-66.
5. Lu Ang, Zhang Chong. The practice and revelation of Guangzhou-Foshan co-location[J]. Macroeconomic Management, 2013(04): 77-78. DOI: 10.19709/j.cnki.11-3199/f.2013.04.031.
6. Chen, Shiao-Ling, Liao, Yuan-Tao, Liu, Song-Ling. Exploring the implementation of co-location under the power constraint--A case study of Guangzhou-Foshan co-location[J]. Modern Urban Studies, 2012, 27(07): 64-69.
7. Li Hui, Qian Huahua. The construction of "Guangzhou-Foshan co-city" and its inspiration to the cooperation between local governments in China[J]. Science and Management, 2012, 32(01): 12-16.
8. Zhu Honglin. Study on the governmental synergy of "Guangzhou-Foshan co-city" [D]. Guangzhou University, 2011.
9. Li Chengmei. Research on the government's role in the process of transport integration in the same city of Guangzhou-Foshan [D]. South China University of Technology, 2010.

Technological-progress and accident-drive: the presentation of the topic of "medical beauty" in the domestic media——the media reports of " People's Daily", "Southern Metropolis Daily" and "China Women's Daily" as examples(2000-2022)

*Kaiyin Zeng

South China Normal University, China

Abstract: Based on the framework theory, this study takes "People's Daily", "Southern Metropolis Daily" and "China Women's Daily" three traditional media with different positioning as examples, to have a glimpse of their media discourse presentation and practice of "medical beauty". Through content analysis, it is found that the media presentation of medical and aesthetic topics is mainly driven by events, and its positive aspects are not only reflected in the possibility of providing people with the chance to change their lives and destiny, but also tied to the discourse of technological nationalism. There are significant differences in the framework and topic focus of different media. People's Daily pays more attention to constructive perspectives such as industrial economic framework and governance, while media such as Southern Metropolis Daily lays more emphasis on accidental issues that can draw the attention of the audience.

1.Introduction

As of 2020, the cosmetic surgery market in China reached 256 billion yuan ($36.71 billion), and it is estimated the Chinese cosmetic surgery market will hit 1 trillion yuan in 2025.[1] It can be seen that "medical beauty" has combined with modern social discourse practice, functioning as a window of the public aesthetics and technological evolution

2.Literature review

In today's society, plastic and cosmetic surgery has gain more popularity. However, there are few academic achievements directly related to it in the domestic social science field. Academia place their product "artificial beauties" or the actual "female body" performed by surgery into the research category of consumerism and feminism, and criticize women's options of cosmetic surgery, makeup and other body reconstruction behaviors, which means the subordination to patriarchal culture and capital market consumerism (Jiang Xiuhua (2003)[2], Xu Min and Qian Xiaofeng(2002)[3],Long Hongxiang and Liu Jia(2006)[4]).The existing results are mainly published after 2003.Because at the end of 2003,Hao Lulu's "Freshly Released", "China's No.1 Artificial Beauty", set off a wave of cosmetic surgery among women. Domestic and foreign media are scrambling to report on plastic surgery cases, beauty pageants and other hotevents. As a result, scholars in the fields of jourlism and communication are also discussing. Dong Tiance (2004)[5]and Jin Xia(2004)[6] defines it as a typical case of commercial hype.

What role does mass media play in presenting women's bodies and thus shaping women's aesthetic culture? Views are mainly divided into two categories. On the one hand, some scholars argue that "advertising" plays a pivotal role in consolidating patriarchal culture and manipulating women's bodies (Jiang Xiuhua, 2003[2]; Xu Min and Qian Xiaofeng, 2002[3]). Similarly, Tongxin (2014)[7]commented on posts that clamored for attention to the "body", "They constantly reinforces women's body anxiety". So does the social media, "Viewing images of females who have undergone cosmetic enhancements affected young women's desire for cosmetic surgery." (Walker, Krumhuber, Dayan, & Furnham, 2021[8])

However, there are also studies that challenge this view, arguing that the media has become a tool to liberate women's rights by empowering them. For example, Yang Xianshun and Pan Yingyao (2012)[9] found that feminist advertising can "subvert the patriarchal tradition" to some extent. Findings in fashion magazines echo this sentiment. Tracing back the important historical stages in the development history of Chinese fashion magazines, Zhao Yunze and Gan Guangqian (2008)[10]asserted that the medium was "a powerful tool for the liberation of women's bodies and minds, and even the designer and communicator of women's political discourse". It can be seen that the academia generally believes that women's bodies and appearances are"shaped", which has some connection

zengkaiyinzxy@qq.com

with the mass media. Ke Qianting (2015)[11]found that" plastic advertisements use information such as aesthetic principles, technological myths, and doctors'authority to establish a new discourse on beauty: beauty does not have to be innate, but can be obtained through consumption and medical care intervention". " The prevalent and positive media representation of cosmetic surgery may induce a desire to directly model this behaviour" (Slevec&Tiggemann,2018)[12],and "those who consumed more media believed that cosmetic surgery brought more benefits." (Furnham&Levitas, 2012)[13] . All-aboved shows that we can comprehend the factors and changes that shape women's body and aesthetics in contemporary China by analyzing media discourse, with "cosmetic and medical plastic surgery" providing a good case.

Wenhua(2010)[14]classifies the perspectives of the mass media on cosmetic surgery into three types: one is approval, regarding it as women's "right" and "liberation"; the second is criticism that cosmetic surgery is just a means for the colonization of women's bodies(Morgan,1991)[15] and a conspiracy of the patriarchy and capitalism as the " beauty myth " (Wolf ,1991)[16]";the last voice comes from the state-owned Chinese media, reckoning the pursuit of body beauty as an result of economic development. For a long time, the body has been the core concern of gender studies. Women's struggle for further liberation in modern society is bound to be accompanied by a new understanding of their own bodies. Therefore, plastic surgery, as an important means of transforming the female body, inevitably contains multiple connotations.

To sum up, first of all, scholars have thought the guiding effect of mass media on aesthetic culture to varying degrees, but combined with the basic blank of the traditional media form of "newspaper" and the content presentation method of "news report". Second, the mainstream studies stress the feminist framework, selecting "the female body" or "the presentation of the female body in the media" as the research object, while temporarily neglect the significant body modification technology of "medical aesthetics and plastic surgery". Furthermore, the above results have a certain time distance from the present. Taken all-above mentioned in to account, in the modern society where appearance and body anxiety are increasingly rampant, this paper intends to use the content analysis method, to make up for the existing gaps by analyzing the presentation methods of "medical aesthetics and plastic surgery" in the traditional mainstream media in China. Specifically, the research questions of this paper are mainly:

(1) What are the characteristics of the quantity and trend of mass media reports on medical aesthetics?

(2) Are there differences in the coverage of medical aesthetics by different media, and what are they?

(3) What aspects do different media focus on in medical aesthetics?

3.Research methods

This paper selects three traditional paper media, "People's Daily", "China Women's Daily" and "Southern Metropolis Daily" as the research objects for their reports on medical aesthetics and plastic surgery. There are two factors. First, traditional media reports are a stable carrier of social attitudes and changes. Compared with Internet texts, they can form a more stable frame of samples. Second, the market positioning of media determine media reports to a certain extent. Therefore, this paper selects the types of central official media, industry newspapers and market-oriented urban newspapers, in an attempt to more comprehensively understand the attitudes and reports of different types of media on the phenomenon of medical beauty. I conducted a search in the period from January 1,2000 to April 21,2022 with the keywords of "medical beauty" and "plastic surgery" through the WiseSearch Database, and obtained 89 articles in People's Daily and" Plastic Surgery"304 articles in China Women's Daily and 812 articles in Southern Metropolis Daily respectively. After manual review, repeated and less relevant reports were eliminated, and a total of 713 valid samples were finally determined.

According to the hypothesis, we set the dominant frame and topic type as the research dimension and coding category. The coding work was done by two coders together, and the final coders reliability kappa was 0.741.

Dominant Frames: According to discourse tendencies, reports are divided into progressive frames, risk frames and others.

Types of topics: All reports are classified into five categories: Rampant medical beauty disorder, adjustment and governance of the medical beauty industry, emerging cosmetic surgery trends, emerging cosmetic surgery technologies, and others. Among them, the category of "emerging plastic surgery trends" includes not only new changes aimed at plastic surgery recipients, such as students' plastic surgery fever, job seekers' plastic surgery fever, and the increase in the number of male beauty seekers, but also summer plastic surgery fever, going to Korea for plastic surgery, and plastic surgery at a younger age. change, appearance anxiety, etc

4.Research findings

4.1 Number of reports:Market media mainly focus,event-driven reports

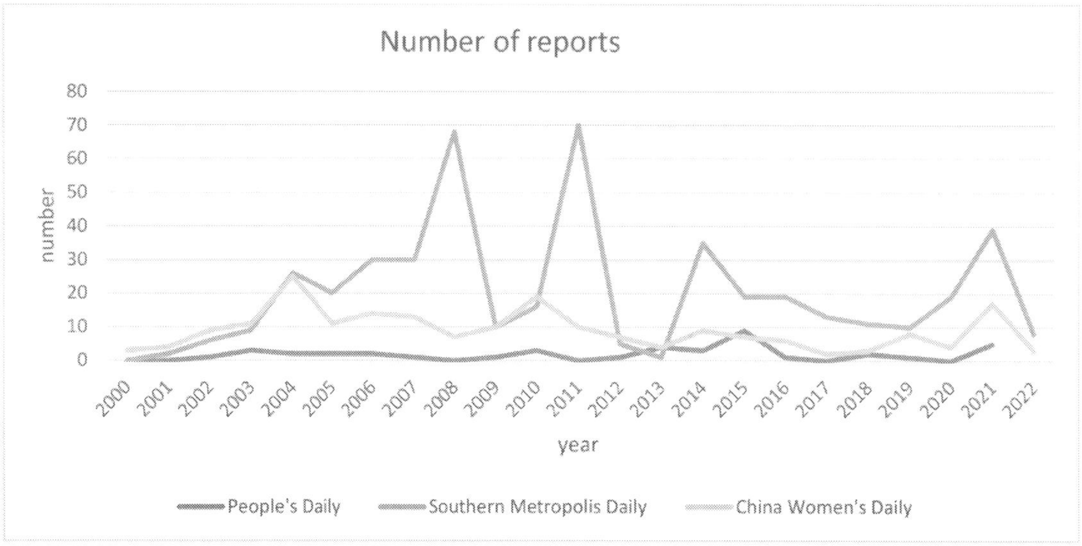

Fig. 1. Number of reports

After sample cleaning and statistics, from January 1,2000 to April 21,2022, there were 41 reports on medical beauty and plastic surgery in People's Daily,206 in China Women's Daily, and Southern Metropolis Daily up to 466 articles. The Southern Metropolis Daily paid the highest attention to the topic of medical beauty, peaking in 2008 and 2011.Southern Metropolis Daily, as a market-oriented media, mainly focuses on the supply and demand side of emerging markets for medical and aesthetic issues, and has a strong timeliness. For example, also in 2011, during the Spring Festival, it was reported," Give me the whole' lucky nose'"(2011.02.14); on Women's Day, "the whole' prosperous nose' for a good marriage" (2011.03.07); During the National Day, it became" Lucky Micro-Shaping is very popular nowadays" (2011.09.30). It can be described as "advance with the times".

As can be seen from broken line chart 1, the issue of medical aesthetics has not received sustained attention in the media discourse sphere,so the number of reports fluctuated dramatically and was driven by important events. This finding also echoes the presentation of media reports on other issues, and media discourse is often influenced by issues such as policies, disasters, and emergencies. The media presentation of medical beauty topics is mainly influenced by the beauty pageant schedule. For example, the number of relevant reports in China Women's Daily peaked at 25 in 2004.Among them, the number of keyword reports containing" artificial beauty" is 14, because it is at the stage of "China's first artificial beauty" Hao Lulu undergoing plastic surgery. At this time, medical aesthetics reports mainly focus on the impact of medical aesthetics on an individual's life path. A few reports claim its positive side, claiming that it can"make the career path more confident and add value to professional quality"(2004.06.18);but most of them adhere to the angle of criticism and reflection, such as "make beautiful women or make selling points?" (2004.01.10); "Is Free'Renovation'a Gospel or Another Trap?" (2004.06.08). In order to increase the credibility and influence of the views, many reports quoted authoritative sources. Zhou Xiaozheng, a sociology professor at Renmin University, believed that "'artificial beauties' are modern feet" (2004.01.14)

4.2 Reporting framework: There are differences in the reporting framework among the media, and the risk framework dominates

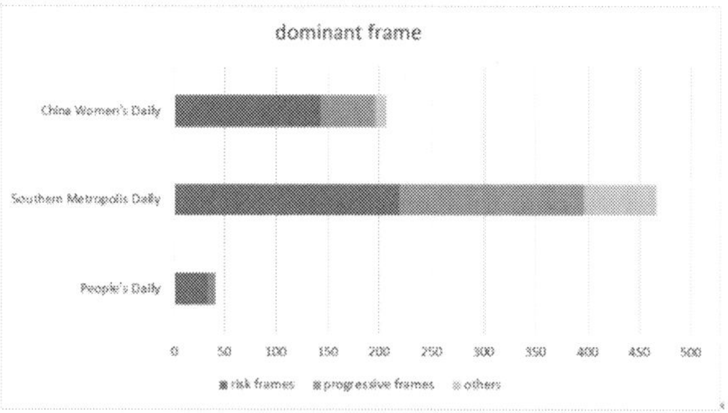

Fig. 2. dominant frame

In media reporting, framing is a process of re-"selecting" and "emphasizing", making an aspect of intended emphasis more prominent in the text(Pan&Kosicki,1993)[17].In response to the second research question, this paper found that the three media selected the risk framework as the dominant framework when reporting medical and aesthetic-related issues, as shown in Figure 2,accounting for 47%,68.9%and 80.5%,respectively.Relatively speaking, however, Southern Metropolis Daily's use of the progressive framework is much higher than that of the other two newspapers, accounting for 38%,much higher than the others.

The risk here mainly emphasizes not the risk of medical beauty technology itself, but the abuse of medical beauty technology by unqualified subjects. In other words, even in the risk framework, medical beauty technology is not completely denied, but focuses on the issue of safety qualifications. The headlines of "People's Daily" often use rhetorical techniques such as rhetorical questions, contrasts or metaphors to strengthen the tone, thereby deepening the reader's impression and reminding the reader to reflect. For example, "Destroy 200,000 Faces in 10 Years, Beauty or Disfigurement?" (2003.11.21)," "Face Project" Can't Buy" Job Tickets"" (2005.08.03).

What is worth highlighting is that the risk of medical aesthetics is mainly demonstrated in the use of medical aesthetics technology as the icing on the cake in the event of "excessive" pursuit of beauty, rather than as a medical means in the restoration of patients in disasters such as burns. For example, "People's Daily" once reported the political news that "courageous actions lead to damage to the appearance, Shanghai citizens can get free plastic surgery"(2007.01.15); "China Women's Daily" also reported the case of children injured in the Wenchuan earthquake who regained their beauty due to medical beauty surgery(2008.07.15).Such concept may also be related to traditional Chinese concepts such as "parents who receive body and skin", explianing why the media has a negative attitude towards "plastic surgery" as a whole.

When reporting on positive events such as "the use of medical aesthetic plastic technology to repair congenital or acquired facial defects that hinder the normal life of patients", the three newspapers all chose the progressive framework and showed a relatively positive reporting attitude. This also reflects the progressive discourse of medical aesthetics as a "technology", echoing Wenhua's (2010) statement that medical aesthetics technology is bound up with the discourse of national progress to reflect China's desire to be accepted by the international community. For example, the words of "China Women's Daily" linking Chinese medical beauty technology with the world, "Injection beauty technology has attracted much attention from the medical beauty industry at home and abroad,which is called the future development direction of medical beauty(2007.05.22)."Previous studies on technological discourse in China's cyberspace have also found that technology is often tied to the country's political and economic goals(Wang Qin&Ouyang Guohua,2020)[18]. For example, the media discourse on artificial intelligence is highly related to China's manufacturing and technological dreams(Zeng et al,.2020)[19].The research on medical aesthetics in this paper also confirms this view.

In addition to the technical level, the progressive framework of medical aesthetics is also reflected in taking it as an "opportunity" to change the fate. On Women's Day, a report in Southern Metropolis Daily (2011.03.08) said: "Beauty is the passport of mankind, and beauty is also the goal that all women pursue. Every woman who knows how to pamper herself will never give up any opportunity to be more beautiful.". It can be said that the "opportunity discourse" is intended to encourage readers to increase their advantages in seeking employment, finding a partner, etc. through medical cosmetic surgery, so as to realize the "transformation" and "renew" of destiny.

In this paper, the chi-square test confirms that the frame difference between different media types is significant($p<0.00$***), and the choice of frame for different media does show differences. This finding is still consistent with the "marketization of urban newspapers and the trend toward entertainment"(Su Linsen,2013)[20],and also reflects the choice of different media to appeal to the state's position and the public's position (Zeng Fanxu et al.,2013)[21].In the special medical field of medical aesthetics and plastic surgery, which is far more profitable than public welfare, private plastic surgery institutions bring more benefits from advertising; Words and pictures such as "beauty" and "model" are also more suitable for the tastes of most readers.

4.3 Types of issues: lack of industry governance issues, and chaos attracts more attention

Through statistical analysis of the types of topics, this paper finds that the medical beauty industry has the lowest proportion of governance and adjustment (9.39%), and the issue that the media attaches most importance to is the rampant chaos in medical beauty (30.7%).It can be seen that the media's attention to constructive reports is slightly less, while accident reports are more able to enter the media's field of vision. This also shows that the media discourse lacks constructive reports

on the medical and aesthetic industry, and the audience is exposed to more news that is a catastrophic frame. From the perspective of emotional and behavioral impact, this may lead to an increase in people's negative emotions and a decrease in motivation to take positive behaviors (Xu Jinghong et al.,2019)[22].

This paper also found a significant difference in the association between media type and issue type through the chi-square test($p<0.00$***), which indicates that different media have significantly different selected issues. Among them, "Southern Metropolis Daily" is more inclined to the topics of "emerging cosmetic surgery trend" and "rampant medical beauty disorder", accounting for 29.4%and 28.5%respectively; "China Women's Daily" is similar to the former. Correspondingly, although People's Daily does not pay much attention to the topic of medical beauty, it pays more attention to the perspective of industry governance. In recent years, the huge economic volume of the medical beauty industry has made it impossible to be ignored by the mainstream media. And just as People's Daily reported on video games, "In the report of the 'industrial economy' framework, they (here refers to individuals such as medical aesthetic practitioners and plastic surgery recipients)are no longer required to be present as objects. Correspondingly The protagonists in the report are the country, the government, enterprises, capital and entrepreneurs (He Wei&Cao Shule,2018)[23].

5.Conclusion

This article attempts to analyze the media's shaping and presentation of medical and cosmetic discourse by analyzing the content of medical aesthetics and cosmetic surgery technologies in three media: People's Daily, China Women's Daily, and Southern Metropolis Daily from 2000 to 2022.Specifically,it is found that the topic of medical aesthetics carries the progressive side of technological nationalism, and can also be used as an "opportunity" to bring people the possibility of changing their lives and destiny. But on the other hand, the media has not formed continuous attention to the topic of medical beauty, and mainly pays attention to the chaos and qualifications of the medical beauty industry. In terms of presentation methods, there are obvious differences in the presentation frameworks and topics chosen by different media. "People's Daily" focuses more on constructive perspectives such as industrial economic framework and governance, while "Southern Metropolis Daily" and other media pay more attention to accidental issues that can attract the attention of the audience.

This article contributes to the presentation and sorting out of different types of domestic media on medical and aesthetic issues. However, future research can also focus on more diverse aspects, such as actors in reports, the relationship between media presentation and time, etc. These topics are also worth digging into

References

1. China Daily. (2020) Cosmetic surgery market in China worth over 250b yuan. http://www.chinadaily.com.cn/a/202001/03/WS5e0ef315a310cf355825cb.html
2. Jiang Xiuhua. (2003). A Cultural Review of Female Body Reconstruction Behavior. Women's Research Review (03), 37-46.
3. Xu Min, Qian Xiaofeng. (2002). Weight loss advertisements and morbid slender culture: On the cultural control of the female body by mass media. Women's Studies Series (03), 22-29.
4. Long Hongxiang & Liu Jia. (2006). Cultural reflection on artificial beauties. Journal of Nantong University (Social Science Edition) (06), 99-102.
5. Dong Tiance. (2004). How the media should deal with commercial hype -- from the "artificial beauty" phenomenon. Journalist (04), 25-27.
6. Jin Xia. (2004). "Beauty Manufacturing" and the Women's View of the Media. Contemporary Communication (02), 87-89.
7. Tongxin. (2014). "Becoming a Better Self": Fashion Forum and Female Body Consumption——Taking the Fashion Edition of Tianya Forum as an example. Journalism University (01), 140-145.
8. Walker, C.E., Krumhuber, E.G., Dayan, S., & Furnham, A. (2021). Effects of social media use on desire for cosmetic surgery among young women. Current Psychology,40(7),3355-3364.
9. Yang Xianshun & Pan Yingyao. (2012). The spectacle of feminism being stared at: Interpretation of feminist advertising in the postmodern perspective. Modern Communication (Journal of Communication University of China) (02), 25-29.
10. Zhao Yunze & Gan Guangqian. (2008). Fashion Communication and Women's Politics: A Little Enlightenment from the Development History of Chinese Fashion Magazines. International Press (10), 46-49+95.
11. Ke Qianting. (2015). Plastic advertisement, beauty concept and body culture. Women's Research Series (01), 71-77.
12. Slevec,J.,&Tiggemann, M.(2010). Attitudes toward cosmetic surgery in middle-aged women: Body image, aging anxiety, and the media. Psychology of women quarterly,34(1),65-74.
13. Furnham, A., & Levitas, J. (2012). Factors that motivate people to undergo cosmetic surgery. The Canadian journal of plastic surgery=Journal canadien de chirurgie plastique,20(4),e47–e50.
14. Hua, W. (2013). Buying beauty: Cosmetic surgery in China (Vol.1).Hong Kong university press.
15. Morgan, K.P. (1991). Women and the knife: Cosmetic surgery and the colonization of women's bodies. Hypatia, 6(3),25-53.
16. Wolf, N. (2013).The beauty myth: How images of beauty are used against women. Random House.
17. Pan, Z. & Kosicki, G, M. (1993) Framing analysis: An approach to news discourse, *Political Communication*, 10:1,55-75.
18. Wang Qin & Ouyang Guohua. (2020). Understanding Technonationalism: Collective Memory, Identity Recognition and International Competition--Based on the Observation of Lenovo 5G Voting Event. Journal of Northwest University for Nationalities (Philosophy and Social Sciences Edition) (03), 59-67. doi: 10.14084/j.cnki.cn62-1185/c.2020.03.008.
19. Zeng, J., Chan, C.H., & Schäfer, M.S. (2020). Contested Chinese dreams of AI? Public discourse about artificial intelligence on WeChat and people's daily online. Information Communication & Society, (2), 1-22.
20. Su Linsen. (2013). The Represented Other: The Media Image of Chinese Workers. International

Press (08), 37-45. doi: 10.13495/j.cnki.cjjc.2013.08.004.

21. Zeng Fanxu, Dai Jia & Zheng Jie. (2013). Frame competition, resonance and diffusion: Analysis of media reports on PM2.5 issues. International Press (08), 96-108. doi: 10.13495/j.cnki. cjjc.2013.08.009.
22. Xu Jinghong, Guo Jingyu, You Xinyang & Hu Shiming. (2019). Constructive journalism: concept definition, main features and value enlightenment. International Press (08), 135-153. doi: 10.13495/j.cnki. cjjc.2019.08.008.
23. He Wei & Cao Shule. (2018). From "Electronic Heroin" to "Made in China": Discourse Changes in the Game Report of People's Daily (1981-2017). International Press (05), 57-81. doi :10.13495/j.cnki.cjjc.2018.05.004.

Hegemonic Gaze: Tyranny, Media Chauvinism and the Image of Women - Based on Chinese Mainstream Media's News Coverage of Ailing Gu's Winter Olympics Victory

Tieyu Zhou[1,a], Ranjun Hua[2,b], Han Wang[1,c] and Linyi Gao[1,d]

[1]Faculty of humanities and arts, Macau University of Science and Technology, Taipa, Macau, China
[2]School of Media Science, Northeast Normal University, Changchun, Jilin, China

Abstract. During the 2022 Winter Olympic Games in Beijing, Chinese freeskier Ailing Gu received much media attention. Past literature and journalistic practice on the media and sport suggest that sport has been an area of absolute male hegemony since ancient times and that the media is constructed around hegemonic male values that help to reinforce 'historical gender norms. This paper examines the construction of Ailing Gu's image of the 'other' in the media frameworks and discourses based on news stories about her in the official and commercial media. The study finds that the deconstruction of Ailing Gu's image as an athlete in the news coverage, which appears to reinforce the discourse of female athletes, still serves the ultimate goal of male domination and remains, in essence, a chauvinistic and hegemonic gaze.

1 Introduction

The 2022 Beijing Winter Olympics has received widespread attention at home and abroad. As a rising sports star of domestic and international attention and the most famous athlete on social media during the Winter Olympics, Ailing Gu has become the focus of media coverage. By the end of the 2022 Winter Olympics, Ailing Gu's Weibo super talk had been read by more than 9 billion people and was widely discussed and followed by netizens.

At the same time, the media coverage agenda in the Chinese context has shaped the public's perception of traditional female athletes. However, is the media portrayal of Ailing Gu a different kind of male gaze? Is the political propaganda and image construction of patriotism still essentially subordinated to male domination? Based on this, this study will explore the news construction of Ailing Gu's image of women in the Winter Olympics in the Chinese media today in an attempt to expose the gender discursive construction behind the news texts.

2 Research ideas and methodology

2.1 Research Questions

1. Is there a tendency in the contemporary mainstream media to report male domination as a value?

2. Is the political aim of patriotic propaganda achieved through media propaganda? Is its essence a chauvinistic and hegemonic gaze?

3. Is there a deconstruction of the image of women in the relevant news coverage?

2.2 Research Methodology

2.2.1 Sample selection

This study used a Wisers search to retrieve the reports on Ailing Gu's victory in the Winter Olympics throughout 2021-2022. A sample of 36 feature articles with a view count of over 100,000 was selected by hand. Sixteen of these articles were in the official media and 20 in the business media. The names of the media are as follows.

2.2.2 Content analysis method

This study analyzes the media's portrayal of Ailing Gu's female image during the Winter Olympics based mainly on comparing the official and commercial media. By organizing the news reports of Chinese mainstream media on Ailing Gu's Winter Olympics, the study examines how the female image is built or deconstructed in the news reports. Through the content analysis of the articles, the author coded the content according to four dimensions (male gaze, objectification of women, expression of male and female discourse power, and identity representation). The coding table is as follows.

[a] tieyuchou@163.com, [b] 2010592153@qq.com
[c] 2282564215@qq.com, [d] haydeeg01@outlook.com
Correspond author: Linyi Gao, haydeeg01@outlook.com

Table 1. Category of the code

Main categories	Category Coding	
Male gaze	V1. Presence or absence of male gaze in reports	None = 1 Viewing fetish (yes) = 2 Gaze fetish (yes) = 3 Object worship (yes) = 4
Objectify woman	V2. Presence of objectified women in reporting	None = 1 De-selfing (yes) = 2 Instrumentalization (yes) = 3 Commoditization (yes) = 4
Male discourse expression	V3. Presence of male discourse power expressions in the report	None = 1 Discourse rights = 2 Discourse power = 3
Female discourse exprssion	V4. Presence of female discourse power expressions in the report	None = 1 Discourse rights=2 Discourse power = 3
Identify shaping	V5. Reported expressions of Gu Ailing's status	None = 1 Z-era youth(yes)=2 Genius(yes)= 3 Athlete(yes)= 4 Champion(yes)=5

3 Review of the literature

3.1 The coupling and embedding of gender and media in the field of sport.

The new concept of "gender," which emerged in the 1960s, is a product of the same period as the second wave of feminism and has evolved in academic circles from a single term to a category of analysis used to explain the social constitution of femininity and to study the reasons for the maintenance of a male-dominated society and the ownership of power, gradually It became the new vocabulary of Western feminism.

In the 1970s, Ann Oakley drew a clear line between gender and sex[1]. In the same period, the American scholar Tuchman made the first connection between the media and the image of women in his book Fireplace and Home: Images of Women in the Mass Media, in which he referred to the 'symbolic annihilation', where women are censured, trivialized or even ignored by the media in society and culture[2]. Furthermore, nowhere is this 'symbolic annihilation' more prominent in society than in sport. Tracing back to the development of sport, it can be said that the development of the sport has been the process of the awakening of women's consciousness, a field that since its inception has been the domain of absolute male hegemony and has served to construct and reinforce a patriarchal society dominated by men. The role of the media in sports has been more to reinforce the absolute male hegemony, and the struggle for women's power in sports has been chiefly downplayed or even ignored by the media. For example, the Olympic Games, a symbol of equality, have only been used for men, and the father of the Olympic Games, Alexander Kubitan, fought for women's access to sport throughout his life, which has undoubtedly put up much resistance to women's equal rights in sport[3].

Later, the study of the role of the content of media texts in the image of women entered the academic perspective, and Rosalyn Gill mentioned in her book Media and Gender that different types of media have different effects on the construction of the image of women[4]. However, in the Chinese context, sports journalism has always been predominantly male, portraying male athletes primarily as "brave, strong and full of blood," while female athletes are mostly portrayed as "soft, feminine and physically coordinated." At the same time, as female athletes compete in less competitive sports, the media's portrayal of women is also predominantly negative.

On the topic of the construction of women's social images in the Chinese media, Professor Liu Liqun mentions in his book A Tutorial in Media and Women's Studies that in the Chinese context, the translation and interpretation of gender theory are "tightly focused on the advocacy and protection of women's subjectivity, women's social and political rights [5]." Professor Liu Liqun argues that the Chinese media is concerned with women's subjectivity and discourse rights, is committed to balancing gender power between men and women, and has a positive role in promoting equality of gender and political rights. However, we are afraid we have to disagree with Professor Liu Liqun's utopian idea. We believe that the construction of women's image in the Chinese media is still in the service of patriarchal domination but only changes the way of communication and does not achieve gender equality in the media. Based on this, the author proposes a new concept of "media chauvinism" in this paper, arguing that today's media is still the absolute domain of male hegemony and that the focus on women only serves to consolidate men's domination of the discourse. In this paper, the author uses the example of "Ailing Gu," the female athlete representative of Chinese sports in 2022, as an example to support the view that male hegemony is still absolute in the media sphere and that the promotion of women serves the patriarchal society of male domination to a greater extent.

3.2 Agenda Setting

The concept of agenda setting dates back to the 1920s with Lippmann, who, in his book Public Opinion referred to the concept of a mimetic environment in which we all live in a virtual reality constructed for us by the media [6]. His famous 'searchlight' metaphor, in which a light is shone on a dark wall, is akin to how the media present us with the idea that we can only look at the lighted area and assume that this is our actual reality. This concept is the theoretical basis of agenda-setting, which was introduced by McCombes and Shaw 1972, who argued that although the media cannot determine audience perceptions, they can influence them by adjusting the order and weight of stories[7]. In 1997, McCombs and Shaw introduced a second concept of agenda setting, "attributive agenda setting," which suggests that the media can change audience perceptions through smear and other means[8]. In the era of intelligent media, Guo Lei et al. proposed "online agenda setting," which suggests that online agenda setting can link audiences to different issues or attributes through a network of relationships, forming a social network of relationships on the Internet level. In this paper, the author will take agenda setting as the basis and introduce Entman's framework theory, i.e., from "defining the problem," "taking the pulse of the cause," "moral judgment," and "suggestions for countermeasures [9]." In this paper, we will analyze the media effects of the news coverage of Ailing Gu by different Chinese media and reveal the real purpose of media communication based on the four news frames of "defining the problem," "taking the pulse of the causes," "moral judgment" and "suggestions for countermeasures," in conjunction with the attributes of news agenda setting. This is an attempt to support the concept of "media chauvinism" proposed by the author.

4 Data analysis

4.1 Consistency of the code

Data were processed and analyzed using SPSS tools for the coding results. For the reliability test of this coding scale, this study was a coding exercise completed by three social science researchers with quantitative backgrounds after understanding the research context and coding requirements. The author calculated a Krippendorff coefficient of 0.7 based on this coded data, and the data largely converged between 0.8 and 1.0 with solid reliability.

Consistency was measured using Kendall W values for each of the five question outcomes, and the results were as follows.

Table 2. Presence or absence of four dimensions

	Male gaze	Objectify woman	Discourse expression		Identify shaping
			Male	Female	
Kendall's W	0.59	0.786	0.648	0.408	0.422
Chi-Square	61.93	82.528	68.065	42.839	44.277
df	35	35	35	35	35
Asymp. Sig.	0.003	0	0.001	0.17	0.135

For Male gaze, p=0.003<0.05, original hypothesis denied, coding results are consistent. W=0.590, the degree of consistency is moderate.

For Objectified females in the report, p= 0.000010<0.05, original hypothesis rejected, coding results consistent. W= 0.786, the consistency is strong.

For Male discourse representations in reports, p= 0.001<0.05, original hypothesis rejected, coding results consistent. W= 0.648, the consistency is strong.

For Female discourse representation in the report, p=0.170>0.05 Original hypothesis accepted, coding results not consistent. W=0.408, the consistency is moderate.

For Identity of Ailing Gu, p=0.135>0.05 Accept original hypothesis; coding results not consistent. W=0.422 Moderate consistency.

The calculated measures revealed strong consistency in the coding results for questions 2 and 3 and moderate consistency for questions 1, 4, and 5. The three coders formed a relatively unified opinion after a discussion for those coding results that were not significantly different.

4.2 Empirical analysis of content

4.2.1 Male gaze

Table 3. Male gaze type of distribution of different media

	Official media	Commercial media	Total
Viewing fetish	10.42%	10.00%	10.19%
Gaze fetish	18.75%	41.67%	31.48%
Object worship	29.17%	13.33%	20.37%
None	41.67%	35.00%	37.96%
$\chi 2$=8.047			

In the judgment of the male gaze, in order to comply with the cardinality test for the expected number of times, the author firstly made a dichotomous judgment of "yes" and "no"; secondly, based on the specificity of the study, the sample type of male gaze was divided, and the original data was recorded(no=0, yes=1). The number of samples in which the male gaze was present (67) represented 62% of the total sample; the minimum expected count was 1<4.89<5, p=0.041<0.05, and the presence or absence of the male gaze differed significantly between media types ($\chi 2$=8.047, df=3).

According to the above table, we can see that the types of the male gaze are also different, with the highest percentage being "gaze fetish" (31.8%). However, the difference is that some commercial media are biased towards commodified fetishization, i.e., the "commodified" category that corresponds to the objectification of women.

4.2.2. Objectifying Women.

Table 4. Objectifying Women type of distribution of different media

	Official media	Commercial media	Total
De-selfing	12.50%	6.67%	9.26%
Instrumentalization	18.75%	23.33%	21.30%
Commodification	8.33%	21.67%	15.74%
None	60.42%	48.33%	53.70%
$\chi 2$=4.328			

Objectification is a form of "dehumanization or depersonalization." According to the coding results, 58 samples stated that there was no objectification of women, and 50 samples stated that there was the objectification of women, with a similar proportion, close to 50%. Those who felt that "objectification of women exists" were concentrated in the category of "instrumentalization" (21.3%). "Instrumentalisation" is a representation intended to magnify specific physical characteristics of women and their functional utility, to separate them from the woman as a whole and treat them purely as tools, and to generalize or describe the image of women in this way. In the texts in question, this is often expressed in terms such as Ailing Gu's "good physical qualities" and "bridging the gap in female skiing in China." The different media types do not differ significantly in specific biological categories ($\chi 2$=4.328, p=0.224, df=3), with the official and commercial media showing a more uniform trend. However, according to the word frequency ratio, the commercial media focused more on the expression of "commercialization," such as "money-earning ability," "spokesperson," and "fashion. The emergence of labels and phrases such as "money-earning power," "spokesperson" and "fashion" amplifies the commercial value of Ailing Gu as a female athlete.

4.2.3 Expressions of discourse

Table 5. Expressions of discourse type of distribution of different media

		Official media	Commercial media	Total
Discursive Rights	Male	10.42%	65.00%	40.74%
	Female	37.50%	35.00%	36.11%
Discursive Power	Male	35.42%	36.67%	36.11%
	Female	27.08%	15.00%	20.37%
None		44.79%	64.17%	55.56%
$\chi 2$=15.627				

In the table with the statistical distribution of discourse power, the representation of male discourse power did not change according to the media subject (p=0.269>0.05). Nearly 60% of the men in the sample were represented as having 'discursive power in the news stories about Ailing Gu. Similarly, due to the singularity of the event subject, there was no significant difference in the representation of women's discourse power across media subjects (p=0.196>0.05). However, the proportion of women's representation of 'discourse power' was as high as 60.2%, in contrast to men's 'discourse power' .By comparing the differences in coding frequencies and the chi-square test, the summary concludes that the differences between the categories of discursive expression power of men and women are significant ($\chi 2$=15.627, p=0.000<,0.05, df=2);. At the same time, the official and commercial media are more consistent in shaping gender discursive power, showing no significant differences or effects.

Concerning the results of the analyzed data, we can understand that Ailing Gu, as a woman with a certain social status, has external empowerment of her in the media system. The male voice is not weakened in this process but somewhat shifted to a group 'power' output. According to Foucault's theory of power, 'discursive power' is a coercive norm that, by defining and operating on something, has the power to dominate it. Therefore, it can be concluded that even though Ailing Gu has a certain social status as the subject of the narrative, she is still suspected of having a 'weakened voice' in the official media, which is dominated by male voices. In contrast, the commercial media gave more balanced data on the representation of male and female discourse, and the commercial media was more inclined to open gender attitudes and equal rights practices of freedom of expression than the official media.

4.2.4 Identity Shaping

Table 6. Different distribution of Identity Shaping

Categories	Frequency	Proportion
Z Generation Youth	12	11.11%
Genius	22	20.37%
Athlete	14	12.96%
Champion	53	49.07%
none	7	6.48%
Total	108	100%

The focus is on Ailing Gu as an individual, a young and talented 18-year-old athlete whom the media have given various names and labels. According to the data results, "champion" was the most frequent word, accounting for about 50% of the total. This phenomenon is presented in the official and commercial media, with no distinction between the sexes.

The identity labels given to Ailing Gu are not only "champion" but also "Z-era youth," "athlete," and "genius. " According to the results of the descriptive statistics, in addition to the specific identity representation given by the word champion in a specific context, Ailing Gu is mostly described as "someone else's child" or "talented girl," which is mostly reflected in the text as "talented The term "gifted" and "all-round learner" is used in the text. The two categories "Z-era youth" and "athlete" have a general connotation and are not specifically reflected in the text. The media portrayed both the public and private image of Ailing Gu - the 'world champion' in the public eye and the 'academic genius' in the private sphere - with the intention to create a representative female figure utilizing 'deification,' implicitly using her as a benchmark or yardstick against which to compare or measure the women of the new era.

5 Factors and Measures

5.1 Factors influencing the construction of the image of women in sports journalism

5.1.1 Gender bias entrenched in the world of sports media

The intense nature of sport and the pursuit of power and speed make the sport seem to be a natural no-go area for women. In all types of media, women are often "silent" or "voiceless" due to the absence of media. Men have long dominated sports, and the imbalance between the sexes is particularly pronounced. As the number of female athletes and their achievements have increased rapidly, the number of women in the media has increased, and their perspectives have become more diverse, but the gender bias in sports media coverage has not disappeared.

In the 2022 Winter Olympics in Beijing, Chinese female athletes, represented by Gu Ailing, gained the attention of the media in China with their outstanding achievements. Through a multidimensional reading of the selected sample of news reports, we found that both official and commercial media, in the articles focusing on Gu Ailing, repeatedly emphasized her status as a "female athlete" and described her around the more distinctive aspects of femininity, such as physical characteristics and dressing. As one of the most talked about athletes at the Games, Gu Ailing was portrayed in the media as much more than a female athlete on the field. In addition to her achievements and recaps of the games, the coverage focuses more on her off the field, with more attention given to her family, education, and business value than to her contribution to the sport.

As a male-dominated society does not automatically change its gender power structure, women have to change their position in sport by challenging and changing the male domination, as in the case of Gu Ailing. Some media coverage has focused on Gu Ailing's behavioral displays of female empowerment, such as the deliberate display of bangs on her forehead during games and interviews about Gu Ailing's attitude towards women in sport. However, this seeming emphasis on the representation of 'her power' in sport is, in fact, further entrenchment of the structural characteristics of the male-dominated and female-dominated field, masking the longstanding passive subjugation of women with the struggles of the female athletes themselves, and is an implicit reflection of the entrenchment of gender bias in the world of sports media.

5.1.2 The male gaze in the confines of the traditional gender order

Influenced by the traditional "male superiority over women" gender order, Chinese men have long been in a repressive position, holding the power of social discourse and perpetuating the "gaze" on female groups as a manifestation of their gender power. The term "gaze" was first coined by film critic Laura Mulvey in her essay "Visual Pleasure and Narrative Cinema," where it refers to the core ideas of feminist theory and the "male gaze." The "gaze" in this context refers to the central idea of feminist theory and the "male gaze[10]."The "gaze" is a way of seeing with a will to power, concealing a dichotomy between the subject and the other. In the male gaze, men often act as the viewer and women as the watched, achieving self-actualization in the male act of watching.

In the case of the Winter Olympics, a major sporting event, the media coverage of Gu Ailing is a concrete manifestation of the male gaze's desire to probe and control female athletes and even objectify their bodies.

According to the author's analysis, in the selected sample of news reports, the male gaze tendency was present in both official and commercial media, with the highest percentage of gaze fetish among the coding types of viewing fetish, gaze fetish, and worship fetish. The texts were more likely to direct readers to Ailing Gu's outstanding appearance, surging business value, and mythical growth story than to her sporting achievements, which perfectly matched men's expectations and imagination of the perfect woman.

Under the influence of the traditional gender order, the construction of women's image in news reports is mainly to highlight male power, satisfy the desire of the male gaze and weaken women's independence and contribution. In the coverage of Gu Ailing, various media outlets have not failed to portray her abilities positively. However, they have inevitably and intentionally led to the exploration of the reasons for Gu Ailing's achievements in various fields such as sports, academics, and business, specifically the exploration of her father's identity and family background, as well as her appearance, figure, and dress. This reflects that the media in China has fallen into the habit of expressing the male gaze discourse under the long-term confinement of the traditional gender order.

5.1.3 Objectified female symbols in a consumer society

In the media, women's many characteristics are easily the object of media communication and are seen as symbols to be consumed. In today's consumer society, women are often 'labeled' by the media. As in the previous analysis of the 'male gaze, the degree of attractiveness of women is closely linked to what men 'see,' and the degree of attractiveness to men becomes an essential measure of the value of women. For example, female celebrities and anchorwomen are often labeled as "soft porn" in current media coverage, and such stereotypes and social perceptions "are bound to affect the construction of serious social events in the media[11]."

The emergence of Gu Ailing as a confident, healthy, independent, successful woman seemed to be a break with the conventional image of women and prompted collective praise in the domestic media. Some scholars have pointed out that symbolic capital as a relational concept, a value recognized and identified by social groups, can give women a certain legitimacy in sports journalism and help them be recognized more quickly[12]. More symbolic capital and a surge of attention have made more women fully 'visible' in sport and able to reap the benefits beyond sporting awards.

However, behind this seeming acknowledgment of women's power is an orgiastic consumption of the rich symbolism of women in Gu Ailing by the media, especially in the commercial media. According to the author's analysis of the texts on Gu Ailing in both the official and commercial media in China, the commercial media show a more pronounced tendency toward commodification and fetishization than the official media, which places more emphasis on seriousness. In today's traffic and profit-driven media environment, the media exploit women's special symbols and use them as capital for consumption in order to increase attention and indirect income and Gu Ailing, who has become the most popular after the 2022 Beijing Winter Olympics, has allowed the media to do so.

5.1.4 Unbalanced discourse under media chauvinism

Some researchers have examined the three prominent roles of discourse, one of which is that "discourse is a vehicle for structural power[13]." Discourse and its power are not just ordinary linguistic expressions but deep socio-political phenomena with cultural and ideological underpinnings in constant flux. It is thus clear that discourse and power are closely linked, and that discourse is, in fact, the externalization of power, dominated and controlled by it. In the media field, the power-controlled reporting discourse is a concrete manifestation of media chauvinism.

From the author's analysis of Gu Ailing's reports, it can be seen that at the level of discursive power, the proportion of men's discursive power in official media reports is close to 50%, which is a reflection of men's long-term dominance in discursive power. In the news reports focusing on Gu Ailing, the greater power of women's discourse is, in fact, due to the social status of Gu Ailing as the subject of the narrative, while more women are often caught in the dilemma of losing their voices in the news reports, losing more benefits than men in the disparate power relations between the sexes.

In the case of Gu Ailing's report, men still dominate the channels of expression, and the official media is still suspected of undermining women's voices. The media still serves the patriarchal domination of China's media to construct the image of women, and the media still focuses on the absolute hegemony of men. The discussion of Gu Ailing's nationality and other content repeatedly emphasizes her identity as a Chinese athlete, which is subordinated to the ideological and political position held by the media. Unlike the ideologically charged official media, the commercial media have achieved a relatively balanced representation of gender discourse, with no significant difference in the proportion of power and rights of men and women. This shows that the Chinese commercial media have shown more openness to gender inclusiveness than the official media in their coverage of Gu Ailing, a practice of pursuing gender discourse equality in media chauvinism.

5.2 Measures to reshape the image of women in sports news

5.2.1 The content of the communication is based on the concept of gender equality

In the development of women's sports, female athletes have always strived for the right to speak for themselves. Gender differences and the logic of discrimination based on nature should be discarded, and the media should pay more attention to gender issues in sports, giving more voice to female athletes and advocating for them to fight for their rights more positively. At the same time, the media should be more sensitive to the gender issues that already exist in sports reporting, strengthen gender awareness in media presentation, and, in addition to the general principle of "journalistic professionalism," be sensitive to gender issues in reporting, adopt terms that contribute to fairness and equality between the sexes in specific contextual variables, and frame the reporting in a more The framing of the story is a more multidimensional demonstration of female power rather than a condescending male gaze.

At the same time, the structural gender issues behind the content of media coverage should also be considered. According to the 2020 China Journalism Development Report published by the China Journalists Association, "the gender ratio of journalists in China is equal, with 50.62% of men and 49.38% of women[14]. While female media workers occupy half of the industry, men still dominate the decision-making power in the industry, and men almost monopolize the sports field. The loss of voice of female media workers inevitably leads to the strengthening of the male gaze and the media chauvinism that serves patriarchal domination, and the media image of female athletes is trapped in the male imagination and construction for a long time.

5.2.2 Media writing returns to the value of sport itself

Lippmann pointed out in his "mimetic environment" theory that people's knowledge of the external objective world cannot be based on their practice and that most of the impressions of objective reality formed in people's minds depend on the media's reflection of them, and that if media communication is accurate and comprehensive, people can obtain the closest understanding of the real world. In sports reporting, the media's portrayal and construction of female athletes is both a reflection of ideology and social culture and a reaction to the formation of sports culture and gender structures, which profoundly impact public psychology and social consciousness.

Therefore, the media should make reasonable use of the function of 'agenda setting,' return to the exploration of the value of sport itself in sports reporting, increase the attention and coverage of female athletes, and guide the audience to understand the value of sports and female athletes themselves objectively. Instead of catering to the market's 'aesthetic' complex and focusing too much attention on the construction of female athletes' images outside of the field, we should

Instead, they should return to the true nature of sports journalism and disseminate information about women's sports in objective, fair and balanced manner so that the purpose of sports "nurturing" can be implemented.

5.2.3 Open and tolerant attitudes that play down gender differences.

The media's coverage of female athletes, in which information about women's bodies, families, and private lives is scrutinized, identified, and extracted, dramatically demonstrates the gender power order that is still maintained today, with the diverse symbols of female athletes being subjected to constant ideological discipline and social psychology in the logical control of consumer society[15].

Of course, within this long-established structure, the media should also capture the patriarchal system and the looser parts of the ideology, as well as the dilution and blurring of gender and body concepts in the field of sport. In sports such as weightlifting, there is a growing number of female athletes with a masculine perception of the body and a growing social acceptance of them. It is evident that perceptions of gender are fluid rather than static, "subject to revision and reshaping according to the ideological direction of the dominant society and the circulation of values in cultural events." The media should start by being more open and inclusive in their approach to reporting on gender-related topics. In their reporting, they should try to balance the two sexes, reduce the emphasis on gender differences, and avoid creating sensitive topics. Whether in the official or commercial media, this is an inescapable problem that the media must face.

6 Conclusion

As a phenomenal female athlete in the 2022 Beijing Winter Olympics, Ailing Gu has become the subject of intense media coverage in China. By examining and analyzing the image construction and discursive practices of female athletes in China's official media and mainstream commercial media, it is easy to see that gender power is still present in news texts in an overt manner. In a male-dominated discourse system, the media are accustomed to imagining and presenting female athletes through a hegemonic gaze, abusing and extracting female symbolic capital to serve the intentional political propaganda and the male-dominated gender structure. Whether it is Ailing Gu, who has a certain social status, or the many female athletes who remain 'invisible,' their image and power may not seem to be a central part of the sporting sphere, but in fact, they have a deeper political meaning, which is indicative of a country's social psychology and political civilization. Using Ailing Gu as an example, the author intends to lift the haze of media chauvinism from female athletes, not only in the hope that our media will make objective analysis and forward-looking guidance on gender issues but also to focus on the significance of the media writing of female images for the grand "modernization of national governance."

 https://doi.org/10.1051/shsconf/202315503003

7 Appendix

Appendix 1: Coding result

Coding table

V1. Presence or absence of male gaze in reports None = 1 Viewing fetish (exist) = 2 Gaze fetish (exist) = 3 Fetishism (exist) = 4
V2. Presence of objectification of women in reporting None = 1 De-selfing (exist) = 2 Instrumentalisation (exist) = 3 Commodification (exist) = 4
V3. Representation of male discourse in reporting. none=1 Discursive rights=2 Discursive power=3
V4. Presence or absence of female discourse in the report. no = 1 Discourse rights = 2 Discourse power = 3
V5. Representation of Gu Ailing's identity in the report: none=1 Z Generation Youth (exist)=2 Genius (exist)=3 Athlete (exist)=4 Champion (exist)=5

Coder: a=Linyi Gao b=Tieyu Zhou c=Han Wang				
		V1		
Central media:		a	b	c
1	Why is Ailing Gu so popular? The answer lies beyond the field of play	3	1	1
2	"Besides talent, there's love" - Ailing Gu's journey	1	1	3
3	Ailing Gu, the flower grower's boy!	4	4	4
4	A leek box is a testament to the fact that Ailing Gu is as good as gold	4	4	1
5	Embracing love to reach the "boiling point" of everyone's life	3	4	4
6	How do you celebrate "three dunks for one"? Ailing Gu's answer is very "Ailing Gu"	4	4	4
7	Special feature: first time! Gu Ailing enjoys the "victory slide"	1	1	1
8	Ailing Gu: I have less than 1% of the talent	1	1	1
9	Watching women take flight on the ice and snow! An old man's memories of watching the race (4)	1	1	1
10	Gu Ailing: How to "bridge" things with skiing?	1	1	3
11	China's new art review: acknowledging the ordinary and "creating a flower in the ruins"	4	4	2
12	The most fashionable and skinny sportswear this year, should be worn like this	2	2	3
13	Gu Ailing: My confidence is not based on my looks, but on skiing!	3	1	1
14	What should I do if I meet Ailing Gu in the street? Waiting online, it's quite urgent	4	2	1
15	When the light of sport shines in every day	1	2	3
16	What are you waiting for when people who are better than you are also more disciplined than you?	3	4	4
Business media:				
17	Ailing Gu: "Ordinary talent" for the Winter Olympics	3	3	3
18	Feature article / Gu Ailing takes a selfie in Winter Olympics fashion	3	1	1
19	Ailing Gu shares the secret to her success: she sleeps 10 hours a day and doesn't compete to beat other	1	4	1
20	Ailing Gu's Winter Olympics win and the surge in brand endorsements for the talented girl	4	3	1
21	Will Gu Ailing become the next Yao Ming and Liu Xiang?	3	3	3
22	As China's winner has left the American media with some nagging doubts, Gu Ailing's response is bol	3	3	4
23	Gu Ailing: I'm in cram school too, I just passed too, I'm just like everyone else!	1	1	1
24	Who is Gu Ailing after all the buzz?	3	4	1
25	The greatest gift to Chinese sport from slash girl Gu Ailing is not a gold medal	3	4	3
26	Liu Shengjun: Gu Ailing's success in the ring is a result of 'business school thinking'	3	4	4
27	Gu Ailing wins silver against pressure!	1	1	1
28	Gu Ailing wins again, but what does her father have to do with it?	1	2	2
29	The image of the "dry rice man" is deeply rooted, but Gu Ailing's silver medal is based on eating buns	3	1	1
30	Gu Ailing: Stronger all the way, crossing new peaks all the way	1	3	4
31	The real identity of Gu Ailing's father revealed? It's never a single parent that destroys a child	1	2	2
32	Enjoying the process/Valley Ailing: Back to school looking forward and nervous	1	1	3
33	I'm amazed at how much of a cyber-violence Gu Ailing can get	3	3	3
34	I have no power to change two countries	3	3	1
35	Gu Ailing responds to controversy: Representing China in the Winter Olympics for these reasons	1	3	2
36	Ailing Gu: No regrets about representing China, will be Salt Lake City's Olympic bid ambassador	3	3	2

	V2			V3			V4			V5		
Central media:	a	b	c	a	b	c	a	b	c	a	b	c
1	1	1	1	1	1	1	2	2	1	5	5	5
2	1	1	1	1	1	1	2	3	3	3	3	3
3	3	4	2	3	1	1	2	2	1	3	2	3
4	3	1	2	1	3	1	3	1	1	4	3	2
5	3	3	2	2	3	3	1	3	3	3	3	3
6	1	2	1	3	3	3	2	2	2	5	5	4
7	1	1	1	1	1	1	2	3	2	2	5	2
8	1	1	1	1	1	1	3	3	1	2	5	5
9	1	1	1	1	3	3	2	3	3	4	5	4
10	1	1	1	2	1	1	2	2	1	1	5	5
11	3	3	3	3	3	3	2	1	1	3	3	3
12	4	4	4	2	3	3	1	1	1	3	5	3
13	3	1	1	2	1	3	2	3	1	3	1	5
14	3	1	1	3	1	1	2	1	1	1	5	5
15	1	1	1	1	1	1	2	3	1	2	5	5
16	1	2	2	2	1	3	2	1	3	5	5	5
Business media:												
17	1	4	4	1	3	3	2	2	2	3	5	2
18	1	1	1	2	1	1	1	2	2	5	4	4
19	1	1	1	1	1	1	2	1	1	2	4	4
20	4	4	4	2	3	1	2	1	1	3	5	3
21	4	4	4	2	1	1	1	1	3	5	4	4
22	3	3	3	2	1	1	3	1	1	5	5	5
23	1	1	1	1	1	1	2	1	2	2	5	5
24	1	1	1	1	1	1	2	3	1	2	5	5
25	3	3	3	2	1	1	1	1	1	3	5	2
26	2	4	4	2	3	3	1	1	1	1	5	5
27	1	1	1	1	1	1	3	1	1	4	5	5
28	1	1	1	1	3	3	3	3	2	1	5	1
29	1	1	1	1	1	1	3	1	1	1	5	4
30	1	4	4	1	1	1	2	1	1	4	5	5
31	1	2	2	2	3	2	2	3	2	3	3	3
32	1	1	1	1	1	1	2	1	1	2	5	5
33	2	4	3	3	3	3	1	1	1	5	5	5
34	1	3	3	2	3	3	2	3	2	5	5	5
35	1	3	3	1	1	1	2	2	2	5	5	5
36	3	3	3	1	1	1	2	1	1	4	5	5

Reference

1. Oakley, A. (1985). Sex, gender and society (1972). Londres Temple Smith.
2. Tuchman, G., Daniels, A. K., & Benét, J. (Eds.). (1978). Hearth and home: Images of women in the mass media. Oxford University Press.
3. Zheng Yanfang. (2014). Gender bias and social construction: The changing history of women's sports rights. Sports and Science ,35(2),92-95.
4. Gill, R., & Gill, R. M. (2007). Gender and the Media. Polity.
5. Liu Liqun, & medium. (2004). Gender and Media Communication. Communication University of China Press.
6. Lippmann, W. (1965). Public opinion. 1922. URL: http://infomotions. com/etexts/gutenberg/dirs/etext04/pbp nn10. htm.
7. McCombs, M. E., & Shaw, D. L. (1972). The agenda-setting function of mass media. Public opinion quarterly, 36(2), 176-187.
8. McCombs, M., Llamas, J. P., Lopez-Escobar, E., & Rey, F. (1997). Candidate images in Spanish elections: Second-level agenda-setting effects. Journalism & Mass Communication Quarterly, 74(4), 703-717.
9. Paletz, D. L., & Entman, R. M. (1980). Presidents, power, and the press. Presidential Studies Quarterly, 416-426.
10. Li, Yuan & Jing, Xuemin. (2021). An exploration of new gender power in micro-political communication based on social media. Journalism (10), 33-40+60. doi:10.15897/j.cnki.cn51-1046/g2.20210716.003.
11. Zheng Junyuan. (2020). The media presentation of women involved in power-related sexual assault (Master's thesis, Dalian University of Technology).
12. Peng, Huaxin & Zhao, Huiying. (2022). Perceptions of "her" identity: the "gender
13. segregation" of the sports journalism profession. Journal of Xiangtan University (04), 182-188. doi:10.13715/j.cnki.jxupss.2022.04.027.
14. Xiong, Guangqing. (2021). The change of discourse in the Internet era and its impact: An analysis based on the dual perspective of power and rights. Journal of Renmin University of China,35(4), 19
15. China Journalists Association. China journalism development report [R]. Beijing: China Journalists Association, 2020: 20-21.
16. Wang Runbin & Xiao Libin. (2015). The Olympic Games in the light of media spectacle theory:A cultural critical perspective. Sports and Science, 36(2), 26-33.

https://doi.org/10.1051/shsconf/202315503004

Psychology and Solutions of Perpetrators in Cyber Violence

Zhenming Chen

Hanvos Kent international school, Ningbo, Zhejiang Province, China, 315700

Abstract. As of December 2021, the number of Chinese Internet users reached 1.032 billion, an increase of 42.96 million from December 2020, and the Internet penetration rate reached 73.0%. In the Internet era, people's freedom of speech has been greatly released, and many insulting and inflammatory remarks and pictures posted on the Internet have brought great harm to people's personal health, especially teenagers [1]. Therefore, this paper studies the psychology of online abusers and tries to provide corresponding solutions. It can be concluded that the psychology of abusers is divided into subjective and objective parts, and the corresponding solutions are to improve the relevant legal construction, strengthen the platform management, and deepen legal education.

1 Introduction

Cyber violence refers to the violence committed by citizens against other people on the Internet, which is "defamatory, slanderous, infringing on reputation, damaging rights and interests, and inflammatory", and is an extension of social violence on the Internet.

The perpetrators often cause damage to others' reputations, rights and spirits, including through inappropriate comments and comments on the Internet that are extreme or contain overly personal views; spreading rumors or malicious rumors to lower the positive evaluation of the person at the time; and spreading private information such as photos and videos without the person's permission, causing disturbance to the person's daily life and bringing unnecessary harassment [2].

With the increasing development and power of technology, the Internet has entered millions of households. People's daily lives cannot be separated from the convenience brought by the Internet; some social apps have attracted a large number of Internet users through their efficient communication performance and information dissemination performance.

While creating benefits for society and bringing convenience to citizens, the freedom of the Internet has also resulted in the emergence of cyber violence, which has the widest range of victims and the largest number of victims. Cyber violence has a negative impact on society and youth psychology. Therefore, this paper examines the psychology of the perpetrators of cyber violence and provides corresponding solutions.

This study provides a reference for the purification of the Internet environment and the better use of the Internet.

2 Psychological analysis of perpetrators

Due to the wide coverage of the Internet and the large number of people involved, people who think differently from the general public are often encountered on the Internet.

They may be real "patients" who have been traumatized; they may be people who have been abused on the Internet and want to find an outlet through Internet violence; or they may be teenagers who are not yet involved in the world and lack proper guidance.

In short, the abusers vent their emotions through the Internet and want to express some things they cannot say in real life through the Internet in a radical way through the anonymity of the Internet. Generally speaking, abusers can be divided into "subjective abusers" and "objective abusers".

2.1 Subjective violence

The so-called "subjective" is to vent their negative emotions or to gain attention through online violence. Subjective violence is the progression of objective violence without proper guidance. The main causes of subjective violence are psychological trauma, stress reactions, and mental illness.

Victims are unable to face these behaviors head-on, so they will only do so by extreme means - harming others to achieve their own psychological balance or to relieve psychological stress. The victim does not have a good way to vent their emotions, and the perpetrator continues to get away with it, so this strong gap leads to the victim becoming the perpetrator.

Abusers with mental illnesses are the source of this behavior; because of their different ways of thinking from

chenzhenmingnb@gmail.com

the general public, they will express their opinions on the Internet, which contain their own personal emotions and often contain extreme ideas and sensitive words, causing irreversible harm to some ordinary people.

2.2 Objective Violence

Objective violence is consistent with a psychological activity - "herd mentality." Herd mentality is when an individual receives the influence of the behavior of the outside crowd and behaves in his or her own judgment in line with public opinion or the majority [3]. Such abusers do not have their own subjective awareness of violence and show an ignorant and blind attitude toward online violence. Such behavior is inseparably related to the environment of the Internet.

Some social news with a reversal is the best example. For example, Luo Xiang, a well-known professor, doctoral supervisor, and director of the Institute of Criminal Law at China University of Political Science and Law's School of Criminal Justice [4]. Because of his excellent language organization ability and humorous and funny speaking style, he attracted many viewers and got a very good response on the Internet. On March 9, 2020, he was invited to be in the beeper, with more than one million fans in two days and more than ten million fans in six months [5]. But soon, because of the excessive speculation and excessive reasoning of some network abusers, biting the bullet and pointing out that some of his published words are full of connotations, hints and other styles, the network stirred up opposition to Luo Xiang, and this trend slowly expanded over time, finally leading Luo Xiang to withdraw from the network. Because of the herd mentality, more and more netizens will believe others' words because of some one-sided words and join them as abusers.

3 Solutions to reduce cyber violence

3.1 Strengthen the construction of relevant laws

As a carrier of the rapid emergence of the new era, the speed of change and form change of the Internet are very rapid, and some related laws may lag behind. Therefore, it is necessary to keep pace with the times, respond to the rapid changes of the Internet and formulate or strengthen the relevant laws. Relevant legislative bodies and management departments should revise more targeted laws and regulations to keep up with the times as much as possible.

At the same time, increase the coverage of the law, and refine it to avoid legal loopholes or targeted deficiencies so that every citizen is aware of the importance of compliance with Internet norms. Relevant law enforcement agencies should also strengthen law enforcement efforts to achieve a timely crackdown on serious inappropriate speech, which can curb online violence by increasing penalties.

3.2 Strengthen the supervision of platforms

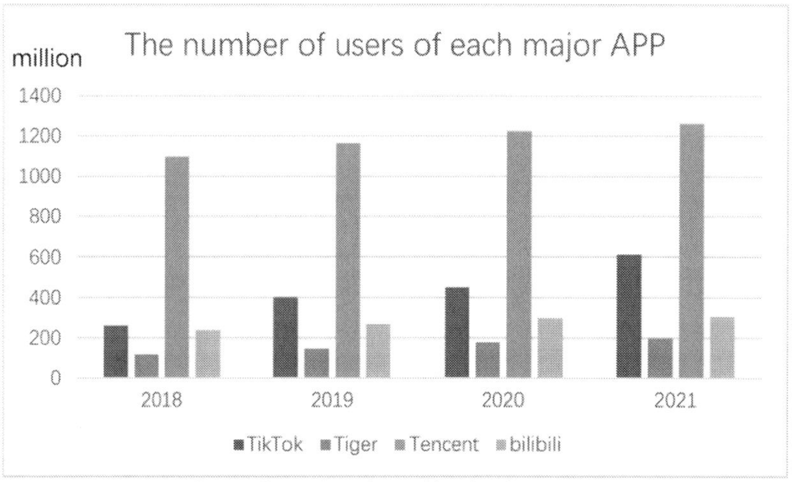

Fig. 1. The number of users if each major APP[7]

At the same time, relevant network software should strengthen the management of users, such as mandatory real-name registration to achieve the purpose of one person, one account, and provide personal information to law enforcement agencies to warn online abusers. Since February 2022, a short video app, ShakeYin, has implemented a strategy of displaying IP addresses for real user comments.

The aim is to reduce online violence, to reduce the "virtual nature" of the Internet by displaying IP addresses, and to reduce the freedom of speech and "anonymity" of Internet users. Once this policy was introduced, some extreme and negative comments were significantly reduced, and this further highlights the inextricable relationship between strengthening personal information management and online violence [6].

In addition, APPs can screen users by means of tests, through which they can effectively reduce users who are not yet mature in their thinking and have little legal awareness. A well-known video creation APP - bilibili uses this method to improve the overall user quality. Compared with the same type of APP like Racer, Tiger, Tiktok and Tencent, Bilibili users' feedback indicates that the APP is more transparent with comments and contains a lower level of vulgar and violent words [7].

3.3 Strengthen the popularization of education

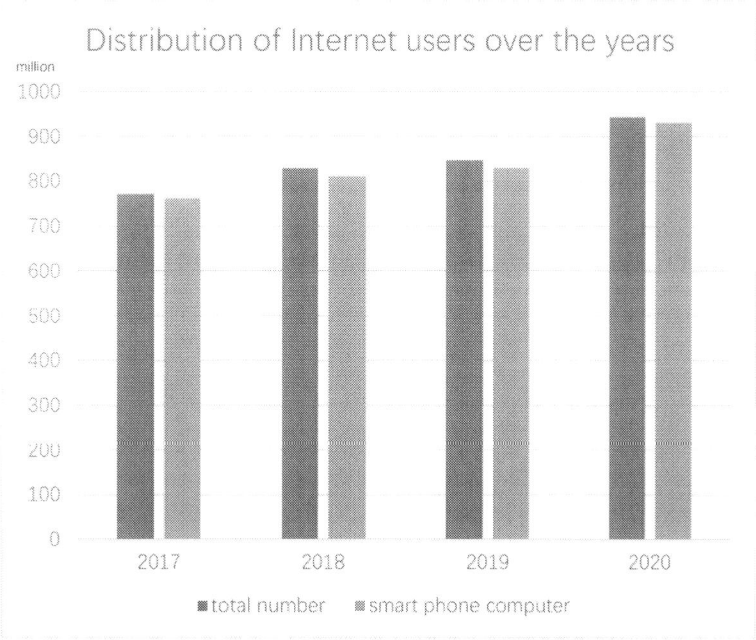

Fig. 2. Distribution of Internet users over the years[8]

As of June 2020, the number of China's Internet users reached 940 million, and the number of cell phone Internet users reached 932 million. On June 29, 2020, the China Internet Network Information Center (CNNIC) released the 46th Statistical Report on the Development Status of the Internet in China.

The report introduced in detail the "Scale and Structure Distribution of Internet Users in China". This survey report points out that 940 million Chinese Internet users have no undergraduate education, and over 40% of them have only junior high school education [8]. This data shows that the overall quality and awareness of Chinese Internet users are low, so there is a long way to go to strengthen the awareness of Internet users.

First, the government can add the curriculum about Internet-related knowledge into the nine-year compulsory education to help users establish good Internet awareness and fundamentally improve the quality of Chinese Internet users. At the same time, it can promote legal education on some online APPs and invite some famous bloggers or net celebrities to promote knowledge related to the Internet. For example, Mr. Luo Xiang is a very good example. His humorous and interesting lecture style and strong legal background attract a large number of viewers, so some social media can achieve the purpose of law popularization by strengthening the promotion of such video creators.

4 Conclusion

This paper has studied the psychology of online abusers and tried to provide corresponding solutions, and concluded that there are two kinds of online violence: "objective" and "subjective." Society should strengthen legal education, increase people's legal awareness, and improve relevant laws as the main theme to maintain the network environment. Although this process is very long, some ideas have been ingrained in people's psychology, it is difficult to improve. But with the increase of social living standards and the improvement of people's knowledge reserves, the eradication of the Internet environment is sure to be achieved.

References

1. Internet violence_Baidu encyclopedia. Baidu Encyclopedia. (n.d.). Retrieved September 5, 2022, from https://baike.baidu.com/item/%E7%BD%91%E7%BB%9C%E6%9A%B4%E5%8A%9B/4707152
2. Cyberviolence. Cyberviolence | Tech Without Violence. (n.d.). Retrieved September 5, 2022, from https://techwithoutviolence.ca/cyberviolence#:~:text=Cyberviolence%20is%20defined%20as%20online%20behaviours%20that%20criminally,happen%20online%2C%20through%20smartphones%2C%20during%20Internet%20games%2C%20etc.
3. Crowd psychology. Psychology Wiki. (n.d.). Retrieved September 5, 2022, from https://psychology.fandom.com/wiki/Crowd_psychology#:~:text=In%20social%20psychology%2C%20Crowd%20psychology%20is%20an%20aspect,established%20due%20process%2C%20they%20have%20also%20provoked%20controversy.
4. Wikimedia Foundation. (2022, July 14). Xiang Luo (jurist). Wikipedia. Retrieved September 5, 2022, from https://zh.wikipedia.org/zh-cn/%E7%BD%97%E7%BF%94_%28%E6%B3%95%E5%AD%A6%E5%AE%B6%29

5. Luo Xiang, the legal "duanzi player", was "cyber violence" for three months and withdrew from the network_Tencent News. (n.d.). Retrieved September 5, 2022, from https://new.qq.com/rain/a/20210616A00CCA
6. Advantages and disadvantages of intellectual property: Everything to know. UpCounsel. (n.d.). Retrieved September 5, 2022, from https://www.upcounsel.com/advantages-and-disadvantages-of-intellectual-property#:~:text=There%20are%20many%20advantages%20to%20IP%2C%20including%20the,financing%20for%20your%20business%206%20Greater%20export%20opportunities
7. Bilibili user group characteristics analysis, bilibili user crowd introduction_Tencent News. (n.d.). Retrieved September 5, 2022, from https://new.qq.com/rain/a/20220512A059FG00
8. Name, E. (n.d.). Official report: 940 million Chinese Internet users more than 90% did not go to undergraduate school, more than 40% of junior high school education. Retrieved September 5, 2022, from https://www.guancha.cn/politics/2020_09_29_566746.shtml

The Cognition Trend of Chinese Traditional Media on Feminism and the Underlying Reason for Existing Negative Reports

Yining Cao

Beijing No.4 High School International Campus, Beijing, 100003, China

Abstract. With the positive development of the media, various mainstream cultures and ideologies have also changed. It is difficult for the two to accommodate. "Feminism", as a progressive thought, spreads in the media space and has a great impact on the cognition of social equality between men and women. This paper studies and analyses about 200 articles from two traditional Chinese print media, exploring the cognition trend of Chinese traditional media on feminism and the deep reasons for existing negative reports. Conclusions can be drawn that the general attitude of traditional Chinese media to feminism is neutral and positive, while both the traditional dualism of patriarchal society and the reform of social economic structure have influenced the cognition and inclusiveness of traditional media to feminism.

1 Introduction

In the year of 2005, the Law of the People's Republic of China on the Guarantee of the Rights and Interests of Women passed the constitutional amendment. In the past, Chinese women have lived in the framework of Confucian ethics, which is essentially masculine, patriarchal, and hierarchical [1]. With the coming of the reform era, women opened the way to find their self-identity under the influence of the market economy. Feminism has become a prominent issue since the 21st century. "Gender-related perspectives have become increasingly diverse and have entered the public discourse space, but most media, especially mainstream media, still follow the norms of Chinese women's inferiority to men in love, marriage, and family". It is not difficult to find that there are still some traditional mainstream media and new social media gradually stigmatizing and misinterpreting "feminists", which can be regarded as a sign of the media's lack of inclusiveness. The stigmatization of feminist is reflected on the denial of feminist identity, the failure to advocate radical feminist movements, and the weakening of the idea of gender inequality. The term feminism has fallen into a false dilemma and is used as an offensive symbol in various contexts. It is judged to be non-inclusive, exclusionary of men, and even disturbing to social stability, and the meaning and group rationality behind it is being ignored.

This paper takes the texts and headlines from Beijing Youth Daily and Global Times in 2019, 2020, and 2021 as the main objects of investigation. By searching "feminism" keyword samples in WiseSearch Database, the author selects the relevant samples that directly illustrate the action or characteristics of feminists or express views about feminism. In conjunction with these findings, this paper investigates the inclusiveness of Chinese traditional media to feminism and its causes from a social perspective. The media's portrayal of feminists directly affects the readers' and popular culture's perception of feminists, so studying this topic is of great significance in the context of contemporary China.

AliceCaoYining@163.com

2 Literature Review

Feminism, a term that first appeared in France, refers to a social theory and political movement created and initiated to end sexism, sexual exploitation, discrimination, and oppression, and to promote equality among sexual classes. Feminism was introduced to China in 1919 during the May Fourth Movement. The pioneers of the May Fourth Movement used the issue of women's rights as a breakthrough to advocate individual rights and rebel against traditional culture. The issue of women's rights was then seen as the yardstick of modern civilization and a prerequisite for human liberation [2]. The Chinese women's liberation movement preceded the West, and the slogan "women can hold up half of the sky" at the founding of New China in 1949 "meant that what the men could do under the leadership of the Party, the women could also do." The Chinese Communist Party led the debate on the "new woman" with more ambition for personal progress and patriotism, turning women into part of the national workforce. In 1995, the Chinese government made it clear that "equality between men and women" is a basic state policy. The related laws and regulations are made to ensure women's participation in the implementation of state, social, and economic affairs, women's employment rights, equal income rights, and labor protection rights, as well as women farmers' rights to land, etc.

Women's legal equality with men is clearly guaranteed, but in reality, traditional patterns of gender relations are still deeply rooted. With modernization, women's empowerment has become a hot topic in the media, and countless feminist scholars have explored whether women can obtain equal rights in the media context. The emergence of media such as Weibo and Zhihu has provided women with the opportunity to enter the public sphere and gain a voice, and on these platforms, women are able to express their views on current events and politics in the same way as men. But in the process of interaction, feminism's struggle against patriarchy can hardly avoid the social stigmatization of feminism. According to the research of Wang Lan, women are able to express their views on current affairs and politics on the new media platform as men do, and are no longer confined to the trivial matters in daily life. Many women have become opinion leaders through the new media. But at the same time, the number of media reporting on feminism is very small, and some of them are negative. Zeng Changqiu states in his article that the development of the Internet does not necessarily promote the harmonious development of gender in society. The online media's focus on women's bodies constructs the image of women "seen" on the Internet, which is a derogatory and distorted image of women [3]. Not only in the new integrated media, but also in the traditional print media, women have gradually shed their passive status and become feminists who actively fight for their rights. However, the stigmatization of feminists in the media is like a misogynistic orgy. Yang Yuke conducts a textual analysis of more than 200 news media reports, exploring the process of feminism's stigmatization. It shows how feminism has been commercialized, misinterpreted, and taken out of context as radical egoism [4].

3 Methodology

In this paper, the content analysis was used to search articles from Beijing New Youth Daily and Global Times. Using the WiseSearch Database, the headlines and articles that talked about "feminism" from 2019 to 2021 were searched. By analyzing the content of the articles, their reporting plots, and the characteristics of the words used, the articles were categorized into three types, namely neutral, negative, and positive attitudes toward feminism. And the author specifically analyzed the nature of the negative texts to explore the story paradigms, stereotypes, or other underlying causes of the negative attitudes.

4 Result and Analysis

4.1 Beijing Youth Daily

There are 93 reports on feminism in Beijing Youth Daily from January 1, 2019 to December 31, 2021, and 65 of them meet the requirements after screening. As can be seen from Figure 1, Beijing Youth Daily's attitude toward women's rights is neutral and positive, with less and less bias from 2019 to 2021. Among the reports in 2021, negative feminist reports are no longer present. They are replaced by praise and promotion of the works of feminists that continue to emerge. For example, in the article "Ueno Chizuko - A Name That Cannot Be Ignored," the author praises Ueno Chizuko's feminist ideas. The author writes, "Chizuko Ueno's words are worth remembering: feminism is never the idea of the weak trying to become strong; feminism is the idea of seeking respect for the weak as well."

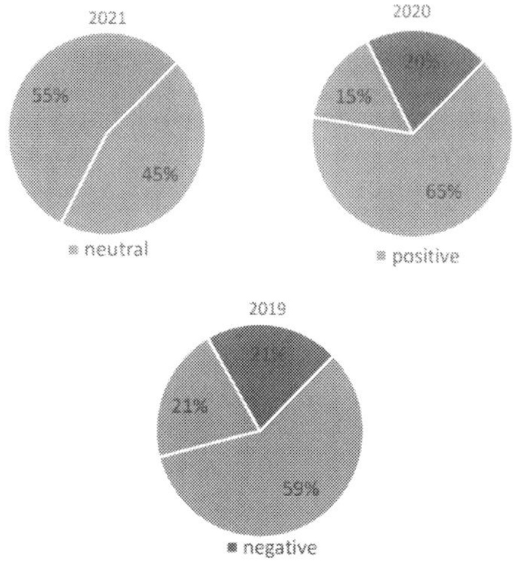

Fig. 1. Beijing Youth Daily's attitude toward feminism (original).

4.2 Global Times

The Global Times had a total of 105 reports in which feminism appeared in the full text between January 1, 2019 and December 31, 2021, and 72 of them were screened for compliance. According to Figure 2, the Global Times is generally neutral in its portrayal of feminism, and is less righteous and aggressive in its promotion of feminism than the Beijing Youth Daily.

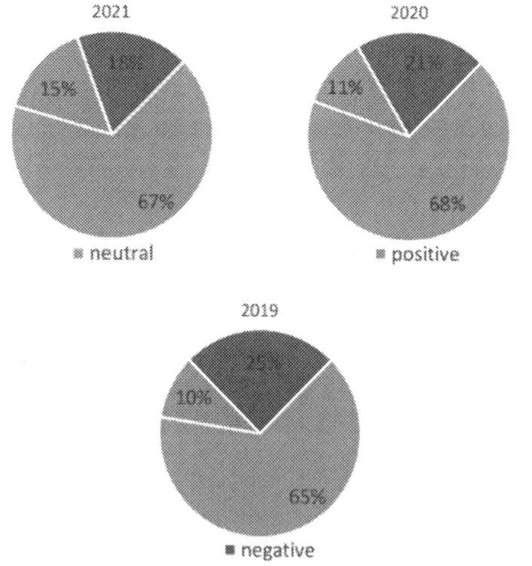

Fig. 2. Global Times's attitude toward feminism (original).

5 Discussion on the Underlying Reasons for Negative Reports

Among the articles in these three years, there are direct depictions of feminists or feminist events, as well as indirect depictions of feminists in film and television and political activities. Basically, there are two types of negative attitudes toward feminism. One is the misinterpretation of feminism and the other is the denial of one's feminist identity.

5.1 Misinterpretation of feminism under the influence of patriarchal society

Most of the misinterpretations of feminism are false critiques of feminists from the nature of gender binary and masculine centrism. On April 26, 2019, in an article called "Woman, When Your Name is Power," the author points out that the historical pseudo-propositions of feminists are "if I were a man" and "unfortunately, I am not a man". In a nutshell, men are regarded as a standard and reference point. The essence of this statement is that feminists want to be men, which is not possible. And this is not at all the intention of feminists, but the plight of women tossing and turning and rebelling under the patriarchal culture and language. Mulan is a hero on the battlefield when she disguises herself as a man. In other words, women are only equal beings when they possess androgyny. This is the stereotype under the gender dichotomy and the emphasis on masculine centrism [5].

In the article "what are they against when they are accusing Papi?" the author discusses the feminist position with the online event of the influencer Papi's right to a surname. There are three reasons for the "surname" debate: the convention of the patriarchal family system, the continued market for patriarchal inheritance and male superiority, and the advancement of women. To solve this problem, the most urgent task is to dilute the concept of "surname". Surname and first name, in the end, is a symbol used to distinguish different people, it does not have much special meaning. Nowadays, people are fighting over it, but it is just a kind of male superiority and female inferiority ideology and the concept of succession that is at play [9]. The author of the article, however, attributes this argument to feminist fallacies, portraying feminists as elite, arrogant, and inconsiderate of the lower classes.

In fact, Professor Chizuko Ueno once expressed, "feminism is never the idea of the weak trying to become strong; feminism is the idea of seeking respect for the weak as well." Women have long been an oppressed group, and patriarchy and misogyny are widespread phenomena in society. However, in fact, it is the feminism preached by society that has somewhat abandoned the average women. The essence of slogans like "women can hold up half of the sky" and "whatever a man can do, a woman can do it too" is to encourage Chinese women to participate in productive work and contribute economically to the country. In fact, however, they still follow the patriarchal family code in the context of feminism propagated by the state. Women in the new era are walking into a historical paradox. Contemporary Chinese women, while they are allowed to share in the power of discourse, have lost their gender identity of discourse in the culture. Society does provide support for women's economic independence, yet women must still act in accordance with traditional social norms and cultural values [6].

And the pro-patriarchal, elite population that the author describes as excluding ordinary women is the very picture of the refined egoist. At the same time, the Internet is rife with a phenomenon where people who are essentially individualists are regarded as feminists. They have their own names, such as "microblogging women" and "garden feminism", but the common feature of these names is that they bind egoists, who have no connection with feminist claims, to the category of feminists. The term "Chinese Feminism in the Field" should be literally interpreted as feminism of Chinese origin or feminism with Chinese characteristics. Other views point out that the word garden means romanticized, and feminism of Chinese origin is a variant of feminism that has been romanticized or alienated in China, meaning that feminism with Chinese characteristics is not feminism in the true sense of the word. Therefore, in the online environment, it is not appropriate to simply label a group as stigmatized without analyzing its origin and the mechanism behind it [7]. From these points, it seems that the "feminism" that is often rejected in the media corresponds to romanticized feminism. The fact that some self-interested discourse unrelated to feminism is tied to feminism reflects the misuse of the term feminism and the erasure of the historical image of feminists. When egoism is pretended to be feminism, the real idea of feminism will be erased, and what is weakened is their suffering and struggle.

5.2 The denial of feminist identity in the process of economic transition

In the article "The Huge Risk Behind the High Divorce Rate" reported on March 30, 2019, the author indirectly points to the rise of feminism as the cause of imbalanced gender relations and increased divorce rates: "This has led to, on the one side, the rise of feminism, the high status of female and low status of male, and the imbalance of yin and yang, thus leading to gender inequality and easily causing an imbalance of gender relations in marriage. On the other side, some unmarried women are more likely to be tempted by material things, such as houses and cars. They pursue materialistic marriages and this has caused a lot of pressure on the male partner, leaving a poor foundation for their marriage and serving as the root cause of conflicts after marriage."

With the economic and social development, the education and employment of modern women have been improved in general. The improving economic status of women in the family has rapidly increased their sense of independence. Such changes have made a great number of married women, on the one hand, feel more confident in self-sufficiency, and on the other hand, more difficult to tolerate an unsatisfactory marriage [8]. This is one of the many reasons for the increase of the divorce rate. What this phenomenon has led to is an open mind where marriage is just one of the life choices for women [9]. The article not only implies that the rise of "feminism" has led to a new kind of gender inequality, but also objectifies unmarried women and describes them as "for sale" and materialistic. In the face of the rapid change in the class status of women and men in the process of economic transition, the anxiety and pressure accumulated by the division between the lower class and the powerful class are finally released in the new media [10]. However, the "gender equality" between men and women is only valid in a society where all people are married.

6 Conclusion

In conclusion, the data showed that the attitude of Chinese traditional media toward feminism is, to a large extent, positive and neutral. The misunderstanding and stigmatization of feminism can be attributed to the paradoxical identity of women in the process of modernization and the dregs of patriarchal society. The dual demands of patriarchal society on women have become more pronounced after the reform and opening up. To the present day, it is evolved into the objectification of women and the radicalization of the stigma of feminism. Women have both the status and opportunities to advance, but still find it difficult to escape the demands placed on them by a patriarchal society. Additionally, the rising status of women and their rising control over marriage and economy have also caused rising anxiety among a certain proportion of men. When feminism is demonized and restrained by institutions with certain authorities, gender equality in society is difficult to be achieved all over. The change of the media's cognition on feminism has a direct effect on the change of people's perceptions, so its inclusiveness and correctness are extremely important.

References

1. Han, X.: Exploring the Relationship between the Internet and Women's Empowerment in China (2021).
2. Wang, L.: Feminist Communication in the New Media Environment (2017).
3. Zeng, C.Q.: Gender and Communication—The "Seen" Female Image in the Network Media (2015).
4. Chen, Y.: Radical Feminist Label—How Feminism is Stigmatized in the Media Platform (2014).
5. Dai, J.H.: Invisible women: women and women's films in Contemporary Chinese films (1994).
6. Zhang, Y.: "Surname", how much confusion—Perspective on the struggle for children's surname right (1996).
7. Zhang, G.N.: A study on the meaning of Chinese rural feminist words—the stigmatization of women's rights in the network environment (2018).
8. Song, Z.Y.: A psychosocial analysis of the phenomenon of "rising divorce rate" (1998).
9. Ueno, C.: Onnagirai: Nippon no misojinī (2010).
10. Cao, J.: New media, new rhetoric and class relations in China in transition—Take "green tea bitch" as an example (2015).

https://doi.org/10.1051/shsconf/202315503006

The Analysis of Why Women Kill Season 2 Based on the Development of Feminism in China

Zhen Guo[1,*]

[1]North China Electric Power University, English Department, Baoding, Hebei, China

Abstract. The second season of Why Women Kill has been widely disseminated and discussed in China. By combining an in-depth viewing of Why Women Kill II and the current development of Chinese feminism, this paper adopts the textual close reading to analyze the connotation of beauty, the awakening of female self-awareness, and the abnormal development of Chinese feminism. The work hopes to promote the further development of Chinese feminism, raise the level of gender consciousness in the whole society and accumulate experience and provide reference for similar feminist drama series and related researches in China and abroad.

1 INTRODUCTION

Tracing the development of Chinese feminism with the change in economic and social systems and the continuous communication between Chinese and Western cultures, Chinese women have gradually improved their social status from absolute obedience to the patriarchy to self-awakening and daring to fight for their rights. However, women are still in a disadvantaged position, bearing family and social pressures and facing unceasing social problems concerning the violation of women's rights. Chinese feminism in the new era is facing great challenges.

Nearly 150,000 people rated Why Women Kill II on Douban and over two million people searched for it on Baidu, causing widespread discussion on the Internet. Why Women Kill II explores the meaning of beauty, what lies beneath the surface and what a woman is willing to do to find her destination. Compared with the first season of Why Women Kill, the second season has a more innovative design and deepens the reflection on feminism, hitting the pain of most viewers in China.

2 CLOSE READING AND RESEARCH OF THE TEXT

2.1. Interpretation of Beauty

Women on TV are often young, beautiful and slim as if old, plain or fat women are frumpy. But the three main female characters in Why Women Kill II are far from attractive. Rita Castillo is aging; Dee Fillcot is obese; the heroine Alma Fillcot is neither young nor beautiful. If beauty is power, in the battle of it, those who have beauty become the strong ones and those who don't have are naturally reduced to the disadvantaged.

However, who defines the standard of beauty? Beauvoir states in The Second Sex: "In a dressed woman, though nature present, it's captured and reshaped by the desires of men"[1] Beauty is what men want to see. That is why from commercials to films there are few women with mediocre bodies and common faces. The nature of male-oriented visual culture is to place men and women in a "see/be seen" mode. [2] Thus, females' beauty is defined and judged by males. Women in a patriarchal society are unable to reject this "male gaze" and therefore acquiesce to the standards of beauty they set.

Moreover, men view women as objects and deny women's self-actualization through the objectification of women, as if women's self-confidence comes from the male gaze.[3] By placing too much emphasis on the external and measuring women against indicators as objects, men ignore the spiritual interior of women as "human beings", as if the only value of women is in their external appearance not in their role as men who can think and contribute to the world through the power of their minds.

The patriarchal definition of beauty and the objectification of women lead to women's inexorable desire to be complimented on their appearance by men. Alma, decades after being irresponsibly hurt by her high school hunk, is overjoyed by his compliment on her appearance. While Catherine Castillo is suspicious of Scooter's (Dee's boyfriend, Rita's lover) "raging love", only because of the man's nonsense words of love, she puts aside the doubts, and gladly accepts the "unsolicited love". As the moth to the flame, knowing that the outcome is doomed, women still determine to shape themselves into the way men want to see them.

* Corresponding author: GZCH20020118@163.com

The male-oriented visual culture model subtly internalizes its own value judgments and aesthetic interests into the way women evaluate themselves, i.e., men see, women are seen, and women see themselves seen by men. In this way, the beauty that meets the standard of men's appreciation becomes the standard of women's beauty in society and constantly transforms women's aesthetics and their perception of themselves. [1] The yardstick set and compared by the opposite sex, rather than women who no doubt know women better, creates a gargantuan and unrelenting anxiety, low self-esteem and psychological burden for women, especially those who do not meet the so-called "beauty standard". Women unconsciously compare themselves to it, believing that if they do not meet it, they are defective, and then women began to constantly self-doubt and denial.

Ultimately, women are misled by the patriarchal "beauty standards" and use them to judge and hurt each other. " Even though Rita knows it is offensive to call a girl "fat" and has also been frustrated after being called "old" by her husband, she insists on using "aesthetic standards" as a sharp sword to stab Dee and Catherine, who were also women. Women do not unite against men as blacks do against whites and the proletariat against the bourgeoisie. Instead, they are working with men and hurting other women.

Today, white, young and thin images are all the rage. From social media to clothing shops, women are constantly being taught that you can only be considered beautiful if you are as white as snow, thin as a bone and look younger than your age. Thus, appearance anxiety has been planted in every woman's heart. Bloggers who are 170 and weigh 110 describe themselves as overweight. More and more people pursue the square shoulder or Scapula Gyrus Syndrome in medical term, which causes breathing difficulties, get extremely long nails regardless of the damage to nails and even health, let alone the restrictions on life, and claim to make peace with the makeup-free while the first thing to do after taking a selfie is to open Photo Apps and beautify according to the standard. While more and more people on the Internet are dieting to obtain the “A4 waist”, "glasses waist" appears...... Customs and fashions are often dedicated to isolating a woman's body from her transcendence[3], in order to achieve the "beauty" recognized by the patriarchal society, women hurt their bodies cruelly, which is similar to the "foot-binding" in the Qing Dynasty and the girdling in the Middle Ages, which were abandoned by history. The corrupt customs we think we have got rid of just change different forms. Defining women by their appearance alone anytime is superficial, ridiculous, and sensational.

"Undefined girls" is an Internet buzzword, an outlet for women’s appearance anxiety. However, contrary to the original intention, many companies have found new commercial opportunities. Whoever hires a plump girl without make-up first to endorse is likely to see a year-on-year increase in revenue. The true meaning of "undefined" is regardless of height and weight, beauty and ugly, and not only girls but also women should not be defined by the external. There is no absolute standard of beauty. Each woman has her own shining point, and is worth loving and being loved. Inner beauty is taught from childhood, but nowadays, external appearance, rather than talent, has become the most important criterion for judging women revealing the holes in current aesthetic education, which teach people to understand, appreciate and create beauty. On the other hand, this in no way means that beauty is the most important and the only. Society should not be biased and overemphasize superficial beauty. Cultivating women's confidence, independence, bravery and perseverance is equally important and worthy of recognition.

2.2. The Awakening of Female Self-awareness

The underlying reasons and path behind Alma's progression from inconsequential housewife to infamous club president with bloodstained hands are worth pondering. The constant negative reviews since the airing of Why Women Kill II may be based on a stereotypical logic about women: if a woman commits a crime, she must be forced to do so; if a woman kills someone, it is highly likely that the person killed was her long-time abusive husband. Most women portrayed in literature have been soft and weak for generations, even in the first season of Why Women Kill. But who says that women must be the victims, not the perpetrators? Alma is a female serial killer, symbolizing a declaration of women's autonomy to control their own destinies and the awakening of female consciousness. Kill is the biggest thrill of Why Women Kill. Women kill for themselves rather than for men, which is a transformation and rebirth from the consciousness level, showing the progressive consciousness of women who pursue self-liberation and realize their inner desires. The virtual realization of the protagonist's control over her destiny, reversal and outburst of desire inspires the real viewers and increases their satisfaction. Viewers take on the role to make up for the regret and dissatisfaction of real life and complete the ideal. [4] Feminist theory has a basic premise: women are in an oppressed and discriminated class worldwide, not only because they belong to a certain stratum, but also because of being a woman.[5] It is the desire of almost every woman to escape the oppressed and submissive status and become the real director of her own life.

The status of Chinese women is steadily rising, the proportion of women with higher education has surpassed that of men, and women have the right to participate in almost every field. However, compared to men, ordinary women without background and beauty still have fewer opportunities for work and promotion, because the concept of "woman is the womb" is engraved in the DNA of the male-dominated society and women are automatically classified by employers and society into families before making their own choices. The essence of Alma's journey to the top is the struggle of ordinary women to pursue their dreams and fulfill their desires. The shared "second sex" status of women makes the Chinese female viewers of Why Women Kill II able to put themselves in Alma's shoes even though they have political and cultural ideological differences with the U.S., experience the control over their own destiny and succeed, which are unattainable in reality. Women hope to get pleasure and the inspiration that they can control their own future from Why Women Kill II.

Secondly, the female characters can wake up and leave men to exist as independent individuals. Alma goes from a life of husband and daughter, with no friends and no social circle, to successfully making it into the fancy club; from initially having to ask her husband's opinion on buying a dress to finally coercing him to kill; from only being the veterinarian's wife to being a notorious murderess on the news. Rita succeeded in marrying up when she was young and beautiful, but ultimately she is willing to leave Carlo's millions to run away with true love she thinks. Dee is able to keep awake when Scooter blandishes and finally offers to break up with this playboy.

The sixth national census shows that 2.47% of women over the age of 30 are single, about two times higher than the results of the fifth; according to 2019 civil affairs statistics, China's divorce rate reached 3.36‰. Getting married and having children seems no longer to be the only choice for Chinese women. The pursuit of true love, economic independence and ideological independence are three main reasons. Firstly, more women are unwilling to improvise, putting quality before quantity. In order to meet true love and if there is no so-called true love, they prefer to be alone. The most important thing to achieve this is the economic base. There is a famous line in Little Women from Aunt March, women must work hard. If they don't, they will get married. I don't have to because I am wealthy. As women can earn a high income on their own, they do not need to be confined to the family and rely on husbands and sons. More and more women choose to invest in themselves, making themselves more powerful. Men are no longer the sole source of income at home and women can also be the backbone of the family. At this time, the average man and their wealth are nothing to the successful women. Last, Chinese women are generally educated and are reluctant to settle down in the family as unpaid nannies. With more choices and broader views, women become economically and spiritually independent and possess the ability to judge and choose their own life goals independently, rather than being pressured by family and society. Marriage for women, especially for highly educated women, may mean sacrificing more time and energy, career stagnation or dual pressure on family and career, which reduce their inner happiness. [6]

Throughout Why Women Kill II, the theme is different from the first season of marriage and gender. The second season focuses on women themselves, creating the Elysian Park Garden Club, a women's scene which highlights the characteristics of the female group. The content is more free from males, and male characters no longer serve as the driving force for the development of the story, nor do they constitute the motivation for all the actions of the women, which accord with the theme of women's awakening.

2.3. The Lopsided Development of Chinese Feminism - The Generalization of the Circle of "Socialite

Similar to Gu Jia's desperation to join the circle of wealthy housewives in the Chinese drama "Nothing but Thirty", Alma is desperate to join the Elysian Park Garden Club. Ironically, what the club does is merely a group of bored and pretentious rich wives spend their afternoons drinking and gossiping. They are only with people in their own class, and a poor, plain woman like Alma would never fit in. But from the moment Alma decides to join the club and begins to erase her original look, she craves more, becomes insatiable and the image she creates is unveracious and contrary to her original intention. In order to possess a fancy and fashionable dress as the only way to be remembered and to enter the celebrity circle, Alma, who receives money from her husband, enters a high-end clothing store for the first time and is shocked by the sky-high price, so she memorizes the details of the dress and buys cloth for making the dress, which is also known as fake. In order to have face, Alma steals antiques and paintings from her late neighbor Ms. Yost's house to decorate her home and wears Ms. Yost's high-end clothes, becoming The Man in Case.

The term debutante is originally used to describe those elegant women in authority who often attend various fashionable social gatherings. However, the current definition of celebrity seems to infinitely enlarge appearance and wealth but discard morality and education.[7]

Nowadays, there are numerous indistinguishable celebrities online, Hermes in hand, Rolex on wrist and Bentley behind the wheel. The number of people who can have luxury goods (forgeries perhaps) in photos that are endowed with a social value is increasing with the help of group buying or sharing. Besides beauty, high level and consumption of clothing, food, housing and transportation and superior social circle have become the touchstone of successful women. To gain access to powerful resources and social contacts, some women squeeze their wallets or even violate the bottom line of morality and challenge the authority of law, which is ultimately vanity. The generalization of the celebrity circle, a phenomenon of group disguise that widely exists in social media, reflects the identity anxiety, psychological imbalance and identity-seeking demands of some people [8]. It is society that writes a blank check for this vanity. People will be amazed, praise and admire your social status because they default it is a leap of status accomplished by your own efforts and real talent and they default rich people are naturally superior, which is actually a distortion of values.

When more and more people find that they can realize the leap by creating a false persona, the public's trust is depleted, and pretentious posers are incessantly suspected. After years of hiding behind the mask, it is even difficult for oneself to distinguish. If not rely on one's own efforts, daydreams will be cracked and one ends up being a joke and gossip.

3 CONCLUSION

This paper innovatively takes the development of Chinese feminism as a starting point and adopts the textual close reading approach to explore the profound reasons why the second season of Why Woman Kill has gained a large Chinese audience and heated discussions which can be concluded as the connotation of beauty, the awakening of women's self-awareness and the abnormal development of Chinese feminism.

Through research and analysis, this paper finds that there are still many people judging women only by their appearance and many people pretending to be socialites, smearing women and feminism in the current society. The development of Chinese feminism is at its highest level in history while facing unprecedented but unavoidable problems. Why Women Kill II catches these pain points and reaps a large number of fans, which not only shows that feminist TV series in China and abroad need to pay great attention to social reality, reflect social problems and provide possible solutions, but also shows that the benign development of Chinese feminism urgently needs social attention and joint forces. This work hopes to accumulate experience for similar feminist-themed TV dramas in China and abroad, provide a reference for future studies of Chinese feminism or literary works using Chinese feminist perspectives, and jointly promote the further development of Chinese feminism and raise the level of gender awareness in society.

References

1. Simone de, B. (2011) The Second Sex. Shanghai Translation Publishing House, Shanghai.
2. Jiang H. Mass Media and Gender. [J]. Journalism & Communication.2000, (03): 15-22+93-94.
3. Wang K. Advertising: More than "attractive"? -- The deep cultural connotation behind the female image in commercial advertisements. [J] Guangdong Sea Breeze (2000), (03): 52-53.
4. Chen RX. Progress or Confinement? -- The "Thrill Point" construction of the American TV series "Why Woman Kill" and the reconsideration of female expression. [J].Audio-visual ,(2021),(02): 36-38.
5. Song X. A Feminist interpretation of Kate Chopin's Story of an Hour. [J].Journal of Shenyang Agricultural University (Social Science Edition),(2010),(01):110-112.
6. Zhao W, Liu XY. The "Happy Singleness" of Chinese highly educated women: An empirical study on singleness and subjective well-being.[J]. Research of Finance and Economics, (2019),(04):42-53.
7. Zhang M. Reflections on the changing meaning of "famous yuan". [J].Chinese Character Culture , (2021),(12):124-125.
8. Wang WC. Media ritual and Symbol consumption: User self-presentation in social media -- A case study of "SH Yuan" event. [J].Southeast Communication . (2021),(07):109-112.

Development of Internet of Things in the field of smart home security

Guoyi Shi

International College, Chongqing University of Posts and Telecommunications, Chongqing, 400065, China

Abstract. With the rapid development of the Internet of Things and the constantly breakthrough of sensing technology, the Internet of Things has been further applied to the field of smart home security. The progress of the Internet of Things in the field of smart home security not only improves the efficiency of civil home security, but also improves people's life experience. This paper first introduces the security range of the Internet of Things technology applied in the field of smart home security, and then introduces Zigbee, a typical communication network of the Internet of Things. Then it introduces the sensors involved in environmental monitoring and intelligent door lock in the Internet of Things. In the end, it summarizes the content of this paper.

1 Introduction

Intelligent security on the basis of traditional security, the integration of Internet of Things、 cloud computing、 artificial intelligence、 big data and other new generation of information technology, improve the image perception、 data processing、 information analysis、 correlative fusion capabilities, to achieve pre-prevention、 in-process detection and post-processing[1].The application of Internet of Things technology in the field of smart home security has many advantages and a wide range of coverage, which can greatly protect people's lives and property.

2 The range of smart home security based on Internet of things technology

2.1 Environmental Monitoring

Environmental monitoring mainly provides real-time environmental monitoring and alarm processing for home environment. Data of gas、 flame、 smoke、 temperature, humidity and so on, which is related to living safety and comfort are collected through sensors. Dates are collected and are transmitted to the remote control system in real time. Then the remote control system analyzes the data and compares it with the safety data values. If abnormal data is found, the alarm will process.

2.2 Smart home door lock

With the continuous development of Internet of Things technology, the traditional key unlocking has been unable to satisfy people's requirements for convenience and security. There are much unlocking ways based on intelligent locks now, such as fingerprint recognition、 face recognition、 wireless remote lock and unlock password, people can according to their own needs to choose a right way of unlocking, or combine several unlocking ways. It can solve the problems of traditional key security and is more conducive to preserve people's life and property security.

3 Zigbee Communication Network

3.1 Introduction to Zigbee wireless network technology

Zigbee is an open、 low cost、 low power、 low delay、 large network capacity、 high reliability、 high safety factor of the new short-distance wireless communication technology. Zigbee generally adopts the combination of IEEE802 transceiver and Zigbee protocol station, it form a wireless data transmission network platform. Within the range of the entire network, each Zigbee network data transmission module can communicate with each other. These Zigbee nodes need very little energy to work. Data transmission between nodes is carried out through radio waves in the way of relay, so Zigbee communication efficiency is very high. Zigbee is a good communication network solution for smart home security technology and the technical characteristics of Zigbee will be analyzed below.

3.2 Zigbee Layered protocol

Zigbee protocol from bottom to top are the physical layer (PHY)、 media Access Control layer (MAC)、 network

Email:1244061494@qq.com

https://doi.org/10.1051/shsconf/202315503007

layer (NWK)、application layer (APL). The hardware layer is PHY and MAC defined by IEEE802.15.4, and the software layer is NWK and APL defined by Zigbee Alliance. See Figure 1 below. In addition, it can be intuitively seen from Figure 1 that the Data transmitted by the user-defined APL can pass through two relatively independent but actually closely related wireless channels, namely Data Server and Manage Server [1].

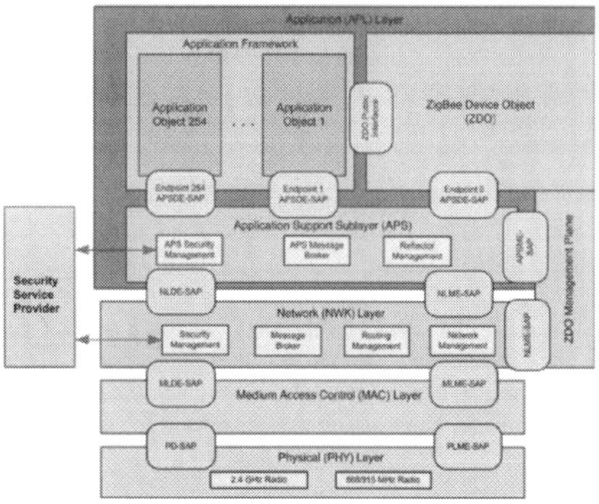

Fig. 1. Zigbee protocol architecture

3.2.1 Physical Layer (PHY).

The physical layer offers the ports between MAC and the wireless physical, the channel mainly provides management services and data services. For example: checking and evaluating whether the channel is free and the channel connection status, which directly determine whether the signal can be transmitted successfully. The CSM-CA algorithm can be used for idle channel evaluation (CCA), which simply means that the underlying hardware detects whether a certain carrier exists in the space, and whether its energy is higher than a defined value. There are three methods as follows: first, energy detection (ED) is carried out in the current channel. If the energy of a channel exceeds a certain value, the channel is considered occupied; Second, carrier detection (CS). CS is more intelligent than ED monitoring. ED monitoring can only monitor whether the channel can be used, while CS monitoring can judge whether the signal of the current channel is the same as its own IEEE802.15.4 signal according to the carrier characteristics, if not, the channel will be considered busy. Third, it is determined by the above "and" "or" "nor" logic of ED monitoring and CS detection. According to different countries and regions, three different frequency bands are provided, 868MHZ, 915MHZ and 2400MHZ, which provide 1 channel, 10 channel and 16 channel respectively. The center frequency of each channel can be calculated by Equation 1:[2]

$$\begin{cases} f_c = 868.3MHZ & k = 0 \\ fc = 906 + 2(k-1)\ MHZ & k = 1,2,\ \ldots,\ 10 \\ fc = 2405 + 5(k-11)MHZ & k = 11,12,\ldots,26 \end{cases} \quad (1)$$

3.2.2 Media Access Control Layer (MAC).

As can be seen from Figure 1 Zigbee protocol stack structure diagram, MAC is located between PHY and NWK and mainly provides data transmission service and management service. MAC Common part Molecular layer Service Access point (MCPS-SAP) is mainly responsible for receiving data from network layer、 transmitting data through peer entity keys and reporting data from other devices to the upper layer. The MAC Management Entity (MLME) provides MAC management functions such as channel scanning、 network association、 disassociation、 beacon synchronization and so on. It is also responsible for maintaining MAC management object database, namely the PAN Information Base (PIB).

3.2.3 Network Layer (NWK).

NWK is a protocol layer between APL and MAC, which is the most important part of Zigbee protocol stack. The main task is to provide functions to ensure the correct operation of IEEE802.15.4-2003 MAC, and to provide appropriate service interfaces for APL. The NWK consists of two service entities, namely Network Layer Data Entity (NLDE) and Network Layer Management Entity (NLME). The NLDE provides the following data transmission services: 1. After receiving the APL protocol data unit, the NLDE adds the NWK layer frame header to generate the NWK layer protocol data unit (NPDU).2. The NLDE can send data to the target device or the next-level device on the path to the final target. NLME provides management services: 1. Configure a new device.2. Start a new network.3. Connect or disconnect from a network;4. Search and record services such as valid transmission of information.

3.2.4 Application Layer (APL).

The APL is the top layer of the Zigbee protocol stack, It includes Application Support (APS)、Zigbee Device Objects (ZDO) and Application Framework (AF). APL is mainly developed by users according to specific applications to maintain functional attributes of nodes, discover the work of other nodes in the workspace of this node, and make multiple nodes communicate with each other according to service and demand. AF is the working environment of device application objects. About maximum of 240 application objects can be defined. APS provides an interface between the NWK and the APL, which contains a set of objects that can be agreed upon by ZDO and devices for use through the Service Access Point (SAP). The Zigbee device object ZDO sits between APS and AF and provides an interface for both. Customized by different manufacturers according to different needs. AF interacts with ZDO through ZDO public interface.

4 Hardware

4.1 Sensor

Sensor or Transducer is a monitoring device. It can sense and measure information. Sensor also can convert and output electrical signals or other required information forms. Sensor has miniaturization、digital、intelligent、multi-functional、systematic、network and other characteristics. Basing on perception function, sensor can be divided into: thermal sensitive element、 photosensitive element、gas sensitive element、force sensitive element、magnetic sensitive element、sound sensitive element and so on. Here are two typical smart home detectors.

4.2 Environmental Detecting Sensor

4.2.1 Gas sensor.

Gas is a general term for gas fuel, which plays an indispensable role in modern household life. It can release heat through combustion for the use of residents and enterprises. Common fuel gas is natural gas, gas, petroleum, biogas and so on. This paper mainly introduces the sensors for monitoring natural gas. The main component of natural gas is alkanes, of which methane accounts for the majority. Methane is a highly toxic gas, mild poisoning, patients will appear dizziness、 headache、 fatigue and other symptoms, severe poisoning, patients will feel chest tightness、breathing difficulties and even life-threatening. So gas leaking can be detected by monitoring the concentration of combustible gases such as methane, such as MC226A catalytic combustion gas sensor. MC226A catalytic combustion type gas sensor bases on the principle of catalytic combustion effects work, it consists of monitoring unit and compensating element matching bridge of one arm. When meeting combustible gas, its monitoring components resistance increases and the bridge circuit output voltage changes, the voltage variation as gas concentration increases in proportion to increase. The MC226A operates at a voltage of 2.5V±0.1, with a sensitivity of 20%LEL, methane: 15-40mV、20%LEL, ethane: 10-30mV, and a measurement range of 0-100%LEL. Beyond this value, the intelligent home control system will send alarm signals, so as to achieve the purpose of protecting family life and property safety.

4.2.2 Fire detection monitor.

In people's daily life, fire is a common major disaster. It often threatens people's safety and social development, which has attracted great attention. The main function of the fire detector is to monitor the home environment. If the fire is detected by the sensor, the alarm signal will be sent immediately. Common fire detectors include smoke sensors, flame sensors, temperature sensors, special gas sensors and so on. In the early stage of a fire, there will generally be a lot of smoke, and the temperature around the ignition point will rise sharply. The combination of smoke sensor and temperature sensor can effectively detect the fire, predict the size of the fire, alarm the fire, and provide information about the fire, and reduce the loss caused by the fire. MQ-2 smoke monitor and DS18B20 temperature detector will be introduced next[3].

MQ-2 smoke detection has wide range、 high sensitivity、 fast response、 high stability、 long life and other characteristics. Its working principle is built-in tin dioxide semiconductor photosensitive material, the material is Surface ionic N-type semiconductor. At 200~300 ℃, tin dioxide adsorbs oxygen from the air and form negative ion adsorption of oxygen, which reduces the electron density in the semiconductor, thus increasing its resistance value, so that the electrode cannot generate an electrical signal. So MQ-2 can judge whether a fire occurs.

DS18B20 is a common temperature sensor and output digital signal. It has the characteristics of small size、 low hardware overhead、 high precision、 strong anti-interference ability and so on. It is suitable for various kinds of narrow spaces. DS18B20 temperature detector operating voltage range of 3.0V-5.0V, measuring temperature range of -55℃ to +125℃, can be set according to the actual conditions of the temperature threshold. For example, assumed the temperature threshold is set to 40 ° C, if the sensor detects that the temperature exceeds this range, it will alarm.

4.3 Smart home door lock

4.3.1 Introduction of smart home door lock.

With the progress of science and technology and the improvement of people's living standards, people attach more importance to the safety of life and property. The traditional security door device has simple design principle and low safety factor, which has been hard to satisfy people's needs. Because of the continuous development of information technology, smart home door locks based on the Internet of things have entered people's daily life [4]. Smart home door lock mainly completes two functions: 1. Monitoring function;2. Unlock function.

4.3.2 Monitoring function of smart home door lock.

As it name implies, the monitoring function is monitoring the surroundings of the user's home. The user's family needs to input biological information, such as fingerprint、 voice and appearance into the database of the smart home system. When an outsider approaches, the surveillance camera will first capture the person's biological information and compare it in the database. If there is no information about the person in the database, the user will be alerted. Users can check the alarm information, decide whether the external person is a visitor and open the door.

4.3.3 Unlocking function of smart home door lock.

4.3.3.1 Unlocked by user Biometric Authentication.

The system collects the user's fingerprints, voice, facial features and other biometric data and stores them in a database. When entering the door, the biological data of visitors will be collected through cameras、 microphones、fingerprint collectors and other devices, then compared in the database. If the comparison result is determined as a family user, the door will be opened. Users can also set the type of authentication mode and choose one or multiple biological information [4] to authenticate.

4.3.3.2 RFID verification unlock.

RFID (Radio Frequency Identification) is a kind of automatic identification technology. It can fast exchange and store information through radio waves without contact. Connecting database through wireless communication and data access technology, it can realize non-connect two way communication. During identification, the electronic label can be read, written and communicated by electromagnetic wave.The access control system uses RFID reader to identify the electronic tag. If the electronic tag matches the set value, it is regarded as successful matching and the door lock is opened [5].

5 Conclusion

This paper first describes the structure of Zigbee protocol stack in the communication network of intelligent home security system. And then introduces how to realize the environment monitoring module and intelligent door lock module in the intelligent home security system. In the environment monitoring module, the sensor used in the gas monitor and fire detection monitor and its working principle are analyzed. In the intelligent home door lock, there are two ways to verify the lock, respectively biometric verification and electronic tag verification.

REFERENCE

1. Jinhui . (2014). Based on ZigBee smart home security system of information security research, a master's degree thesis, hangzhou university of electronic science and technology). http://202.202.43.73:8000/rwt/CNKI/https/NNYHGLUDN3WXTLUPMW4A/KCMS/detail/detail.aspx?dbname=CMFD201402&filename=1014228895.nh
2. yan. (2007). The analysis and design of the ZigBee protocol stack (a master's degree thesis, xiamen university). http://202.202.43.73:8000/rwt/CNKI/https/NNYHGLUDN3WXTLUPMW4A/KCMS/detail/detail.aspx?dbname=CMFD2008&filename=2008057843.nh
3. wen-hai zhang. (2015). Intelligent household security system design based on Internet of things (a master's degree thesis, xi 'an university of electronic science and technology). http://202.202.43.73:8000/rwt/CNKI/https/NNYHGLUDN3WXTLUPMW4A/KCMS/detail/detail.aspx?dbname=CMFD201601&filename=1016006505.nh
4. Yang Haichuan. (2013). Based on the intelligent household security system design and implementation of the Internet of things (a master's degree thesis, Shanghai jiaotong university). http://202.202.43.73:8000/rwt/CNKI/https/NNYHGLUDN3WXTLUPMW4A/KCMS/detail/detail.aspx?dbname=CMFD201302&filename=1013021061.nh
5. Qian Lang, Luo Xiaojuan, Song Lulu & Zhang Xinran. (2021). Internet of Things Technology (03),28-30. doi: 10.16667/j.issn.2095-1302.2021.03.007.

Consumers' Adopt Intention for Contactless Delivery during COVID-19: An Extended Perspective on The Theory of Planned Behavior

Haotian Wang[1,*], Mengxin Chen[2,*]

[1] Logistic Management of Navigation Institute, Jimei University
[2] Communication Engineering, Heilongjiang University of Science and Technology

ABSTRACT.COVID-19 is a severe disease and has now spread around the world as a highly infectious pandemic. COVID-19 has rapidly and deeply affected our daily life, and most countries have taken measures to prevent the spread of this disease, such as restricting entry and centralized isolation. Contactless delivery is a non-face-to-face delivery method which help ensure sufficient distance between the rider and the customers, thus further reduces the pandemic transmission risk. While relatively few people know and adopt this service. This study used the theory of planned behaviour to investigate people's intention to adopt contactless home delivery services during COVID-19. Research results provide both contributions to academics and practitioners.

1 Introduction

The outbreak of the COVID-19 virus caused serious problems that influenced nearly everyone's life. The Chinese government has carried out many controls over various regions, such as restricting entry and centralized isolation. under the severe epidemic situation, the distribution of food and daily necessities become a key issue. Many People are worried about the ways and the impacts of supplies delivery [1].

Traditional delivery method enables a face-to-face delivery way between the riders and consumers which has been adopted by many platforms and consumers. Given the severe pandemic situation, contactless delivery has become a promising delivery way to help reduce disease transmission risk. Contactless delivery means that when placing an order, users can negotiate with riders about the designated placement of goods, such as the company's front desk or at their home doorstep. Through "order notes," telephone calls, and app message systems, customers and riders can communicate with each other. After delivery, the rider will inform the user through a telephone call, an app message, or another channel to pick up the food or goods by himself. Contactless delivery can ensure sufficient distance is maintained between the rider/driver and the customer. With the development of technology, a variety of tools have been created to conduct contactless delivery, like the smart cabinet or an unmanned aerial vehicle (drone). Many platforms and stores have adopted contactless delivery technology, like Meituan and Ele.me in China and Instacart and Postmates in the U.S.

Distribution is now a very important part of people's lives. However, due to the arrival of COVID-19, people's attitudes have been greatly affected. People began to carefully consider the convenience and risk of distribution services. If people can feel the convenience of using contactless delivery, their attitude toward using contactless delivery will be more positive. If people attach great importance to COVID-19 and think it is very harmful, this fear will also have a positive impact on their attitude toward using contactless delivery.

This paper aims to examine people's attitude toward using contactless delivery during COVID-19. Specifically, how do individuals consider the risk and the convenience during contactless delivery? To answer this question, the extended theory of planned behavior (TPB) model was adopted as the theoretical foundation. TPB consists of attitude, subjective norms, and perceived behavioral control.

We collected online survey data by using Wenjuanxing, a popular online survey platform in China. In all, 574 valid responses were collected. Based on TPB theory, this paper explores the influence of the perceived risk and perceived convenience on consumers' attitude. Research results could provide both contributions to academics and practitioners.

In the following parts, we first review the relevant literature and then build a research model and hypotheses. We then introduce the research methodology for research hypotheses testing. Finally, we discuss the research results,

*Corresponding author.
Haotian Wang Email: 2857268839@qq.com;
Mengxin Chen Email: 328771751@qq.com

research implications, and limitations of the study.

2 Theoretical Foundation and Research Model

2.1 Theory of Planned Behavior

Ajzen (1991) developed the theory of planned behavior (TPB) to understand the psychological basis of volitional behavior. TPB includes three key antecedents of behavioral intention: attitude, subjective norms, and perceived behavioral control. It assumes that behavioral intent is a good proxy for actual behavior, as researchers have verified [2].

2.2 Attitude

Attitude refers to the numerous behavioral beliefs about the possible results of a certain behavior that individuals hold. Attitude is a factor that can easily affect people. This argument has been proved experimentally [3]. Attitude is the negative or positive psychological feeling of an individual toward a particular behavior.

H1: The attitude toward contactless delivery is positively associated with contactless delivery adopt intention.

2.3 Subjective Norms

Subjective norms are also a very important factor in influencing people's behavior. Subjective norms refer to the pressure perceived by individuals from others important to them, such as parents and friends to support or oppose a certain behavior. In most studies, subjective norms have the weakest influence on behavioral intention [4], but the influence of subjective norms on non-negative behavior is very significant.

H2: Subjective norms are positively associated with contactless delivery adoption intention.

2.4 Perceived Behavioral Control

Attitudes and subjective norms examine subjective factors, while perceived behavioral control (PBC) involves the evaluation of objective factors. Behavioral intention is affected not only by personal attitude and subjective norms but also by the difficulty of performing specific behaviors. The research of contactless delivery points out that PBC is an effective indicator to predict people's use of contactless delivery. If a person has the ability to choose the distribution method, they will be very willing to try this way.

H3: Perceived behavioral control is positively associated with contactless delivery adoption intention.

2.5 Perceived Convenience

Perceived convenience is the factor to measure people's perception toward convenience [5]. It is an important factor for testing people's satisfaction. This method was also used on our previous paper, investigating perceived convenience's impact on perceived value and adoption intention in predicting public bicycle adoption [6].

H4: Perceived convenience is positively associated with an attitude toward contactless delivery.

2.6 Perceived Risk

In the consumption context, perceived risk can be defined as a subjective perception regarding the uncertainty and negative consequences of the purchase of a product [7]. In the present study, the uncertainty was due to the perceived risk of the COVID-19 pandemic. In this paper, we measured people's awareness of the risk of COVID-19. Those who think that COVID-19 is a very serious concern tend to use contactless delivery.

H5: Perceived risk is positively associated with an attitude toward contactless delivery.

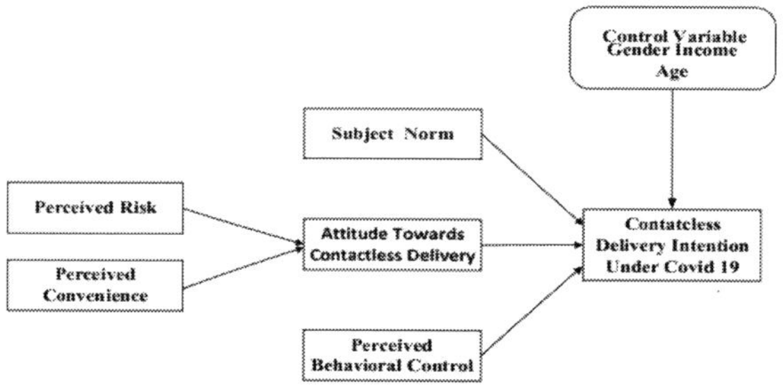

Fig. 1. Research Model (self- organized)

3 Research Methodology

We collected questionnaires through online channels via the widely used Wenjuanxing survey company. The data were collected from July 12 to July 16, 2022. A total of 607 questionnaires were collected. We deliberately conducted a check of the answers to ensure the authenticity and accuracy of the data, and finally 575 valid responses were recorded.

3.1 Questionnaire Development

The items of attitude, subjective norms, perceived behavioral control, perceived convenience, and perceived risk were measured on a 5-point, Likert-type scale from “strongly disagree” to “strongly agree.” The Likert scale is the most widely used and most effective tool for scaling responses in survey-type studies. Measures of attitude (three items), subject norms (three), perceived behavioral control (three), behavioral intention (three), perceived convenience (four), and perceived risk (four) were adapted from the existing literature. To measure attitude, people were asked for their view regarding contactless delivery, as being "good," "smart," "interesting," or not. For subjective norms, people were asked for the opinions of people near them on their use of contactless delivery. Perceived behavioral control focused on the ability of respondents to choose to use the contactless delivery by themselves or not. Behavioral intention inquired into the willingness of the respondents to use contactless delivery in the future. Perceived convenience measured respondents' perception of the ease of using contactless delivery and the perceived risk measuring their awareness of the risk of COVID-19.

3.2 Sample Demographics

The participants were selected from a nationwide sample to participate in the survey. The questionnaire also asked respondents for basic information about their gender, age, education level, and income. Of the 575 respondents, 45.4% were male and 54.6% were female. The median age of the participants was 27 years (M = 30.17, SD = 10.39) and the median monthly household income was 10000¥ to 50000¥. In terms of education level, 64.3% of the participants had received a bachelor's degree, 18.1% held a master's degree or a doctorate, and 17.6% had obtained a high school certificate or a junior school certificate. Among these qualifications, bachelor's degrees accounted for the highest proportion.

Table 1. Demographic Profile of Respondents (self- organized)

	Demographic Characteristic	Frequency	Percentage
Gender	Male	261	45.39%
	Female	314	54.61%
	0-20	75	13.04%
	21-30	282	49.04%
Age	31-40	123	21.39%
	41-50	63	10.95%
	above 60	32	5.58%
Education Level	high school	34	5.91 %
	junior college	67	11.65 %
	bachelor	370	64.35 %
	master	96	16.7 %
	doctor	8	1.39 %
Household Income	below 10000	137	23.83 %
	10000 to 50000	248	43.13 %
	50000 to 200000	123	21.39 %
	above 200000	67	11.65%

Before the regression analysis, we screened the missing data, conducted descriptive statistical analysis on each variable after deleting it from the total sample, and calculated the reliability of each variable (Table 2).

Table 2. Item Descriptive (self- organized)

	Min	Max	Mean	SD	Cronbach's α
Attitudes towards contactless delivery	1	5	4.338	0.743	0.883
Subjective norms	1	5	4.193	0.766	0.867
Perceived behavioral control	1	5	4.009	0.918	0.875
Behavioral intention	1	5	4.266	0.722	0.855
Perceived convenience	1	5	3.905	0.875	0.856
Perceived risk	1	5	3.989	0.878	0.861

4 Results

These data analyses were conducted using SPSS and AMOS. And in terms of the hypothesized relationships between perceived convenience and attitude, the results showed that perceived convenience was positively related to attitude (β=0.285, p<0.001), which supports H4. It also meant that contactless delivery is more convenient, so people will be more willing to adopt contactless delivery. In addition, the results showed that perceived risk was positively related to the attitude (β=0.293, p<0.001), which further supports H4. If a person considers COVID-19 as a serious problem, they will probably choose contactless delivery.

Regarding the TPB variables, attitude (β=0.249, p<0.001) was positively associated with contactless delivery adopt intention, which positively supports H1. It might be easier for those individuals who have a high level of attitude to adopt this behavior. Similarly, the coefficients of PBC are β=0.260 and p<0.001—similar to the results for attitude—and have a significant positive effect on the behavioral intention of using contactless delivery, with a greater impact than the attitude. Overall, subjective norms have a higher relationship to the behavioral intention. If the people who are critical of your behavior suggest you use contactless delivery during COVID-19, your personal willingness will be reduced. Finally, for the controlled variables, age, gender, and educational level all had a significant influence on the attitude and behavioral intention—β=0.061, β=0.048, β=0.09—respectively.

Table 3. Test Results (self- organized)

Hypothesis	Path	Coefficient	T value	Supported
H1	ATT➡BI	0.249***	6.363	Yes
H2	SN➡BI	0.292***	7.361	Yes
H3	PBC➡BI	0.260***	4.524	Yes
H4	PC➡ATT	0.285***	7.182	Yes
H5	PR➡ATT	0.293***	4.752	Yes

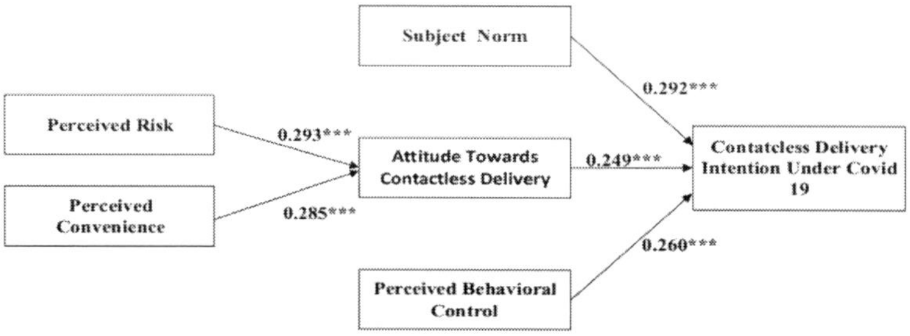

Fig. 2. Model Results (self- organized)

5 Conclusion

With the spread of COVID-19, contactless delivery is becoming ever more popular. In the contactless delivery process, residents can receive their purchased goods more conveniently and reduce the risk of contracting COVID-19. This research is based on TPB theory and enriches the literature on contactless distribution adoption.

Firstly, attitudes toward contactless delivery play an important role in people's contactless delivery adopt intention. Through our research, we found that attitude has a positive impact on contactless delivery intention under COVID-19, and that two of the antedecents—perceived risk and perceived convenience—have a positive impact on attitude consent, with perceived risk having a larger impact. No research study has explored this aspect before. Secondly, we also found that subjective norms and perceived behavioral control also have a positive effect on people's behavior. Therefore, it is recommended that policymakers should pay attention to perceived risk and perceived convenience to improve people's behavioral willingness and pay more attention to people's psychology and in practical applications to formulate better plans.

This study makes certain theoretical contributions and has practical significance. For the theoretical contributions, the research results provide good suggestions for the application of TPB and the exploration of people's adoption intention and the influence of two antecedents, perceived convenience and perceived risk, on attitude. For the practical contributions, this research has provided many suggestions for the non-basic distribution businesses and the government to formulate policies, analyzed the factors that easily affect people's attitudes, and showed how to make this technology more convenient and bring more help to people in trying times.

This study also has some limitations. Firstly, our samples are all from areas with very strict control of COVID-19. Subsequent papers can sample and analyze some areas with less strict control measures. Secondly, this paper also does not consider that in some remote areas, the transportation industry is relatively less developed than other areas, and the perceptions and attitudes of people there may be different from those who participated in this research.

REFERENCES

1. Wang, X., Wong, Y. D., Qi, G., & Yuen, K. F. (2021). Contactless channel for shopping and delivery in the context of social distancing in response to COVID-19 pandemic. Electronic Commerce Research and Applications, 48, 101075.
2. Ajzen, I., & Fishbein, M. (2005). The influence of attitudes on behavior. In D. Albarracin, B. T. Johnson, & M. P. Zanna (Eds.), The handbook of attitudes (pp. 173–221). Mahwah, NJ: Erlbaum
3. Ajzen, I. (1991). The theory of planned behavior. Organizational Behavior and Human Decision Processes, 50(2), 179–211.
4. Lim, H., & Dubinsky, A. J. (2005). The theory of planned behavior in e-commerce: Making a case for interdependencies between salient beliefs. Psychology and Marketing, 22(10), 833–855. https://doi.org/10.1002/mar.20086
5. Jih, W. J. K. (2007). Effects of Consumer-Perceived Convenience on Shopping Intention in Mobile Commerce. International Journal of E-Business Research, 3(4).

6. Hazen, B., Overstreet, R., & Wang, Y. (2015). Predicting Public Bicycle Adoption Using the Technology Acceptance Model. Sustainability, 7(11), 14558–14573.
7. Hussain, Safdar, et al. "EWOM Source Credibility, Perceived Risk and Food Product Customer's Information Adoption." *Computers in Human Behavior*, vol. 66, no.

Intercultural Communication Research about a Foreign Online Celebrity "I am Jerry Guo"

Xinyu Zhao*

Internet and New Media Department of CCNU, Wuhan, Hubei, China

Abstract: In the context of new media, "I am Jerry Guo", a foreign online celebrity, has successfully practiced intercultural communication between China and the United States as a folk individual. This case can provide experience for the effective development of intercultural communication in the wave of globalization. This study argues that the success of Jerry Guo lies in the combination of hard discourse and soft forms. Jerry Guo tells cultural connotations in a firm and objective discourse and presents cultural information in vivid and lively images by videos. This communication path effectively crosses cultural borders between countries, breaking down the prejudicial elements contained in cultural stereotypes, and promoting audiences to build a deeper knowledge of foreign cultures.

1. Introduction

Under the wave of globalization, more and more cultures are spreading around the world across nationalities. On this basis, social media has facilitated the development of intercultural communication with its advantages of openness, broad compatibility, interactivity, and retrieval convenience. It has promoted cultural communication and integration at both official and private levels. However, there are still many problems in intercultural communication, such as prejudice and misunderstanding between countries, and symbolization and fragmentation of national images. In this context, "foreign online celebrities" have achieved more effective intercultural communication by virtue of their individual power. "I am Jerry Guo" is one of the most representative foreign online celebrities. Jerry Guo is an American Jew who became interested in Chinese culture when he went on an exchange program in Taiwan, China, during his college years. As a result, he adopted the authentic Chinese name "Jerry Guo" as his online nickname. He began posting videos on major social media platforms around the world in 2017, telling stories of what he saw firsthand both in China and the US. After nearly five years of posting videos. Guo has gained many fans in China, and has arguably become one of the most well-known "foreign online celebrity" on Chinese online platforms. Therefore, this study explores the key points of Jerry Guo's intercultural communication practice, aiming to provide experience for more effective practice of intercultural communication in the new media era.

2. Literature review

2.1 Intercultural Communication in the New Media Era

As an important branch of communication, the development of intercultural communication has closely followed the changes of the times. The wave of globalization has promoted changes in social interaction and relations. New media platforms have increased in the new media technology environment. Cultural communication has become more frequent, cultural discussion has become more diverse. Invisible intercultural communication subjects have incubated simultaneously ([1]Ji, 2020).

In the current context of globalization, researches on the relationship between new media and intercultural communication focus on three aspects: the influence of national or ethnic culture on the development of new media; the impact of new media on cultural identity; the impact of new media (social media) on intercultural communication, such as intercultural relations, adaptation, and conflict ([2]Chen, 2012). The following section will highlight the current state of research on "the impact of social media on intercultural communication".

The emergence of social media has facilitated the formation of many virtual communities, and individuals' cultural identity presents dynamicity and multiciplity in different virtual communities, seriously challenging the stability of individuals' cultural Structures. In addition, virtual communities also show certain power tendencies, and Western culture breaks the cultural neutrality of

* Corresponding writer: 1759526913@qq.com

cyberspace with its implicit power advantage ([3]Chen, 2012). At the same time, social media gives people a unique way to display their culture, and based on the openness and flexibility of new media, people can form connections and freely form "virtual communities". However, in the "virtual community", there are both resonance of identities and the possibility of division. On the one hand, the Internet enhances the degree of cultural integration, and research has shown that online support communities formed by Chinese students, when fully utilized, can help them gain more emotional support and significantly reduce cultural pressures when fully utilized ([4]Ye, 2005). On the other hand, although intercultural communication has become more fluid and efficient, people do not have a deeper understanding of different cultures. Factors such as the language and geography of the communicator will significantly affect global cultural flows. In addition, the large number of people have involved in online communication. The frequent presence of noise in the process of conversation has led to the loss of meaning, divergence of opinions in some cases of cultural conflicts ([5]Xiao & Hu, 2021). But as Hall says, culture is, at its core, a communication process ([6]Edward, 1966). Social media enables more people to engage in cultural presentations and image performance. As intercultural communication promotes dialogue among more cultures in the midst of conflict, how to avoid cultural conflict and promote adaptation will become one of the key issues in intercultural communication research.

As regards the practice of intercultural communication, the current situation of overseas communication of mainstream media has triggered scholars' research on official foreign communication. Both the communication effect topics, such as image construction, and communication issues, such as cultural bias, have aroused the attention of scholars. Liu Yang and Xu Jia use Facebook and Twitter as the background to study the current situation of the operation of mainstream media in China. They point out that although the operation of mainstream media on overseas social networks is more normatively and highly concerned, it still needs to be improved in terms of indigenization ([7]Liu, Xu, 2013). Intercultural communication mainly reflected in the increased opportunities for individual voices and frequent individual interactions at the civil level. For example, the popularity of Li Ziqi overseas in 2019 is an excellent success of folk intercultural communication practice, which has been studied by many scholars. Cao Xiankun analyzes the reasons for Li's overseas explosion in terms of video content, audiovisual language, narrative techniques, and Chinese cultural elements ([8]Cao, 2021).

Some scholars have also studied the communication practices of foreigners in China from the perspectives of intercultural communication and semiotic theory. But the attention to the successful cases of folk intercultural communication practices is not enough. There is still room for further exploration and research.

2.2 Concept and researches of foreign online celebrities

The term "online celebrity" refers to a person who becomes popular online because of an event or behavior noticed by netizens, or because of long-term and continuous output of expertise. Wang Guohua, Gao Wei, and Li Huifang ([9]2018) borrow the definition of "online celebrity" and define "foreign online celebrities" as "foreigners who have become popular on Chinese online platforms because they have been noticed by the majority of domestic netizens".

In terms of international communication, Du believes that "foreign online celebrities" have the advantages of expanding communication subjects, enriching communication content, and innovating methods. Still, there are also problems such as homogenization and vulgarization of content ([10]Du, 2019). Xia and Pan ([11]2020) analyze the content of videos about China by blogger "Korea Dongdong" on Bilibili, and suggest that Chinese media should create more grounded content to bring them closer to the audience and improve the international communication ability of China ([12]Zheng, 2019).

Zheng and Jin analyzed the video content characteristics of five "foreign online celebrities", who came from different countries. Still, their videos had certain common features, including structure, theme and opinion tendency. Liang ([13]2020) screened out five head "foreign online celebrities" and analyzed three aspects of their videos: the theme type, content characteristics and narrative style. She pointed out that the value of "foreign online celebrities" should be full played in intercultural communication.

The current research on "foreign online celebrities" is not sufficient, and most of them are comprehensive or group studies, lacking detailed analysis of individual cases. This paper analyzes the videos created by Jerry Guo, starting from the specific case of him, using the textual analysis method. This method can not only enrich the theoretical content of "foreign online celebrities" research, but also further broaden the research horizon.

3. Research Methodology

This paper uses the case study method to analyze the intercultural communication practices of contemporary folk, taking video blogger "I am Jerry Guo" as an example. Since releasing videos in 2017, Jerry Guo and his team have produced 472 videos. From 2018 to 2020, Jerry Guo was selected as one of the top 100 Uploaders of the year for three consecutive years on Bilibili. On October 17, 2022, he had 7.055 million fans, 44.141 million likes, and 930 million total video views on Bilibili. Among the many bloggers practicing intercultural communication, Jerry Guo has become one of the most famous "foreign online celebrities" on Chinese and global online platforms, making this case representative and typical.

472 videos posted by Jerry Guo in the past 5 years were selected in this paper. 465 videos with distinctive content and complete structure were screened out as

research objects, excluding 4 videos in which Guo simply interacted with his fans and 3 videos in which he cooperated with CCTV News. By watching these videos, the category construction was completed based on the specific contents, with ten indicators in three dimensions. Then, based on the above data and semiotic theory. I analyzed the content characteristics of videos, exploring the connection between the video symbols and the cultural connotation behind them. Through analysis, I summarized the characteristics of videos in the process of dissemination and the reasons for the successful intercultural practice of the account "I am Jerry Guo".

4. Analysis of this research

4.1 Theme content

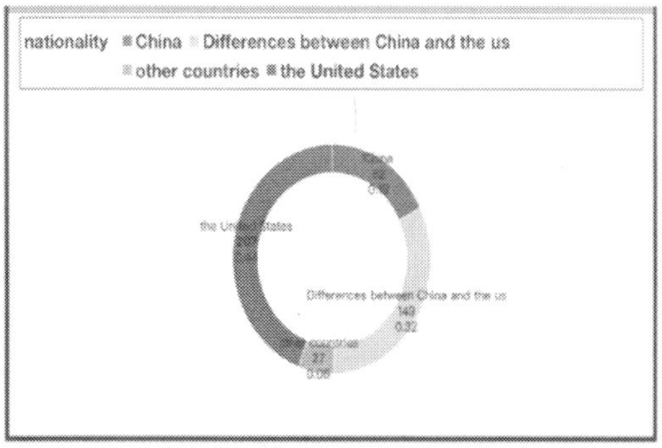

FIG.1 Nationality of selection of Jerry Guo's videos

In terms of nationalities, most of Jerry Guo's videos are about the differences between China and the US. As an American, he exported American native culture in a down-to-earth way, showing the world in the eyes of Americans. As a Chinese student, he explored the novel elements of Chinese culture as an explorer. But in any case, he always faces cultural differences and answers people's doubts. He conveyed the differences he felt through video, which greatly satisfied the public's demand for information from other countries and eliminated misunderstandings.

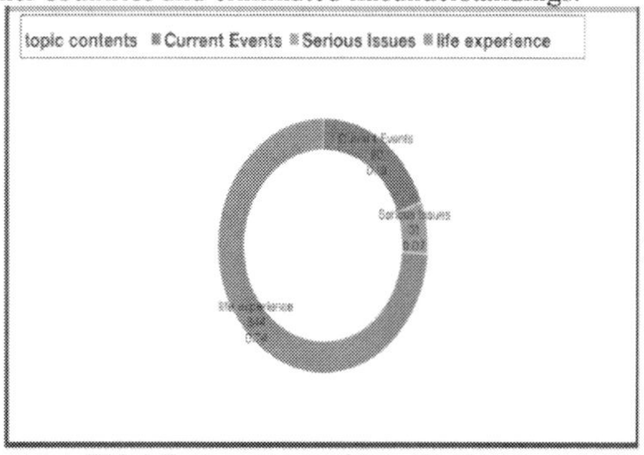

FIG.2 Topic contents of Jerry Guo's videos

In terms of contents, as an intercultural short video blogger, although his main content is about relaxing life experiences, he has increasingly preferred hot topics that are generally concerned by international audiences around the world, such as "U.S. Police Violent Law Enforcement", "Hong Kong Violent Assault on Police" and "Mumbai Wealth Gap". Jerry Guo closely follows the hot topics of current affairs, discussing serious and sensitive issues. As a neutral person, he tries to show and introduce these marginal and sensitive events, making up for the silence in the marginal discourse.

4.2 Values

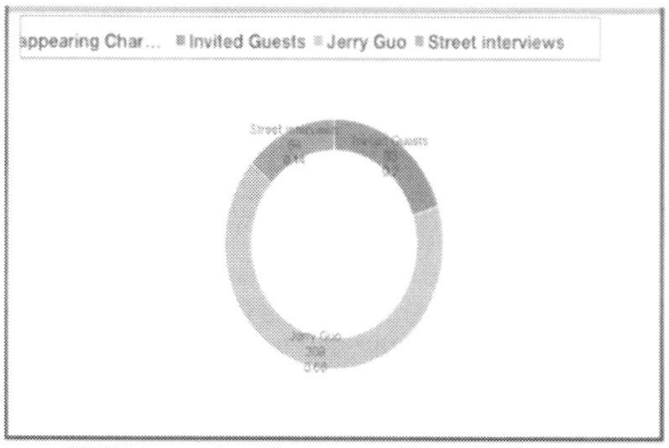

FIG.3 Appearing Characters of Jerry Guo's videos

First, most of Guo's videos adhere to an objective position, rarely pre-set positions and emphasize the principle of truth. Among the 465 videos, 89% of the videos are field visits, and each video is produced by himself. The primary editing method is a one-way, linear time mode, and insists on recording the experience process with a straightforward and clear narrative structure. Based on the above three points, his videos use the subject's "sense of presence" and coherent narrative logic to embody a documentary creative method. Besides, Guo's videos focus on street interviews, while inviting guests as auxiliaries. 20% of them answer questions and explain viewpoints. Taking the perspective of primary position culture as the starting point, the cultural holders speaking out in person can make the content more credible and appealing. 14% of them use the method of inviting guests to lead audiences from different cultural backgrounds to better enter the cultural environment.

Second, Guo's videos try to maintain the value orientation of equality, tolerance and openness. In the process of showing different cultures, he always insists on cultural dialogues and promotes consensus. While exploring Chinese culture, he went to Guizhou, Chongqing and other places to talk face-to-face with the locals and listen to his cultural views. He also cooperated with other Chinese bloggers, such as Jing Hanqing and Huanong Brothers, to promote the formation of the understanding of Chinese and American folk culture, trying to establish a cultural model of "solidarizing the other". Meanwhile, as a communicator, when Guo entered other cultures, he chose to abandon the inherent thinking brought about by his own cultural accumulation, actively considered the experience of the local culture, and expressed his views with a neutral attitude toward the other, showing an excellent cultural transference and empathy ability. For example, he was once challenged to survive a night in an Internet cafe in China, eating snacks and screaming while playing games like other netizens.

Based on the above analysis, I believe that the videos of Jerry Guo adopt a rigid discourse, taking a firm and objective stand to confront sensitive topics and marginal issues, outputting content with the values of equality and tolerance.

4.3 Narrative features

First, as a nonverbal symbol, Jerry Guo constructed the role of a "stranger". For Chinese audiences, he is a stranger from a different culture. For foreign audiences, he is a communicator of different cultures with ethnic closeness. This identity makes Jerry Guo "others" on two levels. The first-person perspective of "cultural other" can quickly shorten the distance between the cultural recipient and itself in the same context, and the communication relationship no longer appears as a binary between subject and object, but transforms into inter-subjectivity. Through the intermediary of "stranger", the self and the other are each other's subjects in the communication process, and can promote cultural development with an equal interactive relationship. The blending has made the audiences in China and the US show more gestures of dialogue when facing culture.

Second, Guo's videos are used as context symbols to create a "communal meaning domain". The flow of meaning in communication activities requires such a domain for both the sender and receiver. For intercultural communication, smooth communication between the sender and receiver requires an intercultural "common domain". Most of Guo's videos are records of real life, covering a large number of daily scenes that everyone has experienced, such as Internet cafes and communities. These scenes easily evoke viewers' memory about their own clothing, food, housing and transportation, and reproduce the "common reality" of the public. In the process of reappearing, many close and life-like scene symbols have opened up the common meaning space of intercultural communication among foreign people. The supplement of appropriate comparative information, such as the comparison of Chinese and American Internet cafes, has broadened the extension of the "common sense domain", providing a broader window for deepening cognition for both parties across cultures, promoting the circulation and sharing of meanings between the two cultures.

4.4 Manifestations

FIG.4 The comments word cloud of Jerry Guo's videos

Jerry Guo's videos are good at using down-to-earth expressions to get closer to the audience. As a long-time "Chinese netizen", Guo is good at making fun of network memes on social media, such as "dogsbody". I randomly selected 10% (47) from 465 videos, sorting the comments of each video according to their popularity. After crawling the top 30 respectively, I obtained 1410 samples, using Python to perform word segmentation and vocabulary statistics, and using FineBI to make the above word cloud map. It can be seen from the figure that the audience is also keen to joke with Guo: "Guo Zhui" for his unclear Chinese pronunciation, and "Not spicy" for him always challenging spicy food. It can be seen that Jerry Guo takes the local cultural symbols as the entry point, the communication effect as the purpose point, and the audience's satisfaction as the foothold. Not only did he not marginalize his "different culture" status, but it strengthened the foreign audience's appreciation for him. The understanding of cultural symbols has also enhanced the local audience's trust in "I am Jerry Guo".

Guo's account focuses on showing individuality and creating a differentiated visual identity. His video cover design is relatively uniform. Generally, a small number of keywords are placed in the middle of the picture, mainly in yellow, and the visual focus is deliberately set to bring a strong impact. At the same time, Jerry Guo always takes the phrase "Hello everyone, I am Jerry Guo" as a fixed opening remark, straight to the point. The features at the end are also prominent. Jerry Guo usually ends the video with the sentence, "Then let me go now". The black background and the fixed expressions of "I am Guo Jerry" in five noticeable white characters are repeated, creating a very engaging scene and symbolic rituals. And this ritualized behavior not only enhances the audience's sense of identity with Jerry Guo, but also makes the main symbol "Jerry Guo" deeply rooted in people's hearts.

Based on the above analysis, this study believes that the video "I am Jerry Guo" adopts a soft form and conducts communication based on the audience's perspective, thus constructing a new communication strategy that is more flexible than serious news.

5. Conclusions

For a long time, there have been problems in worldwide intercultural communication, such as symbolization, fragmentation, etc. These problems are not conducive to international cultural integration. The successful practice of short video blogger Guo Jerry provides a reference for contemporary intercultural communication. Comprehensive analysis of the video content of the short video account "I am Jerry Guo", I found that the key to Guo's success lies in the combination of hard discourses and soft forms.

In terms of "theme content" and "values" of his videos, Jerry Guo adopts hard discourses to respond to sensitive topics and hot current affairs actively. Cultural differences are faced, and the content through on-site experience and interviews are highlighted. It is accurate and objective, showing Guo's own cultural value orientation of "equality, tolerance and openness", bringing a "hardcore" texture to the audience. In terms of the "narrative features" and "manifestations", he adopts a soft form to improve the communication effectiveness through the construction of characters and scene symbols, to gain audience's experience with the grounded personality expression, and to tell culture more softly.

To sum up, as a private "foreign online celebrity", Jerry Guo has explored a more effective path to reduce transmission resistance: "resolutely convey the true content with hard discourse narrative, and make it easy for the audience to accept it with soft communication". Through this path, the role of the "bridge" of foreign online celebrities is activated. They transcode the information of one country and then disseminate it to other countries. By creating coequally shared values among cultures and effectively bridging the cultural gap with soft narratives, the foreign online celebrity may have achieved the maximum effect of intercultural communication under the current situation.

References

1. Ji, D.Q. (2020). A Plum Echo Chamber? Boundary-Breaking and Demarcation of Cross-Cultural Communication in the Age of Social Media. News and Writing. 3, 10-16.
2. Chen, G.M. (2012).The Impact of New Media on Intercultural Communication in Global Context. China Media Research. 8(2), 1-10.
3. Ye,J.(2005). Acculturative Stress and Use of the Internet Among East Asians. Internation Students in the U.S. Cyber Psychology & Behavior.8,154-161.
4. Xiao, J. & Hu, W.T. (2021). Challenges of Intercultural New Media Communication and the Theoretical Construction of Studies. Journalism & Communication Review. 74(1), 101-117. http://qikan.cqvip.com/Qikan/Article/Detail?id=7103620754
5. Hall, Edward, T. (1966). The Hidden Dimension. New York: Doubleday; translated to Japanese in 1970 by Toshio Hidaka and Nobuyuki Sato as Kakureta Jigen. Tokyo: Misuzu Shobo, p.1
6. Liu, Y., Xu, J. (2013). Current Situation and Countermeasures of Chinese Media Using Overseas Social Networks. External Communication.9,42-44.http://qikan.cqvip.com/Qikan/Article/Detail?id=47106966
7. Cao, X. K. (2021). Research on Cross-cultural Communication of Pastoral Food Short Videos--A Case Study of Liziqi Short Videos. China Press. 2, 12-13.10.13854/j.cnki.cni.2021.02.005
8. Wang, G. H., Gao, W., Li, H. F. (2018). Analysis of the Characteristics, Dissemination Function and Management Countermeasures of "Foreign Internet Celebrities"--A Case Study of Ten Foreign Internet Celebrities on Sina Weibo. Journal of Intelligence. 37(12), 93-98. 10.3969/j. ISSN. 1002 – 1965.2018.12.014
9. Du, G. D. (2019). How to Play the Role of Foreign Internet Celebrities in China's International Communication. International Communication.6,67-77. http://qikan.cqvip.com/Qikan/Article/Detail?id=7101098230
10. Xia, F. F., Pan, H. Q. (2020). An Analysis of the International Communication Path Base on the Perspective of Foreign Internet Celebrities--A Case Study of Han Dongdong, a Vlogger of the Bilibili. News Research Guide. 11 (24), 22-24. https://kns.cnki.net/kcms/detail/detail.aspx?dbcode=CJFD&dbname=CJFDLA20
11. Zheng, E. J. (2019). Research on the Dissemination of Short Videos of "Foreign Internet Celebrities" on Bilibili Video Network. Doctor dissertation. Inner Mongolia University. https://kns.cnki.net/kcms/detail/detail.aspx?dbcode=CMFD&dbname
12. Liang. X. H. (2020). Research on the Phenomenon of "Foreign Internet Celebrities" from the Perspective of Cross-cultural Communication. Doctor Dissertation. Central China Normal University. 10.27159/d.cnki.ghzsu.2020.002298

Is There a Barrier between Seniors and Smartphone Use in The Internet Age? A Study of Digital Disconnection among Older Adults

Ruotong Jiao*

Jilin University, Department of Advertising, 2699 Qianjin Street, Chaoyang District, Changchun City, Jilin Province, China

Abstract. With internet and smartphone usage increasing every year, seniors are being exposed to digital media, either actively or passively. To help the elderly better integrate into the information society, clarifing the causes of their digital disconnection and developing appropriate solutions is essential. In this study, semi-structured interviews are utilized to investigate the self-adjustment and external intervention of the elderly in the age of digital media and to explore the ways, causes, and effects of digital disconnection of the elderly. The focus issue of this study is to clear the subjective and objective factors of digital disconnection, especially cognitive problems, group influences, and technological access barriers in the elderly population concerning digital devices and the Internet.

1. Introduction

With advancements in the national system, medical technology and economic development, China is stepping into population aging. The proportion of senior Internet users aged 60 and over was 11.3% as of June 2022, droping 0.2% from the same period in the last years, according to *the 50th Statistical Report on Internet Development in China*, published in August 2022. Only 44% of senior citizens use the Internet, and some argue that aging and informatization are significant characters of contemporary era [1].

The digital divide refers to a situation in which people of different genders, ages, and classes differ in their capability to access and use new information technologies, causing further inequalities [2]. This theory among older people works in three layers: accession, use and knowledge, and the three layers are progressively deeper. Limited knowledge and using skill gaps make it hard for older netizens to integrate into digital products and services [3]. What's more, the aging-friendly design of the Internet still deficient, leading some active seniors to withdraw.

However, there is few researches setting foot on the digital disconnect of the elderly. The empirical studies presented in this article are from explorative research, which used interviews as a context sample to investigate what dilemmas older netizens face and why they are in this predicament, especially in cognitive problems, group influences and technological access. Classifying this question would provide insight into their disconnect behaviors and propose workable solutions, which might also enhance the aging-friendly design.

* Corresponding author: 1220912739@qq.com

2. Literature Review

2.1 A Study of The Subjective Factors of Digital Disconnect among Senior Citizens

First, physical and psychological issues shape elderly adults' Internet use. Physiological factors like reduced brain activity and weakened neuromodulatory control lead to negative learning in older people [4]. The elderly are concerned about privacy and security due to a lack of digital skills [5]. As for emotional motivation, such as the feeling seems like outsiders, the fear of trying new things and making mistakes may play a role in digital disconnection [6]. Furthermore, psychological age impacts older inhabitants' digital connections, the average age of them with media access is lower than those without connections [7]. Some seniors, old but young in mind, are willing to use digital media [8]. In addition, gender differences in the senior population also change digital media consumption. Research finds older female users are more tested to adapt to the use of new technologies [9] and their learning tendencies between older women and men are also different [10]. Thus, we may conclude that taking physical and psychological issues into account should be vital in promoting the senior population's digital connection.

Second, given the senior population's residential status, the family composition has a significant impact on the digital media access of the elderly [11]. When children become the main source of information and financial support for the elderly, their desire for digital feedback on the internet plays a crucial role [12], in other words, the

https://doi.org/10.1051/shsconf/202315503010

living condition of the elderly group directly or indirectly affects the degree of digital media access through emotional factors [13].

Third, the degree of digital disconnect among the elderly is realtive to education and financial status. For example, well-educated people use digital media more frequently than poor ones [14]. So with that, older adults who are informed have easier access to digital using digital media than others [15]. Low-income seniors are more prone to utilize digital media sparingly and carefully [16]. Lack of education, limited finances and an overwhelming fear of emerging technologies, all hinder seniors from having a great experience using digital media.

2.2 A Study of The Subjective Factors of Digital Disconnect among Senior Citizens

First, group and social factors are critical external factors in Internet use of older adults. They are disposed to use digital media in households with good access facilities [17]. If there are many household jobs, fewer patient relatives, or less energy to guide them, they would be reduced or even stop using the digital press [18]. Besides, the digital access of family and friends can affect the choice of older adults [19]. The social circle of the elderly is unfamiliar with digital media, which may be short in external access motivation. Regarding social factors, inadequate social support is also an apparent cause of digital disconnection [20]. The development of regional economies and information societies also act on the distribution of digital media resources and social ideas and shape older people's use.

Second, policies impact the utilization of digital media among older. The unbalanced and inadequate development of "The internet & Pension" may prompt the more senior crowds to leave and turn to pick the traditional pension model [21].

Finally, the Chinese Internet infrastructure, essential resources, and data layer dominance are defined, and there is also a lack of supply of policy tools. Accordingly, this paper argues that the lack of Internet-related policymaking or neglect of older people can sometimes lead to a digital disconnect among older people.

3. Materials and Methods

A research group is an elderly group that takes into account demographic characteristics. Based on the reflection brought by the concept of digital disconnection, the author conducted a semi-structured interview with the survey respondents from October 2020 to October 2022 to investigate the formation of their specific actions and ideas. With the permission of the interviewee, nearly four words of interview notes were compiled.

This study followed the corresponding coding procedure, first extracting 20 open codes, then extracting 5 related codes by using axial shape and selective coding, further refining the interview context and summarizing the phenomenon more concretely. By combing the logical framework, this study determines the primary and secondary relationships between the core categories and different categories and tries to describe and interpret the phenomenon of digital disconnection comprehensively.

This study adopts the combination of online and offline methods, which can transcend the limitations of time and space to collect suitable samples. To ensure the reliability and validity of the research, the design of the interview outline paid attention to avoiding the induced questions, and two interviews were organized to obtain more detailed and in-depth information. Finally, the data analysis adopts the methods of cross-organization and triangle tests, trying to make the conclusion more scientific and normative.

4. Results & Discussion

Owing to the expansion of the Internet and various contributing factors, including environment, education, health, etc. By analyzing the interview contents, the seniors could be divided into two categories based on their initiative and the possibility of participating in smartphones since more senior citizens are now either actively or passively exposed to digital media. However, their frequency of exposure to digital media differs depending on information production, i.e., "completely disconnected" and "partially disconnected." In these two stages, the degree of disconnection reduces while the degree of access grows.

4.1 Completely Disconnected and Partially Disconnected

Completely disconnected is a state in which the elderly do not have any contact with digital media and are far from the Internet. According to the analysis of the interviews, because these older adults have poor hardware and software equipment conditions in Internet contact or are short in literacy, it is difficult for them to complete the operation alone. There are many alternative ways when facing online troubles to boots. They have children, relatives, and friends to solve other situations, so they passively stay away from the Internet or refuse to contact it, which generates a negative attitude toward it.

When operating media among the partially disconnected elderly group, their media use breadth and usage are influenced by their different needs. By compiling and summarizing the interview results, the media use of the interviewed elderly was as follows.

Table 1 Digital Media Usage among The Seniors

Digital Media Usage Behavior	Reason	Example
Connected	Social Interaction	Use social apps such as WeChat to communicate with friends.
	Entertainment	Watch short videos as a form of daily entertainment.
	Electronic Payment	Use WeChat/Alipay for daily QR code payments.
	Epidemic	Show the health

 https://doi.org/10.1051/shsconf/202315503010

	Managemen	code when entering public spaces.
	Information Acquisition	Browse news sites and search for information.
Disconnected	Physical Condition	Poor audiovisual ability or otherwise.
	Knowledge & Skills	Limited knowledge and unmatched skill.
	Economic Restrictions	Lack of smart equipment and relevant network devices.
	Digital Exclusion	Be alert and sensitive to the Internet.

In light of varying degrees of information intake, the partially disconnected older groups can be split into two groups, passive reception and active acquisition. The passive group has fewer practices of digital media, and their usage is in a small range. Interviews revealed that they had relatively low information needs and single-use functions, and the Internet is one of the ways to enrich their daily life. For these oldsters, using the Internet mainly serves purposes of socialization and entertainment. The respondent (S17) said, "I do not want to learn that, I don't know how to shop online and I am not interested. I just want to read the news."

The elderly obtaining information proactively would be strongly willing to integrate into the Internet. They trust the Internet for good experiences in social communication and information acquisition. Meanwhile, their peer social circles are more active in approaching digital media, and the attitude toward internet integration will be more positive. For example, respondent (S16) said, "I can learn more international events, domestic events, and weather forecasts anytime and anywhere," and the respondent (S26) said, "The Internet is an encyclopedia for me."

4.2 Influencing factors of digital disconnect among the seniors

Older users' integration possibility and willingness to consume digital media will be subject to Physical conditions, which are also the first consideration for aging-friendly design. Seniors with healthy conditions, which we define as good hearing and visual abilities and the capability to operate smartphones and other electronic devices, are more likely to join the Internet and are more proactively learning to use smart devices. Those in poorer health, with poorer hearing and vision, as precisely as the respondent (S4), said, "my eyes always tired when watching the phone," and prefer to use large fonts and simple elderly push-button machines.

Moreover, the ability to discriminate,comprehension, and educational are crucial aspects that change prior years' behavior. Seniors who perform well in these aspects are more sensitive. They treat personal and financial matters cautiously to prevent receiving unfavorable exposure. Those weak in these areas will opt to cut off to "defend themselves." Additionally, some of them will consciously control their digital access by disconnecting themselves after a predetermined time.

Given the environmental impact, the Internet access environment strikes at their digital access status. Such as the living environment, peer social circle, and the Internet access status of location are closely related to it. Seniors having encouragement from youngsters are persistent in scooping up smartphone. Most elders in the interview expressed a desire to communicate with their children via WeChat video. It also increased their urge to use the Internet. Respondent (S9) can grasp some skills with the help of his family, and his willingness to access digitally is enthusiastic, "it is my children who teach me, and the young people around me can also learn very quickly."

Group also exerts influence on the initiative of the elderly to drive digital media. They become intrigued and attempt to connect when most of their friends use smartphones because of the spread of contagious emotions like sharing. When a peer circle's Internet access is rare, there is no positive feedback and a generalized negative emotional influence, which can eventually lead to disconnection.

The installation of inter-territorial network infrastructure is another substantial element of the digital disconnect of the aging population. During the interview, some elderly gives a higher recognition to digital management for convenient online experiences and are optimistic about accessing digital media. In contrast, some older adults in towns and remote areas with lower Internet penetration can meet most of their essential needs in life and do not have the conditions to access the Internet.

The experience of intelligent devices concerning usage requirements makes an impression on access. The interviews indicate that older people are driven to have digital access under better device experiences.

5. Conclusion

The experience of intelligent devices concerning usage requirements makes an impression on access. The interviews indicate that older people are driven to have digital access under better device experiences.

The focus issue of this study is to clarify the subjective and objective factors of digital disconnection, especially cognitive problems, group influences, and technological access barriers in the elderly population concerning digital devices and the Internet. However, the inadequate age-appropriate design has caused some active seniors to drop out. This mixed reality requires more investigation and feasible government regulations to bridge the digital divide among seniors, especially in the accession and knowledge gaps.

Undeniably, when people get older, their physical function will decline, making it more difficult to use digital media. It is essential to pay attention to the social benefits of Internet access to the elderly. All social sectors should help them learn how to use digital devices. Only when the digital access of the elderly is effectively promoted, can we build a favorable digital ecosystem that can better serve the public welfare.

What needs to be pointed out is that this study has

limitations and room for future improvement. After all, some interviewees can't participate in the interview independently because of their physical condition and other reasons, which is bound to strike the survey results. How to overcome the factors that may cause data deviation in the process of studying the elderly population deserves further discussion and discussion. Overall, this study may provide some experience and strategic reference for other scholars to study issues such as the elderly's adaptation to the Internet age and the elderly-friendly functional design.

References

1. Zhan, H., & Xizhe, P. (2020). Strategic Changes and Policy Choices in the Governance of China's Aging Society. *Social Sciences in China*. 41. 185-208.
2. Huang, C. X., (2020). Status, Challenges And Countermeasures of The Digital Divide in Older Adults. *People's Tribune*, (29).
3. Zhang, W. P., & Fan, J. H. (2019). The Construction of A Social Support System for The Digital Divide in Old Age. *Scientific Research on Aging*. (02),63-70.
4. Mahncke, H. W., Bronstone, A., & Merzenich, M. M. (2006). Brain plasticity and functional losses in the aged: scientific bases for a novel intervention. *Progress in brain research*, 157, 81-109.
5. Wagner, N., Hassanein, K., & Head, M. (2010). Computer use by older adults: A multi-disciplinary review. Computers in human behavior, 26(5), 870-882.
6. Blažič, B. J., & Blažič, A. J. (2020). Overcoming the digital divide with a modern approach to learning digital skills for elderly adults. Education and Information Technologies, 25(1), 259-279.
7. Hong, Y. A., Zhou, Z., Fang, Y., & Shi, L. (2017). The digital divide and health disparities in China: evidence from a national survey and policy implications. Journal of medical Internet research, 19(9), e7786.
8. Gana. K., Alaphilippe D. & Bailly. N. (2004). Positive Illusions And Mental And Physical Health in Later Life, *Aging & Mental Health*, 8:1, 58-64, DOI: 10.1081/13607860310001613347
9. Liu, Q. Q., Zuo, M. Y., Liu, M. C. (2021). An Empirical Analysis of The Continued Use of Internet Applications by Older Adults Based on Expectation Confirmation Theory. *Management Review*, (05), 89-101.
10. Yan, G. F., Tian, R., Xiong, Z. Y., Sun, L. H. (2019). Examining The "Digital Gender Divide" in The 5G Era: Causes And Solutions: Insights from The OECD Report "Bridging The Digital Gender Divide". *Journal of Distance Education*, (05), 66-74.
11. He, Q., Zhang, X. D. (2020). Factors Influencing The Digital Divide Among The Elderly And Strategies for Social Integration. *Journal of Zhejiang University of Technology (Social Sciences)*, 16(04), 437-441.
12. Zhou, Y. Q. (2014). Digital Generation Gap and cultural Feedback: A quantitative investigation of the "quiet Revolution" in the family. Modern Communication: Journal of Communication University of China, (2), 117-123.
13. Antonio, A., & Tuffley, D. (2015). Bridging the age-based digital divide. *International Journal of Digital Literacy and Digital Competence (IJDLDC)*, 6(3), 1-15.
14. Friemel, T. N. (2016). The digital divide has grown old: Determinants of a digital divide among seniors. New media & society, 18(2), 313-331.
15. Van Dijk, J. A. (2017). Digital divide: Impact of access. *The international encyclopedia of media effects*, 1-11.
16. Nie, Z. Y. (2017). Analysis of The Internet Usage Status of The Elderly Based on Segmentation Representation. *New Media Research*, (19), 40-41. DOI:10.16604/j.cnki.issn2096-0360.2017.19.014.
17. Korupp, S. E., & Szydlik, M. (2005). Causes and trends of the digital divide. European Sociological Review ,21(4) ,409-422.
18. Drori, G. S., & Jang, Y. S. (2003). The global digital divide: A sociological assessment of trends and causes. Social science computer review, 21(2), 144-161.
19. Lan, Q., Lu, X. H. (2019). Exploring The Differences in Internet Use Among Urban Older Adults And Their Influencing Factors - An Empirical Study Based on CFPS2016 Data. *Soft Science*, (01),104-108.
20. Helsper, E. J. (2021). The Digital Disconnect: The Social Causes and Consequences of Digital Inequalities. London: SAGE Publications Ltd.
21. Song, Z., Song, T., Yang, Y., & Wang, Z. (2019). Spatial-temporal characteristics and determinants of a digital divide in China. Sustainability, 11(17), 4529.

Using the Extended Theory of Planned Behavior to Predict Privacy-Protection Behavioral Intentions in the Big Data Era: The Role of Privacy Concern

Mengxin Chen[1,*], Haotian Wang[2,*], Ruobing Zhang[3,*]

[1]Communication Engineering, Heilongjiang University of Science and Technology
[2] Logistic Management of Navigation Institute, Jimei University
[3] Visual Communication Design, Xiamen University

ABSTRACT. This research extends the theory of planned behavior to explore how privacy concerns influence privacy-protection behavioral intentions via attitudes, subjective norms, and perceived behavioral control. Using snowball sampling, 637 participants were recruited in China to complete this online survey. After eliminating the invalid questionnaire, this study received valid samples of 599 respondents. The finding demonstrated that privacy concerns were positively associated with individuals' attitudes, subjective norms, and perceived behavioral control. And these factors also were found positively influence privacy-protection behavioral intentions. Theoretical and practical contributions are discussed as well.

1 Introduction

With the development of science and technology, the amount of data generated and captured in modern society has continued to increase rapidly, and the number of statistics has increased by the PB (1024TB) level, which has entered the era of big data. The concept of big data first appeared in Lenny's Gartner report in 2001 [11], where the term "big data" was defined as large and complex data set that computing facilities were unable to handle at that time [16]. The current big-data environment, especially regarding data mining and machine learning techniques, involve serious threats to privacy [16]. As a result, privacy concerns have attracted scholarly attention. Facebook revealed that "almost 50 million" of its users had their confidential information exposed by a security flaw in September 2018 [7]. Public polls have demonstrated evidence of heightened problems related to privacy leakage online users in the big data era as it has become easier to locate an individual's information [3]. Moreover, data leakage incidents continues to occur, exacerbating privacy concerns [3]. In addition, online service providers may leak users' personal, private data to others without notifying users or obtaining their authorization for providers' interests. Thus, protecting individual's privacy and confidential information has become extraordinarily challenging in the big data era.

Privacy-protection behaviors (PPBs) may help users minimize locating and collecting of individual's information by preventing them from giving truthful and accurate individual's information online. Therefore, this study seeks to explore the factors that predict that an individual will practice privacy-protection behaviors.

The theory of planned behavior (TPB) is one of the most widely used theories to examine behavioral determinants. It assumes that individuals' behaviors are influenced by their attitudes, subjective norms, and perceived behavioral control [2]. A large number of researches have proved these factors play an important role in privacy-protection behavioral intentions [10]. Additionally, previous studies also have extended the TPB by taking privacy concern as an antecedent. Some scholars have noted that privacy concerns could also affect individuals' perceptions of practicing privacy-protection behaviors and even directly predict privacy-protection behavioral intentions. For example, Ho found that adolescents' privacy concerns are important for influencing their future PPBs on SNSs [10]. Moreover, Heirman found that privacy concerns are significant predictors of attitudes toward privacy-protection behaviors, which can be significantly affected by privacy-protection behavioral intentions (PPBIs) [9]. Hence, this study also seeks to explore the direct and indirect effects of privacy concerns on shaping individuals' privacy-protection behavioral intentions.

This study's objective is twofold. We aim to determine (1) whether individuals' attitudes, subjective norms, and perceived behavioral control affect privacy-protection behavioral intentions and (2) whether privacy concerns directly and indirectly affect individuals' privacy-protection behavioral intentions through attitudes, subjective norms, and perceived behavioral control.

*Corresponding author. Mengxin Chen Email: 328771751@qq.com;
Haotian Wang Email: 2857268839@qq.com;
Ruobing Zhang Email: 627360474@qq.com

 https://doi.org/10.1051/shsconf/202315503011

Additional detailed theoretical and practical implications will be provided in a later discussion.

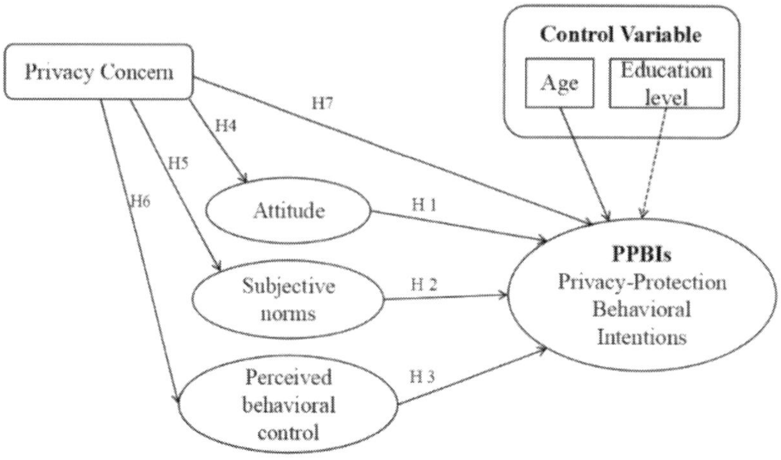

Fig. 1. Conceptual model (self- organized).

2 Literature Review and Research Hypotheses

2.1 Privacy-Protection Behavioral Intentions

Previous literature has documented the link between privacy concerns and privacy-protection behaviors, which has found that people protect online privacy through different ways [15]. Privacy-protection behaviors are defined as "specific computer-based behaviors that consumers take to protect their information safe" [12]. Specifically, individual's privacy-protection behaviors can reduce the collection, use, and share of individual's information to protect online privacy. For instance, individuals can protect online privacy information by restricting their shared information and taking privacy-protection behaviors [6]. Privacy-protection behaviors discussed in this study include providing incomplete individual's information, avoiding providing truthful and accurate individual's information, and refusing to provide actual individual's information online. In this research, we focus on the direct and indirect factors that affect PPBIs.

2.2 Theoretical Framework: The theory of planned behavior

Adopting the theory of planned behavior, this study adopts the theory-driven approach to explain the potential factors that might be related to privacy-protection behavioral intentions (PPBIs) online. The TPB assumes that personal intentions are the closest antecedent of his or her actual behavior. Moreover, individual's behavior intentions are affected by three factors : attitude, subjective norms, and perceived behavioral control [2].

Attitude. Attitude indicates individual's assessment of behaviors. When assessment is more positive, the individuals are more likely to implement specific behaviors [5]. Previous studies have found that attitudes may have positive effect on privacy-protection behaviors. For example, Ho showed that adolescents' attitudes toward PPBs positively predicted their future PPBs [10]. When people perceive privacy protection as important, valuable, and appropriate behavior, they tend to engage in PPBs online. Hence, we posit the hypothesis as follow:

H1: Attitude is positively associated with PPBIs.

Subjective norms. Subjective norms is defined as the perceived rate of behavior and the perception that significant others want the individuals to implement the particular behavior [14]. Through previous studies, subjective norms have been shown to have a positive effect on PPBs. For example, earlier research indicated that, when significant others expected that individuals would not disclose their privacy, those individuals did not engage in sharing their confidential information [9]. In other words, if significant others expect people to engage in PPBs, those persons tend to engage in PPBs online. Hence, we propose the following hypothesis:

H2: Subjective norms are positively associated with PPBIs.

Perceived behavioral control. Perceived behavioral control indicates the perceived efficacy of an individual in the expected behavior [9]. Previous research has indicated that perceived behavioral control might had a positive influence on privacy-protection behaviors. For example, when individuals were able to perceive their behaviors, they were less likely to engage in leaking their private information online [9]. Moreover, when individuals perceive that it is easy to implement PPBs, they tend to engage in the behavior online. Hence, we propose the following hypothesis:

H3: Perceived behavioral control is positively associated with PPBIs.

2.3 An extension of the TPB: Privacy concern as an antecedent

Defined as one's concern about his or her personal data being used, privacy concern could affect PPBIs directly and indirectly through attitude, subjective norms, and perceived behavioral control. Previous study has indicated that a positive association exists between privacy concerns' degree and privacy-protection behaviors: the higher the degree of concern, the more likely to implement privacy-protection behaviors [10]. The relationship between privacy concerns and PPBIs was initially proposed and discussed by Altman. He found that "people try to enforce the required level of privacy by using behavioral mechanisms" [1]. Privacy concerns occur frequently in the big data era. With the continuous growth and adoption of social media services and strategies, it is necessary to understand their role in individual's privacy [8].

Previous study has indicated a positive association between privacy concern and adolescents' attitudes toward PPBs [9]. Based on these, we further explore the degree to which privacy concerns are likely to have an impact on attitudes toward PPBs online. In addition, we also investigate the influence of privacy concerns on the subjective norms and perceived behavioral control. When people have strong privacy concerns, their family and friends expect them to engage in PPBs (i.e., subjective norms). If an individual has a high degree of privacy concern, her or his significant others perceive that individual as someone who wants to protect her or his personal information. As such, they expect people to engage in PPBs, and in turn, to satisfy their significant others' expectations, they are more likely to implement PPBs in order to protect their privacy online. Additionally, a high degree of concern about privacy may lead to a greater degree of attention to privacy issues, which may affect their ability and determination to implement PPBs (i.e., perceived behavioral control) and to become knowledgeable about it. Thus, such individuals can be expected to have a high degree of behavioral control over these behaviors. In summary, attention to privacy issues reflects the level of people's privacy concerns, which may affect their implementation of PPBs. Hence, we propose four additional hypotheses:

H4: Privacy concern is positively associated with attitude.

H5: Privacy concern is positively associated with subjective norms.

H6: Privacy concern is positively associated with perceived behavioral control.

H7: Privacy concern is positively associated with PPBIs.

3 Method

3.1 Data Collection

During July 2022, an online survey was conducted in China to collect data. We employed the snowball-sampling technique and 637 adults were recruited. We asked participants to sign an informed consent before they accessed the questionnaire, and those who were not willing to participate and failed to answer the filter questions correctly were excluded. We acquired 599 valid samples after eliminating the invalid questionnaires. The mean age of these 599 respondents was 28.77 years, with a range of 18–72 years old. About 39.7% had obtained a high school diploma, and 30.6% had obtained a college degree.

3.2 Measures

Based on the research background, this study adjusts the existing and effective questionnaire table. (see Table 1). First, we used Ho et al.'s (2017) attitude scale, subjective norms scale, and perceived behavioral control scale; each of these scales had three items for a total of nine items. Additionally, we adapted Mosteller and Poddar's (2017) privacy concern scale (comprising four items) to our current context. Finally, we adapted Adhikari and Panda's (2018) privacy-protection behavior intentions scale (comprising three items) to our current context. The finding showed that all scales have a good validity and reliability.

Control variables. Demographic variables were used as control variables for this research, including age and educational level.

Attitude. Attitudes were measured by three items using a five-point scale (1 = strongly disagree and 5 = strongly agree) [10]. The respondents were asked about their attitude toward engaging in privacy-protection behaviors online. A higher score indicates the greater degree of attitudes (M = 3.91, SD = 0.96, Cronbach's α = 0.87).

Subjective norms. Subjective norms were surveyed by three items using a five-point scale (1 = strongly disagree and 5 = strongly agree) [10]. The respondents were asked whether most of the people who are important to them thought that they should implement PPBs online. A higher score indicates the greater degree of subjective norms (M = 3.86, SD = 0.96, Cronbach's α = 0.87).

Perceived behavioral control. Perceived behavioral control was measured by three items using a five-point scale (1 = strongly disagree and 5 = strongly agree) [10]. The respondents were asked whether they considered themselves able to implement PPBs online. A higher score indicates the greater degree of perceived behavioral control (M = 3.88, SD = 0.93, Cronbach's α = 0.83).

Privacy Concern. The degree of attention to how to handle personal privacy information on online service providers was surveyed by four items using a five-point scale (1 = strongly disagree and 5 = strongly agree) [13]. The respondents were asked about which personal privacy issues they are concerned about when they are online. A higher score indicates a greater degree of concern about privacy online (M = 3.81, SD = 0.94, Cronbach's α = 0.88).

Table 1. Summary of measurement items (self- organized).

Study measure	Survey items	Cronbach's alpha	M	SD
Attitude [10]	Do you agree with the following statement? (1 = Strongly Disagree, 5 = Strongly Agree)	0.87		

	1. It is important to engage in PPBs online.		4.00	1.01
	2. It is necessary to engage in PPBs online.		3.85	1.12
	3. It is meaningful to engage in PPBs online.		3.87	1.10
Subjective norms [10]	Do you agree with the following statement? (1 = Strongly Disagree, 5 = Strongly Agree)	0.87		
	1. Most people who are important to me, they implement PPBs online.		3.95	1.02
	2. Most people who are important to me, they think I should implement PPBs online.		3.80	1.11
	3. Most people who are important to me, they expect I implement PPBs online.		3.82	1.10
Perceived behavioral control [10]	Do you agree with the following statement? (1 = Strongly Disagree, 5 = Strongly Agree)	0.83		
	1. It is possible for me to engage in PPBs online.		4.01	0.96
	2. It is easy for me to engage in PPBs online.		3.82	1.12
	3. It is mostly up to me whether or not I engage in PPBs online.		3.79	1.14
Privacy Concern [13]	Do you agree with the following statement? (1 = Strongly Disagree, 5 = Strongly Agree) I am concerned that online service providers...	0.88		
	1. may keep my private information in a non-secure manner,e.g. saving my personal information on a public server.		3.91	1.01
	2. may divulge my information to unauthorized parties without my consent,e.g. authorizing my personal information to online shopping platforms and real estate companies.		3.74	1.13
	3. may use my information for other purposes without notifying me or getting my authorization,e.g.analyzing my daily activities to derive information about me and profiteering on big data.		3.80	1.11
	4. may sell my information to others without notifying me or getting my authorization,e.g. selling my personal information to P2P platforms and intermediary organ.		3.76	1.13
Privacy Protection Behavior intention [1]	Do you agree with the following statement? (1 = Strongly Disagree, 5 = Strongly Agree)	0.84		
	1. I will provide incomplete personal information online.		3.88	1.01
	2. I will avoid giving real and accurate personal information online.		3.75	1.13
	3. I will refuse to provide real and accurate personal information online.		3.78	1.11

3.3 Preliminary analysis

For this research, we employed the maximum likelihood estimation for the path analysis on AMOS 26.0 in order to examine hypotheses investigating the association between attitudes, subjective norms, perceived behavioral control, and privacy concerns with PPBIs. Before conducing a path analysis, we examined the study data to ensure no abnormal values, and all measure items were found to be normally distributed (kurtosis and skewness were within the range of ± 1). Moreover, we used SPSS 26.0 to conduct a reliability test on potential variables. The test results indicated that the Cronbach's alpha values of all potential variables range from 0.83 to 0.88, and all potential variables had good reliability.

4 Results

To support or refute the hypotheses, a path analysis was conducted. The path analysis showed a good fitness of the measurement model (CMIN/DF = 0.99, TLI = 0.99, CFI = 1.0, RMSEA = 0.01). The results are reported in Table 3 and Figure 2. These factors may affect individuals' PPBIs that aim to protect their online privacy. Respondents' age (β = -0.01 , p = 0.88) was not significantly correlated to PPBIs, and their educational level (β = -0.10 , p = 0.01) was negatively related to PPBIs. Attitude (β = 0.09, p = 0.03), subjective norms (β = 0.11, p = 0.01), and perceived behavioral control (β = 0.20, $p < 0.01$) were found to be positively related to PPBIs, thereby supporting H1, H2, and H3. Privacy concern was positively associated with attitude (β = 0.34, $p < 0.01$), subjective norms (β = 0.27, $p < 0.01$), and perceived behavioral control (β = 0.31, $p < 0.01$), thus supporting H4, H5, and H6. Finally, privacy concerns were positively correlated to PPBIs ($\beta = 0.23$, $p < 0.01$), thus confirming H7.

Table 2. Hypotheses testing (self- organized).

Hypothesis	Path	β	P	Analysis Results
H1	Att→PPBIs	0.09	0.03	Supported
H2	SN→PPBIs	0.11	0.01	Supported
H3	PBC→PPBIs	0.20	***	Supported
H4	PC→Att	0.34	***	Supported
H5	PC→SN	0.27	***	Supported
H6	PC→PBC	0.31	***	Supported
H7	PC→PPBIs	0.23	***	Supported

Note(s): Att: Attitude; SN: Subjective norms; PBC: Perceived behavioral control; PC: Privacy Concern; PPBIs: Privacy-Protection Behavioral intentions

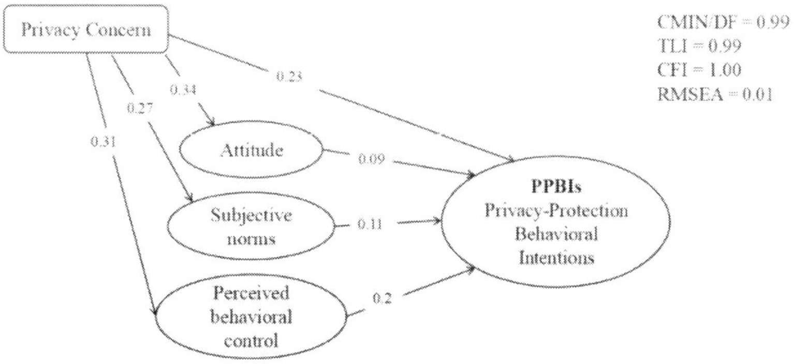

Fig. 2. Full model for PPBIs (self- organized).

Note(s): The rectangles represent observed variables and the ovals represent latent variable.The coefficients in the figure are standardized β coefficients. Age and educational level were treated as control variables.

5 Conclusions

The finding of this research supported the variables of the TPB in predicting PPBIs. Consistent with a previous study, attitudes, subjective norms, and perceived behavioral control positively affects PPBIs [10]. The finding indicated that individuals were more likely to implement PPBIs when they perceived it to be beneficial and important. The results showed that subjective norms were positively with PPBIs. And it indicated that individuals were more likely implement PPBIs if they perceived that others expect them to engage in particular behaviors. Additionally, we found that individuals' perceptions of behavioral control may result in their greater likelihood of engaging in privacy-protection behaviors if they consider themselves able to engage in PPBIs online.

This research extended the TPB by examining the influence of privacy concerns. It also explored how privacy concerns affect PPBIs directly and indirectly via attitude, subjective norms, and perceived behavioral control. The finding indicates that privacy concerns do, indeed, motivate individuals to engage in privacy-protection behavioral intentions directly and indirectly via attitude, subjective norms, and perceived behavioral control. Specifically, we found that privacy concerns influence individuals by generating the positive attitudes toward privacy-protection behaviors and by shaping subjective norms and perceived behavioral control, which lead, in turn, to behaviors that are aimed at protecting their privacy online.

6 Discussion

The findings suggest that this study has theoretical contribution and practical significance. The theoretical contribution is mainly reflected in extending the TPB by taking privacy concern as an antecedent to explore how it influences attitudes, subjective norms, and perceived behavioral control, consequently affecting PPBIs. Furthermore, in regard to the behavioral component of privacy concerns, it was found that attitudes, subjective norms, and perceived behavioral control significantly mediate the association between privacy concerns and PPBIs. Finally, this research found that the positive association between privacy concerns and privacy-protection behavioral intentions if they knew how to handle their private information in relation to online service providers. This will provide a reference for online service providers with privacy-protection measures.

This research has several significance. This study analyzes cross-sectional data to prevent causal inference. In future research, we may increase longitudinal surveys with the aim of clarifying causation. Secondly, this study employed snowball sampling to recruit respondents. Thus, further research utilizing a representative sample is necessary in order to make research more universal and persuasive.

References

1. Adhikari, K., & Panda, R. K. (2018). Users' Information Privacy Concerns and Privacy Protection Behaviors in Social Networks. *Journal of Global Marketing*, 31(2), 96-110. https://doi.org/10.1080/08911762.2017.1412552
2. Ajzen, I. (1991). The theory of planned behavior. *Organizational Behavior and Human Decision Processes*, 50(2), 179-211. https://doi.org/10.1016/0749-5978(91)90020-T
3. Alashor, T., Han, S., & Joseph, R. C. (2017). Familiarity with Big Data, Privacy Concerns, and Self-disclosure Accuracy in Social Networking Websites: An APCO Model. *Communications of the Association for Information Systems*, 41, pp-pp. https://doi.org/10.17705/1CAIS.04104
4. Altman, I. (1975). *The Environment and Social Behavior: Privacy, Personal Space, Territory, and Crowding.* Monterey. https://eric.ed.gov/?id=ed131515
5. Baker, R. K., & White, K. M. (2010). Predicting adolescents' use of social networking sites from an extended theory of planned behaviour perspective. *Computers in Human Behavior*, 26(6), 1591-1597. https://doi.org/10.1016/j.chb.2010.06.006
6. Baruh, L., & Popescu, M. (2017). Big data analytics and the limits of privacy self-management. *New Media & Society*, 19(4), 579-596. https://doi.org/10.1177/1461444815614001
7. BBC (2018, October 1), *Facebook security breach: up to 50m accounts attacked*, http://www.bbc.com/news/technology-45686890?ns_source=Facebook&ns_mchannel=social&ns_campaign=bbcnews&ocid=socialflow_Facebook
8. Brandtzaeg, P. B., Luders, M. & Skjetne, J. H. (2010). Too Many Facebook "Friends"? Content Sharing and Sociability Versus the Need for Privacy in Social Network Sites. *International Journal of Human-Computer Interaction*, 26(11-12), 1006-1030. https://doi.org/10.1080/10447318.2010.516719
9. Heirman, W., Walrave, M., & Ponnet, K. (2013). Predicting adolescents' disclosure of personal information in exchange for commercial incentives: an application of an extended theory of planned behavior. *Cyberpsychology, Behavior, and Social Networking*, 16(2). https://doi.org/10.1089/cyber.2012.0041
10. Ho, S. S., Lee, E. W. J., Lwin, M. O., & Yee, A. Z. H. (2017). Understanding Factors Associated with Singaporean Adolescents' Intention to Adopt Privacy Protection Behavior Using an Extended Theory of Planned Behavior. *Cyberpsychology Behavior, and Social Networking,* 20(9), 572-579. https://doi.org/10.1089/cyber.2017.0061
11. D. Laney. (2001). 3D data management: Controlling data volume, velocity, and variety, *META group research note*, 6(70), 1.
12. Milne, G. R., Labrecque, L. I., & Cromer, C. (2009). Toward an understanding of the online consumer's risky behavior and protection practices. *Journal of Consumer Affairs*, 43(3), 449-473. https://doi.org/10.1111/j.1745-6606.2009.01148.x
13. Mosteller, J., & Poddar, A. (2017). To Share and Protect: Using Regulatory Focus Theory to Examine the Privacy Paradox of Consumers' Social Media Engagement and Online Privacy Protection Behaviors. *Journal of Interactive Marketing*, 39(1), 27-38. https://doi.org/10.1016/j.intmar.2017.02.003
14. Pelling, E. L., & White, K. M. (2009). The theory of planned behavior applied to young people's use of social networking web sites. *Cyberpsychology & Behavior*, 12(6), 755-759. https://doi.org/10.1089/cpb.2009.0109
15. Yap, J. E., Beverland, M. B. & Bove, L. L. (2012), "Doing Privacy": Consumers Search for Sovereignty through Privacy Management Practices. *Research in Consumer Behavior*, 14, 171-190. https://doi.org/10.1108/S0885-2111(2012)0000014013
16. Yu, S. (2016). Big Privacy: Challenges and Opportunities of Privacy Study in the Age of Big Data. *IEEE Access*, 4, 2751-2763. https://doi.org/10.1109/ACCESS.2016.2577036

More Attached, Less Stressed: Viewers' Parasocial Attachment to Virtual Youtubers and Its Influence on the Stress of Viewers During the COVID-19 Pandemic

Yulan Tan

Postgraduate Student, Beijing Foreign Studies University, Beijing, China

ABSTRACT. The COVID-19 pandemic is stressful for people, yet it has witnessed an exponential growth in the number of Virtual Youtubers and their viewers in China. Virtual Youtubers (VTubers) are people represented by virtual 2D or 3D avatars on different platforms. From the perspective of media psychology, an online survey of 669 participants was conducted to examine the intensity of Chinese viewers' parasocial attachment to the VTubers and the influence of such attachment on the stress of these viewers during the pandemic. As a result, positive correlations were found between viewers' parasocial attachment to VTubers and the stress relief of these viewers.

1 Introduction

Many people have been suffering from stress in response to COVID-19 pandemic [1]. This is a time when people are discouraged from having face-to-face conversations and have to adapt to a lifestyle restricted within their homes. To satisfy their emotional and interactive needs, more and more people watch VTubers streaming online. They are fascinated by the Vtubers with charming voices from the actors behind the avatars (i.e. Nakanohito), with attractive avatars that capture Nakanohito's facial expressions and body movements, and with Nakanohito's charisma as well. As a result, the companies related to VTubers developed rapidly during the pandemic in China. A case in point is ANYCOLOR (a Japanese entertainment company with Virtual Youtuber projects) which owns Vox Akuma whose 1.7-hour debut on Bilibili earned approximately 1,100,000 yuan overnight [2].

Previous research about the VTubers mainly focused on the ontological nature of VTubers and VTubers' relationship with the audience. Zhou suggests that the external form of VTubers' virtual bodies is shaped by the consumer society and hyper-modernity [3]. As for VTubers' relationship with the viewers, Lu discovered that the Otaku community thought that VTubers were mentally more distant to them compared with real-person streamers [4], and Huang claims that fans play a crucial role in shaping the multiplicity of VTubers [5]. In addition, Stein conducted an experiment to prove that despite viewers' more intense parasocial interactions with VTubers, these viewers believed that VTubers had less psychological human-likeness [6]. Furthermore, Luo mentions that the emotional connection between the viewers and VTubers is mainly manifested in the attachment of the viewers to the avatars of the VTubers when he investigates the embodiment of VTubers with a phenomenological approach [7].

However, it can be seen that though some critics notice the parasocial relationship and even the attachment built between VTubers and viewers, how these connections may affect the audience during the COVID-19 pandemic has not been thoroughly studied. Therefore, from the angle of media psychology, this paper analyses the parasocial attachment formed by the viewers to VTubers and its influence on the stress of these viewers in the COVID-19 era with an online survey, hoping to supplement the parasocial attachment theory with fresh evidence and enrich the research about VTubers' effects on people in these trying times.

2 Literature Review

In the field of media psychology, the relationship between a public figure and his or her audience is usually explored by scholars with parasocial theory. Horton and Wohl observe that when watching television, the viewers are engaged in parasocial interaction (PSI) with the performers on television, and that the two seem to be in a parasocial relationship (PSR) [8]. This work belongs to the early literature in parasocial theory, in which PSI and PSR are used interchangeably to refer to a usually one-sided and nonreciprocal manner in which the spectators interact with the media personae. Then, the two concepts are distinguished more clearly by recent scholars, with PSI meaning the viewers' one-sided perception of the performers during the time of actual viewing; and PSR meaning the lasting processes which include affective and cognitive responses continuing outside the viewing time

tanyulan9@outlook.com

[9].

Parasocial attachment (PSA) refers to the progression of a PSR to the point when the media figure becomes a source of perceived safety and security [10]. Stever argues that many similarities exist between the attachment to parasocial characters and the attachment in face-to-face relationships, and Stever also believes that forming parasocial attachment to celebrities may help ease people's discomfort and pain [11].

3 Methods

3.1 Platforms

An online survey was conducted among participants recruited through posts on different social media platforms in China to increase the diversity within the sample, including Bilibili (a Chinese video website where many VTubers stream), Weibo (the Chinese version of Twitter), and Baidu Tieba (an online forum like Reddit). The survey was implemented in Chinese in July, 2022.

3.2 Participants

In total, 669 participants completed the questionnaire online. For this study, 4 respondents are excluded from the analysis because they reported that they had never watched the live stream, video clips, or the work of a VTuber during the COVID-19 pandemic, so eventually the answers of 665 respondents are examined in this essay. 21.05 % of the respondents are identified as male (n = 140); and 78.95% as female (n = 525). All of the audience members of VTubers were under 41 years old, and most viewers were 18 to 25 years old (n = 445; 66.92%). The majority reported that they were still attending schools (*n* = 217; 37.27%) or were enrolled in higher education (*n* = 346; 52.03%). About 15.34% of participants reported that they were employed in different sectors or were seeking employment (*n* = 102).

Table 1. Demographic characteristics of respondents ($N = 665$).

Demographic Characteristics	*N*	%
Gender		
Male	140	21.05%
Female	525	78.95%
Age		
Under 18 years old	191	28.72%
18-25 years old	445	66.92%
26-30 years old	22	3.31%
31-40 years old	7	1.05%
Occupation		
Primary school students and Junior high school students	37	5.56%
High school students	180	27.07%
Undergraduates	305	45.86%
Postgraduate students and PhD candidates	41	6.17%
Civil servants	9	1.35%
Business managers and financial industry	9	1.35%
Planting, animal husbandry, construction industry	3	0.45%
Service, culture and education industry	10	1.5%
Internet industry	19	2.86%
Medical industry	6	0.9%
Other	46	6.92%

Note. N is based on the respondents who have reportedly watched the live stream, video clips or work of a VTuber during the COVID-19 pandemic.

3.3 Variables

Items were adopted from existing literature to measure the constructs in this study. The questionnaire was modified slightly to reflect the context of VTubing. Table 2 demonstrates the descriptive statistics of all items.

Parasocial Attachment (PSA). The first two items for parasoical attachment were derived from Stever's research [11], and the remaining three items in this category were adapted from the research of Rubin et al. [12], Tuchakinsky [13], and Wu et al. [14] to measure the intensity of parasocial relationship and the degree of its progression towards parasocial attachment. These items have a 5-point Likert scale that ranged from 1 (strongly disagree) to 5 (strongly agree).

Stress Relief. The three items were adapted from the perceived stress scale [15] and measured on a 5-point Likert scale ranging from 1 (strongly disagree) to 5 (strongly agree). A higher score indicates more obvious effects of relieving stress.

Table 2. Mean Scores (M) and Standard Deviations (SD) of the variables

Items	*M*	*SD*
Parasocial Attachment (PSA)		
PSA1: On a scale from 1 to 5, watching him/her during the pandemic tends to make me feel: (1 means uncomfortable and insecure; 5 means comfortable and safe)	4.51	0.652

PSA2: During the pandemic, his/her presence and companionship is one of the sources of my sense of security.	3.94	0.978
PSA3: I actively search for his/her information.	4.57	0.650
PSA4: I think that he/she could be a friend of mine.	4.04	1.029
PSA5: I often think of him/her, even when I'm not watching him/her.	4.16	0.968
Stress Relief (SR)		
SR1: During the pandemic, when watching him/her, I can temporarily forget about my worries in real life.	4.32	0.814
SR2: During the pandemic, ever since I began watching him/her, I have been more relaxed, and less nervous and stressed, than I was before.	4.22	0.858
SR3: During the pandemic, ever since I began watching him/her, I have been feeling encouraged by him/her and more confident in dealing with the difficulties in real life than I was before.	4.13	0.961

Note: N = 665.
Pandemic here means the COVID-19 pandemic.
Watching him/her means watching a VTuber's live stream, video clip, or work.

4 Results

The Cronbach's alpha of the essential eight items in this questionnaire (α= .904) shows the evidence of reliability among all measures of parasocial attachment and stress relief. As for the validity of these items, the Kaiser–Meyer–Olkin measure verified sampling adequacy (KMO = .914) and Bartlett's test of sphericity reached significance ($p < .001$). Moreover, a correlation analysis (Pearson's r) was conducted to look at the relationship between parasocial attachment and stress relief in general. Results indicate that stress relief was significantly and positively related to parasocial attachment (see Table 3), most notably to the sense of security in parasoical attachment (see PSA2). Generally, the more the respondents were parasocially attached to VTubers, the more they were relieved from stress during the COVID-19 pandemic.

Table 3. Correlations (Pearson's r) between Parasocial Attachment (PSA) and Stress Relief (SR).

	SR1	SR2	SR3
PSA1	.533**	.539**	.518**
PSA2	.577**	.594**	.604**
PSA3	.423**	.461**	.484**
PSA4	.508**	.579**	.593**
PSA5	.545**	.544**	.579**

*Note: N = 665; **p<0.01.*

Although the general results above clearly proved the positive correlations between parasocial attachment and stress relief, the difference of these correlations among various subgroups in the sampling frame is still worthy of further investigation. Firstly, gender may affect these correlations. To begin with, it can be found in the survey that women tended to form a stronger parasocial attachment to VTubers than men, with higher mean scores in every one of the five items used to measure the parasocial attachment. More importantly, the correlation between parasocial attachment and stress relief was stronger in women than men on the surface, with two bigger Pearson's r among the first two items of PSA and the first two items of SR. This means that when women were parasocially attached to VTubers, the positive effect of this attachment was more instant and only stayed on the level of mid-term emotional regulation. However, it seems that the effect of parasocial attachment affected men's stress in a more profound way, because when men considered VTubers as one of the sources of their security during the pandemic, they felt more encouraged than women to overcome the difficulties in real life. Therefore, rather than being influenced by the instant effect and mid-term emotional regulation effect of PSA, men were more affected by its motivating effect to address the actual stressful problems in the harsh reality of pandemic.

Table 4. Correlations (Pearson's r) Between Parasocial Attachment and Stress Relief in Men (n=140) and Women (n=525).

	SR 1		SR 2		SR 3	
	Men	Women	Men	Women	Men	Women
PSA 1	.493**	.540**	.509**	.545**	.467**	.529**
PSA 2	.514**	.592**	.558**	.601**	.650**	.582**

*Note: **p<0.01.*

Secondly, age can influence the correlations between PSA and SR. It is worthy of note that although children and teenagers under 18 had the second highest mean scores in items related to PSA across all age groups, it was in the third group with people from 26 to 30 that the correlations between PSA and SR was the most obvious. This means that for people from 26 to 30, forming a strong parasocial attachment to VTubers significantly helps them ease the stress during the pandemic. In addition, the fourth group with people from 31 to 40 had highest mean scores and the smallest standard deviations in all of the items, but this group does not worth to be discussed in detail because this group had only seven samples and the significance of all of its correlations was lower than the 0.05 benchmark.

Table 5. Correlations (Pearson's r) Between Parasocial Attachment and Stress Relief in People who were under 18 (n=191), 18-25 years old (n=445), 26-30 years old (n=22) and 31-40 years old (n=7).

	SR 1				SR 2				SR 3			
	Under 18	18-25 years old	26-30 years old	31-40 years old	Under 18	18-25 years old	26-30 years old	31-40 years old	Under 18	18-25 years old	26-30 years old	31-40 years old
PSA 2	.552**	.575**	.699**	0.471	.544**	.604**	.742**	0.600	.546**	.600**	.812**	0.710

Note: ** $p<0.01$.

Thirdly, occupation may also be an influential factor in this correlation. There are differences between people who were studying and working. Students had higher mean scores and smaller standard deviations in the items related to PSA, indicating that students tended to build a stronger parasocial attachment to VTubers than people who were working. However, it was the people who were working who experienced more effective and deeper stress relief from their parasocial attachment to VTubers during the pandemic.

Table 6. Correlations (Pearson's r) Between Parasocial Attachment and Stress Relief in People who were studying (n=563) and working (n=102).

	SR 1		SR 2		SR 3	
	Studying	Working	Studying	Working	Studying	Working
PSA 2	.564**	.634**	.574**	.674**	.589**	.665**

Note: ** $p<0.01$.

5 Discussion

5.1 The Intensity of Parasocial Attachment to VTubers

The parasoical relationship between fans and real-person celebrities has been thoroughly studied in the past, yet the viewers' parasoical attachment to the rising VTubers on the Internet during the pandemic has not been paid enough attention in the academia. Hence, the present study conducts a survey to look at how intense the parasoical attachment to VTubers is among their audience, and compares such intensity in different subgroups. It has been found that women in the subgroup of gender, people under 18 in the subgroup of age, and students in the subgroup of occupation tended to form a stronger parasoical attachment to VTubers during the COVID-19 pandemic. This may help identify the groups that have a strong parasocial attachment to VTubers more quickly in the future research.

5.2 The Influence of Parasocial Attachment to VTubers on the Audience

The findings in this paper strongly support the soothing effect of parasocial attachment proposed by Stever [11]. It is clear that there were positive correlations between the parasocial attachment (PSA) to VTubers and the stress relief for the audience of VTubers during COVID-19. Moreover, this positive effect of PSA on stress relief during the pandemic was manifest in different aspects in women and men, and more manifest in people from 26 to 30 in the subgroup of age and people who were working. This can help us see how the parasocial attachment to VTubers may positively influence different types of viewers.

6 Conclusion

This study confirms the positive correlations between the viewers' parasocial attachment to VTubers and the stress relief of these viewers in general. It also provides evidence for Stever's belief that people's parasoical attachment to media figures can help reduce people's discomfort and pain. Moreover, the soothing effects brought by such parasoical attachment to VTubers can vary in aspects and degrees among different subgroups of viewers. These results may inspire the future research related to parasocial attachment and the influence of VTubers, and provide guidelines for the people watching VTubers in the COVID-19 era. Despite these theoretical contributions, the current research has a few limitations that deserve the attention from future researchers. Firstly, since the participants were recruited randomly on the online platforms in China during the pandemic, the sampling profile was inevitably biased to some degree. In the future, it is suggested that a survey using stratified sampling should be conducted to replicate this study in different countries to make sure that there are diversified VTuber viewers from various backgrounds and adequate samples in each subgroups of VTuber viewers. Thus, variations of the correlations between PSA and SR can become clearer and more convincing. Moreover, because this study is based on self-reported data in an online survey, there can be bias caused by dishonest respondents. Therefore, controlled experiments and longitudinal surveys should be conducted to further test and refine these findings in the future.

References

1. C. Wang, R. Pan, X. Wan, Y. Tan, L. Xu, C.S. Ho, R.C. Ho, Immediate psychological responses and

associated factors during the initial stage of the 2019 coronavirus disease (covid-19) epidemic among the general population in China, International Journal of Environmental Research and Public Health, 17(5), 2020, 1729. doi: 10.3390/ijerph17051729.

2. Z. Zhao, J. Zhang, Virtual Youtuber earns a million in a 2.5-hour stream, Ifeng, June 28, 2022, Retrieved July 15, 2022, from https://i.ifeng.com/c/8HBoRXAQ6HF.
3. B. Zhou, Body-technology Interaction: the Example of Virtual Youtubers, Public Communication of Science & Technology, 13(20), 2021, pp. 161–164. doi: 10.16607/j.cnki.1674-6708.2021.20.049.
4. Z. Lu, C. Shen, J. Li, H. Shen, D. Wigdor, More kawaii than a real-person live streamer: Understanding how the Otaku Community engages with and perceives virtual youtubers, Proceedings of the 2021 CHI Conference on Human Factors in Computing Systems, 2021. doi: 10.1145/3411764.3445660.
5. Y. Huang, H.Chen, K. Fang, A Study on the Influence of Fan Participatory Culture on the Development of Virtual Youtubers, New Media Research, 8(09), 2022, pp. 101–105. doi: 10.16604/j.cnki.issn2096-0360.2022.09.012.
6. J.P. Stein, B.P. Linda, N. Anders, Parasocial interactions with real and virtual influencers: The role of perceived similarity and human-likeness, New Media & Society, 2022, doi: 10.1177/14614448221102900.
7. W. Luo, The Embodied Communication of Virtual Youtuber in the context of Technological Philosophy, Public Communication of Science & Technology, 14(05), 2022, pp. 131–133. doi: 10.16607/j.cnki.1674-6708.2022.05.033.
8. D. Horton, R.R. Wohl, Mass communication and para-social interaction, Psychiatry, 19(3), 1956, pp. 215–229. doi: 10.1080/00332747.1956.11023049.
9. J.L. Dibble, T. Hartmann, S.F. Rosaen, Parasocial interaction and parasocial relationship: Conceptual clarification and a critical assessment of measures. Human Communication Research, 42(1), 2016, pp. 21–44. doi: 10.1111/hcre.12063.
10. G.S. Stever, Parasocial Theory: Concepts and Measures, In P. Rössler, C.A. Hoffner, & L. van Zoonen (Eds.), The International Encyclopedia of Media Effects, essay, John Wiley & Sons, Inc., 2017, pp. 1–11.
11. G.S. Stever, Mediated vs. Parasocial Relationships: An Attachment Perspective, Journal of Media Psychology, 17(3), 2013, pp. 1–31.
12. A.M. Rubin, E.M. Perse, R.A. Powell, Loneliness, Parasocial Interaction, and Local Television News Viewing, Human Communication Research, 12(2), 1985, pp. 155–180. doi: 10.1111/j.1468-2958.1985.tb00071.x.
13. R.H. Tukachinsky, Para-romantic love and para-friendships: Development and assessment of a multiple-parasocial relationships scale, American Journal of Media Psychology, 2011, pp. 73-94.
14. Y. Wu, Y. Sun, N. Zhu, X. Shi, The Influence of Otome Video Games on Female Gamers' Views of Ideal Love, Youth Studies, 2020, pp. 50–70, 96.
15. S. Cohen, G.M. Williamson, Perceived Stress in a Probability Sample of the United States, The Social Psychology of Health, 1988, pp. 31–67.

Facebook Cyber Security Evaluation

Jinnan Sun[1,a*]

[1]Columbia University, School of professional Studies, New York, United States

Abstract. This evaluation is focused on conducting a cybersecurity risk assessment on Facebook, which will map their current practices against the NIST CSF Framework's 4 of the 5 functions. The evaluation will first list the research of Facebook's data breach from public information, then will show the analysis of missing control based on the NIST Framework and provide a logical connection between them.

1. Introduction

1.1 Background

Facebook is an online social media and social networking service owned by Meta Platforms. Facebook has had its fair share of scandals in recent years. The latest one on Facebook's list of problems is "Data Breach" [1].

In 2021, the personal data of more than 530 million Facebook users were made public on online forums and this data was already captured in 2019. Since 2010, Facebook has been frequently found to have security vulnerabilities and data breaches.

1.2 Research aims

The evaluation will evaluate and assess Facebook's cybersecurity capacities based on **NIST** Framework's 4 functions (Identity, Protect, Detect, Respond) and identify the missing control.

After assessing the cybersecurity capacities, the evaluation will apply the **CMMI** Cybersecurity model to measure Facebook'smaturity and roadmap

1.3 Risk assessment results

The risks will be identified in 5 levels based on the estimation of their impact and likelihood, which forms a **Heat map**. The numbers of risks and their risk levels are listed in the table below:

Table 1 risk assessment result self-drawn

Risk level	Number of risks identified
Very low	0
Low	0
Medium	3
High	1
Very high	1

*Corresponding author: E-mail: 1280538273@qq.com

The risk assessment is conducted to identify potential risks associated with the missing NIST Framework functions, which are listed in the below table.

Table 2 NIST Framework functions self-drawn

NIST Framework			
#	Risk Category	Description	Risk rating
1	Identity ID.GV-3	Weak legal awareness and didn't comply with GDPR	Medium
2	Protect PR.AC-1	Not only authorized individual has access to data	Medium
3	Protect PR.AT-5	Lack of security and responsibility training on employees	High
4	Detect DE.CM-1	Lack continually monitor of potential vulnerabilities	Medium
5	Respond RS.MI-1	Lack of emphasis on containing known risks	Very high

CMMI cybersecurity platform:

In this evaluation, after assessing the risks based on NIST Framework, the CMMI cybersecurity platform will be used to identify enterprise risks and prioritize a roadmap for building organizational resilience based on the seven determined missing controls and associated risks evaluating maturity level.

1.4 Literature review

NIST framework has become a hot topic in the recent several years as a high-profile data breach is on the rise. The domestic and foreign academics are relatively active in studying the concept of the NIST framework and how to be used [2]. NIST framework was published to help regulators and industry players identify and mitigate cyber risks. The author Shen Lei mainly talks about three components and four different ways of using the NIST framework in the literature [3]. Barbara Krumay also addressed NIST could be extended to other topic areas, he pointed out that NIST should be engaged more in detecting, responding, and recovering from incidents in his paper [4].

Scholars have analyzed NIST concepts in detail, however, combining the NIST framework with real data breaches and using the CMMI framework to help solve the current situation of enterprises has rarely been studied directly.

2. Risk assessment & recommendation

2.1 Identify

ID. GV-3: Legal and regulatory requirements regarding cybersecurity, including privacy and civil liberties obligations, are understood and managed

Facts:

Facebook didn't comply with the GDPR's user privacy policy. So, over 530 million Facebook users' data were posted publicly in August 2021, these data was scraped from Facebook in 2019 [5]. However, it didn't notify these users about this scrap of personal data as it fixed the vulnerability by Sep 2019.

Risks:

Facebook's legal awareness is weak.

Facebook will be fined for violating the laws of different areas and causing serious social impact and reputation damage.

Recommendations:

Facebook should be in contact with lawyers in different regions (especially in the US and Europe) to meet the legal requirements.

Stay on top of regulatory changes and be proactive about checking the updated standards and regulations.

Ensure that company policies and procedures are properly documented. That non-compliance is continuously monitored and corrected through the company's legal department, such as making sure the legal professionals work with compliance officers.

Risk level: Median Likelihood: Remote Impact: Catastrophic

2.2 Protect

PR. AC-1: Identities and credentials are issued, managed, verified, revoked, and audited for authorized devices, users, and processes

Facts:

Facebook doesn't pay attention to the restrictions on employee data access and misses the proper identification and access control to prevent unauthorized individuals from viewing and posting internal information.

Risks:

Third-party developers could view and change the data that they aren't entitled to access.

The internal information would be exposed to the public and hard to track who revealed it.

Recommendations:

The "CIA" of security should be strictly observed. Facebook should ensure that only authorized individuals could be able to access these data and create or change this information.

Facebook should develop a new system to monitor and report the authorized access control, and also should audit the authorized individuals' accounts regularly.

Risk level: Median Likelihood: Occasional Impact: Significant

2.3 Protect

PR. AT-5: Physical and cybersecurity personnel understand their roles and responsibilities

Facts:

These Facebook developers have not gone through rigorous training to make it clear to them that user data should be stored on internal, non-public servers. As a result, about 540 million users' records were captured by Facebook developers and stored in an Amazon public cloudpublic server in 2019.

Risks:

The developers didn't truly understand cybersecurity people's responsibilities. They would be attacked easily (such as phishing attacks) and more internal data would be revealed through their wrong operation.

Recommendations:

Instill the corporate culture and educate employees on cybersecurity before they start their work. Make them clearly understand the serious consequences if they fail to maintain cybersecurity.

Regularly hold relevant training on cybersecurity, through the training to inculcate the importance of cybersecurity and how to achieve it.

Risk level: High Likelihood: Probable Impact: Significant

2.4Detect

DE.CM-1: The network is monitored to detect potential cybersecurity events.

Facts:

Facebook does not continuously monitor for potential cybersecurity vulnerabilities, resulting in vulnerabilities that arise going undetected promptly. So, an internal glitch in the Facebook system prevented the privacy settings from working, so the private posts of more than 14 million users were publicly revealed in 2018.

Risks:

The vulnerability becomes large and affects other operations, causing a major impact on the entire system and leading to disruptions of operations.

Recommendations:

Keep testing the vulnerabilities within the system, hold the internal stress test of it and produce a continuity plan.

Develop a monitoring system to continuously monitor the vulnerabilities

Document these vulnerabilities in detail and report to

specialists who're responsible to solve this cyber trouble and try to avoid this in the future

Risk level: Median

Likelihood: Remote Impact: Catastrophic

2.5 Respond

RS.MI-1: Incidents are contained.

Facts:

Most vulnerabilities identified in Facebook were not mitigated. From 2007, when Facebook first encountered massive privacy problems, to the present day, frequent user privacy attacks have shown that Facebook does not contain themseriously.

Risks:

Cyber vulnerabilities and potential risks will increase. It may be more difficult to deal with therisk afterward

Users lose confidence in Facebook and think Facebook is a company full of lies

Recommendations:

Use monitoring tools and analysis tools to screen and organize network security vulnerabilities, set up specialists or automated systems to help resolve vulnerabilities, and detailed records to avoid recurrence

Facebook could use Archer to analyze operating risks, and input the questionnaires on the platform and these questionnaires would be recorded

Risk level: Very high Likelihood: Frequent Impact: Catastrophic

3. CMMI level analyze

3.1 CMMI's current level & target level

Based on the above five determined missing controls and associated risks, the CMMI Maturity level is measured in seven areas: ensure governance framework, establish risk management, identify and manage risks, ensure risk mitigation, ensure risk mitigation, ensure risk detection, ensure risk exposure, and ensure resilience.

The CMMI Maturity level is scored by the CMMI Scoring methodology in five levels, which are shown in the histogram below:

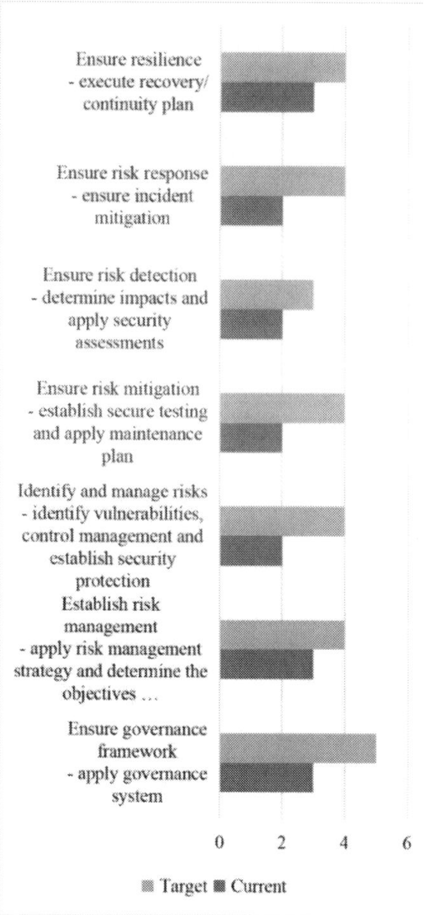

Figure 1: Histogram self-drawn

4. Conclusion

The cybersecurity risk in Facebook can be successfully assessed by the NIST framework was clearly expressed. CMMI model was combined with the NIST framework to address Facebook's current management weaknesses on five levels. Future work will investigate how to use CMMI's road map to formulate specific actions to solve management problems in different stages in an orderly manner, and how to use other frameworks to analyze other important risks within Facebook, such as market risk and financial risk.

5. Appendix

5.1 What is Archer

RSA Archer Suite is a risk management platform that provides solutions in enterprise operational risk management. Archer suites risk data from organizations and outputs risk analytics to provide an integrated risk report. These reports can help enterprises document the known risks and try to avoid them.

5.2 Heat map

This risk heat map offers data visualization for assessing the risks in Facebook through quantifying likelihood and impact. Risks are divided into five severity levels according to the depth of the color, the lightest cells represent the risk level is *very low*, and the darkest is *very high*.

Impact					
Catastrophic (5)	5	10	15	20	25
Significant (4)	4	8	12	16	20
Moderate (3)	3	6	9	12	15
Low (2)	2	4	6	8	10
Negligible (1)	1	2	3	4	5
	Improbable (1)	Remote (2)	Occasional (3)	Probable (4)	Frequent (5)
	Likelihood				

Very Low
Low
Medium
High
Very high

Figure 2 heat map self-drawn

5.3 What is "CIA"

The CIA of security.

C is confidentiality, to ensure that only authorized individuals could view information

It is integrity, only authorized people could create and change information

A is availability, to ensure that the data and system are available when authorized people want it.

5.4 CMMI scoring methodology

Standard definition of maturity

	Level 1	Level 2	Level 3	Level 4	Level 5
	Performed	**Managed**	**Defined**	**Quantitatively managed**	**Optimized**
People	General personnel capabilities may be performed by an individual, but are not well defined	Personnel capabilities achieved consistently within subsets of the organization, but inconsistent across the entire organization	Roles and responsibilities are identified, assigned, and trained across the organization	Achievement and performance of personnel practices and predicted, measured, and evaluated	Proactive performance improvement and resourcing based on organizational changes and lessons learned (internal & external)
Process	General process capabilities may be performed by an individual, but are not well defined	Adequate procedures documented within a subset of the organization	Organizational policies and procedures are defined and standardized. Policies and procedure support the organizational strategy	Policy compliance is measured and enforced. Procedures are monitored for effectiveness	Policies and procedures are updated based on organizational changes and lessons learned (internal & external) are captured
Technology	General technical mechanisms are in place and may be used by an individual	Technical mechanisms are formally identified and defined by a subset of the organization, technical requirements in place	Purpose and intent are defined (right technology, adequately deployed); Proper technology is implemented in each subset of the organization	Effectiveness of technical mechanisms are predicted, measured, and evaluated	Technical mechanisms are proactively improved base on organizational changes and lessons learned (internal & external)

Figure 3 CMMI scoring methodology self-drawn

References

1. Franklin D. Azar & Associates, P.C. (2022) Facebook (Data Breach) Retrieved August 8, 2022, from https://www.fdazar.com/practice-areas/class-action/facebook-data-breach/.
2. Heiligenstein, M. X., Jodi, Robert, & Smith, O. (2022, March 21) Facebook data breaches: Full timeline through 2022. Firewall Times.
3. Shen, Lei. (2014) The nist cybersecurity framework: overview and potential impacts. J. Scitech Lawyer, 10(4): 16-19.
4. Krumay, B., Bernroider, E. W., & Walser, R. (2018) Evaluation of cybersecurity management controls and metrics of critical infrastructures: A literature review considering the NIST Cybersecurity Framework. J. Secure IT Systems, 369–384.
5. Holmes, A. (2021). 533 million Facebook users' phone numbers and personal data have been leaked online. J. Business Insider.

The Effect of Algorithmic Recommendations on Female University Students' Willingness to Engage in Romantic Behaviour

1[st]Conghui Guo [1*,a], 1[st]Xuexin Guan[2,b]

[1]Institute of Media and Design, Beijing Technology and Business University, 102488, Beijing, China
[2]Institute of Applied Arts and Sciences, Beijing Union University, 100191, Beijing, China

Abstract. With algorithmic recommendations becoming an important way and means of information distribution, social media platforms are developing rapidly, and students in colleges and universities today have become the main audience for their information access. In this paper, through the application of TPB model in the intention to fall in love of young female groups in colleges and universities, four variables, namely attention, attitude, subjective norm and perceived behavioural control, were studied in depth, and a research model of factors influencing the relationship behaviour of young female groups was constructed based on the theoretical model of planned behaviour. According to the results, the factor of attention did not have a significant positive effect on relationship behaviour intentions, and behavioural attitudes, subjective norms and perceived behavioural control (PBC) had a significant positive effect on relationship behaviour intentions.

1 Introduction

1.1 Background

Algorithmic recommendations are based on tracking a user's online behaviour and using mathematical algorithms to calculate personal characteristics, environmental characteristics and other relevant information, and thus to deduce what the user might like [1]. In terms of trust in sources, women trust community forums and algorithmic news apps more; the younger the user, the higher the trust in Weibo [2]. The younger the user, the higher the level of trust in Weibo [2].

According to the "2020-2021 China Mobile Social Industry Research Report" published by Ai Media Consulting, as the public's awareness of mobile social networking increases, the proportion of female users and their activity level gradually increases, and their initiative in mobile social life also tends to increase [3]. In the process of seeking positive relationship practices and developing their own view of marriage, in a subtle way, social media quietly influences the willingness of some female university students' relationship behaviours.

1.2 Research Analysis

According to the existing literature, many scholars at home and abroad have conducted relevant research on the psychology of university students' behaviour. The Theory of Planned Behavior (TPB) has been applied to research [4], but the focus has been mainly on university students' job search and entrepreneurship, examination-based education and teaching research. Therefore, on the basis of the applicability of the TPB model in the study of algorithmic recommendation on female university students' intention to engage in romantic behaviour, this study proposes a hypothesis that the four dimensions of attention, attitude, subjective norms and perceived behavioural control will influence the behaviour based on the cognitive mediation model as well as the attention dimension. The influence of information in new media on personal thinking and value orientations has, to a certain extent, influenced the establishment of students' views on relationships [5].

1.3 Problem statement and objectives

The general structure of this paper is as follows: Firstly, we review and summarise the previously available literature and research. Secondly, based on the existing research, we present a theoretical model and our hypothesis. After that, analysed the survey process and the sampling data. Finally, an empirical test of the model and hypotheses of this study and the sharing of the findings.

*Corresponding author: [a]1908010211@st.btbu, edu.cn
[b] 2019010310010@buu.edu.cn

2 Literature Review

2.1 Algorithmic recommendations

Algorithms, as a new way of understanding communication and changing it, are a way of disrupting and transforming the existing status quo of communication [6]. In the 21st century, where social media and internet technologies are constantly evolving, this power is presenting itself in a more insidious and widespread way [7].

2.1.1 important results

In the age of Internet media, as a comprehensive ability of media cognition and utilization, has become a basic quality necessary for the public [8]. Wang Yuchen believes that the young people are dependent on the new media for their thoughts and consumption, and use the messages conveyed by the new media as a reference point for moral reference and values, transforming them into internal spiritual beliefs and behavioural intentions over time [9]. At the same time, based on Li Qiaowen's view that personalised recommendations can lead to an imbalance in the values of the audience, which can eventually lead to extremes in thought and behaviour [10].

2.2 Theory of Planned Behavior

The Theory of Planned Behaviour (TPB) is widely used to explain and predict human social behaviour. "Intention to act" is "the main direct determinant of 'behaviour'". Behavioural intentions are "influenced by both internal and external motivational variables, with the internal motivational variable being "behavioural attitudes" and the external motivational variable being "subjective norms"". Behavioural attitudes" are "individuals' positive or negative evaluations of the outcome of their behaviour"; "subjective norms" consist of "the level of motivation to align with the opinions of those who have a significant influence on them".

2.2.1 Important results

Li Yongqiang, Bai Xuan and Zeng Zheng made a study on the applicability of the Theory of Planned Behaviour in the study of college students' entrepreneurial intentions, illustrating the applicability and good predictive power of TPB in the college student population from different perspectives [11]. Meanwhile, Shaohui Liu has also studied the formation mechanism of college students' entrepreneurial intentions under the perspective of TPB theory, which gives us a new perspective [12].

1.1 Attitudes

According to the Theory of Planned Behaviour, attitude is the degree to which an individual's cognitive and affective evaluation of a particular goal-directed behaviour, positive or negative, towards performing a particular behaviour, reflects the actor's willingness to perform that particular behaviour [13, 14, 15]. Therefore, we have the following conjectures.

H1: Attitudes towards algorithmic recommendations related to relationship information pushing are positively correlated with behavioural intentions

1.2 Subjective norms

Subjective behavioural norms reflect an individual's willingness to perform a behaviour with the perceived influence of social pressure to behave or not to behave in a certain way [13]. therefore, the following conjecture was formulated.

H2: Subjective norms of willingness to fall in love behaviour are positively related to behavioural intentions

1.3 Perceptual behavioural control

Perceived behavioural control refers to the judgement of an individual's ability to control. In this study, which will examine perceived behavioural control of willingness to act in love, our conjecture is as follows.

H3: Perceived behavioural control of relationship intentions is positively related to behavioural intentions

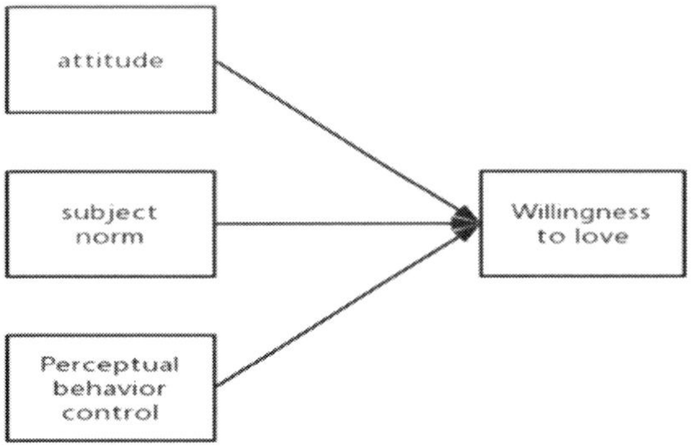

Fig. 1. Conceptual Model
Source. Self collection of pictures

2.2.2 Summary.

Currently, new media networks are quietly changing our thinking patterns, behaviours and lifestyles, and have become the "third space" after classrooms and extracurricular activities [16]. The algorithmic recommendation is based on the accurate "portrait" of the user to achieve the accurate distribution of information needs of the user, and its logic has three main mechanisms, namely content-based recommendations, collaborative filtering recommendations and recommendations based on association rules [17].

In the new media environment, not only do college students' perceptions of love differ, and different individuals' relationship choices, motivations and attitudes vary widely, but various factors also influence college students' relationship patterns and attitudes at different levels and levels [18]. Nowadays, The bad talk of some online celebrity anchors and their close to "ruinous" view of love are like "poisonous tumours" that subconsciously influence the ideological and moral quality and behaviour of university students [19]. Therefore, exploring the relationship between algorithmic recommendations and the willingness of female university students to fall in love is an important topic in the current new media era and beyond.

3 Methodology

3.1 Research design

In this paper, the TPB Theory of Planned Behaviour was used to study the respondents' willingness to engage in romantic behaviour, and a questionnaire was used to collect data. In order to ensure the scientific validity of the questionnaire, this study strictly follows Ajzen's questionnaire model for questionnaire design and theoretical modelling, and on top of this it also consults questionnaires from existing research on relationship behaviour for question setting.

3.2 Data collection

The research subjects of this paper are female university students who go on and read algorithmic news pushing content about relationship topics, and the online questionnaire survey was conducted through the questionnaire star platform in the first stage, and the data was statistically processed using spss26.0 software in the later stage. A total of 382 questionnaires were collected, excluding 28 invalid questionnaires (short response time, respondents who have used algorithmic recommendation media, etc.), the number of valid questionnaires was 354, with a valid return rate of 92.6%, according to which, this paper carried out data analysis and model testing.

3.3 Data analysis

After the questionnaires were returned, the data collected were screened to exclude those who did not meet the age requirement and those who had never used social media to view relationship-related topics, and the data were analysed and processed using SPSS 26.0 software (Table 1).

Table 1. Summary of measurement items

Measurement	Projects
Concerns	1. You often browse relationship topics
	2. You are interested in the topic of relationships
	3. Have you read relationship-related tweets
Attitude	4. How often do you get tweets about relationship-related content?
	1. How much do you like the algorithm's recommendations for reading information on relationship topics?
	2. What do you think about big data pushing relationship-related topics
Subjective norms	1. You believe that relationship-related content can give you some positive guidance on emotional issues
	2. Publicity from online media influences you to pay attention to the content of relationship topics
	3. Positive feedback from friends and family will influence you to use the algorithmic medium to read content on relevant topics
Perceptual behavioural control	1. You prefer to use big data recommendations to get content about relationship topics than other ways
	2. I think the content pushed by Big Data can help me with my emotional problems
	1. I would like to read about relationship topics and gain experience and knowledge from them
Willingness to act	2. When faced with relationship difficulties, you would consider talking openly with your online friends
	3. Consider seeking advice and help from online friends when you are facing relationship difficulties
	4. In your future life, you will draw on relevant relationship tweets to fall in love

Data collection and statistics by authors

The study found that the majority of the effective age group were in their undergraduate studies, accounting for 69.9%, with 137 female university students (36.9%) having had 1-2 relationship experiences. The percentage of female students who like to browse relationship content is 33.9%. 34.46% of female university students were tweeted once a day and 33.05% of female students were tweeted multiple times a day (Table 2).

Table 2. Status of individual projects

	Projects	Percentage of
Grade Level	Undergraduate studies (freshman and sophomore)	30.63
	Undergraduate studies (junior and senior year)	39.27
	Studying for a Master's degree	17.28
	Doctoral studies	12.83
Relationship experience	Single	27.75
	1-2 relationships	36.91
	3-4 relationships	20.16
	5 paragraphs or more	15.18
Push frequency	Multiple times a day	33.05
	Once a day	34.46
	Once a week	19.12
	Once a month	13.28
Likeness	Very unlike	6.78
	Dislikes	9.6
	General	16.1
	Likes	33.9
	Loved it	33.62

Source. Self collection of pictures

4 Results

The reliability of the current questionnaire was examined using the Cronbach alpha coefficient, and the results showed that the reliability of the total questionnaire was 0.943, which was above 0.9, indicating that the questionnaire was very reliable. The reliability of the attitude, subjective norm, perceived behavioural control and behavioural intention among the subscales was above 0.85, and these four subscales also had a certain degree of reliability and stability (Table 3). In terms of construct validity, the KMO and Bartlett's sphericity tests conducted showed a KMO value of 0.971 > 0.5 and a Bartlett's significance of 0.000 < 0.001, indicating a good fit for factor analysis (Table IV).

Table 3. Reliability statistics

Name	Kronbach	Number of items
Concerns	. 058	2
Attitude	. 860	3
Subjective norms	. 853	3
Perceptual behavioural control	. 872	3
Behavioural Intentions	. 864	3

Source. Self collection of pictures

Table 4. KMO and Bartlett's test

KMO Number of sample suitability measurements.		.971
Bartlett Sphericity Test	Approximate cardinality	3754.746
	Freedom	91
	Significance	.000

Source. Self collection of pictures

In terms of correlation, SPSS 26.0 was used for correlation analysis, and the results showed that there were significant two-by-two correlations between attitudes, behavioural norms, perceived behavioural control and behavioural intentions, with all correlations above 0.80 and all positive correlations. This indicates that the theoretical model constructed and the research hypotheses proposed are all reasonable (Table 5).

Table 5. Relevance

	Concerns	Attitude	Code of Conduct	Perceptual behavioural control	Behavioural Intent
Pearson Correlation	. 704**	.889**	.863**	.862**	1
Significance	.000	.000	.000	.000	
Number of cases	354	354	354	354	354

Source. Self collection of pictures

For the regression analysis, R-squared, ANOVA and coefficient table tests were used, with an R-squared of 0.832 and an adjusted R-squared of 0.830, which showed a small deviation from the adjusted R-squared and more stable data (Table VI). The significance of ANOVA was 0, which was less than 0.05, indicating that the regression equation was significant (Table VII). In the coefficient tables, the significance of attitude, subjective norms and perceived behavioural control was less than 0.05, indicating that the variables had a significant effect on the dependent variable. The significance of the level of concern was greater than 0.05, indicating that the two variables had a non-significant effect on the dependent variable (Table VIII).

Table 6. Summary of the model

Models	R	R Square	Adjusted R-squared	Errors in standard estimates
1	.912[a]	.832	.830	.460593487965111

a. Predictor variables: (constants), perceived behavioural control, attention, behavioural norms, attitudes

Source. Self collection of pictures

Table 7. ANOVAa

Models		Square and	Freedom	Mean Square	F	Significance
1	Return to	365.935	4	91.484	431.229	.000[b]
	Residuals	74.039	349	.212		
	Total	439.974	353			

a. Dependent variable: behavioural intentions

b. Predictor variables: (constants), perceived behavioural control, attention, behavioural norms, attitudes

Source.Self collection of pictures

Table 8. Coefficientsa

		Unstandardised factor		Standardisation factor		
Models		B	Standard errors	Beta	t	Significance
1	(Constant)	.196	.148		1.331	.184
	Concerns	-.081	.043	-.041	-1.860	.064
	Attitude	.217	.055	.208	3.921	.000
	Code of Conduct	.450	.054	.435	8.276	.000
	Perceptual behavioural control	.329	.050	.311	6.563	.000

a. Dependent variable: behavioural intentions

Source. Self collection of pictures

The adjusted R-squared of the independent variables on behavioural intention in the whole model is 0.830, which means that the degree of explanation of attitude, behavioural norms and perceived behavioural control on behavioural intention is 83.0%. At the same time, the significance p-values of attitude, behavioural norms and perceived behavioural control were all less than 0.05, indicating that attitude, behavioural norms and perceived behavioural control can all influence behavioural intention and are positive. In summary, hypothesis H1 H2 H3 is valid.

5 Conclusion

"Algorithmic recommendation technology has brought dividends and opened up a Pandora's Box of questions about the invasion of privacy and the bias of data [20]. In the context of algorithmic recommendation technology, new technologies are influencing the thinking, perceptions and perceptions of the world of post-90s university students at an accelerated pace [21]. Through reviewing the literature and reading previous studies, the main focus of these studies has been mainly on students' guidance and reflection on psychological problems, and very few studies have looked at the issue from the students' own perspective [22].

In this study on the impact of algorithmic recommendations on female university students' intentions to engage in romantic behaviour, it explores whether the level of attention paid to algorithmically recommended networks by the female youth cohort affects their consumption intentions, and examines the impact on intentions to engage in romantic behaviour from four perspectives: information attention, attitudes, competent norms, and perceived behavioural control. In addition, the survey population for this paper's study was female university students at school, covering only a small social group. What could be considered in future research is to expand the research subjects and further analyse the potential influencing factors of willingness to engage in romantic behaviour. In the course of the study, this research again validates the TPB theory and hopes that more excellent studies will emerge in the future.

References

1. Xu Youfei. (2018). The "information cocoon" effect of algorithmic recommendation and its reflection: the example of personalized news recommendation of "Today's headlines". Journal of Journalism Research (14), 42-43+49.
2. Yu, G. & Fang, K. Ren. (2020). Does algorithmic content pushing lead to information cocooning? --An empirical analysis based on media diversity and source trust. Shandong Social Science (11), 170-174+169. doi:10.14112/j.cnki.37-1053/c.2020.11.026.
3. Hu, Yu-Lan. (2021). A study on the impact of social media use on users' survival status during sudden public health events: An example of Sina Weibo use during the New Crown Pneumonia outbreak. New Media Research (18), 49-55. doi:10.16604/j.cnki.issn2096-0360.2021.18.012.
4. Zhang Min, Meng Die, & Zhang Yan. (2020). A conceptual model of social media users' continuous use behavior in the framework of "use-satisfaction" analysis. Journal of Information Resource Management, 10(1), 92.
5. Qiu Lu. (2019). An analysis of the impact of the new media era on the new era of college students' view of love. Chinese and foreign entrepreneurs, 4.
6. Yu, G. & Zhao, W.Y.. (2020). Algorithms as a new view of communication: the future of communication and the reconfiguration of communication science. Journal of Southwest University for Nationalities (Humanities and Social Sciences Edition) (05), 145-149.
7. Yu, G., Yang, Y. & Yan, Qiaomei. (2018). Algorithm as power:The power revolution of algorithmic paradigm in news communication. Editor's Friend (05), 5-12. doi:10.13786/j.cnki.cn14-1066/g2.2018.5.001.
8. Jue Wang. (2020). Public media literacy in the post-epidemic era. Journalism and Writing (08), 4.
9. Wang Yuchen. (2020). A study on the influence of new mainstream media on the values of youth groups. Journalism Research Guide (05), 58-60.
10. Li Qiaowen. (2021). Research on the influence of algorithm recommendation news on audience values and its guidance mechanism (Master's thesis, Shandong University of Finance and Economics).
11. Li YQ, Bai Xuan, Mao Yu, & Zeng Zheng. (2008). An analysis of factors influencing students' entrepreneurial intentions based on TPB model. China Soft Science, (5), 122-128.
12. Liu Shaohui. (2012). Research on the formation mechanism of college students' entrepreneurial intention from the perspective of TPB theory improvement (Doctoral dissertation, Harbin Institute of Technology).
13. Ajzen, I. (1991). The theory of planned behavior. Organizational behavior and human decision processes, 50(2), 179-211.
14. Fishbein, M., & Ajzen, I. (1977). Belief, attitude, intention, and behavior: An introduction to theory and research. Philosophy and Rhetoric, 10(2).
15. Swan, J. E., & Trawick, I. F. (1981). Disconfirmation of expectations and satisfaction with a retail service. journal of retailing.
16. Su Wenming, Wang Zhengshu, & Feng Jiao. (2015). Research on the alienation of college students' psychological characteristics and countermeasures in the new media era. Educational Review, (1), 75-77.
17. Lin Saifa. (2022). The negative impact of algorithm recommendation on college students' thinking and response. Journal of Fujian Jiangxia College.
18. Wang Cuntong. (2011). An empirical study of contemporary college students' relationship identity and behavioral choices. Youth Studies, (3), 22-29.
19. Nayla Yarbaiti & Zhang Zhen. (2020). The negative influence of various mobile phone social software on contemporary college students' view of love and the investigation of countermeasures. Modern Business Industry (04), 74-75. doi:10.19311/j.cnki.1672-3198.2020.04.034.
20. Zhou Min. (2021). The negative impact of "algorithmic recommendation" and strategies to cope with it. Journal of Hubei Second Normal College (12), 98-105.
21. Gu Quan. (2017). Exploring the thought and behavior patterns of post-90s college students in the context of big data. Collection, 7.
22. Li T., Zheng Y. X. & Yan Yuteng. (2022). Have marriage and childbearing been de-institutionalized in China? --Findings and discussion based on a survey of Chinese university students' views on marriage and childbearing. Women's Studies Series (03), 85-102.

The Historical Transition of All-China Women's Federation

Muye Tian*

Beijing National Day School

ABSTRACT: All China Woman's Federation (ACWF) is the sole official institute dedicating to gender equality in China, and had served for more than 76 years of history. Recently, however, the recognition of the organization among the public has been noticeably falling. In hope to answer the question "why and through what process has the influence of ACWF fell", the researcher explored the development of Chinese public's perception over ACWF's role during the 1946 - 1978 period, in which the organization first took shape, by chiefly looking into the A Brief History of Contemporary Chinese Women's Movement, a document written by a sub department of ACWF, largely academic and functions more as an historical research, and a list of most frequently used words by the largest, government funded Chinese newspaper People's Daily in describing ACWF during the period. The researcher finds notable discrepancy between the two sources, which helps to explore the power dynamic of the ACWF and its direct supporter, the CPC government, as well as the image that CPC government has propagandized its people to think of the ACWF. The research functions as part of a longer research that examines the change of Chinese public's recognition over ACWF for the entirety of its history.

1 Introduction

There are many institutions that have participated the course of gender equality and women's right. However, few are like All-China Women's Federation (ACWF), more frequently referred to as "Fu Lian" by Chinses people. Fu Lian is considered as the largest women's organization [1, 2]. The uniqueness of Fu Lian can be found in its nature as a semi-government department. Being a government department, ACWF has more powerful measurements to change women's lives than other civil organizations. The power is, however, largely given by the state government, and ACWF's decisions are often influenced by the state government [3].

As the engraving scenes in The Gender of Memory: Rural Women and China's collective past presented, Fu Lian played a crucial role in the collective era of China as a progressive force that pushed the live quality of women forward. However, by investigating people's awareness of Fu Lian in 2022, in a sample size of 100 people, merely 16 percent of them knows how to contact the organization and only 9 percent of the subjects reorganized it as a "very important" organization. This no wonder raised the question: why and through what process has the influence of ACWF fell?

The paper aims to examine the development of Chinese public 's perception over ACWF's role in contemporary China, specifically from 1946 – 1978, a stage when the institution first took shape.

By chiefly reviewing the titles of reports on ACWF on People's Daily, the largest new paper under the direct control of the central party of the Communist Party of China (CPC), on a ten year basis and comparing it to a document, A Brief History of Contemporary Chinese Women's Movement, by Women's Studies Institute of China (WSIC), an official research institution of women's right and gender issue serving both the ACWF and the central party committee, the research discovers remarkable discrepancy between the actual actions conducted by ACWF and the reports presented by People's Daily which shapes people's perception of ACWF.

The paper is part of a larger study that examines a larger time span of ACWF's development and answers three questions, "what ACWF was, what it is, and what it would be in the future", of which this paper corresponds to the first question: "What ACWF was".

2 Primary Sources & Methodology

The main source recognized as the "official document" of ACWF's action is A Brief History of Contemporary Chinese Women's Movement, which will hereafter be referenced as "the WSIC document" or The Brief History [4]. The WSIC document is one of the most important sources of ACWF's history, as WSIC is a research department serving directly to ACWF, therefore making it ACWF's "official historical record". Though the WSIC document, as the its name implies, is a general documentation on the modern women's movement in China from 1949-2000, its central subject is still the development of ACWF, as it was the main push force behind most of the major progress of the women's

*Corresponding author. Email: muye_tian@163.com

movement during the 1949–2000 period.

The writing style of the documentation is described by the summary as "an organic combination between macro and micro history" and is compiled with "solid references" [4], yet the book's emphasis on many of ACWF's actions are insufficient, generally documenting the history of ACWF in the way that the CPC preferred it to look like. Most notably, it lacks specific evidences that reflect the public's reaction to ACWF's policy. Sequencing wise, the book divides the development of women's movement as 5 stages from 1949 – 2000, primarily according to important events in the history of CPC [5]. Additionally, the publication date of the book is also worth of mentioning. Though the book is a documentation on the history of ACWF from 1949 – 2000, the book is published only in 2017.

The second main source that the research references is a sheet of titles of reports on ACWF by People's Daily from 1946 – September 2021, created by the researcher. The reports on People's Daily are different from the WSIC documentation because as a newspaper under the direct control of CPC, its reports served the purpose of national propaganda, demonstrating the state power of CPC. The image of ACWF presented by People's daily is thus the direct reflection of what CPC expect people to know and to think about ACWF and would help to explain the decline of ACWF's influence among the public of China through the lens of central government's propaganda on ACWF to its people. Though the WSIC document is also filtered through the political lens of the CPC, it should be noticed that the document is most likely not targeted towards the mass public as average Chinese citizens would have little incentive to purchase a documentation surrounding the very specific topic of women's movement. Additionally, the news reports are filtered through a set of dynamically changing political lens from the 1946 to 1978 while the WSIC publication is very likely filtered through merely the political lens of the 2017 CPC government. Another detail to consider is WSIC's major role as a department of ACWF, which allows it to be affected less directly by the central government. Thus, A Brief History of Contemporary Chinese Women's Movement is at some degree a self-documentation of ACWF, while the reports presented on People's Daily is more directly the reflection of CPC's characterization on ACWF.

To have a bird view of the mass data counts, the author first extracted the report titles from People's Daily and then separated them as ten-year periods from 1946 to 1978. Noticeably, this division is not made based on the historical events, therefore the periods divided in the investigation cannot fully overlap with the records on the WSIC document. After separating the titles on a ten-year basis, the author filtered the titles of every ten-year period through an online word-frequency-analysis site called "weiciyun" (may be briefly translated as Micro Word Cloud). With weiciyun's built-in dictionary, key words used in the reports are sequenced by the frequency they are used on several Excel sheet. The most frequently used terms of the decades are used to represent the state's characterization on ACWF during these periods. While importing the titles to the weiciyun filtration, the author removed certain words that may reflect little about the image of ACWF. Incomplete words and words that may refer to the name of ACWF are removed from the word frequency ranking lists. The word frequency ranking lists may be found at the end of the paper as appendixes.

Apart from the use of the two main sources, the author also referenced the work of Gail Hershatter, The Gender of Memory: Rural Women and China's collective past to examine the local impact of ACWF's acts, as the two main sources are broad official records on the macro effects of ACWF, and introducing Hershatter's work would allow the effects of ACWF to be better evaluated at a micro scale as the book include valuable record from the local gazetteers regarding the effect of ACWF actions.

3 1946-1956 Development Analysis

1946-1956, the first ten years of People's Daily's reports and the records on The Brief history showed a preference correlation. Noticeably, reports on the international voicing of ACWF significantly outnumbered reports on their domestic efforts. The most frequently used terms like "democracy" (Minzhu), "international" (Guoji), and "meeting" (Huiyi) generally all reflected the theme when traced back to the original report titles.

The theme of international voicing is also presented in the WSIC document. The WSIC document included the subtitle of "the opening of the new status for new-China's women's international conversation" as one of the major themes of Chin's women's movement in the 1946 – 1956 period, which largely overlaps with the report period and the phenomenon it presented. The "international conversation" presented by ACWF generally tends to be "international support", as many of the international voicing during the periods are supports to communist parties in other countries, chiefly Communist Party of Greece and the North Korea Communist Party during the 1946-1956 period. The subjects of international support during the era demonstrated ACWF's place as a central supporter to the central government's political decisions, including its diplomatic attitude.

Noticeably, however, when making direct conversation with certain institutions of the foreign countries, the organizations directly contacted by ACWF are often the countries respective women's right organizations than with diplomatic departments. This tendency may indicate the slight differentiation between ACWF's international image and CPC's, rendering ACWF as fundamentally a women's right institution even though their supports are often upon matters barely relevant to the subject of women's right, chiefly oppositions against worldwide wars.

Upon the theme of "international support", the second most notable trend of the decade is "meeting". Considering ACWF's creation in merely 1949, the first ten years is the rime that the organization refined itself as a proper national political institution.

Trending words of the decade provided firm support to the conclusion as the word such as "meeting" (Huiyi), "Representative" (Daibiao) and "big conference" (Dahui) The WSIC depicted the process as "the organization and refinement of women's working institution and women's organization, as well as the beginning of women's work" [5]. This trend essentially reflects ACWF's early-stage development through the form of hold meetings of various sizes and topics. The topics of the meetings, according to the report topics presented on People's Daily, include both international topics such as Asia women's conference (Yazhou funü huiyi) and local topics such as the propaganda of the new marriage law of 1950.

Overall, when referring to the result presented in the appendix, the reports of ACWF during the 1946-1956 period focuses more on the organization's international voicing efforts than their local efforts. Discrepancy is observed when People's daily's depiction on the image of ACWF is compared to the documentation of WSIC in their 1949-1956 period document. The WSIC document focuses chiefly on ACWF's local efforts, generally mobilization of women to participate various new policies passed by the central government. While the policies are often released by the central government, and some are simply general policies that treat women and men population with little difference, such as the land reformation policy, it is notable that the majority of policies propagandized by ACWF targets specifically women. Some of the most specified policies includes the abolition of prostitution, the new marriage law and the popularization of the new-style midwifery. Furthermore, it is noticed that even when dealing with the general policies, the ACWF would work form the lens of gender equality. In the matter of land reform, for instance, The Brief History documents that ACWF members focused on the education of rural women on their civil rights so that those who are long suppressed by the feudalistic norms may participate in the process of land reform [4].

Therefore, the image of ACWF presented by People's Daily reports and the image of ACWF presented by the WSIC document presented a preference correlation, a partially accurate propaganda on the effort of ACWF focusing chiefly on their international voicing and ignoring many of the domestic efforts of ACWF. Generally, however, both the sources highlighted ACWF's support to the central party's decisions but also showed its unique nature as an women's welfare institute.

4 The 1957 – 1967 analysis

In the 1957-1967 period, the high frequency words could barely reflect the efforts of ACWF in the WSIC document.

The Brief History, in its 1956-1966 chapter, recorded various efforts of ACWF on local topics. The efforts includes both involvement in government agenda and discussion of gender issue.

The involvement in the government agendas, such as the anti-right (Fanyou) movement and the great leap forward (dayuejin) movement. The involvements of ACWF, however, shows little about its identity and image. The ACWF essentially acts much similar as any other mobilization government agency would in involving the agendas. In the great leap forward movement, for instance, ACWF mobilized women to participate more intense labor, similar to what a people's commune would do in that period. Notably, however, though ACWF demonstrated high support to the central government's decision, the relationship between the two tightly connected groups deteriorated slightly in anti-right movement. The WSIC document pointed out that by the end of the movement, "the progression of women's movement was damaged" [4]. In September 1957, 6 of the women's representatives was identified as right wings in the third national women's representative meeting. According to the WSIC document, as the anti-right movement spread in the country, various civil women's organizations ceased to be active. The event ended with ACWF becoming one of the few organizations working on women's interests, yet, as the forementioned, the political activism of the central party has nonetheless made its relationship with ACWF more intense.

"Making the visible invisible" is one of Hershatter's main criticisms towards the communist party's revolution in women's social role. Hershatter argues that in the process of attempting to liberate women from housewifery, the communist party wrongfully ignored the agency of family and treated it as a remnant of the federal "old society", leading to depreciation and reflectance of domestic works [6]. Interestingly, there has been a discussion regarding the appreciation of domestic labor lasting from 1956 – 1957. The discussion was initially triggered by a letter published on the "Chinese Women" magazine in 1956, titled "is it ok for employed women to quit work in order to participate housewifery?", regarding the story of a women who was employed but find it more economically efficient if she stays at home and take care of her children than working while hiring a baby sitter. The publication led to intense civil discussion following the publication of later issues of "Chinese Women" on how employed women can resolve such problem. The discussion was so popular, to the point where many important political figures including the vice president of ACWF, head of personal bureau of the state council and vice-chairman Nenlai Zhou made statements in 1957, addressing that domestic works should be equally considered as any other forms of labors [4]. Domestic labor, however, as described by Hershatter, remains to be invisible after the declaration of government leaders. Two hypotheses can be made. The first being that reflectance of domestic labor is simply the result executional failure —— the proposal was somewhat successfully performed as a subsection of the "great leap forward", but eventually failed with the movement, as Hershatter described that the "great leap forward" movement aimed to socialize traditional domestic labors, but the collapse of the movement marked the end the attempt. Relating the statement to document of WISC, the sequence of events matches as the "great leap forward" movement was initiated in 1958, after the addresses were made, which explains how the attention to family realm was eventually lost. The second explanation would be that any attempts to further concern women's domestic labors were probably ceased by the criticism towards the proposal of "be industrious and thrifty in managing a household" (Qinjian chijia), one of the slogans made to encourage domestic labor in the "cultural revolution" activism starting from 1967.

The Brief History also documented ACWF's

participation on various agendas that aims to improve the health of women during the period. Among them, ACWF's participation in the childbirth reform (jiesheng gaige) and the birthrate control (jihua shengyv) agenda are the two most important ones [4]. The Brief History documented that by the December of 1962, the party central and the state council have together released the "instruction regarding the serious popularization of birth control plan" in response to the dramatic increase in birth rate following the economic and welfare improvement in the same year. The record indicated that by 1965, the birth rate control policy has achieved significant effect by dropping the birth rate from 7.5 from 1963 to 6.0 in 1965 [4].

The theme of the ten-year efforts is thus described in the publication as still political synchronization with the central government, as the executive agency of its proposal regarding women's interest. The newspaper report HFWs however, continues to focus on Fulian's function as an international speaker of the central government on international affairs. Various terms such as "Zhichi" (support), "Guoji" (international), and Zhidian (call) occupy the top of the frequency list. Even terms usually lead to the image of domestic revolutionary struggles in the 1950s and 1960s such as "Renmin" (people), "Douzheng" (struggle) leads to texts that refer to Fulian's efforts in supporting the agenda of foreign countries against the American political expansion at the time. Suggested by the HFW list, Japan and Cuba were most frequently mentioned by Fulian in its international voicing. Tracing back to the original text, on Japan, Fulian chiefly supported its feminism movements, and left-wing movements, more specifically Japanese people's act against the US-Japan security treaty; on Cuba, Fulian chiefly supported its effort against America's military invasion and political sanction [4]. Though increasingly tough supports on foreign political struggle are presented during the period, feminine tone is still presented in the voice of Fulian, as the word "Furen", referring to the first lady diplomacy, and "Chahuahui" (tea party), were both used for 34 times, merely 9 times less than the usage of the word "Riben" (Japan) and "Guba" (Cuba), the major political subjects of the ten years.

5 1968-1978 Analysis

The 1968-1978 decade is one of the darkest periods in ACWF's development. The first half of decade marks ACWF's major conflict with the central government and its period of complete standstill and by the latter half, its reconstruction.

In May 1966, the infamous "cultural revolution", a political purge that eventually cause national disturbance, took place, leading to cruel political conflict within the party. Many political and administrative activities were sabotaged by the political insurrectionists "the gang of four". ACWF suffered great lost in the event. The Brief History described the period as "The paralyze of Women's organizations and the standstill of women's work (work to improve the life quality of women)".

According to the WSIC document, by the second half year of 1966, the local women's federation of Tianjin, Shanghai, Beijing, Anhui, Heilongjiang, Hunan and Xinjiang have ceased all their work [4]. From the second half of the 1966 to 1967, ACWF ceased all its activities aside from formal mobilizations with the central union and central of the Communist Youth League, as well as necessary international affairs [4], however, according to The Brief History, even the international affairs of ACWF were disturbed as the number of affairs declined significantly. ACWF did not send representatives to the World Children's conference held by the Women' s International Democratic Federation (WIDF) in October 1966 and ceased to contact with the organization. By the November of 1967, ACWF stopped all its international communications. In June thirteenth 1968, ACWF was put under the "military monitor" of military representatives from the central government. In the same year's September, all ACWF organizations were dismissed by the newly formed "revolution council" [4].

The restriction of ACWF started at before the end of the cultural revolution, in 1973. A social comment from People's Daily in March of 1973 suggested that women women's organizations should be reorganized, started the reconstruction process. The reconstruction was quickly preformed in 1973, and by April 1975, all the districts of China had their women's representative meetings held [4]. Kang keqing was voted as the new president of ACWF, after Cai chang. She will lead the ACWF to redefine itself in the next decade.

Politic wise, ACWF displayed discrepancy with the central government, rejecting its radical leftwing agenda. Many officers of ACWF sacrificed their lives in the process, and some more wrongfully accused by the central government for unjust reasons [4]. Comparing to the earlier decades, the discrepancy between the institute and the party reached a pinnacle in during the cultural revolution.

The international voicing of ACWF also recovered, following the reconnection of China with UN in 1971 and the involvement in the UN's Commission on the Status of Women in 1974. The most remarkable achievement in the reconnection is probably the present of ACWF in the status commission's meeting in 1975 [4].

The recovery of medical system was also an important agenda in the recovery process of ACWF. Like the other recovery agendas, the process started in 1971. The two major achievements that falls into the category are the continuation and reinforcement of birth control policy, achieving even lower birth rate by the end of 1976 than the birth rate of 1968, and the recovery of medical facilities dedicated for women and children, which, unfortunately failed to recover to its 1966 state by 1976.

People's Daily's reports in the 1968-1978 period are extremely rare, merely 48 (including several reports on women's federations in other countries rather than ACWF) reports comparing to the 381 reports during 1957-1967, and 513 reports during 1946-1956. Despite of the scarcity of reports, the image of ACWF presented generally correspond to The Brief History, as most reports of the time reflect the reconstruction process of ACWF, as the most frequently used terms are generally related to the women's representative meeting as well as the election of Kang, such as "Zhaokai" (hold; holding meeting),

"Xuanchu" (elected), "Zhuxi" (president), "Shida" (tenth(referring to the 10th women's representative meeting held at the time)) and "Huiyi" (meeting) .

Noticeably, however, there are only three reports before the reconstruction of ACWF in 1971, which indicate that all the criticisms and abolitions actions conducted by the CPC government upon ACWF were not reported. It is unknown whether this is result of media focus of the time was on other events or that the CPC concealed related repots due to certain political concerns.

In general, the 1968-1978 period marks a time of great tension between the CPC and the ACWF, which concludes the early exploration stage of ACWF in its identity and agency as an axillary mobilization and diplomatic organization for the central government, specializing in gender issues. After the reconstruction of ACWF, the institute lost many of its early, experienced members and received many proletarian farmers instead , which led to a long restoration process of the institute's executive ability, as well as a redefinition on the institute's identity in the next decade.

6 conclusion

While the agency of ACWF in the 1946-1978 period includes both domestic mobilizations and international voicing, the research discovered that the CPC government seemed to prefer reports on its international activities. It is, however, no clear answer to the preference; the preference is arguably strange, given that the main influence of ACWF to the public would be on their local mobilization. It is possible that their mobilizations are displayed elsewhere, such as revolutionary mural or local newspapers, and that the international reports acted as a supplemental note on the organization's work. The research nonetheless has no evidence for the explanation and further research is required.

In the contemporary China, when feminism is again an active agenda in the political discourse of China, we need to rediscover the history of ACWF --- feminists under the supervision of the central government, in hope that we can learn more about the dynamic of civil right agendas and governmental intervention in the Chinese political environment.

REFERENCES

1. Zhang, Tong. Power, women and revolution: The role of the All-China Women's Federation in China's revolution. State University of New York at Binghamton, 2003.
2. Tsui, Justina Ka Yee. Chinese women: active revolutionaries or passive followers?: a history of the All-China Women's Federation, 1949 to 1996. Diss. Concordia University, 1998.
3. Yihong, Jin. "The all China women's federation: challenges and Trends." Chinese Women Organizing (2020): 123-140.
4. Jiang Xiuhua et al. ed, Dangdai zhongguo funü yundong jianshi [A Brief History of Contemporary Chinese Women's Movement] (Beijing: Zhongguo funü chubanshe, 2017).
5. Dang Dai Zhongguo Fu nü Yun Dong Jian Shi (1949-2000) (Beijing Shi, Beijing: Zhongguo fu nü chu ban she, 2017), i-v.
6. Gail Hershatter, Gender of Memory: Rural Women and China's Collective Past (Asia Pacific Modern; 8) (Berkeley and Los Angeles, California: University of California Press, 2011), 184-186.

A Study on the Media Framework during Beijing Winter Olympic Games from a Comparative Perspective:Taking People's Daily Online and CNN as Examples

Zhuqi Yu[1,~], Nuo Cheng[2,~], Chunfu Zhang[1,*]

[1]School of Statistics and Data Science, Zhuhai College of Science and Technology
[2]School of Foreign Languages, Tongji University
~Co-First Author
*Corresponding Author

Abstract. Upon the conclusion of the 2022 Beijing Winter Olympic Games, the media coverage of this sports event of worldwide prestige has piled up as countless. However, the coverage of the Beijing Winter Olympics varies across countries, and it is up to different media to tailor and package the information to be delivered to the public according to their value standards, political tendencies, ideologies, etc. Based on a review of existing literature, this paper is devoted to the exploration of the underlying media framework of China and United States through a comparative study with a certain theoretical basis by taking their practices as examples to distinguish the choices made by these two countries on conducting both negative and positive reporting and their divergence in pursing either the "universal solidarity" or conflicts and divisions.

1 Introduction: Reports on the Beijing winter Olympics by Chinese and US media

1.1 Introduction

The Winter Olympics column launched by People's Daily in 2022 is an exclusive reporting sector dedicated to the Beijing Winter Olympics, covering the opening and closing ceremonies, competition events, participants, Olympic venues and other topics. Under the concept of creating an online version, People's Daily pays close attention to the conversion of the written texts of People's Daily into online reports to free viewers from limited reading scenarios. Therefore, a collection of reports themed on the opening ceremony of the Beijing Winter Olympics or related figures reports, readable in a single sit, and many with fabulous photos, is presented in this paper. Compared to most media reports, the reports from People' s Daily Online are more focused and in line with the visual dwelling pattern of viewers: the content is more precise and concentrated on a key news spot or practice, making itself a reference for many media workers to conduct reports in recent years. Unlike U.S. media that is featured by unscrupulous pursuit of traffic, or most domestic news media focusing on factual matters, People's Daily's coverage of the Winter Olympics is characterized by its more in-depth reporting of the same event. For viewers who want to know more about the athletes and events they are interested in, this type of coverage is more comprehensive and tailored to their taste, driving them to explore the stories behind the news.

The current series of reports on the Winter Olympic Games on the People's Daily Online has exceeded one million, so how many of them are focused on the opening ceremony? What are the characteristics of these reports in terms of content presentation? What are the implications for news media workers? How to carry out theoretical research based on the reports of People's Daily Online? The attention to the above questions is relatively scarce. To this end, based on a review of existing literature, this paper is devoted to the exploration of the underlying media framework of China and United States through a comparative study with a certain theoretical basis by taking their practices as examples to distinguish the choices made by these two countries on conducting both negative and positive reporting, so as to shed light on future media reporting works.

1.2 Theoretical Basis: The Framework Theory Perspective

The framework theory was proposed by psychologist and anthropologist Gregory Bateson, and later introduced and further developed by Entenmann in the studies of communication. According to Entman's theory, "'framing' includes 'selection' and 'highlighting'... ...constructing

*Corresponding author: cfzhang@zcst.edu.cn
[1,~]First Author: stafi77@163.com
[2,~]Co-First Author:ChengNuo001206@163.com

reality rather than reflecting it when news is reported." [1] In other words, different media would tailor and package the information to be delivered to the public according to their value standards, political tendencies, ideologies, etc., and the way the media choose to frame the facts is "framework". This paper will take Chinese and American media as examples to analyze the media frameworks adopted by Chinese and US media in their coverage of the Beijing Winter Olympics with corresponding cultural explanations.

2 Literature Review

2.1 Based on the News Report

Since the end of the Beijing Winter Olympics, there is an absence of articles comparing the media frameworks of China and the U.S. taking the Beijing Winter Olympics as an example for the time being, but the coverage of the Winter Olympics by the U.S. media has drawn the attention of the domestic media, especially some inaccurate reports, which have all shed light on the present research. Earlier, many researchers have observed the differences between Chinese and American media frameworks and conducted studies on international hot spots as well as major ongoing events, such as the Sino-US trade war, the COVID-19 epidemic, and the World Health Assembly. Among them, the Sino-US trade war has received the most attention from researchers. In her master's thesis, "A Comparative Study of China-US Trade War Coverage Frameworks between China Daily.com and New York Times.com", Cao Ying argued that in the course of the Sino-US trade war, the two networks each have their own coverage frameworks, reflected mainly at three levels: high, medium and low, and each has its own strengths and weaknesses. [2] Lu Yizhou and Ma Deyong in "The Media Framework Effect and its Continuity - an Experimental Study on the Issue of China-US Economic and Trade Friction" devised four media framework scenarios: pro-China framework, pro-US framework, competition framework, and competition framework + free discussion, with which the changes in audiences' attitudes toward the US and the trade war in different contexts were observed. [3] In "Media Framework Effects in Complex Public Opinion Issues: An Experimental Study of the U.S.-China Trade Dispute as a Case Study", they conducted an empirical and comparative analysis of different media frameworks and their political consequences in the U.S.-China trade friction through surveys, experiments, and statistics with Chinese netizens as respondents. In the aftermath of the COVID-19 epidemic sweeping the world, Chinese and foreign measures to prevent the epidemic are distinctively different, and some scholars underpinned the different frameworks adopted by the Chinese and American media to report the COVID-19 epidemic. [4] In Yuanxin Wang and Dehuan Liu's "Analysis of the LDA Thematic Model of Western Mainstream Media Coverage of the COVID Epidemic from the Perspective of Framework Theory - Taking New York Times and The Guardian as Examples", the framework theory was employed with content analysis to study the content of Chinese and US mainstream media coverage in terms of headlines and language features, and to summarize the respective framework characteristics. [5] In addition, other researchers have focused on the discrepancies between Chinese and US media when covering the Olympic events, and Sun Xiaosheng and Zhang Yining in "Comparison of Chinese and US Media Coverage of the Rio Olympic Games from the Perspective of Evaluation Theory - Taking the China Daily and the Washington Post as Examples" revealed the different positions of Chinese and US mainstream media about news coverage and their differences in leveraging attitudinal resources. [6]

In conclusion, Chinese and American media frame facts in different ways. The news framework of the media is a conscious act of constructing social reality and shaping public perceptions, and the production of news and the dissemination of information will affect the perceptions of their audiences, and there are great differences between China and the United States in terms of geopolitical, institutional, and cultural aspects, as well as the Chinese and American media stand on different positions to report on this Winter Olympics in Beijing.

2.2 Based on Existing Literature

Guo Qing et al. argued that international public opinion is a synthesis of the international community's views on the organization and preparation of the Beijing Winter Olympics and related Chinese affairs, and suggested that (1) differences should be narrowed in the light of Olympic culture; (2) cross-cultural mindset should be adopted to communicate sports stories. [7] Zhang Lei et al. divided the significance of the Winter Olympics communication into two kinds - intrinsic and extrinsic, i.e., bringing ice and snow to the public and China to the world, and he also proposed that the culture of ice and snow can be distinct from intelligent mobile terminals, sports events-based educational promotion [8], Yang Zesheng et al. delved into the issue from the perspective of media professionals and suggested that the reporting should be based on Chinese elements with international vision; grounded on the scenes, but go deep behind the scenes; training professionals and introducing the three main parts as the inspiration of media coverage. [9] Du Heng et al., through the analysis of the Beijing Summer Olympics, proposed: (1) it is important and urgent to guide international public opinion on media response to the Beijing Winter Olympics. (2) theoretically, the concept and goal of public opinion guidance should be established; a new government-media model should be introduced; and the government's international public opinion guidance should be innovated. (3) a brilliant, extraordinary and excellent Winter Olympic Games can strongly cancel out the prejudice and defamation of the West against our country. [10] Li Jinbao et al. analyzed by statistical methods in the form of questionnaires, divided the audience into five categories: need-driven, participatory, selective, opinion-based and segregated audiences to investigate media response measures in each category, such as facing up to the current situation of Winter Olympics communication; amplifying the leading role of superior projects; bridging the multiple compartments of sports fans; and playing the advantages of all-media communication. [11] Huang He et al. conducted an analysis of the media agenda network existing in Western media Olympic coverage through the net agenda setting theory, and briefly examined the foreign communication countermeasures that China could carry out in terms of strategy, tactics, vision, path and pattern. [12] People's Daily's official website also launched a special section to cover the Beijing Winter Olympics, [13] which highlights with the opening ceremony as an example: the determination of China's flag bearer for the opening ceremony; the security force behind the Beijing Winter Olympics; Zhang Yimou's detailed explanation of the concept of the Winter Olympics; 95% of teenage actors/universities supporting the Winter Olympics; foreign audience's praise of the opening ceremony; the view that the lighting ceremony was designed as "unexpected but reasonable"; and the discussion related to the rehearsal, the silent contributor to the Winter Olympics, and the pilot film.

In summary, the focus of existing literature is on theoretical analysis of the public opinion of the international community on the Beijing Winter Olympic Games, with little emphasis on specific news coverage by domestic and overseas media. This paper will conduct a comparative analyze over the framework and agenda setting of the Chinese and US media based on their coverage of the Beijing Winter Olympic Games.

3 Distinctive coverage of the Beijing winter Olympics by Chinese and US media

3.1 Positive Publicity and Negative Reporting

While the whole country is celebrating the arrival of Chinese New Year, the 2022 Beijing Winter Olympic Games are about to kick off and Beijing will become the world's first "City of both Summer and Winter Olympics". Ahead of the opening of the Winter Olympics, Chinese media spotlighted the opening ceremony, such as the People's Daily's "Ice and Snow! The Pilot Film for the Opening Ceremony of the Beijing Winter Olympics goes online" greatly raised people's expectation for the opening ceremony, and the 2022 Spring Festival Gala was featured by the festive atmosphere of the New Year with the program "Dating with Ice and Snow". What the media conveyed was a positive message. From the "Three Olympic Questions" a century ago to the "City of the Two Olympics" today, the 2022 Winter Olympics in Beijing carries the expectations of the whole nation. In the course of the Beijing Winter Olympics, Chinese athletes made breakthroughs one after another and finally reached a historical milestone of nine gold medals. During this period, the young athlete stars such as Gu Ailing and Su Yiming received a lot of attention from the media, labeled with the key words "a promising future", "hopeful" and "a new generation", which were all with positive connotations. People's Daily's report "Raise the national flag, play the national anthem! Gu Ailing Rises to the Top of the Podium Again" and other online reports have all connected Gu Ailing to the country and the Chinese nation. In addition, Chinese media made continuous efforts to encourage and recognize all athletes whoever struggled, though failed in some cases, but showcased the spirit of the striving, and thus deserves to be recognized, for example, People's Daily's "Imperfect, yet gorgeous". These reports all serve as positive publicity with simple but piercing headlines and concise descriptions.

In contrast, the Beijing Winter Olympics were negatively reported by the US media. While the COVID-19 epidemic is sweeping across the globe, and it is China's epidemic prevention policy that made it possible for its people to live in a relatively safe environment, the strictness of which was evidenced in the Beijing Winter Olympics where athletes were protected tightly within the security line. However, given all these efforts, the U.S. media repeatedly speculated that the Beijing Winter Olympics were too strict and tried to attack the Games on the grounds of "human rights". Yet undoubtedly, the strict epidemic prevention measures of the Beijing Winter Olympics truly protected the safety of athletes and coaches. In addition, Gu Ailing's choice to represent China as a "second-generation immigrant" was hyped by Fox News as a "betrayal of the United States" and fabricated some missing events of athletes at the Beijing Winter Olympics in order to mock Gu Ailing for making that decision. The negative reports were unfounded and poorly received, as Aaron Blunck, a U.S. athlete who lived in the Olympic Village, publicly stated that he had seen various irresponsible reports about the Beijing Winter Olympics in the United States. "Those were actually false. In fact, everything was remarkable. Everyone, from the staff to the nucleic acid testers, to the accommodations, it was the best Winter Olympics we've ever attended." The consistent hype of these American media outlets only added to their declining credibility.

3.2 "Universal Solidarity" versus Conflicts and Division

The Winter Olympics was supposed to be an event for countries around the world to put aside conflicts and compete in friendship. The Beijing Winter Olympics adheres to the concept of universal solidarity and a friendly and inclusive attitude. Chinese media also reported with the principle of speaking well about Chinese story and conveying Chinese voice. The "Folded Willow Sends Love, This is China's Romance to the World!", "I've eaten about 200 dumplings since the Winter Olympics", "Last night, the 'welcoming pine' lit up the starry galaxy", all these reports contain the concept of "harmony" and the expectation of "harmony". During the Beijing Winter Olympic Games, the diet of athletes from different countries in the Olympic Village appeared frequently on the "hot search" list, with some enjoying dumplings and others obsessed with sesame balls. Chinese media paid close attention to the everyday routine of athletes, showing the warmth of the Beijing Winter Olympic Games in their daily lives and conveying the beautiful vision of "universal solidarity". Athletes of different skin colors and languages competed on the same arena with great vigor and enthusiasm, building friendship in the pursuit of excellence and realizing ideals in fairness and justice. The epidemic has separated people and distanced countries from each other, but sports can bring people closer together, as the Beijing Winter Olympics lighted people's hearts with ice and snow in the festive atmosphere of new year, and dispersed division with tolerance and openness.

The U.S. media has never given up on creating conflict and division. Before the Beijing Winter Olympics, the U.S. media, represented by CNN, reported that the Biden administration would announce its decision on not designating government officials to the Beijing Winter Olympics, citing "human rights issues" in China. It also disseminated rumors of "genocide" and "forced labor" in Xinjiang to discredit China and compromise its international image. In the course of the Beijing Winter Olympics, CNN smears China from various angles such as artificial snowfall and site selection. "Fake it when there is no snow", "Snow and ice manufacturing companies have gone to the extreme", "This is dangerous!" The headline was particularly negative, claiming that "the artificial snow-making technology used at the Beijing Winter Olympics will not only consume 'a lot of energy', but some elite athletes claim it will make the event 'far from safe'". In addition, CNN "checked the satellite photos and official maps" which showed how "the ski center in Beijing's Yanqing district was built on the ruins of a national nature reserve", implying that the Chinese side contradicted the "Green Olympics" principle. After the closing of the Beijing Winter Olympics, CNN published a report titled "2022 Beijing Winter Olympics Divides the World, but Represents a Success for China," claiming that the Winter Olympics was a "deeply polarizing event" for the world. It even claimed that "China is taking advantage of the Winter Olympics to confuse the public and dodge criticism of the so-called 'human rights issues' from the international community", which was truly a distortion of facts. When it comes to Gu Ailing's representation of China, CNN barely showed its narrow-mindedness, as Gu Ailing is a "second-generation immigrant" who is eligible to apply for Chinese citizenship, and such comments that Gu Ailing betrayed the U.S. only brought to light the hypocrisy of "freedom" that the U.S. has been preaching. The above study of the media framework of China and the United States in the coverage of the Beijing Winter Olympics shows that the media in the two countries hold different positions and value standards, but the Olympic Games is a stage to put aside political conflicts for friendly exchanges, and sports are designed to unite people together. The "reality" constructed by the American media certainly has its own value stance, but it goes against the spirit of sports.

4 People's Daily's Agenda Setting on the Beijing Winter Olympic Games

Huang He et al. argued that the best response to Western media is to fully understand their ways and patterns of constructing Olympic issues, and that there is a great difference in the way Western media report on the Olympic Games in non-Western countries. [12] Traditional U.S. media coverage overemphasizes its own values as a way of evaluating the outside world, focusing on the political and humanistic information of the country hosting the Olympics, rather than reflecting a true picture of the performance of the host country.

4.1 Focus on the Needs of the Audience for a Better Access to Information

The live streaming of the Winter Olympics is just one of the many communication efforts. In fact, viewers often have a curiosity about the work of the Winter Olympics and develop a desire to learn more about the Games. To this end, mainstream websites such as People's Daily Online provide real-time updates from different angles and perspectives to meet the needs of different viewers. The so-called "real-time update" means that the public is served with timely text reports on specific projects.

For example, the audience would like to take a look at the behind-the-scenes preparation of the opening ceremony, and the report "Behind the opening ceremony of the Beijing Winter Olympic Games is the silent dedication of these universities" [14] explains to us how the universities have participated in the event with professors leading students to participate in scientific and technological innovations in the opening ceremony performance. On one side, there are experts from different disciplines who contribute to the screen presentation in terms of culture and science and technology for the opening ceremony, and on the other side, there are students from the universities who participate in the performance; the professors are more capable than the students in generating greater social benefits and are more dedicated to the opening ceremony, but the report does not rub off the students' contribution however. As another example, the audience wondered how director Zhang Yimou, the conductor behind the opening ceremony, perceived the opening ceremony, and the report "Zhang Yimou gives full marks to the opening ceremony presentation"[15] shared his views: more cordial, more relaxing, and more minimalistic, when taking the comparison with the 2008 Beijing Summer Olympic Games into account.

It is evident that People's Daily pays great attention to the needs of different audiences on the Winter Olympic Games series, associating the reports with people or events related to the Winter Olympic Games. These real-time updated news are comprehensive and well prevent one-sided media reports by delivering intuitive and concise stories of memorable moments in the Winter Olympic Games. Meanwhile, a connection was established between the key paper reports in People's Daily and its rolling reports online, promoting viewers to gain a more in-depth understanding of the news events. In real cases of news reporting, a reasonable migration and utilization of this kind of reporting model can be made to help viewers have a deeper understanding of the news event, provide viewers with the access to first-hand information about the news event (e.g., live photos of the winning moments, animation of difficult action clips), and promote the formation of the reporting system. It is noteworthy that, unlike the U.S. reporting approach of gaining exposure by fabricating facts, People's Daily uses more declarative sentences in its headlines, which is more attractive to viewers and generates a stronger sense of cohesion and patriotism.

4.2 Help viewers to understand Olympic events through simulated scenarios

Coverage that is complex, boring and abstract often makes it difficult for viewers to relate to the Winter Olympics events, which is a dilemma common in news coverage. If the viewers sitting in front of the TV can be immersed in a specific scene and placed in a specific competition site, the competition scene on the screen could be as vivid as to strengthen the viewers' understanding of the sports event. For example, viewers may consider curling to be less spectator-friendly than alpine skiing, and in the report "What the curlers are shouting" [16] the athletes' shouting "hahaha" was mimicked, which is actually the communication word in curling, "hurry", that is used for requiring intensified friction of the curling brush against the ice so as to give the curler a speed boost.

These reports are like microfilms that portray in depth the whole picture of the Winter Olympic Games by reproducing the performance of the athletes. What the implication this brings for media professionals is that a similar simulation could be introduced into the agenda setting process, leading viewers to put themselves into the games as they read the reports. This is important for enhancing the audience's enthusiasm for the Winter Olympics. Second, attention should be paid to the history and stories behind the Winter Olympics. For example, the story "The network security force behind the opening ceremony of the Winter Olympics" [17] describes how security personnel stationed at the main media center to ensure a safe watching experience for the world.

4.3 Make the Winter Olympics penetrate into millions of households

Despite the brevity of the rolling news on the Winter Olympic Games, the concept of "from the field to the audience, from the audience to the world" is deeply embedded. For example, prior to the introduction of short track speed skater Wu Dajing, the main body of the report started from topics familiar and interesting to the audience, such as the report "Four memorable moments in the mind of Wu Dajing" [18], which introduced the short track team winning the gold medal in the mixed relay; Ren Ziwei winning the gold medal in the men's 1000m; the women's team winning the bronze medal in the women's 3000m relay; and the scene where Fan Kexin kissed the ice race. The report ends with "I hope that the team will hold the momentum to fight shoulder by shoulder with me, by which it is expected to raise the audience's confidence in the Chinese short track skating team.

Another remarkable feature is the special focus on the peripherals of the Winter Olympics. This is reflected in two aspects.

First, by taking into account the perceptions of foreign audiences about the opening ceremony, it meets the audience's preference chain from domestic to foreign affairs and helps to understand the quality of the Beijing Winter Olympics in all aspects. For example, the report "The focus of the Russian mainstream media on the opening ceremony of the Beijing Winter Olympics and the Russian audience's admiration" [19] describes the focus of the Russian media on the opening ceremony and the Russian netizens' wishes for the success of the Beijing Winter Olympics.

The author not only states in concise language the technological elements and modernity of the opening ceremony, but also proclaims the spirit of innovation and dedication to the country of the university professors and research institutes. These reports can be used as materials, interspersed with reports of people's deeds, or refined and transferred to cases for the publicity of universities and the country. This is helpful for raising the patriotic consciousness of contemporary youth, cultivating their correct values, and shaping ideal, capable and responsible characteristics.

In addition, the report "Xinjiang elements shine in the opening ceremony" [20] mentions that the cross-country skier from Altay, Xinjiang, had her mother in tears when she placed the torch into the big snowflake on the main torch stand, igniting the fire of enthusiasm among the cadres and masses of Xinjiang. This also expresses a core idea: the people of Xinjiang are an inseparable part of China. It also indicates to us how powerful the media coverage is in depositing national unity, for example, while benefiting the audience, it is also constantly coalescing the close relationship between us and our compatriots of all ethnic groups. The values of journalists conveyed in these reports are exactly what are currently in short, therefore the resources behind news media should be exploited wisely and prudently.

The value behind the rolling news coverage of People's Daily is not only reflected in its diverse connotations, unique and clever conceptions, and clear and exquisite on-site photos, but also in the fact that it derives a new way of media, which can be observed in the content section on the People's Daily website. In addition to the opening ceremony of People's Daily Online, there are comment sector set for viewers to exchange opinions on the news and generate a collision of thoughts through the National Progress Forum, as well as a photo and video sector, providing abundant first-hand photos and videos of the Winter Olympic Games. These sections open another window for viewers to understand the hot news of the Winter Olympics in all aspects and expand their scope of the news. More crucially, People's Daily Online provides a platform to create personal accounts, allowing individuals to discuss related reports and read the parts where they find themselves satisfied. Currently, People's Daily has over 600,000 rolling news reports on the opening ceremony, and more and more news media workers are still adding to this volume.

5 Conclusion

Through the comparison between Chinese and US media based on Beijing Winter Olympics coverage, this paper draws on the framework theory to identify the similarities and differences between Chinese and US media in structuring the factual hierarchy in their coverage, as well as summarizing the existing literature on the agenda network of Western media, giving a review to the the literature on agenda setting for domestic media coverage of the Beijing Winter Olympics, and initiating a discussion on the different Chinese and US media coverage of the Beijing Winter Olympics and the agenda setting of People's Daily Online. As the different frameworks for media coverage of the Beijing Winter Olympics in China and the United States have become the object of comparison for researchers, through the analysis of literature, it has been found that the domestic media responded to the U.S. coverage through a process from "U.S. coverage - Chinese response" to "Chinese preparation - U.S. coverage - Chinese response", and although considerable results have been obtained through this research practice, certain limitations should be addressed. In general, the response is mostly grounded on empirical theories, such as the analysis of the significance of the communication of this Winter Olympics, the media or means that are available, etc. There are relatively few research and empirical studies, and it is still worth examining whether the research methods have observed scientific principles. It is not advisable to judge the response of our media to U.S. media coverage solely on the basis of experience, but rather on the scientific evidence to assess its performance. Therefore, the diversity and scientific nature of the research approach is what needs to be addressed, and a more extensive research approach should be adopted to evaluate the differences in Chinese and American media coverage beyond just the Winter Olympics to make our coverage more discursive in the world, which requires research efforts to develop appropriate framework theories and agenda setting based on the reality of the Chinese society.

References

1. Zelizer, B. Covering the body: The Kennedy Assassination, the Media, and the Shaping of Collective Media, and the Shaping of Collective Memory, University of Chicago Press,1982, p.8.
2. Cao Ying, Fu Jicheng (supervisior), A Comparative Study on the Reporting Framework of China-US Trade War Between China Daily and New York Times, MA thesis of Xiangtan University. 2020. DOI:10.27426/d.cnki.gxtdu.2020.001677.
3. Yizhou Lu, Deyong Ma, Media Framing Effects and Their Durability in Contemporary China: An Experimental Study on the Issue of China-US. Trade War, Journalism Research,vol.175,2020(11),pp.50-65+126
4. Deyong Ma, Yizhou Lu, Media Framing Effects in Foreign Policy Issues: An Experimental Study on China-U.S. Trade War, Chinese Journal of Journalism & Communication, vol.42,2020(05), pp: 99-120. DOI: 10.13495/j.cnki.cjjc.2020.05.007
5. Yuanxin Wang, Dehuan Liu, LDA Topic Modeling Analysis for COVID-19 News of Western Mainstream Media from the Perspective of Framing Analysi: An Analysis of New York Times and Guardian Publications, Advertising Panorama (Theoretical Edition), 2020(03), pp:76-89
6. Xiaosheng Liu, Yining Zhang, Comparative analysis of Chinese and American media coverage of Rio Olympic Games from the perspective of evaluation theory -- Taking China Daily and Washington Post as examples, Journal of Changchun University of Science and Technology (Social Science Edition), vol.34,2021(06), pp:163-167
7. Qing Guo, Xi Yang, Mission, Opportunity and Challenge: Reflection on the International Communications of the Beijing 2022 Winter Olympic Games, China Sport Science, vol.40,2020(11), pp:3-9.DOI:10.16469/j.css.202011001
8. Lei Zhang, Jun Xie, Research on the Communication Strategy of Ice and Snow Culture under the Background of 2022 Beijing Winter Olympics, Journal of Xi'an Phsyical Education University, vol.38,2021(01), pp: 55-62. DOI: 10.16063/j.cnki.issn1001-747x.2021.01.008
9. Zesheng Yang, Han Gao, Research on Olympic TV communication from the perspective of news field——Taking the Beijing Winter Olympics program of CCTV as an example, Journalism Lover,2021(02), pp:82-84. DOI:10.16017/j.cnki.xwahz.2021.02.020
10. Heng Du, Xiaoyi Zhang, International Public Opinion Guidance and Media Countermeasures for Beijing Olympic Winter Games, Journal of Being Sport University, vol.44, 2021(1), pp: 52-61. DOI: 10.19582/j.cnki.11-3785/g8.2021.01.007
11. Jinbao Li, Leilei Ye, Gang Hong, Winter Sports Communication Strategies: A Case Study of the Beijing Winter Olympics, Media Observer, 2021(12),pp:41-47.DOI:10.19480/j.cnki.cmgc.2021.12.015
12. He Huang, Zheng Li, Xiaohan Yang, Agenda Network Characteristics of Olympic Reports in Western Media and Communication Strategies for Beijing 2022 Winter Olympic Games: Semantic Network Analysis Based on News Coverage of the Recent Six Olympic Games, Journal of Shanghai University of Sport,vol.45,2021(05),pp:10-20.DOI:10.16099/j.sus.2021.05.002
13. People's Daily Online, XXIV Winter Olympic Games, http://ent.people.com.cn/GB/436846/441848/index.html
14. People's Daily Online, Behind the opening ceremony of the Beijing Winter Olympics, there is dedication of these universities (2022.2.8) [2022.2.26], http://ent.people.com.cn/n1/2022/0208/c1012-32347539.html

 https://doi.org/10.1051/shsconf/202315503016

15. People's Daily Online, Zhang Yimou gave full marks to the opening ceremony (2022.2.6) [2022.2.26], http://bj.people.com.cn/n2/2022/0206/c82846-35123175.html
16. People's Daily Online, What are the curling players shouting? (2022.2.17) [2022.2.26], http://ent.people.com.cn/n1/2022/0217/c1012-32353656.html
17. People's Daily Online, The Cyber Security Forces Behind the Winter Olympics Opening Ceremony (2022.2.6) [2022.2.26], http://bj.people.com.cn/n2/2022/0206/c14540-35123368.html
18. People's Daily Online, Four unforgettable moments in Wu Dajing's eyes (2022.2.17) [2022.2.26], http://ent.people.com.cn/n1/2022/0217/c1012-32353893.html
19. People's Daily Online, Russian mainstream media focuses on the opening ceremony of the Beijing Winter Olympics, and Russian audiences express their admiration to the opening ceremony (2022.2.5) [2022.2.26], http://ent.people.com.cn/n1/2022/0205/c1012-32345771.html
20. People's Daily Online, Xinjiang elements shine at the opening ceremony (2022.2.5) [2022.2.26], http://xj.people.com.cn/n2/2022/0205/c186332-35122927.html

*References 2-20 are in Chinese

A Contrastive Corpus Study on Lexical Features of the English Translation of the Report of the 20th and 19th CPC National Congress

Yiyu Zhang[1], Lin Wang[1], Wenming Hu[1], and Xiajing Yao[1,*]

[1]School of Foreign Languages, China University of Geosciences, Wuhan, China, 430074

Abstract. As the channel spreads the Chinese voice and promotes Chinese culture, the status of political document translation is self-evident with the deepening of global communication. Taking the English translation of the reports of the 20th and the 19th CPC National Congress as examples, this paper makes a contrastive study of their lexical features in a corpus-based way. With the assistance of corpus analysis software such as AntConc, TagAnt and WordSmith4.0, the author investigates TTR (type-token ratio), lexical density, average sentence length, high-frequency words, and keywords between two reports. It is found that the report of the 20th CPC Nation Congress has more diversified and native expressions with fewer words used, which provides a guiding significance for the English translation of political documents.

1 Introduction

Delivered by President Xi Jinping, adopted at the 20th CPC National Congress and translated by the Central Institute of Party History and Literature, the document titled "*Hold High the Great Banner of Socialism with Chinese Characteristics and Strive in Unity to Build a Modern Socialist Country in All Respects*" (hereinafter referred to as "the report of the 20th CPC National Congress") further charted the way forward for the cause of the Party and the country, focusing, in particular, on the Party's strategic tasks in the coming five years.

As an authoritative document that serves as a political manifesto, the report of the 20th CPC National Congress played a vital guiding role in uniting and leading the Chinese people of all ethnic groups to uphold and develop socialism with Chinese characteristics on the new journey in the new era. To ensure the key points and content are accurate and understandable to target readers with different cultural backgrounds and language habits, China has invited foreign experts to help edit translated versions of the work [1].

It is worth noting that considerable attention has been paid to corpus-based political document translation research, to name a few, Jia Hui and Sun Minwei discussed translation innovations in the report of the 19th CPC National Congress [2], Wu Yana, Xie Yating, and Xia Mengjun studied explicitation techniques in the English translation of the report of the 19th CPC National Congress [3], and Shang Wenbo focused on the image of the CPC through corpus-based critical translation studies [4]. However, the research object needs to be updated over time and studies on lexical features are still insufficient.

Given that the importance of its English translation is universally recognised, the paper takes the English translation version of the reports of the 20th and the 19th CPC National Congress as research objects, exploring differences between them based on a corpus approach with the help of computer software, then discussing their lexical features by contrasting the research data. Through qualitative and quantitative analysis, it is hoped that some references can be provided for the English translation of political documents.

2 Research method

In this research, the report of the 20th and 19th CPC National Congress from *China Daily* (titled with *Full text of the report to the 20th National Congress of the Communist Party of China* and *Full text of Xi Jinping's report at 19th CPC National Congress* respectively) are selected as research objects [5-6].

The first step in the building process of corpora is text pre-processing, where noise reduction and text normalization take place. Rich text downloaded from *China Daily* websites is cleaned manually and converted into plain text. Then TagAnt is used for POS (Part-of-Speech) tagging so that AntConc, by which lexical density is calculated, can be further used. Apart from the functions mentioned above, WordSmith4.0 is also used to analyse the lexical features of both English and Chinese text in the current study. It is an integrated toolkit of modules comprising three functions: WordList, Concord, and KeyWord. The first module, WordList, as the most advocated program for data extraction, provides characteristics observed in the untagged text, including tokens, types, Standardized type-token ratio (STTR),

* Corresponding author: yaoxiajing@cug.edu.cn

mean sentence length, and a list of all words in frequency order. Noting that all text imported into AntConc and TagAnt utilises the UTF-8 character encoding method while all text imported into WordSmith utilises the UTF-16LE character encoding method for the best compatibility.

3 Statistics and analysis

3.1 Type-token ratio

Type-token ratio (TTR) is the ratio obtained by dividing the types (the total number of different words) occurring in a text or utterance by its tokens (the total number of words). A high TTR indicates a high degree of lexical variation while a low TTR indicates the opposite [7]. TTR scores were calculated as follows:

$$\text{TTR} = \frac{\text{Number of different words (types)}}{\text{Total number of words (tokens)}}$$

Standardised type-token ratio (STTR) calculates the mean TTR for every N-words (N is generally set to 1,000 by default but can be changed according to different needs). Those two kinds of ratios reflect the degree of word occurrence in the corpus. However, the basic problem of TTR is that it is affected by the length of the text sample. For larger text, TTR is rather meaningless in most cases. Therefore, the author uses the standardised type-token ratio to analyse two reports. The higher the standardised type-token ratio is, the more easily the word varies, and vice versa. Noting that "NCCPC" in the table is the abbreviation of "the National Congress of the Communist Party of China".

Table 1. Tokens and types

	The report of the 20th NCCPC (target text)	The report of the 19th NCCPC (target text)
Types	3,095	3,029
Tokens	25,346	25,059
Type-Token Ratio	12.21%	12.09%
Standardised Type-Token Ratio	39.77%	39.22%

From table 1, the TTR of the English translation of the report of the 20th and the 19th CPC National Congress is 12.21% and 12.09% respectively. However, the two reports are quite different in length, making TTR have less reference value. Therefore, the paper concentrates on the STTR of the two materials. The STTR data of the report of the 20th CPC National Congress is 39.77%, and the STTR data of the report of the 20th CPC National Congress is 39.22%. The standardised type-token ratio of the English version of report of the 20th CPC National Congress is higher than that of the 19th CPC National Congress report. This shows that the use of English vocabulary in the 20th CPC National Congress report has more variability than that of the 19th CPC National Congress translation, and the scope of vocabulary change is wider than that of the 19th CPC National Congress translation report. This also indicates that the translated version of the 20th CPC National Congress report is more expressive, which is conducive to readers' understanding.

For example, there are four different translation versions of Chinese word "*guan1jian4*" in the report of the 20th CPC National Congress:

(1) It takes place at a **critical** time as the entire Party and the Chinese people of all ethnic groups embark on a new journey to build China into a modern socialist country in all respects and advance toward the Second Centenary Goal.

(2) We have grown stronger in basic research and original innovation, made breakthroughs in some core technologies in **key** fields

(3) The next five years will be **crucial** for getting our efforts to build a modern socialist country in all respects off to a good start.

(4) Our Party **has a pivotal role** in building China into a modern socialist country in all respects and in advancing the rejuvenation of the Chinese nation on all fronts.

However, for the same Chinese word "*guan1jian4*", there are only two translation versions in the report of the 19th CPC National Congress:

(1) The 19th National Congress of the Communist Party of China is a meeting of great importance taking place during the decisive stage in building a moderately prosperous society in all respects and at a **critical** moment as socialism with Chinese characteristics has entered a new era.

(2) We have grown stronger in basic research and original innovation, made breakthroughs in some core technologies in **key** fields.

As is demonstrated by the comparison, vocabulary changes of the 20th CPC National Congress report translation are more diverse.

3.2 Lexical Density

Lexical density is defined as content words (lexical words) divided by the total number of words, which estimates the amount of information in the text [8]. Content words in English include nouns, verbs, adjectives, and adverbs; Function words include pronouns, conjunctions, numerals, mood words, and interjections. Content words indicate the difficulty and amount of information in the corpus. Lexical density is calculated as follows:

$$L_d = \frac{\text{Content words (lexical words)}}{\text{Total number of words}}$$

The more varied a vocabulary a text possesses, the higher lexical diversity. For a text to be highly lexically diverse, the writer has to use many different words, with little repetition of the words already used [9]. In this part, the author studies the lexical density of the translation text to explore lexical features.

https://doi.org/10.1051/shsconf/202315503017

Table 2. Lexical density

	The report of the 20th NCCPC (target text)	The report of the 19th NCCPC (target text)
Nouns	7,581	7,472
Adverbs	674	600
Verbs	3,958	4,003
Adjectives	3,164	3020
Content words	15,377	15,095
Function words	25,346	25,059
Lexical density	60.67%	60.24%

According to table 2, the lexical density of the translation material of the 20th CPC National Congress report is 60.67%, and the 19th CPC National Congress report is 60.24%. The lexical density of the English translation of the 20th CPC National Congress report is larger than the 19th CPC National Congress one. It means there are more content words in the 20th CPC National Congress report translation text. In other words, the 20th CPC National Congress report translation text is more concise and more readable.

3.3 Average Sentence Length

Average sentence length is calculated by dividing the total number of words in the text by the total number of sentences. The average sentence length is positively correlated with the syntactic structure of the translated text. Standard deviation reflects the degree of concentration and discreteness of the data, and the larger the standard deviation, the greater the variation in sentence length of the translated text. The data in table 3 is based on WordSmith calculations.

Table 3. Average sentence length

	The report of the 20th NCCPC (target text)	The report of the 19th NCCPC (target text)
Sentences	930	975
Mean (in words)	26.81	25.87
Std.dev.	14.21	17.37

In accordance with the data in table 3, the standard deviation of average sentence length in the 20th CPC National Congress report translation is larger than the translated text of the 19th CPC National Congress report. It indicates the length of sentences in translation material of the 20th CPC National Congress report has a richer variety of forms. The result is also corroborated by the previous statistics of type-token ratio: the vocabulary in the translation of the report of the 20th CPC National Congress is more varied and abundant.

For the same Chinese sentence "*qiang2diao4 jian1ding4 dao4lu4 zi4xin4 li3lun4 zi4xin4 zhi4du4 zi4xin4 wen2hua4 zi4xin4*" there are two different translations in two reports:

(1) ...strengthen our confidence in the path, theory, system, and culture of socialism with Chinese characteristics. (Abridged from the 20th CPC National Congress report translated text)

(2) It highlights the importance of fostering stronger confidence in the path, theory, system, and culture of socialism with Chinese characteristics. (Abridged from the 19th CPC National Congress report translated text)

After comparing the different translations of the same sentence in the 19th and 20th CPC National Congress report, it can be concluded that in the same sentence, the translation of the 20th CPC National Congress report is shorter than that of the 19th CPC National Congress report, which can also be concluded that the translation of the 20th CPC National Congress is more concise.

3.4 High-frequency words

Table 4 below shows the top twenty high-frequency words in the English translation of the report of the 20th CPC National Congress and that of the 19th CPC National Congress, which indicates that high-frequency words in two reports are "and", "the", "of", "to" and "in". In addition, two reports, both of which focus on Party, people, China, development and so on, maintain the consistency of politics. All high-frequency words are closely related to contents, highlighting the topic of these reports.

Table 4. High-frequency words

	The report of the 20th NCCPC (target text)		The report of the 19th NCCPC (target text)	
Rank	Freq	Word	Freq	Word
1	1832	and	1700	and
2	1541	the	1600	the
3	866	of	951	of
4	702	we	693	to
5	684	to	573	we
6	517	in	483	in
7	493	will	424	a
8	388	a	344	will
9	280	for	327	party
10	260	party	272	people
11	248	s	262	our
12	238	people	258	for
13	220	our	239	with
14	209	with	218	s
15	182	china	207	that
16	181	development	195	chinese
17	179	that	184	is
18	172	have	173	china
19	165	new	162	development
20	156	all	158	must

It worth noting that the frequency of the word "development" in the report of the 20th and 19th CPC National Congress are 181 and 162 respectively, showing that the report of the 20th CPC National Congress is more concerned with the development issue. Especially considering the ongoing COVID-19 global outbreak, policymakers in China are addressing the pandemic's lasting effects and making efforts to spur green, resilient and inclusive growth while safeguarding macroeconomic sustainable development. There are 149 extra modal verbs "will" in the report of the 20th CPC National Congress, demonstrating that the CPC is positively seeking opportunities to provide guidance for the public and serve the country's overall interests.

3.4.1 Modal auxiliary verbs

Modal auxiliary verbs reflect the distance, social relationship, and status between the two sides of communication, revealing different power relations behind the utterance. Halliday and Hasan (1989) divided modal auxiliary verbs in terms of their pragmatic values [10]. High-value modal auxiliaries include "must", "ought to", "need", and "have to"; median-value modal auxiliaries include "will", "would", "shall", "should"; and low-value modal auxiliaries include "may", "might", "can", and "could". Different groups of modalities are related to different politeness degrees of speech. Modal auxiliary verbs of high value indicate impolite speech, which is liable to cause the reader's disfavour, whereas low-value modals suggest a most polite use of language.

In this research, the tagged text is imported into the concord tool of WordSmith, by which frequencies of modal auxiliaries are calculated. To simplify the analysis of results, all modal auxiliary verbs in two reports are classified into three value categories according to Halliday's theory. The following table illustrates the frequency of modal auxiliary verbs in two reports. Noting that frequencies are calculated per 1,000 words.

Table 5. Modal auxiliary verbs

Values	Modal auxiliary verbs	The report of the 20th NCCPC (target text)		The report of the 19th NCCPC (target text)	
		Hits	Freq (per 1,000)	Hits	Freq (per 1,000)
Low	can	16	0.64	21	0.85
	may	5	0.20	1	0.04
	could	1	0.04	0	0.00
	might	0	0.00	0	0.00
	dare	2	0.08	1	0.04
	have to	0	0.00	1	0.04
	Total	24	0.97	24	0.97
Median	would (wouldn't)	0	0.08	1	0.04
	should (shouldn't)	35	1.41	93	3.77
	is/was to (isn't/wasn't to)	5	0.20	9	0.36
	Total	40	1.69	103	4.18
High	must (mustn't)	108	4.35	158	6.41
	ought to (oughtn't to)	0	0.00	0	0.00
	need	5	0.20	12	0.20
	can't	0	0.00	0	0.49
	couldn't	0	0.00	0	0.00
	mayn't	0	0.00	0	0.00
	Total	113	4.55	170	7.09

The analysis of the results indicates a lower frequency of median-value and high-value modal auxiliaries in the 20th CPC National Congress report translation, with a total of 40 median-value modal auxiliaries and 113 high-value modal auxiliaries compared to the 103 median-value and 170 high-value found in the 19th CPC National Congress report translation. Besides, low-value modal auxiliaries, the quantity of which is 24, are consistent in the two reports.

High-value and median-value modal auxiliary verbs are less used in the 20th CPC National Congress report translation in number and frequency, mainly because the Chinese government is committed to building its own internationally friendly and positive image. On the one hand, less modal auxiliary verbs can be gentler and more comfortable for the audience, and on the other hand, using modal auxiliary verbs is a way to call for or express strong feelings. Given that most policies in the 20th CPC National Congress are sticking to prior trends and calls for change appear less than before, as a result, fewer modal auxiliary verbs are used.

3.4.2 Delexical verb "make"

The macro indicators discussed above only scratch the surface of lexical features. Little attention has been paid to specific linguistic items in the translated texts. Thus, this paper further discusses lexical features from the micro level of the delexical verbs to remedy this situation.

Many common words have lost their semantic meaning as a result of the complex mechanism known as delexicalization, and now they must rely on their co-occurring partners to provide meaning [11]. Together, the delexicalised words and their co-occurring partners communicate a common meaning by influencing each other's semantic content.

The author discovers that "have" and "make" have startlingly high frequencies by hesitantly exploring the occurrences of the most prevalent lexical verbs, such as have, get, give, take, make, and do, in WordSmith WordList and Concord modules. Additionally, a more thorough examination (i.e., searching "have/made + noun" structure) reveals that "have" acts as a delexical verb far less frequently than "make". In other words, the word "have" usually has a concrete meaning rather than a phraseological one. Therefore, only the more typical delexical is focused on.

To further instigate, the original Chinese texts of instances where delexicalised "make" is used in interpreted texts are analysed. Instances where the Chinese delexicalised verbs "*zuo4(1)*", "*zuo4(2)*", "*gao3*", and "*jin4xing2*", are interpreted into English delexicalised verb "make" are collected, with the result shown in table 6-1, 6-2 and 6-3:

Table 6-1. Typical Chinese delexicalised verbs in source text

Typical Chinese delexicalised verbs in source text	Number of them in the report of the 20th NCCPC (source text)	Number of them in the report of the 19th NCCPC (source text)
zuo4(1)	18	23
zuo4(2)	130	145
gao4	3	4
jin4xing4	3	12
Total	154	184

Table 6-2. Total number of delexicalised "make" in target text

Total number of delexicalised "make" in target text	
The report of the 20th NCCPC (target text)	The report of the 19th NCCPC (target text)
64	69

Table 6-3. Frequency of delexicalised "make" interpreted from typical Chinese delexicalised verbs in target text

Frequency of delexicalised "make" interpreted from typical Chinese delexicalised verbs in target text	
The report of the 20th NCCPC (target text)	The report of the 19th NCCPC (target text)
41.56%	37.50%

The three tables above show clearly that the delexical verb "make" is more frequently used in the 20th CPC National Congress report translation. It is found that the frequent use of Chinese delexicalised verbs in source text inevitably influences the diction of target text, resulting in the high frequency of delexicalised verbs in both two English translations. The author also believes that the heavy use of delexicalised verbs with collocates of abstract noun phrases is a significant feature of English diction, translators use delexicalised verbs frequently to collocate with different noun phrases to meet the readers' expectation of fully understanding the message. Consequently, target text of the 20th CPC National Congress report tends to be more native than the previous one.

3.5 Keywords

The Brown corpus, the full name of which is Brown University Standard Corpus of Present-Day American English, was the first text corpus of American English. The original Brown corpus was published in 1963-1964 by W. Nelson Francis and Henry Kučera (Department of Linguistics, Brown University Providence, Rhode Island, USA). Taking the Brown corpus as a reference, the words in the two English translations are retrieved.

Table 7. Keywords

	The report of the 20th NCCPC (target text)		The report of the 19th NCCPC (target text)	
Rank	Type	Freq	Type	Freq
1	and	1832	and	1700
2	we	702	we	573
3	will	493	will	344
4	party	260	party	327
5	s	248	people	272
6	people	238	our	262
7	our	220	with	239
8	development	182	s	218
9	china	181	chinese	195
10	have	172	china	173
11	new	165	development	162
12	chinese	156	must	158
13	all	156	have	149
14	system	130	all	136
15	improve	117	new	125
16	must	108	system	118
17	country	93	has	105
18	national	85	work	100
19	work	79	country	94
20	security	78	improve	93

In table 7, the top 20 keywords are selected for comparison, which illustrates that there are 16 same keywords including "and" "we" "will" "party" "people", "our", "Chinese", "China", "development" "must", "have", "all", "new", "system", "work" and "country". The similarity in the two translations is up to 80%, the result having a lot to do with the continuity of the reports and the relevance of their themes.

Compared with the report of the 19th CPC National Congress, the keywords "we" and "will" in the report of the 20th CPC National Congress occur more frequently, which illustrates the CPC is inclined to rally people of all ethnic groups to make efforts for the great rejuvenation of the Chinese nation.

In the top 20 keywords, the keyword "improve" occurs in the translation of the report of the 20th CPC National Congress but not in that of the report of the 19th CPC National Congress. It shows that due to the hit of the COVID-19 pandemic on the economy, China increasingly focuses on economic recovery.

4 Conclusion

Based on a corpus research method, the paper makes a contrastive study of the lexical features between the English translations of the report of the 20th CPC National Congress and that of the 19th CPC National Congress from both macro and micro levels. Lexical features are explored from five perspectives: TTR, lexical density, average sentence length, high-frequency words, and keywords.

The research reveals that two reports have relevance in themes and are a continuum in their major tasks. Specifically, the author found that high-value and median-value modal auxiliary verbs tend to be less used in the 20th CPC National Congress report, the reason for this is that the Chinese government is committed to building its own internationally friendly and positive image. Besides, the frequency of using delexicalised verbs is much higher in the 20th CPC National Congress report, resulting in more native expressions. It is also found that words in the report of the 20th CPC National Congress are more diversified and readable as a consequence of shorter sentence length. The paper hoped to provide some thoughts and ideas on the English translation of political documents.

However, only a limited range of perspectives are examined as an example in this paper due to time constraints. The author remains confident that more lexical features will be found in the future with the growth of corpora and more in-depth research. Additionally, it is proposed that more emphasis be placed on diction, modal auxiliary verbs and delexicalised verbs in translator training because the discussion demonstrates how valuable they are in the translating process due to their high potential of creation.

References

1. CGTN. (2022, October 17). *Foreign Experts Invited to Help Copy Edit Translated Versions of the 20th CPC National Congress Work Report.* https://news.cgtn.com/news/2022-10-17/Foreign-experts-help-translate-the-20th-CPC-National-Congress-report-1ed3T5oXm5G/index.html
2. Jia Hui & Sun Minwei. (2018). Innovation in English Translation of Political Documents -- A Corpus-based Comparison of English Translation of the Reports of the 19th and 18th National Congresses. *Shanghai Translation*, (05), 35-40.
3. Wu Yana, Xie Yating & Xia Mengjun. (2020). A Corpus-based Study on the Manifestation of the 19th National Congress Report. *Overseas English*, (02): 24-25+64.
4. Shang Wenbo. (2020). Research on the Image of Our Party from the Perspective of Corpus Criticism Translation Studies -- Taking the English Translation of the Report of the 19th National Congress as an Example. *East Journal of Translation*, (03): 38-45.
5. The Xinhua News Agency. (2022, October 25). Full Text of the Report to the 20th National Congress of the Communist Party of China. *China Daily*. https://global.chinadaily.com.cn/a/202210/25/WS6357e484a310fd2b29e7e7de.html
6. The Xinhua News Agency. (2017, November 04). Full Text of Xi Jinping's Report at 19th CPC National Congress. *China Daily*. https://www.chinadaily.com.cn/china/19thcpcnationalcongress/2017-11/04/content_34115212.htm
7. Kettunen, K. (2014). Can Type-Token Ratio be Used to Show Morphological Complexity of Languages?. *Journal of Quantitative Linguistics*, 21(3), 223-245.
8. Ure, J. (1971). Lexical density and register differentiation. *Applications of linguistics*, 23(7), 443-452.
9. Johansson, V. (2008). Lexical diversity and lexical density in speech and writing: A developmental perspective. *Working papers/Lund University, Department of Linguistics and Phonetics*, 53, 61-79.
10. Halliday, M. A. K., & Hasan, R. (1989). *Language, Context and Text: Aspects of Language in a Social-semiotic Perspective* (2nd ed.). Oxford: Oxford University Press.
11. Bonelli, E. T. (2000). Corpus Classroom Currency. *Darbai ir Dienos* 24, 205-243.

Visual Truth and Image Manipulation: Visual Ethical Anomie and Reconstruction of Digital Photography

Xia Wang [1*] and Qianqian Hu [2]

[1] Department of Journalism and Communication, Xinjiang University, Urumqi, China
[2] Department of Journalism and Communication, Xinjiang University, Urumqi, China

Abstract: Digital photography has impacted the visual truth in terms of image operation technology. The paper analyzes the performance of image manipulation in digital photography, and asserts that the operation image's resolution of visual truth can be described by three aspects: the fact deviation, the reconstruction of the situation, the trust crisis, the filter survival, the second reprint, and the value change. Consequently, there are three visual ethical problems associated with the manipulation of images by technology, the siege of images on people, and distortions of social order caused by the manipulation of images by technology. Thus, it is suggested that visual ethics be reconstructed from three perspectives: reconstructing the dominant role of human beings in the digital era, returning to the original significance of visual images, and constructing multiple visual structures.

1 INTRODUCTION

The digital camera is not a simple technical replacement for traditional photography, and there are several significant differences between them. The image processor in a digital camera is a microcomputer. In digital photography, photoelectric images are represented by data, which can be displayed through data manipulation. Hence, the formation of digital images is accomplished as a result of the "operation" of the computer. Moreover, the development and popularization of image editing technology have made it easy and simple for people to manipulate images. In spite of some scholars' belief that digital photography is an objective image, it is similar to human visual images [1]. However, the objectivity and authenticity of digital images are greatly reduced by a series of technical means used in the image processing process. Hence, this paper attempts to discuss the technical aspects of digital image manipulation. In this article, the following questions are posed:

(1) What are the effects of image manipulation on visual truth?

(2) How does this result in visual ethical anomie?

(3) How to reconstruct the visual ethics of digital photography?

2 Image manipulation as a means of digital photography

2.1 What is "image manipulation"?

Digital imaging technology has transformed photography from film to digital. Unlike film photography, which only allows limited adjustment of the tone of images, digital photography allows the user to easily change the tone and even the content of images. These photography and image modification behaviors brought about by technology can be understood as the operation of images, and are also considered as the operation mode and technical means of digital photography. As a result, William Mitchell believes that computational tools are the basis for digital imaging skills among digital artists, as they facilitate the transformation, composition, modification, and analysis of images. [2] Liu Canguo pointed out in his book Digital Photography that digital photography involves knowledge in many fields and is a new aggregation of aesthetics and technology, and digital image recording is a natural process in the development history of photography technology. [3] The term image manipulation is used in a wide range of fields, including computer software, automation technology, biomedicine, and physical geography. This paper refers specifically to the process of using digital equipment to produce and process images. While the concept of "manipulation" has opened up an unprecedented range of image creation possibilities, it is also a game involving both technology and concept.

2.2 Image manipulation in digital photography

In contrast to traditional photography, today's image processing technology is more like a digital "bright room" where operators have more creative options.

Email: 956211090@qq.com

2.2.1 The discrete pixel composition of digital images causes the image to lose coherence

Digital images are composed of pixels, which are represented by the numbers 0 and 1. Contrary to the continuous image space of a film image, a digital image is comprised of discrete pixels, resulting in a continuous gradient of color. Pixels contain a limited amount of information. After enlarging an image, it is only possible to see the microstructure of the pixels, and the updated content cannot be obtained. Additionally, image processing software can be used to perform image compression, enhancement and restoration, image change, and segmentation.

2.2.2 Copying, splicing, synthesizing, and altering the semantics of digital images

Photographic equipment and imaging software are constantly being updated and iterated as a result of the advancement of digital technology. As a result, traditional photography completely breaks down the distinction between images and represented objects and creates entirely new concepts and expression languages for creating images. Using Foucault's terminology, "shift, reposition, morph, imitate, disguise, copy, overlap, and apply special effects to a photograph. This method is new to the image of the new, flexible, and without scruples." [4] The replicability of electronic collages reduces the cost of photography, leading to an explosive increase in the number of images.

2.2.3 Digital Virtual Image and the production of surrealism

Computers can do everything from computer-generated images to virtual scenes, especially in creating virtual reality. [5] In contemporary image creation, a kind of image artwork has emerged beyond the existence of the real world. With image editing software, you can alter, edit, and reorganize digital image codes, resulting in a virtual image that is as real as the real one. A montage technique is used to superimpose images, and different image elements are synthesized in a picture using Photoshop. Technological intervention transforms the image into a form of digital information that can be programmed, creating a realm of imagination separate from reality.

3 The impact of image manipulation on visual truth

As Wittgenstein pointed out, "the image cannot be used to judge the truth anymore." [6] With the advent of "image manipulation" in the Internet age, the relationship between image and truth has become more complicated and distant. So how does "image manipulation" contribute to the decline of truth from "indispensable" to "declining"?

3.1 Masking reality and recreating the situation

Heidegger states that the world is visualized as an image. Changes in how images are presented and perceived are redefining our social life and cultural structure. With the aid of beautified software, it is possible to stitch facts from different times and places together. As a result, the image no longer serves as a medium to comprehend reality, but rather serves as a mask, resulting in a scenario in which "life itself is revealed as a huge accumulation of landscapes." [7] "The difference between an image and reality no longer matters. These images substitute for real experience, and for many people, they are hard to distinguish from reality itself." [8] The replicability of images makes it easy for images to proliferate, and social platforms are flooded with beautiful images. On the one hand, it provides a psychological boost to the viewer, but on the other, it obscures the fact that reality is also being eroded by the "manipulation means", and the visual truth is becoming increasingly ambiguous.

3.2 Beautifying the fact and filtering the survival

In the shooting and post-production processes, digital filters are inserted into the camera equipment and image editing software by algorithm engineers in order to achieve instant-beauty function and style change. This makes the filter a form of photography beautification. In this manner, the essence of traditional photography is deconstructed, the authenticity of the image is compromised, and it is increasingly difficult to distinguish between the virtual and real landscapes of society. Generally speaking, the higher the "moisture" of a beautified picture, the lower its "tolerance" is. The image cannot serve as a replacement for the real social face of society until people lose trust in these image landscapes. By beautifying the image, users can feel more present in the social environment. Through the development of beautified image technology, the average person has the ability to display more expression power, expression capital, and social capital on social platforms. In this sense, a beautified image also serves as a form of technical empowerment. A hidden concern lurks behind the rise of filters used by more and more people online to present a "perfect" image of themselves.

3.3 Making trouble and changing the value

Images that are manipulated no longer represent the reality that already exists, but create a "reality" that does not exist or even create much of nothing. In the post-truth era, images encourage the spread of rumors and undervalue the value of truth. Since truth is difficult to distinguish between true and false in the online world, there are neither value-neutral facts nor importance of factual basis. The transmission of manipulated images has also become absurd and unnecessary. It is no longer important to determine whether the image content reflects the real content, but rather the image symbol itself. The identification of image rumors has also become more challenging in recent years. It is difficult to verify the authenticity of images because there are so many image resources, and due to the versatility of digital technology, not every image can be found. Changing the rumor from "there is truth in the picture" to "there may not be true in the picture" simply proves that the picture no longer represents the truth, but only produces confusion.

4 Visual ethical issues caused by manipulating images

In digital photography, image manipulation technologies have invaded the truth, beautification and forgery have caused problems of visual ethics anomy, including image surplus, single aesthetics, and Internet rumors. Afterward, the mechanism of visual ethics anomy resulting from "image manipulation" is examined from three perspectives.

4.1 The manipulation of images by technology

Zhang Jingfang believes that in the new governance era, under the control of the invisible hand of technology, people are literally living in a digital cage. [9] Technology is tampering with the truth while ignoring ethics. More and more "de-incarnation" and "de-face" manipulation image crises have caused people to reflect on their true selves.

4.1.1The absence of reality and the emergence of "hyper-reality"

The danger of manipulating images lies in the conflict between "authentic expression" and "faithful recording". Digital images can be seamlessly connected from production to communication with the help of media such as mobile phones. As a "social tool" for establishing interaction between the two parties, viewers still try to reach the real world of the publisher with images. The real is missed from the screen. In the new media era, the symbolic information world constructed by the media using images is covering the actual reality. Through image manipulation, a kind of quasi-image may be created according to the model, which means the mixing of imagination and reproduction. "Hyper reality is a reality without archetypes and authenticity, created using a real mold." [10]

People publish photos on social media as images-like symbols. From the view of image producers, the manipulated images with beauty and filters may be more like themselves than the real themselves. Technology has also created a hyper-real symbol world in cyberspace. Simulation and reality are entwined under the prevalent use of manipulation methods, and manipulated images are more real than real ones. The abandonment of real simulation and reproduction technology separates the relationship between people and the objective world, and it moves from the actual real world to the image world created by science and technology.

4.1.2Image empowerment

An image's empowerment power is derived from the competition for visibility. By using visual media, the public can express their protest words and make their demands visible. Repeated and secondary creations of the same theme image will result in the echo chamber effect and initiate the Internet carnival. The "Operability" of digital images coincides with the self-empowering mechanisms of Internet users, and the free space and flowing resources enable Internet users to exploit their creativity in unprecedented ways. There has been a growth in new cultural phenomena and practices associated with image transmission and interaction, such as selfie culture and emoji, which indicate the rights and freedoms of users' image discourse.

Visual display often has a distinctive feature of power. Martin Jay once pointed out that we are in a "visual regime", and vision has a strong disciplinary power. [11] Display is the premise of viewing and gaining attention, and attention is what guarantees the power demands of image producers, consumers, and the platform itself, so that power and capital can be linked. Operational means may enable individuals to intercept portions of images and distort the facts in accordance with their own interests. Social platforms without rational management systems are lacking in media literacy, and users tend to copy the network world; therefore, the camera has become a powerful weapon in the hands of every Internet user.

4.2 The siege of people from the image

Although visualization provides us with a rich visual experience, we are losing control of this trend. In visual culture, images have become a means of social power operation, as well as a form of controlling the siege of people's power in the sense of social control.

4.2.1The turning and alienation of viewing

Michelle called the problem of the 21st century "the image turn". Imagery, visual stereotypes, Facebook, hallucinations, copying, copying, imitation and illusion dominate our culture. [12] There is a significant increase in the dominant consciousness of the subject, as well as a certain degree of liberation in the viewer's "viewing" behavior. The modern visual mechanism has not only changed the traditional viewing habit of "do not gaze at what is not polite", but has also created specific visual and ethical concepts, focusing vision into a field that is full of competition and production, and peeping has become a normalized act of watching.

As visual perceptions have changed, the boundaries of graphic representation have expanded, from daring celebrity portraits to cartoons that threaten privacy. Consequently, the images related to personal privacy become the material chosen to satisfy the public's peeping psychology, which is no longer limited to the "presence" of viewing behavior, and the individual's discomfort with moral admonitions decreases. Additionally, there are many factors that contribute to the ethical dilemma associated with modern peeping, in addition to the abundance of viewing objects and the development of viewing mechanisms. More and more image producers and broadcasters have ignored the rights of the objects they present in order to pursue a more exciting look and feel. Without restraint, the private body is listed as an object for public browsing, which blindly satisfies the audience's desire to peep, but establishes an imaginary mirror image of the body image in the invisible.

4.2.2 The proliferation and disappearance of aesthetics

Digital photography has led to a new visual aesthetics, where passive acceptance has been replaced by active participation in aesthetics; aesthetic objects from the past to today's elegant and vulgar appreciation; aesthetic experience is progressively becoming virtual and interactive. Originally, aesthetics was a concept derived from material activities, but now technology and aesthetics are interacting with each other. Currently, operational images benefit from the overall visual aesthetic atmosphere, and their prevalence is nothing more than the performance of aesthetics and technology to expand their own ideologies. In recent years, photography beautification technology has evolved from a tool to represent beauty into a medium for aesthetic expression. Social software thus grasps the traffic password "brushing is reasonable and good-looking is justice", and social platforms reflect the discipline problem in image production and processing. Nowadays, most people accept copies that have been reprinted many times after beautification rather than originals due to the improvement in image transmission speeds and easy access to reprinting. Oftentimes, this leads to stereotypical popular aesthetic practices and aesthetic concepts, and individuals begin seeing the world through others' eyes rather than their own. The aesthetic personality is also eliminated in the subtle influence.

4.3 The reconstruction of social order

When the production and dissemination of images are within reach due to the popularity of mobile phones, images with strong impact effects replace the profound ornamental works, and the worship value of images begins to give way to the display value. The complex visual landscape constantly impinges on people's visual perception, and various image representations constantly challenge the visual limits of viewers, thus tampering with conventional visual concepts.

4.3.1Visual carnival and visual consumption

Technology-driven visual carnivals have already outlived their aesthetic value due to the deep participation of consumer consciousness in the social transformation of the 1990s. Visual consumption has become dominant in image production due to the close connection between visual culture and consumer society. Through the digital medium, visual consumption and image production are mutually beneficial. "The greatest effect of electronic devices is to emphasize and accentuate a need that is already well established." Bowman once stated. [13] As more visual objects are provided to users, the means of image display become more extreme. Some images that indulge low desires and stimulate visual pleasure constantly challenge the established social order and ethical paradigm. The aesthetics of daily life makes experience and entertainment become the guidelines of visual culture.

Modern image production and display technologies not only affect the production and consumption of contemporary social culture, but also influence the standard to judge the attractiveness of an image. Therefore, image producers continue to create visual wonders, capital is deeply invaded, the image becomes a commodity, and personal display usurps ethical constraints of social norms. In order to suppress the initiative of the subject, image manipulation suppresses the influence of visual symbols on other media forms, which is seen as a declaration of war between text and image, as well as constructing a social order dominated by symbols and models.

4.3.2The Visualized "Data Prison"

Online images can easily be copied into countless copies, and they are almost impossible to forget. In light of the fact that data owners are able to access the past data of users at any time and anywhere, we face the threat of a data dictatorship. In addition to personalized accurate push and advertising based on data resources, we can also obtain users' shopping habits by analyzing data and predicting their purchasing power as well as conducting awareness penetration through social media marketing.

With the rapid development of capitalism's digitization, big data applications have already penetrated all walks of life, including the photography and beautification industry. Data has become a capital that can be controlled and consumed. The "2020 shooting beautify Industry Research Report" released by Aurora Big Data found that four out of ten Internet users who use mobile phones possess shooting to beautify apps. With the use of big data algorithms, image manipulation is possible, and users appear to use a variety of image beautification software to display themselves on the platform. However, it is the Internet companies that control the preferences of users to cater to their self-exhibition, and to create a "visual feast".

5 The Visual Ethics Construction of "Manipulation Image"

5.1 Reconstruction of the subject position of the human in the digital photography age

Harari believes that in the 18th century, humanism moved from a god-centered worldview to a human-centered one, pushing God aside. While in the 21st century, dataism may move from a human-centered worldview to a data-centered one, pushing people aside. [14] Image operation is intimately linked to the algorithms that support it. When images are beautified using "portrait presets" and "automatic beauty makeup", these steps are completely hidden from their image attributes by the image retouching software. In this way, the final image presented to the public actually implies a lot of digital "algorithm aesthetic".

The subjects of work creation and aesthetic appreciation generally lack depth, chase flow, and blindly follow the trend. The algorithm creates a seriously homogeneous image environment, which undoubtedly weakens our deep thinking on beauty and deprives us of the possibility of multiple aesthetics. With the widespread involvement of visual texts in social life, operational images gradually affect the formation and dissemination of social values, especially in the field of ideology. Therefore, on the one hand, we should emphasize the humanistic care behind the image, take into account the rational reflection derived from vision, consider the rational expectations associated with the technical picture, and take responsibility for regulating visual ethics on our own. However, while satisfying the fragmented visual carnival in social software, we should pay attention to the spiritual dimension of people and grasp the concept of boundaries and social responsibilities.

5.2 Returning to the original meaning of visual images

Through digital "operation", users create widely disseminated visual texts almost all the time. These images are no longer simple reproductions, but self-presentation with individual aesthetics. The viewer and the viewed compete on the same stage, resulting in the public's attention to pleasure stimulation beyond the original significance of the image. For Plato and later thinkers, there is always a higher value and meaning concern behind sensory entertainment, and people must be wary of the destructive power caused by purely emotional stimuli.

The use of visual images has become a mainstream form of expression in modern society, however, digital technology has provided images with a plethora of possibilities, and aesthetic carnivals which exceed moral norms will adversely affect social development. In response to the "attention economy", technology has opened up new capital markets on new continents. New media profit cannot be separated from the visual prospect that attracts eyeballs and clicks. In order to stimulate people's senses, an increasing number of obscene, ugly and pathological images have been created by individuals. In order to prevent the formation of the image over the content, image producers must develop a conscious ethical character, eliminating vulgarity, superficiality and blindness, and establishing an ethical character. It is imperative, first and foremost, that we reflect on the rationality of using image manipulation techniques; second, we should build and improve media literacy among citizens in moderate "viewing". Additionally, the platform should ensure that the images of the media platform are accurate and position the audience appropriately.

5.3 Constructing multiple visual structures

Unlike print culture, visual culture constructs social subjects through the use of visible symbols and images. It is in this process that society, culture, and self are cognitively shaped. People living in the real world produce their own visual experiences in a variety of ways. Visual construction includes visual construction in the social field and social construction in the visual field. [15] Both emphasize the practice of social visual expression. Since images are more intuitive and expressive than words, the consciousness, emotion and action of the subject are more likely to be affected by images, and visual discourse is more profound and powerful in constructing the cognition and ideology of the subject.

All visual behaviors and expressive practices in the visual field are social, and in the social field, vision is the most important perception channel of the subject. Visual structures are produced by the visual regime using rhetoric and representation to influence how people see and how they see. Technically, the first level of viewing includes elements such as composition, color, and light, and the appropriate use of these parameters facilitates the formation of a comprehensive and objective understanding of society. A viewer should be able to comprehend visual language, understand events through images, infer the motives and intentions of the image publishers, as well as build social cognition from the content level. As a matter of ethics, mass media image information should be conducive to social progress. It is also important for the audience to cultivate good visual literacy, think critically about images, and rationally handle the communication and collision of different visual cultures, so that a diverse and inclusive social world can be constructed that conforms to acceptable ethical norms.

6 Conclusion

The new communication environment is a complex environment in which opportunity and challenge coexist. The public is easy to fall into the dilemma of post-truth because of the visual picture. Due to the development of operating technology, the image is unable to convey its true value and significance to the viewer. The combination of image and virtual reality extends "viewing" into an interactive, experiential, and traversal experience. Through the proliferation of manipulated images, not only is the viewer's rational cognition challenged, but also his or her living autonomy is compromised, and the possibility of a multi-cultural perspective is destroyed. There are many aspects of cognitive power and public order challenges brought about by the digital photography era, including visual ethics. Considering the ethics of digital photography and pursuing a good digital life is an issue that requires further reflection and investigation.

References

1. Zhong Jianming. Analysis of digital photography form -- Rediscuss photography classification. Journal of Hundred Schools of Art, Issue 8, 2011 Total issue 123.

2. Mitchell • William. Reorganizing the Eye: Visual Truth in the Post-Photographic Age. Translated by Liu Zhangbolong. China National Photography Art Publishing House,2017:74.
3. Liu GuoChan. Digital Photography [M]. Zhejiang: Zhejiang Photography Press,2006.
4. Foucault. Foucault Collections [M]. Shanghai: Shanghai Far East Publishing House,2004:257.
5. Zhou Xian. Image technology and aesthetic concept. Journal of Literature, History and Philosophy. Issue 5, 2004 (Total Issue 284).
6. Wittgenstein. The Complete Works of Wittgenstein (Volume 1). Translated by Chen Qiwei. Shi Jia Zhuang: Hebei Education Press,2003:191.
7. Guy DeBo. Image Society. Translated by Wang Zhaofeng. Nanjing: Nanjing University Press,2006:3.
8. Dennis McQuail. McQuail's Theory of Mass Communication. Translated by Cui Baogou, Li Kun. Beijing: Tsinghua University Press,2005:79.
9. "Face" in the Digital Age and the Crisis of Self-Authenticity: Based on Truth capture. Journal of Cultural Research,2021(02):225-238.
10. Ma Xiaoru. On the Concept of Hyper-reality. Journal Philosophical Analysis, Vol.9, No.5, October 2018
11. Martin•Jay, In the Empire of Gaze: On Foucault and the Demeaning of Vision in Twentieth Century French Thought. Translated by Chi Qingli. Guangxi: Lijiang Publishing House, 2002:110
12. W.J.T. Mitchell. Theory of image. Translated by Chen Yongguo. Beijing: Peking University Press,2006:2. •
13. Sigmund Baumann. 44 Letters from the Liquid Modern World. Translated by Bao Lei. Guang Xi:Li Jiang Publishing House.2013:8
14. Yuval • Harari. A Brief History of the Future. Translated by Lin Junhong. Beijing: CITIC Publishing Group.2017.
15. Zhou Xian. Visual Construction, Visual representation and Visuality----an investigation of the three core concepts of visual Culture. Journal of literary review.2017(03):17-24.

Research on the Development of Voice Assistants in the Era of Artificial Intelligence

Yuqi Huang[1*]

[1*]Beijing Etown Academy, No.12 Sihe Road. Daxing District. Beijing, China.

Abstract. Voice assistants have gradually occupied an important position in the products of many electronics companies. Artificial Intelligence voice assistants are able to interpret human speech and respond. Users can ask their assistant questions and manage other essential tasks such as email calendars through verbal commands. This paper analyzes the artificial intelligence voice assistant through the method of comparative analysis. The author studies the development situation of intelligent voice assistants, and compares the differences between Chinese and foreign voice assistants, and finally discusses the relationship between voice intelligent assistants and people's lives. The author found that users in different countries have different functional preferences for using voice assistants, but they can help people's work and life to a great extent. In other words, voice assistants play an important role in contemporary society. Therefore, people need to better understand the relationship between humans and machine

1 Introduction

In recent years, with the gradual development of science and technology, people have changed the lifestyle of finding information in the library when they have a problem. Now people can directly know the answer to the question by using a search engine or artificial intelligence voice assistants. For example, when people use iPhone mobile phones, they often use the siri function. Users only need to activate Siri's voice assistant function and use some simple voice commands to query information, make calls, send information, get directions and play music. It can be seen that siri, as an artificial voice assistant, has completed the work of personal assistant to a certain extent. The data that used to be calculated by the human brain has been superseded by machines, and artificial intelligence is a machine that can imitation human responses. In fact, artificial intelligence voice assistants are based on applications, which often using voice commands to answer users' questions and perform tasks. This interaction is mainly achieved through voice assistants and automatic speech recognition systems that output a set of words or sentences for text by recognizing and transforming the input speech fragments [1][2].

There are also many devices in people's daily life, such as Apple's Siri, Xiaomi's Xiao Ai classmate, home assistant, map navigation assistance and so on. Awakened by means of voice response voice assistant to be in a position to better help people's lives conveniently. People can make their lives better by utilizing the functions of voice assistants. Obviously, the relationship between artificial intelligence voice assistants and humans is getting closer and closer, and it has also led people to reflect on whether voice assistants can replace humans and whether humans will have feelings for voice assistants. This article will explore the types of imitation voice assistants and their impact on people's lives through the research methods of literature review and comparative analysis, and further explains whether humans will develop emotions for artificial voice assistants. The author will first introduce the development and classification of voice assistants, compare the differences between Chinese and foreign voice assistants, and then describe the advantages and disadvantages of voice assistants. By studying the related performance and differences of voice assistant, it can help people better understand the relationship between artificial voice assistant and human.

2 Development of artificial intelligence voice assistant

2.1 Definition of artificial intelligence voice assistant

Artificial intelligence voice assistant is a type of speech semantic application technology which is mainly applied to speech recognition, semantic understanding, speech synthesis and other intelligent speech semantic application technology in electronic devices. Users no longer need to manually operate, and can use many operations simply by giving instructions to the voice assistant. The voice assistant understands the user's needs from the user's language instructions and completes the corresponding tasks for the user. By utilising the voice assistant, the user's experience of using electronic

*Corresponding author Email: 1324538333@qq.com

products such as mobile phones is better. In recent years, voice interactive technology has become the key development direction of artificial intelligence technology. Since Apple launched Siri in 2011, various electronics companies have imitated Siri's voice assistant mode and added the voice assistant function to the company's products, such as Xiaomi's Xiaoai classmate, Google Assistant, Baidu Map voice. These companies are also promoting the implementation of voice interaction technology in people's lives. According to the survey data of China Telecom Terminal Research and Test Center, the user awareness rate of voice assistant is as high as 95.6%, and the use rate is 58.6% [3]. More than half of mobile phone users use voice assistance in their daily life. Obviously, artificial voice intelligent assistant is playing an increasingly important role.

In the past 60 years, the technology of voice assistants has developed rapidly. From the combination of three robust acoustic models at the beginning, the accuracy rate of voice assistants has reached the level of human stenographers, reaching 95% accuracy. Users often use the speech-to-text feature, which allows people on the go to send messages and take shorthand. Amazon launched smart speakers, which connect apps on smart phones with furniture in daily life, opening up a broader space for voice interaction and occupying a larger market. The reason for the success of Amazon speaker is the fact that it can truly interact with dialogue, which is the most requested feature of today's social users [4]. Definitely, the understanding of natural language does not only require the voice assistant to understand the basic meaning of words, but also to understand the deep meaning of sentences through in-depth research and learning, and to connect sentences with each other.

2.2 The development of intelligent voice assistants in China and foreign

2.2.1 China

In China, Internet companies have gradually formed a complete industrial chain of voice intelligent assistant products through intra-industry cooperation. Among them, Baidu, Alibaba and Tencent occupy a sizable market share. These companies also have many products for the application of intelligent voice technology. Companies for users in the daily life of the things need in-depth research, and make efforts to transform these products. In the field of home furnishing, these companies also have certain development. Baidu has launched Baidu IHome, and Alibaba has also launched Tmall series. In addition to these companies, there are many other Chinese companies that have developed products related to intelligent voice assistants[5]. For instance, Haier's voice assistant connects 10 types of smart home appliances under Haier's air conditioners, sweepers, ovens, and water heaters through voice interaction. Users only need to send commands to Haier's intelligent voice system, and Haier's smart home products can operate at any time[9]. Figure 1 shows that the central voice-controlled device can be used to accept commands and link other home devices.

Fig.1. Haier Smart Home Voice Control Device Demonstration[9]

In recent years, people cease to be satisfied with single intelligent voice technology. Intelligent companies gradually began to transform and develop towards technical services. Technology service providers represented by IFlytek began to promote new products. IFlytek advocates innovative and original product development. The current research and development direction are in the two directions of perceptual intelligence and cognitive intelligence. Many online application platforms have also used IFlytek, such as AutoNavi, Didi Chuxing and Ctrip. Recently, IFlytek has continuously increased investment in R&D funds for innovation[6]. In the past five years, IFlytek's product revenue has tripled, which also proves that IFlytek's innovative products are increasingly in line with people's needs.

2.2.2 Foreign

In foreign countries, four major intelligent voice assistants of Amazon, Apple, Google and Microsoft occupy the main market. Among them, Amazon's Alexa and Google's

Google assistant is developing more rapidly. These voice assistants can support and accurately apply to the current mainstream demand functions, speech recognition, semantic analysis and so on.These four voice assistant companies are also constantly changing the core of their product concepts according to different user needs, striving to find suitable market groups and occupying leading positions in diverse markets[5].

As a well-known company in the smart phone market, Apple's products are primarily aimed at its own product series. Siri is mainly applied in Apple's built-in, which is conducive to Apple's promotion to users with the help of Apple's series of products. This kind of intelligent voice assistant is more suitable for users who buy Apple's companion device. Microsoft's Cortana is more associated with the PC. Cortana lets users experience when their phone and computer work together. For people who use the voice assistant for office work, Cortana will help them have a better experience at work than Siri, which is more used in daily life.

Google, as the world's largest search engine, has a strong technical hand in Google assistant. Google Assistant is a virtual assistant software application designed by Google, primarily for mobile and home automation devices. Based on artificial intelligence, Google assistant can hold two-way conversations, unlike Google Now, the company's previous virtual assistant. Google assistant has the most powerful technical advantage over the previous two voice assistants, making it easier to search for information. Google assistant will analyze the user's preferences built on the user's search frequency and provide users with more valuable information. For example, a new "humming search" feature will develop by Google in 2020 will allow users to find songs simply by humming, whistling or singing.

Amazon holds the largest share of the facetious voice market. In 2014, it developed Alexa, a sarcastic voice recognition system, which leads all smart voice assistant companies to become the market leader. Amazon has combined the company's appliances with a voice assistant to make a feature-rich product. Amazon no longer keeps Alexa's voice technology private, but instead provides it to third-party platforms[5]. Through cooperation with other platforms, Alexa has mastered multiple skills in a short period of time, occupying an important position in intelligent voice assistance products.

2.3 Difference analysis of voice assistants in different countries

The Chinese company's products are more aimed at users' personal daily use, such as smart home and map navigation. And foreign intelligent voice assistant companies pay more attention to office use, such as the arrangement of some company meetings. In general, the voice assistants in each country are different, and there is no uniform evaluation standard. The user experience of using a voice assistant is also different. In Germany, most users rated voice assistants lower than in the UK, Italy and Spain. Reviews from these users also reflect the level of development of a country's voice assistants [7]. In contrast, the usage rate of Chinese language assistants is relatively high, and users' evaluation of voice assistants is also relatively high. On the one hand, China has a relatively large base using mobile devices, and the number of users is also relatively large. On the other hand, Chinese companies attach great importance to the publicity of the use of voice assistant functions, which has promoted more people to use convenient software such as voice assistants.

3 Discussion

Voice assistants have begun to integrate into people's lives. It has a positive effect on the society as well as on the individual. For office, artificial intelligence voice assistant can help people work efficiently. For individuals, voice assistants can not only provide convenience in life. To a certain extent, it can make up for the lack of people's emotions. Some people are willing to chat with a voice assistant after they start using it, which also has a lot to do with the faster pace of life. While voice assistants are not as emotionally rich as people, voice assistants can be a qualified listener. It can also be seen that people have emotional dependence on voice assistants. In addition to people's reliance on voice assistants, the development of voice assistant technology also involves emotional issues. For example, the fact that Apple's voice assistant Siri married Amazon's voice assistant Alexa, although a marketing ploy, has also raised questions about whether voice assistants have feelings[8]. Although it is known that artificial intelligence voice assistants are generated by computer programming. But with the development and progress of science and technology, it will be a question worth thinking about whether artificial intelligence will appear human emotions autonomously.

4 Conclusion

In conclusion, with the gradual development of voice assistant, voice assistants are increasingly affecting people's lives. This paper analyzes the current status of the booming and broad development status of artificial intelligence voice assistant. This paper compares the voice assistant products of different companies in China and foreign countries. The author found that China pays more attention to voice assistant products related to life, while foreign customers prefer to use voice assistant products at work. With the development of technology, the relationship between artificial language assistants and humans will become closer and closer. The research method of this article is relatively limited. In the future research, the author will use more research methods such as questionnaires and interviews to better understand the views of artificial voice assistants in people's minds.

Acknowledgements

I would first like to thank my thesis advisor. She consistently allowed this paper to be my own work, but steered me in the right the direction whenever she thought I needed it. Also, I want express my very profound gratitude to my parents and to my partner for providing me with unfailing support and continuous encouragement through the process of researching and writing this paper. This accomplishment would not have been possible without them.

References

1. Sinha Y, Siegert I. (2022) Improving the Accuracy for Voice-Assistant conversations in German by combining different online ASR-API outputs[J]. Human Perspectives on Spoken Human-Machine Interaction, 11.
2. Albuquerque, F. A. D. (2022). "Hey Siri, how much do you know about me?": Intelligent Virtual Assistants and the dilemma between commodity and privacy (Doctoral dissertation).
3. Li S.W., Cheng G.F., He S.W., Zhang D. (2019) Research and Trend Analysis of Voice Assistant Capability Evaluation [J]. Guangdong Communication Technology,39(12):52-56.
4. Xu L.Q. (2019) "The development and future of mobile phone intelligent voice assistants" [J]. Communication World, 26(04): 262-263.
5. Gan Y.J., Su J.G., Lin J., Hong B.Y., Lu W., (2019) "Status and Prospect of Intelligent Voice Products" [J]. Guangdong Communication Technology, 37(12):66-68.
6. Liang C.G., (2021) IFlytek: Smart Voice List TOP20: IFlytek, Baidu, etc. top, more than half of the start-ups, Retrieve from: https://www.sohu.com/a/472474383_115565.
7. Klein A M, Hinderks A, Rauschenberger M, et al. (2020) Exploring Voice Assistant Risks and Potential with Technology-based Users[C]//WEBIST. 147-154.
8. Dommu R., (2019) Siri and Alexa are Lesbians and They Just Got Married, Retrieve from: https://www.out.com/tech/2019/4/29/siri-and-alexa-are-lesbians-and-they-just-got-married.
9. Haier, Home intelligent solution, (2022), Retrieve from https://www.haier.com/business/smarthome/

Influences of Women in the Sports Field from the Sociology Perspective

Dingxu Shi

Address: Xi'an Tieyi High School ICC

Abstract: With the development of the society, the gender issue is getting higher degree of emphasis. In the field of sports, the issues is even with more attention. In this paper, the aim is to discuss the participation and influences of women in the sports field from the sociology perspective, from the aspect of opportunity cost, women participation and interest towards sports

1. Introduction

Ever since the emergence and development of modern sports in the United Kingdom in the 18th century, many scholars have investigated into the development status and history of athletes and practitioners of the sports fields, laying a basis for the studies on the sports field from the sociology perspective (LaVoi and Baeth, 2018)[8]. When the sociology develops as the discipline, an increasing number of stakeholders are involved and considered to examine and evaluate the influences on its development. The gender issue has always been a striking point in the research and development of the sociology, and it is inevitable to be discussed in the context of the sports field. In fact, the participation and influences of women in the sports filed have aroused widespread concerns and attention. As most of the social orders and values are affected by the values, beliefs and interests of males, more and more people are aware of the significance and necessity of researching the participation and influences of women in the sports field in terms of the rights, equality and consequences. In this essay, it will discuss the participation and influences of women in the sports field from the sociology perspective.

2 Theories of women' influences on the sports sociology

After the second wave of feminist in the 1960s to 1970s, feminist has expanded from the social movements to academic researches and studies by penetrating into different fields. The combination of female's participation and influences with different fields and contexts has been prevailing across the world, which is derived from the fact that increasing social awareness has been aroused on the legal rights and interests of women in the daily life (Evans and Pfister, 2021)[7]. The issue of gender relations is so fundamental to sociological analysis that it cannot be simply relegated to a subdiscipline of sociology. This shows that feminist sociology is not only an important branch of sociology, but also that feminism ought to penetrate into other branches of sociology to enrich the theories of related branches of sociology and make sociological research closer to the real social life. Some scholars have pointed out that, as one of the development trends of contemporary western sociological theory, feminism will further affect the way of observation, thinking and problem consciousness in the field of sociology.

The issue of focusing on the participation and influences of women into the sports field started to attract widespread attention and discussion in the 1970s in the North America. The thriving feminist movement has expanded its perspectives from the political, cultural, educational and other fields of the past to the fields of sports that was not taken seriously. In view of the many problems faced by women's sports, such as their sports power, status equality and remuneration, many scholars have carried out in-depth explanations and interpretations. Liberal feminism believes that women are also rational people, so women should enjoy the same rights as males. In other words, it is expected that women should achieve equality between fertility, employment, education and leisure sports through the choice of rights (Litchfield and Osborne, 2013)[9]. As for social injustice, it should be improved in the form of law, which means that the concept and reality of gender discrimination should be changed by the coercive force of the law. From the perspective of social justice, efforts made by women in occupying more important positions and exerting more profound influences in the sports field have contributed to the equal opportunities of women in taking the role as athletes to take part in sports, serving as coaches at all levels of competition and gaining a position in the power structure of the sports field (Elsey and Nadel, 2019) [6]. Through the unremitting efforts of liberal feminism, women's participation in sports has gradually been recognized by the society and confirmed in the form of laws. This is reflected in the participation rights of female athletes and

Email: shidingxu2023@163.com

the emergence of female managers in the modern Olympic Games. In addition, the participation of women in the sports field has also affected the formation of relevant documents and regulations such as the Universal Declaration of Human Rights, the Brighton Declaration on Women and Sport, the International Covenant on Economic, Social and Cultural Rights and the Convention on the Elimination of All Forms of Discrimination against Women, as well as some extended documents (Matthews and Channon, 2016)[10]. All of these documents and regulations are the result of women' active participation into the sports filed and the consequences of feminist sports view achieved in the real society.

Marxist feminism believes that the root cause of women's oppression in different aspects including the sports field lies in the accumulation of capital and profits. In view of the phenomenon of non "equal pay for equal work" between men and women under the current system, women's rights to work and education have been harmed to varying degrees. At the same time, due to the increase of women's domestic labor services, the idea of the expansion of patriarchy under the class logic was formed. With the continuous reflection of feminism, feminism began to move towards a more comprehensive perspective and view of point in interpreting the participation of women in the sports field. In the sociology of sport, a feminist perspective goes beyond sexism, arguing that gender inequalities in sport are often linked to class and racial inequalities (Mays et al., 2010)[11]. In the field of sports, there is a phenomenon of unequal pay for equal work between men and women due to men's possession of capital and power. In addition, there is a lack of opportunities for women to participate in sports and other activities. For example, there are fewer facilities used by women in society, fewer women instructors as well as unequal project development investment in men and women sports. The reason for this situation is that Marxist feminists believe that men occupy the space of power in the world, and relevant public policies are all products of this class logic. In order to make changes and improvements in terms of such situations and reality, the reliance on the participation of women in social decision-making and to realize the expression of women's sports rights and awareness should never be underestimated. These differences and inequalities have been noticed and aware by more and more people, affecting the perspectives and awareness of women in striving for the rights and interests in the sports field.

3 Rational choice: opportunity cost for women in participating into the sports field

There is a concept called opportunity cost in economics, which can be understood that a person's behavior is the result of the choice of one of the many possible behavioral possibilities that can occur. In view of the unidirectional nature of time, the maximum of the possible benefits of the other behaviors that are abandoned is the opportunity cost of a certain behavior, which is also called the cost of alternative cost (Cooky et al., 2013)[5]. Simply put, the cost of a person's behavioral choice is not only money and energy. In fact, it also includes the benefits of other possible choices. Participation in sports is a specific behavior, which means that participation in sports necessarily forgoes some other behavioral choices, and the possible benefits of these other behavioral choices constitute the opportunity cost of participation in sports.

The benefits of participation in sports are fixed: a certain degree of improved physical and mental health as well as social interaction, while its opportunity cost is the benefit of other options that women give up while participating in sports. The higher this benefit, the greater the cost of participation in sports and the relatively smaller the benefit of participation in sports itself. Prior to the modern era, women were less productive with unequal access to production resources, so they had to depend on men for survival resources. The opportunity cost of participating in sports becomes the acquisition of survival resources, which is obviously a greater and more urgent benefit than the benefits of sports. Therefore, before women's labor force was liberated, the huge opportunity cost of participating in sports was generally unacceptable to women, which means that women's participation in sports at that time was not a good choice from the rational perspective.

In the modern society, women's productivity has been liberated. As the social productivity begin to change qualitatively, resources for survival become abundant and human values gradually become more prominent. As a result, more and more women no longer need to obtain means of survival through dependence on men. At this point, the opportunity cost of participating in sports is much lower than before the modern society and becomes completely acceptable (McClearen, 2018)[12]. Even at some points participation in sports may be the most cost-effective option for women. However, since this essay focuses on the participation of women in the sports field, it is inevitable to discuss the differences between different genders. Although the opportunity cost of female participation in sports is gradually decreasing today, it is still at a higher level in comparison with men, and this is clearly evident in professional competitive sports. The tendency can be inferred from the interest of women in sports across two generations.

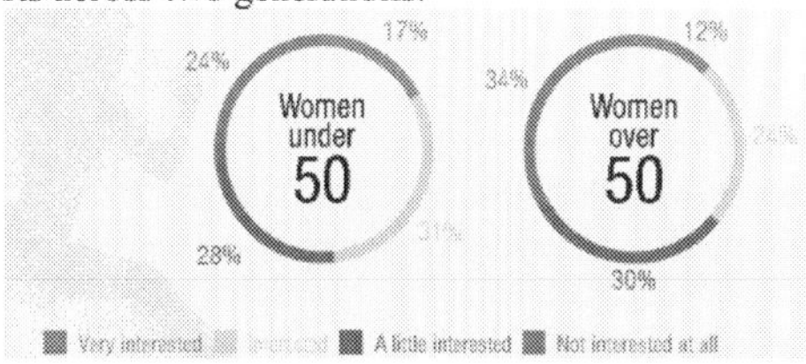

Figure 1 Women' interest in sports across two generations (McDonald, 2015)[13]

The current income level of female professional athletes is generally less than that of men, with only two female athletes ranking the top 100 superstars in the latest Forbes Sports Stars Earnings Ranking as had been released in 2016, ranked 40 and 88 respectively. Meanwhile, more than half of the top 30 Chinese acting stars ranked in the same period were women (Coche and Tuggle, 2018)[4]. On the other hand, the world is currently receiving a lot of attention. On the other hand, most of the professional leagues that have a large and regular fan base in the world are male professional leagues, such as the United States. In addition, most of the current professional leagues with high attention and large regular fan base are male professional leagues, such as the NBA, the English Premier League, and so on. In China, men's games in various leagues are significantly more popular than women's games. In China, the interest in men's games in various leagues is also significantly higher than that of women. Therefore, from the perspective of participation in professional sports, the benefits for women are significantly lower than those for men. Therefore, from the perspective of the participation in professional sports, the benefits for women are significantly lower than those for men, and their opportunity costs are higher as well. As a result, it is natural for the evaluation on the potential opportunity costs of participating into the sports field for women, and the declining opportunity costs may account for the increasing popularity of women' participation into the sports field in more and more occasions. The tendency of female's participation into the sports field also inevitably results in profound influences on a wide range of elements and perspectives.

4 Influences of women by participating into the sports field

Modern researches have proved and verified the benefits of sports activities and participation into the sports for health, including improvements in the physical health and mental health, as well as better adaption and performance in the workplace. Reflecting on the history development process, males occupied a dominant position in the sports field ranging from the coaches, professional sports development and sports field growth direction. However, with the increasing awareness of women of their rights of participating into the sports field and the potential influences, they can gain access to increasing opportunities to participate into the sports field and exert influences on a wider range of people (Whiteside and Hardin, 2011)[21]. Even though more women are participating in collegiate sports today, fewer women than men still participate in professional sports after their graduation. Some may argue that this phenomenon might be resulted from women's lack of interest or choice in professional sports. Arguably, increasing women's professional sports opportunities will not only allow women to further advance their careers, but also bring more positive social and economic outcomes to society. For example, women who raise their profile through professional sports careers can demonstrate leadership, strength, and the value of sports and physical fitness to younger generations (Mirehie and Gibson, 2020)[14]. Therefore, as role models, women in professional sports can inspire girls to participate in sports and experience the many benefits of an active lifestyle. To further discuss the influences of women by participating into the sports field, there are several perspectives that worth further elaborating and discussing.

From the social gender perspective, it mainly emphasizes the expectations and norms posed by the society on males and women in terms of their characteristics, roles, activities and responsibilities. The concept of social gender is different from the distinguishing standard of biological sex (Bum et al., 2018)[3]. Gender is the product of culture, and is caused by thousands of years of patriarchal society's different expectations for men and women. These feminine traits such as gentleness, virtuousness, weakness, etc. are not absolute characteristics of physiology but the result of patriarchal culture. For a long time, people have ignored the social construction of sports. Most of the researches on women's sports in the sociology of sports are conducted from the perspective of functionalism, reflecting the positive role of women's sports development in social development and the promotion of the Olympics, while ignoring other social issues in relevant to women's participation into the sports field (Sherry et al., 2016)[20]. In most sports organizations, men occupy the dominate position with leading power, responsibilities and decision-making sectors except in traditionally all-female events. As have been proved by many facts and evidence that sports are organized for and by men and male domination is self-preservation. One of the most notable pieces of evidence is the major international games and the Olympics, which have always been characteristic of the participation of men throughout the history, and the participation of women is the result of their struggle. According to the Brighton Declaration on Women and Sport states that without women leaders, decision makers and decisive figures in sport, equal opportunities for women cannot be achieved. The "Program of Action" adopted at the 4th World Conference on Women held in Beijing in 1995 proposed to mainstream gender awareness into decision-making in the sports field as well (Pape, 2020)[15].

Reflecting on the participation of women into the sports field with more and more efforts and opportunities, it can be summarized that these practices have challenged the stereotyped views on women and paved diversified paths for their rights and voices in the society. In spite of the stereotyped impression about the social gender, the wider range of participation of women can remind more people of the potential of women in different fields even including the sports field that is used to be connected with features of males. These female pioneers participating into the sports field have set a positive and inspiring example for others in fighting for the legal rights and breaking prejudice on women, thus promoting and encouraging the participation of more women into the sports field (Schmidt, 2013)[19]. The efforts of breaking through stereotyped social gender features have also exerted influences on other fields, shedding light on the active role of women in different walks of life and social activities in regardless of the restrictions posed by stereotyped social gender.

The participation of women into the sports filed can also pose influences on the perceptions of social justice. The social justice perspective focuses on the phenomenon of gender inequality in the sports field. Gender inequality manifests in the sports field as unequal access to power and participation in decision-making in relevant to the sports development and growth process (Barclay, 2020)[2].

Incorporating the social justice perspective into sport development means giving women more equal access to power and authority in the field of sports. As women tend to be in a disadvantaged position for a long time, they are often more sensitive to sports issues, and more likely to feel the unreasonable and inhumane aspects of sports management. However, due to the lack of appropriate channels, their feelings cannot be expressed and transmitted to sports management. The reflection on the issue of women' participation into the sports field from the social justice perspective tends to arouse awareness on the equal access and opportunities of women to the management and participation of the sports field.

Owing to the participation of female's participation into the sports field, there are some improvements and consequences that may be brought by the participation from the social justice perspective. The first one is the awareness of the existence of potential challenges and inequality between males and women in terms of the sports field. People used to be accustomed to the male dominant position in the sports field in terms of the rule settlement, participation rights as well as other regulations in relevant to the sports participation and management. With the participation of more and more women, their influences on the sports field will increase, thus arousing widespread awareness on the existence of inequality for women in the sports field (Piepiora and Petecka, 2020)[18]. In addition, an increasing number of women can gain access to social justice and equality in the sports field as their voices and opinions can be noticed and handled, which is resulted from the participation of women in the sports field. Equal access to the sports facilities, more opportunities of taking part in the sports field and other rights can be guaranteed as a dispensable part of the social justice for women as the participation of them has increased into the sports field. Influences of women' participation into the sports field can also be analyzed from the perspective of female development. At present, the study of women's development in a harmonious society has become a hot issue in the sociology in different countries across the world. Because of the existence of inequality between different genders, women are excluded from serious business, public affairs, and they have long been rooted in domestic spaces and in activities related to the reproductive and social reproduction of offspring. In order to achieve the goal of women's development, the protection of their rights and interests should be achieved in the real world. In the 1960s, women began to take part in sports activities in accordance with the rules and patterns that were developed under the male dominant sports world (Pope, 2017)[17]. But this situation was a reflection of the political field that was not a manifestation of the harmonious development of males and women in the sports field. In different countries all over the world, there is a common problem that women's sports time and quality lag behind men's. How to understand and guide the development of women's sports has become an important task of current sports sociology research.

With the participation of women in the sports field, more and more experience and models can be summarized and concluded from the practice, thus setting a good example for the effective promotion of female development in the sports field. Based on the experience, it can be summarized that the professional development model in different sectors and fields can contribute to the high quality development of women. The accumulation of experience got the professionalized development model can guarantee the efficiency and effectiveness of women participation into the sports field across the world with extending influences on women in different sectors and regions.

The increasing participation of women into the sports field might have also affect the discourse system in the society, which can be considered as the potential and profound influences in the sociology sphere. Discourse has become a hot topic in the study of linguistics and philosophy of language. According to Burr's explanation, discourse refers to a set of meanings, metaphors, representations, mirror images, stories or statements (Al-Ansari, 2011)[1]. Somehow these forms pod discourses together generate a particular view of things. Discourse system refers to the system that forms the overall expression of meaning through discourse. Through the investigation of the women's discourse system in the modern society, it can be found that there is a suspicion that the media and the outside world deliberately magnify the reality tragedy of women's sports. Through this discourse, the fact that women are weaker than men in the society is reinforced. In society, it is undeniable that the words of female endorsers carry the traces of the other's speech, which is more obvious in media reports. For example, when individual media reports on women's sports, they often use discriminatory language, try to highlight their appearance, publish erotic photos, pay attention to the emotional performance of female athletes, and despise women in a roundabout way to show their low status.

To occupy a favorable position in the discourse system, the participation of women in the sports field can present the forces and capabilities of themselves in the strong and effective way. People will be encouraged and inspired by achievements and breakthroughs made by women in the sports field as the successful challenge towards human extreme and the efforts of never giving up. Achievements and records made by women in the sports field can be considered as effective tools and evidence to prove the strength and potential of women in the society, thus contributing to the influential position of women in the discourse system. With the help of the advancements and achievements, the stereotypes of women can be overcame while the voices of women can be expressed and conveyed to a wider range in the discourse system. As a result of the beneficent cycle, more and more women will benefit from the discourse system in seeking for higher positions and more influential discourse power in the society, during which the rights and demands of women can be noticed, catered and protected. The bright future of women' participation into the sports field can be inferred from the increasingly influential position of women in the discourse system while more and more women can exert influences and perform their capabilities in the sports field as a way of expressing themselves.

The public sphere perspective can also be used to

illustrate the influences of women participation into the sports field. Public sphere also known as public policy, is a concept proposed by Habermas in Germany's The Structural Transformation of the Public Domain (Pfister, 2010)[16]. It has been defined as a venue and institution for equal and rational debate on issues of common concern. However, due to the gender blind spot of "de-gendering" in Habermas's theory, the issue of women's public sphere has caused widespread controversy. However, it should not be ignored that the continuous emergence of female groups and female sports venues in sports, and the existence of female sports public spaces has been recognized in reality. Women's sports public space provides a venue and possibility for women to participate in decision-making, improves women's sports skills, enriches women's spiritual life, acting as a tool for women's sports awareness, power and legitimacy, and forms a third space outside the workplace and home. Owing to the cultivation and promotion of the public sphere created by the positive participation of women into the sports field, other women can be encouraged to arouse their awareness on their sports potential and the legal rights of taking part in the sports field. It can create a platform for women to present themselves with enough opportunities and possibilities.

5 Conclusion

In conclusion, the prevailing tendency for the increasing number of women in participating into the sports field has been attracting attention in the global range. An increasing number women have performed with extraordinary achievements and performance in the sports field that might have gone against the stereotypes of women in the history. In fact, many women are still taking a positive role to participate into the sports field to achieve their personal value and exert influences on the society. They have broken through the stereotypes about the gender as women by exceeding records and achieving better performance in the sports field, while the social justice can be maintained and the discourse system can be managed with the active participation of women. Benefited from their participation, the legal rights of women can be integrated and protected, and their voices and demands can also be catered and satisfied through the countless efforts of women participating into the sports field.

References

1. Al-Ansari, M. (2011). Women in sports leadership in Bahrain. Muslim women and sport, 79-91.
2. Barclay, V. M. (2020). Status of Black women in sports among selected institutions of higher education (Doctoral dissertation, University of Iowa).
3. Bum, C. H., Mahoney, T. Q., & Choi, C. (2018). A comparative analysis of satisfaction and sustainable participation in actual leisure sports and virtual reality leisure sports. Sustainability, 10(10), 3475.
4. Coche, R., & Tuggle, C. A. (2018). Men or women, only five Olympic sports matter: A quantitative analysis of NBC's prime-time coverage of the Rio Olympics. Electronic News, 12(4), 199-217.
5. Cooky, C., Messner, M. A., & Hextrum, R. H. (2013). Women play sport, but not on TV: A longitudinal study of televised news media. Communication & Sport, 1(3), 203-230.
6. Elsey, B., & Nadel, J. (2019). Futbolera: a history of women and sports in Latin America. University of Texas Press.
7. Evans, A. B., & Pfister, G. U. (2021). Women in sports leadership: A systematic narrative review. International review for the sociology of sport, 56(3), 317-342.
8. LaVoi, N. M., & Baeth, A. (2018). Women and sports coaching. In The Palgrave handbook of feminism and sport, leisure and physical education (pp. 149-162). Palgrave Macmillan, London.
9. Litchfield, C., & Osborne, J. (2013). Women in the sports pages: A brief insight into Olympic and non-Olympic years in Australia. In Fourth International Conference on Sport & Society.
10. Matthews, C. R., & Channon, A. (Eds.). (2016). Global perspectives on women in combat sports: Women warriors around the world. Springer.
11. Mays, D., DePadilla, L., Thompson, N. J., Kushner, H. I., & Windle, M. (2010). Sports participation and problem alcohol use: A multi-wave national sample of adolescents. American journal of preventive medicine, 38(5), 491-498.
12. McClearen, J. (2018). Introduction: Women in sports media: New scholarly engagements. Feminist Media Studies, 18(5), 942-945.
13. McDonald, M. G. (2015). Imagining neoliberal feminisms? Thinking critically about the US diplomacy campaign,'Empowering Women and Girls Through Sports'. Sport in Society, 18(8), 909-922.
14. Mirehie, M., & Gibson, H. J. (2020). Women's participation in snow-sports and sense of well-being: a positive psychology approach. Journal of Leisure Research, 51(4), 397-415.
15. Pape, M. (2020). Gender segregation and trajectories of organizational change: The underrepresentation of women in sports leadership. Gender & Society, 34(1), 81-105.
16. Pfister, G. (2010). Women in sport–gender relations and future perspectives. Sport in society, 13(2), 234-248.
17. Pope, S. (2017). The feminization of sports fandom: A sociological study. Routledge.
18. Piepiora, P., & Petecka, A. (2020). Personality profile of women practising contact sports using the example of karate kyokushin competitors and handball players. Ido Movement for Culture. Journal of Martial Arts Anthropology, 20(1), 23-29.
19. Schmidt, H. C. (2013). Women, sports, and journalism: Examining the limited role of women in

student newspaper sports reporting. Communication & Sport, 1(3), 246-268.

20. Sherry, E., Osborne, A., & Nicholson, M. (2016). Images of sports women: A review. Sex Roles, 74(7), 299-309.
21. Whiteside, E., & Hardin, M. (2011). Women (not) watching women: Leisure time, television, and implications for televised coverage of women's sports. Communication, Culture & Critique, 4(2), 122-143.

Research on the application status of image recognition technology in textile and clothing field

Mingzhen Ren[1,a*], Yajie Liu[2,b], Xinmin Ma[3,c]

[1] Qingdao University of Technology, Qingdao, Shandong, China
[2] Qingdao University of Technology, Qingdao, Shandong, China
[3] Qingdao University of Technology, Qingdao, Shandong, China

Abstract. With the development of the times, computer, information technology gradually penetrated into all walks of life. Textile and clothing design, production, consumption integration of machine vision, graphic image recognition technology development have become a trend. At present, the intelligence degree of the textile and clothing industry has reached an unprecedented new height. In the production and identification of fabrics, the machine vision gradually replaces the manual work, and realizes the automatic and accurate production line. In the design, manufacturing, consumption and other aspects of clothing, image recognition technology has assumed the historical responsibility. Combined with the Internet of things, cloud computing and other technologies, it has improved the production efficiency and achieved intelligent design and production.

1 Introduction

In the Textile Industry "14th Five-Year" Guidance on Science and Technology[1], Fashion and Green Development issued by China National Textile Industry Council, it is clearly pointed out that textile science and technology adheres to innovation-driven development, and actively promotes the research and development and application of key generic technologies and equipment such as intelligent manufacturing and advanced equipment, green manufacturing and advanced textile products, so as to promote the high-quality development of the textile industry.With the continuous improvement of the intelligent level of the textile and clothing field, the integration of computer, information technology and the textile and clothing industry has promoted the process of the intelligent industry. Image recognition technology has gradually become an important tool to obtain and transfer information in this field, such as the identification of fibers and fabrics, textile defects inspection, human body identification, fabric retrieval, clothing style diagram and sample Image recognition, clothing image retrieval and classification, the above aspects are inseparable from the acquisition, analysis, transformation, recognition and other technologies.

The commercial value of fabric image and the public use rate is higher. Both consumers and manufacturers have a need to select fabrics that match their expectations. Fabric image color, texture, shape and other underlying features are relatively simple and clear, easy to be extracted and distinguished by computer. In addition to the retrieval of similar fabrics by comparing fabric images, the computer can also define the perceptual attributes of some fabrics with obvious characteristics to realize the subjective emotion labeling of fabrics. Such as the analysis of fabric color saturation, contrast, color relationship and texture, pattern, etc. Manually summarize objective indicators into perceptual cognition for fabric retrieval and recommendation.

The key of garment image processing is to extract the target information of garment image from garment image by means of digital image processing technology. Through the effective combination of neural network, support vector machine, deep learning and other methods, clothing image recognition and classification can be realized. Digital image processing technology is used to process clothing images to realize the detection of clothing defects, clothing classification, clothing appearance evaluation, etc. Its significance lies in that the technology makes the acquisition of clothing information no longer limited to the actual clothing, replaces the tedious manual operation, and is more objective and efficient, providing a new method for clothing detection, style recognition, modeling evaluation and so on.

Clothing shopping limited to text retrieval cannot meet the needs of users. Therefore, content-based image retrieval CBIR plays a very important role in clothing sales retrieval and has also been rapidly developed. In recent years, with the development of deep learning, a large number of fabric recognition methods based on neural network learning have appeared.

Image recognition is a technology practice application based on deep learning algorithm[2], its definition can be understood as the use of computer technology for image processing, analysis and

* Corresponding author: [a] maxinmin@qut.edu.cn
[b] 2665337173@qq.com
[c] 29669824@qq.com

https://doi.org/10.1051/shsconf/202315503021

understanding, in order to identify various forms of targets and objects. With the proposal and application of the concept of deep learning, the technology of using computer technology to analyze and understand graphics and images, and to simulate the way of thinking of people, to extract and classify image information, and to complete recognition has been greatly developed. Image recognition is the process of extracting effective information from the image by simulating human visual function, analyzing and processing the information, and finally forming the standard of detection, measurement and control. The process of image recognition mainly consists of four parts. Firstly, the image acquisition and the pre-processing of the image materials are carried out, then the image segmentation and information extraction are carried out, and finally the recognition and display are carried out. Based on the above steps, the corresponding algorithms and operators are used to identify and display the image data.

After intelligent manufacturing, machine vision technology, Internet of things, cloud computing and other advanced science and technology into the public vision and social life, the textile and garment industry in the production and sales of unprecedented new challenges, technological transformation and upgrading or integration of advanced new technology, to cater to the big data process, organic processing and combination of all kinds of production information data. In the context of technological innovation, it is the general trend to improve production quality and operation efficiency and realize high-quality intelligent production[3].

2 Monitoring function of image recognition technology in textile production process

2.1Application of machine vision in intelligent spinning process

Through machine vision products instead of human eyes to do measurement and judgment, we call it machine vision, it is mainly to detect the system. The process of machine vision is to use industrial cameras to capture the image of the device we need to be detected. Then the professional image processing software will be used to collect the information into a digital signal, and it is calculated, output results[5].

The process of spinning mainly includes the main processes such as bag catching, flower clearing, carding, combining, roving and spinning. At present, the machine vision detection technology is mainly applied to the non-fiber detection of raw cotton, online detection of carding cotton net, fine yarn processing detection and barrel yarn processing detection and many other key processes. It has not yet involved the process of bag catching, combining and roving.

Machine vision detection technology involves artificial intelligence, neurobiology and other disciplines. In the process of textile production, this technology mainly uses industrial cameras to obtain real-time image information in the process of production, so as to simulate human visual function. Combined with computer algorithm technology, the image is identified and classified, and the detection data is connected with MES, ERP and other systems to realize real-time data sharing and monitoring, so as to achieve accurate and timely processing of problems in the production process, improve the quality and efficiency of yarn production, reduce the workload of artificial resources, reduce the consumption of artificial resources. The online detection technology based on image recognition was researched and applied earlier in foreign countries. Germany's Trutzschler Company, Switzerland's USTER Company and RIETER Company, etc., took the lead in developing intelligent identification equipment and systems such as isofiber detector, intelligent regulating carding machine and yarn detection system. The carding machine and winding machine developed by Qingdao Hongda Textile Machinery Co., LTD., and the non-fiber sorting machine launched by Beijing Daheng Image Vision Co., LTD., all closely follow the development of machine vision detection and image processing and analysis technology to realize the intelligent real-time detection in the spinning production process, and the key process and real-time data monitoring and judgment of the equipment operation process, in order to achieve high quality Effective combination of production and intelligence. The application of image recognition technology in intelligent spinning production shows that the technology has unique advantages such as non-contact detection, high precision, high efficiency and continuity. It has become an indispensable part in intelligent spinning production and a key condition to guarantee the high quality and efficient intelligent production of intelligent manufacturing and advanced technology equipment.

2.2Progress in the application of image recognition in fabric defects detection

Fabric defects detection can be divided into process detection and product detection, because the fabric weaving process has the existence of complex process, fast speed, high cost, loom environment danger phenomenon, as well as the diversification of fabric, the complexity of the defects and other conditions, it is not suitable for manual detection in the process. Since the 1980s, foreign countries have carried out research and development on the defects detection system, such as the automatic detection system developed by BARCO (Belgium), EVS (Israel) and USTER (Switzerland), which has become a mature product on the market. At present, after decades of development in the field of fabric defects intelligent detection system research and development, it has fully equipped with the basic functions of defects identification, segmentation, classification and a wide range of applications in production, so that the production efficiency has been rapidly improved, the accuracy and accuracy of discriminating defects has also been basic protection[4].

The image processing method based on machine vision technology can divide the defect detection algorithm into five categories: structural algorithm, statistical method, spectrum analysis method, model method and learning method. The analysis results show that the multi-level detection algorithm usually has better results in practical application [6]. The future development trend will extend and develop towards the direction of timeliness, universality, individuation and precision, so as to realize the predictability of intelligent detection of fabric defects in different directions, different scales and different backgrounds.

3 Application of image recognition technology in clothing design and manufacturing

With the deepening of the research on image recognition technology, its application field is constantly expanding [7]. In the process of garment design and production, human body shape recognition, sample recording, style recognition, automatic cutting and sewing, material arrangement, defect inspection and other links are also applied. The use of embedded control platform, the computer through feature extraction, image segmentation, image matching, clothing design and production of the corresponding links to achieve automatic intelligent management, so that the production process to reduce the workload of staff, and work efficiency and quality have been improved. The outstanding performance is the intelligent cutting and sewing, digital graphics processing, fully automatic discharge technology, intelligent hanging system. Intelligent production of clothing based on machine recognition by learning massive clothing production data and mining the information of intelligent production of clothing, intelligent production of clothing more and more industrial and academic attention through intelligent production, aiming at the characteristics of intelligent and combined with clothing production data, the basic method of intelligent production analysis of clothing based on machine recognition is proposed.

In the field of textile and garment, the graphic recognition system in the process of garment production is mainly composed of two forms: manual recognition and computer vision recognition. The measurement of human body shape and the recognition of garment style drawing and garment pattern drawing are mostly manual recognition. It is completed by the researchers of garment universities and clothing majors. The identification, retrieval, segmentation, classification and labeling of the physical picture of clothing, clothing picture and related images are completed by the researchers of computer identification technology and information identification technology.

Influenced by external conditions such as retrieval efficiency and subjective influence, Human body shape recognition, clothing style map and sample layout recognition, as well as traditional artificial clothing detection, classification and evaluation methods generally have many uncontrollable problems. Therefore, the application of digital image processing technology in clothing appearance quality detection, clothing style, fabric recognition and classification, and clothing modeling evaluation should be combined to analyze image recognition technology and traditional recognition methods The paper points out its development trend and application value in the field of clothing.

4 Identification and retrieval of textile and clothing products

At present, with the popularization of the Internet and intelligent electronic products, e-commerce and online shopping have become the inevitable products of social development in the new era[8]. Remote customization of textile and clothing and online purchasing have also become the trend. However, different from other daily necessities, textile and garment products are diversified in color, style, size, fabric and style, which brings difficulties to cloud production and consumption. The image recognition technology arising at the moment brings good news to cloud customization and online consumption of textile and garment.

For a long time, fabric recognition and judgment generally rely on artificial touch and observation to obtain subjective evaluation, the results are easy to be affected by subjective judgment and the experience and physical and mental state of the detection personnel, and the identification efficiency is low and the accuracy is relatively unstable. With the rapid development of the garment industry, the traditional fabric recognition method has been unable to meet the urgent need of the textile industry to pursue low cost labor and high efficiency production [9]. To a large extent, the application of advanced computer image recognition technology to achieve automatic fabric recognition has realized this need. Therefore, in 2002, Lai et al. established KES-FB fabric evaluation system, which comprehensively measured 16 kinds of mechanical properties of fabrics, and then obtained the comprehensive evaluation value of fabrics. In 1998, Fan et al. selected some key characteristic parameters of fabrics based on the physical parameters measured by KES-FB system, and used BP convolutional neural network to realize fabric recognition and evaluation. For the identification and evaluation of fabric, this is undoubtedly an ideal objective evaluation method, but the experimental equipment is expensive, the testing process is tedious, can not achieve a convenient and fast effect. With the rapid development of electronic information technology, the following attempts are made to use image processing method to extract the underlying features of fabric images, and combine fabric attributes to carry out fabric image recognition method, which greatly improves the recognition efficiency and accuracy. Fabric is different from other items because of its anisotropic physical and mechanical properties, and its image texture and shape are slightly different. Therefore, compared with conventional universal images, fabric images have higher similarity, which is difficult to distinguish from a single image recognition. In recent years, based on the research results, researchers have proposed to identify fabrics through visual properties and tactile detection. Generally, tactile images of fabrics are obtained by using tactile sensors, and the underlying characteristics are obtained. By using training convolutional neural network, parameters of fabric properties and extracted underlying characteristics are learned, so as to construct fabric recognition mode. Through the identification of fabric, the quantitative description and identification of fabric image features are completed, and the problem of automatic identification of fabric image features is solved.

From the technical point of view, the correct identification and analysis of various fabric types and organization structure can provide valuable scientific basis and technical parameter data for fabric design, reference, innovation and imitation. Most of the existing fabric recognition technologies are based on individual fabric feature extraction and recognition. The application of these techniques and methods has made some achievements in theory and practice. But from the function, adaptability and correctness of the existing system, there are still a lot of improvements[10].

From the market point of view, the production of textile industry must revolve around the market. In order to develop new products that better meet the market demand, it is necessary to study and re-innovate existing fabrics at home and abroad. Effective identification and analysis of fabric and its organization is a very important link. The automatic identification and analysis of fabric parameters and structure also provides technical support for learning excellent design ideas from others. To fabricate or improve the produced fabric, it is necessary to analyze the fabric and master the technical data such as the fabric structure and the technical conditions of the machine. It is of great significance for our country to learn foreign advanced fabric design technology and produce textile products suitable for domestic and foreign market.

5 Conclusion

With the proposal of the 14th Five-Year Plan and the vision of China's Intelligent manufacturing 2035, advanced manufacturing technology has become a hot research direction, and the textile and clothing fields are no exception. Computer and information technology are gradually integrated into the design, production, consumption and other links in this field. Computer vision, image recognition, image processing and other related technologies are implanted like bamboo shoots after rain. Textile production quality is an important lifeline for the production and survival of textile enterprises. Mastering efficient and fast production methods is an important factor for the development of enterprises in the front. At present, the research and application of relevant technologies in the textile industry production is mainly focused on automatic detection, defect recognition, defect recognition, fabric recognition, style evaluation, quality assessment, fabric density measurement and automatic identification of fabric structure, and has made great progress and results. The application prospect of computer image processing and big data analysis technology in textile and garment industry is very broad. If it is further developed and more widely applied, it will bring higher development benefits to the intelligent application level of Chinese textile and garment industry.

Based on the current digital image processing technology in the field of clothing research. Realize the virtual simulation of clothing fabric touch. The construction of intelligent clothing collocation recommendation system is the research focus of realizing intelligence and integration in the field of clothing. Therefore, it is necessary to combine the existing research with the new theory and technology to make the clothing field develop towards a more intelligent and integrated direction.

References

1. China National Textile Industry Council. Release of the Full text of the 14th Five-Year Guide on Science and Technology,Fashion and Green Development of Textile Industry [J]. Textile science research , 2021(8): 28-44.
2. Bruno Roberto,CENTENO PEREZ,Jiang C.Y. (2018)Research on Image Recognition Based on Deep Learning . Modern Economic Information, (23):419.
3. Xue J.C (2017)Discussion on Intelligent Promotion Scheme and Application of cotton Textile Industry . Cotton Textile Technology, 2017, 45(8): 1-9.
4. XU Weifeng. Application Status of Machine vision in Intelligent Spinning Production [J]. Cotton Textile Technology, 2022,50 (5) :71-74.
5. Xie Y.J, Zhi H.N. (2018)Research Review of Image Recognition Technology Based on Machine vision. Science and Technology Innovation ,(07): 74-75.
6. Yu Y.T. (2016) Research and Application of Garment Image Style Recognition Based on Multi-feature Fusion. Beijing: Beijing University of Technology,45-48.
7. Qin Junju, Cao Xuanping. Thinking about intelligent Clothing Production Technology Based on Machine Recognition [J]. Science and Technology & Innovation, 2022, (3) :47-50.
8. Fu G.W, He Z.P, Liu F,K. (2019)Research on Intelligent and Intelligent Development of Textile and Clothing Industry. Wool Textile Science and Technology,47(8):1-6(in Chinese)
9. LIU Hai. Discussion on Intelligent Production Technology of Clothing Based on Machine Recognition [J]. Textile Industry & Technology, 20, 49 (4) : 58-59.
10. Zhang Zhihua, Xu Yaping, Sun Yuchai. Intelligent Collection and Practice of Garment Production Information Based on RFID Technology [J]. Knitting Industry, 2020 (8): 63-66.

Co-working with AI is a Double-sword in Technostress? An Integrative Review of Human-AI Collaboration from a Holistic Process of Technostress

Mengting Xia[1]

[1] University of Southern California

Abstract. An accumulating body of research has demonstrated that human-AI (Artificial Intelligence) collaboration is an indistinguishable future of work. The study of employee-perspective, affect-related responses, and job-related consequences from the adoption and use of AI is gaining attraction in the field of business and management. Some research has examined the technostress creators or negative responses related to human-AI collaboration, however, a holistic process of technostress that considers both positive and negative impacts has not been adopted. This study integrates the findings to examine the relationship between technostress and human-AI collaboration, based on the technostress creators (challenge technostressors and hindrance technostressors), positive and negative psychological responses, and outcomes in work-related well-being and intention to use AI. Future research opportunities on technostress generated by human-AI collaboration are discussed alongside implications to theory and management practices.

1 Introduction

Artificial Intelligence (AI), a symptom of technological advancements and breakthroughs, is different from previous information technology applications and products. When the machine mimics human thinking and actions through data analysis, it can be called AI as the word "intelligence" represents that the machine can learn and solve problems [1].

Although AI has been studied since the 1950s, the power of AI was yet to be discovered and adopted by organizations to boost business performance in the past twenty years, with the development of machine learning, deep learning, and neural networks [2]. AI has been initially adopted to accelerate the production line in the manufacturing field and subsequently to perform as a coworker in supporting human decision-making in various business fields, changing the future of work [1]. For example, predictive analytics that can help markets generate dynamic and competitive pricing strategies has been a critical marketing effort in the service and retail industry [3]. Chat-box and virtual assistants that use advanced natural language processing and voice recognition have also been applied as communications tools to enhance work productivity [4]. Therefore, the question of whether AI will replace humans in the workplace has been intensely discussed within academia. Yet an agreement on AI job replacement has not been reached. On the one hand, the pessimistic view presented that AI would cause a large proportion of people to lose their jobs due to job replacement theory [5]. On the other hand, some researchers argued that AI, freeing people from repetitive and low-efficiency jobs, would foster employee innovation and generate more new job opportunities [6].

Considering the characteristics of tasks, much literature suggests the development of human-AI collaboration (e.g., [1] [4] [7]). According to AI job replacement theory, tasks in organizations can be divided into four types: mechanical tasks (such as administering daily routines and tracking attendance), thinking tasks (such as analyzing customer preferences and scheduling logistics), intuitive tasks (such as thinking creatively and adjusting effectively to novel situations) and empathetic tasks (such as empathizing with customers and advising therapies to patients) [5]. When it comes to most knowledge work in the business context, it involves both thinking and empathetic tasks, such as stock pitching requires both data analysis and the part of selling to others, and recruiting involves resume screening and communication with the candidates [8]. While organizations have already adopted AI to perform not only mechanical tasks but also thinking tasks, humans can provide an intuitive and emotional approach that AI can not [7]. Therefore, the human-AI collaboration that allows humans and AI each to leverage their advantages in the business processes will be the future for knowledge workers.

Despite the increasing research attention on human-AI collaboration, several literature gaps remain. First, previous literature has identified technostress as one of the negative determinants in human-AI collaboration from the

Email: xiamt0703@gmail.com

individual level, but they have not explored technostress creators in a synthesized way [9]. Second, though previous research has suggested considering a holistic process of technostress, research has yet to examine both positive and negative sides of technostress related to co-working with AI in the workplace [10]. Third, most literature has investigated different models of human-AI collaboration and its relation to job outcomes through team and organizational levels, but research has not examined how technostress influences an individual's job outcomes [11]. This review aims to fill the gap by synthesizing the findings of previous studies and answering the questions: *Which, why, and how do technostress creators influence human-AI collaboration?*

The rest of this paper is organized as follows. The first section presents a review of technostress research and clarifies a holistic process of technostress as the framework of this paper. The next section presents a review of previous studies in three parts: technostressors, psychological response, and outcome. The following section outlines the theoretical and practical implications of human-AI collaboration. Conclusions and future research directions are discussed in the final section.

2 Conception of technostress

Technostress is first proposed by Craig Brod as "a modern disease of adaptation caused by an inability to cope with new computer technologies in a healthy manner" [12]. With the increased use of ICT (Information and Communications Technology) in the professional work environment, a majority of studies have investigated the phenomenon of technostress (e.g., [13] [14] [15]). Tarafda et al. conducted one of the most recognized studies to explain the antecedents of technostress in the organization environment by adopting socio-technical theory and role theory [16]. They identified five technostress creators: techno-overload, techno-invasion, techno-complexity, techno-insecurity, and techno-uncertainty [16]. Later, Ayyagari et al. argued that the previous conceptualization blurred the boundary between technology characteristics and stress [13]. Building upon the person-environment fit model, their work showed that the rise of stress was due to two types of misfits between the characteristics of the person and the environment: the supply-value misfit and demand-ability misfit [13]. The study constructed technostress as five work stressors, which are work overload, role ambiguity, invasion of privacy, work-home conflict, and job insecurity, and found three technology features as the antecedents of technostress: usability, intrusiveness, and dynamism [13].

Aligning with the perspective of the "dark side" of technology, many researchers studied the negative impact of technostress, e.g., decreased job productivity and hindered job satisfaction [14] [16]. However, Califf et al. argued from an organizational stress perspective and posited that a holistic technostress process was constituted by both the techno-eustress subprocess and the techno-distress subprocess [10]. According to the holistic technostress process, an individual cognitively evaluates whether a stressor is positive or negative and then views the stressor as either a challenge stressor (beneficial to workplace tasks) or a hindrance stressor (harmful to work-related tasks), the process of which is called appraisal process [10]. After that, the individual responds to a challenge stressor or a hindrance stressor with a positive or negative psychological reaction, respectively. Finally, the individual performs or changes his or her behavior in the workplace [10].

This study uses the model from Califf et al. as a reviewing framework to synthesize the findings on the relationship between co-working with AI in the workplace and technostress.

3 Technostress generated by human-AI collaboration

This section presents a review of the relationship between human-AI collaboration and technostress. It initially discusses research on technostress creators. Subsequently, studies that elaborate on both positive and negative psychological responses are presented. Final part reviews studies of work-related well-being and behavior outcomes.

3.1 Construct of stressors

When an individual passes through the appraisal process, the individual classifies environmental stimuli into either hindrance technostressors or challenge technostressors [10]. This section reviews literature related to hindrance technostressors (unreliability, complexity, uncertainty, insecurity, overload) and challenge technostressors (usefulness, tech support), respectively.

3.1.1. Hindrance technostressors

Several studies identified the challenges while working with AI, which were mostly aligned with the hindrance technostessors (e.g., [9] [17]). For example, through a qualitative analysis of 24 frontline service employees who worked with robots, Meyer et al. confirmed the issues of work overload, job insecurity, unreliability, and complexity [18]. Arslan et al. also summarized key challenges of human workers-AI team interaction: fear of losing jobs, complexity, overload, and unreliability of the potential of AI-powered team members [19].

The below review presents research that focuses on two of five hindrance technostressors, namely, insecurity and complexity.

3.1.1.1. Insecurity

Among those five hindrance technostressors, job insecurity has been the most studied work stress when investigating the negative influence of AI adoption (e.g., [20] [21]).

Job insecurity describes the degree of an individual's perceived uncertainty about job stability and continuity [20]. When organizations adopt AI, humans experience a mix of hope and fear that contributes to the feeling of uncertainty [22]. On the one hand, people are worried that AI will replace their current jobs in the future, leading to either layoffs or a decline in the employee treatment [20]. On the other hand, AI handles tedious and repetitive tasks, calling for talent upskills [6].

Despite the above research mainly focused on employees' perspectives, one survey investigated whether AI would reduce the workforce in the next five years from the managers' view; the result showed that 47% of the managers agreed, representing an even mix of perceptions [22].

3.1.1.2 Complexity

The complex feature of AI is one of the barriers for employees to understand and perform their job duties [10]. Thus, employees may experience feelings of role ambiguity [13]. In addition, Bedué and Fritzsche suggested that the underlying rules of AI seemed unapparent to the users, thus the results produced by AI-generated can not be interpreted, which was considered the obstacle to developing trust in AI [23].

3.1.2. Challenge technostressors

Two challenge technostressors have been investigated through previous studies, which are usefulness and tech support.

3.1.2.1. Usefulness

Usefulness is the most recognized challenge technostress in the previous studies [17] [24]. For example, Malik et al. found the perception of overall enhancement in job performance when using AI in the workplace [9]. Similarly, Sowa et al. found that a majority of people agreed that cooperation with AI will enhance their job productivity [4]. In the research that focuses on a specific function, the perception level of AI's ability to change human resources management was high because AI could help save the time spent on mundane tasks [25].

3.1.2.2. Tech support

Tech support refers to the degree of organizational support for employees in using AI [10]. Cao et al identified that tech support had an insignificant influence on behavioral intention [24]. Providing employees with the knowledge to work with AI can also be viewed as tech support, which was one of the determinants to increase trust in AI [23].

3.2 Psychological response

An individual generates a positive or negative psychological response as the reaction to a challenge stressor or a hindrance stressor [10]. Since a holistic process consists of two subprocesses, both negative and positive psychological responses are reviewed.

3.2.1. Negative psychological response

Previous studies have successfully investigated negative psychological responses, including fear, concern, threat, etc. (e.g. [24] [4] [26] [27] [21]). This section reviews the findings accordingly.

3.2.1.1 Fear

Numerous studies have investigated whether people perceive the fear of losing jobs and discovered contradictory findings. For example, a piece of qualitative research on 32 working professionals in multinational firms with an average work experience of 7.6 years confirmed job insecurity was the second-highest work stressor through the thematic analysis [9]. However, a recent study analyzed the perceptions of 821 human resource employees and managers in large 500 companies, over half of whom demonstrated "not worried about being unemployed" [25].

To further dive into the different findings of fear of job replacement, researchers have found individual differences, job type differences and task type differences. For example, young generations experienced a little higher job insecurity than senior workers [4]. The differences also existed in job types: office and operation workers, business analysts, and experts and specialists are more likely to experience job insecurity caused by AI [22]. Additionally, the relationship between fear of job replacement and type of job tasks is also investigated. Recently, Vorobeva et al. performed a series of studies on employees in the service industry and discovered that people experienced a higher level of fear of job replacement when engaging in thinking tasks than feeling tasks [28]. Furthermore, when AI was salient, thinking tasks had a similarly negative effect on employees' fear of replacement [28].

Therefore, based on the above literature on negative psychological responses to job insecurity, age differences, job type differences and task type differences have been found when deploying AI in the workplace.

3.2.1.2 Concern

Some studies suggested that when employees used AI, they could face two concerns: personal concerns and technology concerns. Cao et al. identified two personal concerns: personal development concern and well-being concern [24]. Personal development concern is referred to the use of AI that may hinder personal skill development and career prospect; well-being concern is considered another issue due to some possible consequences like depression and cynicism if using AI [24]. In addition to personal concerns, technology concerns were found related to privacy concern, and technology-reliance concern [29] [4].

3.2.1.3 Threat

Cao et al. hypothesized that AI threat was positively influenced by perceived susceptibility to bad decision-making due to the use of AI, and the perceived severity of negative consequences [24]. The result supported the hypothesis and also suggested a negative relationship between AI threat and an individual's attitude towards AI.

3.2.2. Positive psychological response

Trust is the only psychological response that has been discussed when employees working with AI as trust can be viewed from both cognitive and emotional perspectives [30]. Some studies have successfully identified antecedents of trust in AI through technology features and personal features. In a review paper that covers approximately 200 publications in the year between 1999 and 2019, the important antecedents of cognitive trust were AI's tangibility, transparency, reliability and immediacy and one specific determinant of emotional trust was AI's anthropomorphism [30].

3.3 Outcome

Psychological responses lead to the actions, which is the outcome. The outcome can be positive or negative regardless of its psychological responses [10]. For example, positive psychological responses can result in a negative outcome.

This section further classified negative and positive outcomes into two streams: well-being and behavior. Firstly, building upon dimensions of well-being, this section reviews the impact of AI adoption on employees' well-being, including emotional well-being (e.g., anxiety, stress, burnout) and professional well-being (e.g., job engagement) [31]. In addition, the reviews of behavior outcomes mainly focus on the acceptance and resistance of AI.

3.3.1. Negative outcome

3.3.1.1 Negative impact on well-being.

Negative impacts caused by AI technostressors have been found on affective well-being (e.g., job burnout, emotional exhaustion), and professional well-being (e.g., job engagement, change readiness). Firstly, a study on call center employees indicated that perceived stress from working with AI decreased affective well-being in the service industry [27]. Additionally, Kong et al. presented that the higher level of AI awareness contributed to job burnout based on the definition of AI awareness as employees' perception of job insecurity [21]. In this study, burnout was measured in three dimensions: emotional exhaustion, depersonalization and reduced personal accomplishment. Similarly, Wu et al. discovered that job insecurity had a significant effect on emotional exhaustion [20]. In addition to affective well-being, research on professional well-being focused on work engagement and change readiness. Braganza et al. investigated AI adoption as a moderating factor that decreased the positive relationship between psychological contract and job engagement [32]. The same study also argued that the negative effects of AI on job engagement may be caused by job insecurity [32]. Moreover, a study on HR managers confirmed that AI anxiety negatively influenced change readiness and a high-performance work system moderated the effect [26].

3.3.1.2 Negative impact on behavior

Several negative psychological responses have been confirmed the negative influence on AI adoption. For example, both well-being concern and personal development concern negatively affected an individual's intention to use AI [24]. Similarly, data privacy concern was another negative factor in adopting AI [17]. Choi discovered the moderating effect of privacy concern between role clarity and employees' willingness to AI adoption, which also illustrated the negative effect of concern on behavior intention [29].

3.3.2. Positive outcome

3.3.2.1 Positive impact on well-being

Literature has found that working with AI generates a positive impact on professional well-being, including change readiness, and meaningful work. For example, Smids et al. proposed that when AI takes over the boring and repetitious tasks in the workplace, employees can have an increased feeling of pursuing meaningful work [33]. Additionally, challenge technostressors like beliefs about AI have also been found to positively influence change readiness for AI adoption [26].

3.3.2.2 Positive impact on behavior

Challenge technostressors and positive psychological responses have both been studied in the relationship with AI adoption. Perceived usefulness and tech support as challenge technostressors were found to have a positive effect on an individual's intention to use AI [24] [17]. Additionally, trust, as one of the positive psychological responses related to the use of AI positively, could influence employees' willingness to accept AI technology [29] [17].

4 Implications

This paper provides several literature and practical implications. In terms of literature contributions, this paper provides a synthesized review base on Califf et al.'s model of both positive and negative sides of technostress when co-working with AI [10]. As a result, this paper provides clear pathways to understand an individual's technostress process when co-working with AI in organizations. Second, this paper also delivers practical implications for facilitating human-AI collaboration strategy and the related training and support within organizations. A series of dark and bright side variables and outcomes can be combined to minimize the negative influence associated with employees' technostress when they collaborate with AI. For example, tech support could be a challenge technostressor that generates a positive psychological response and a high-performance work system could moderate the negative effect on change readiness.

5 Future research

There are three possible future research directions.

Firstly, this paper suggests that future studies should shed light on balanced research on both the bright and dark sides of technostress related to human-AI collaboration. For example, in terms of psychological response, much literature investigated negative emotions like fear, threat and anxiety. More research on the eustress subprocess will facilitate a clearer understanding of the effects of technostress related to using AI in the workplace on employees' positive organizational behavior, such as employee innovation, and organizational commitment [34].

Another interesting domain for future research is the exploration of technostress related to AI-aid decision-making in a broad of industries and job functions. The development of AI has stepped into supporting decision-making in various business domains, including marketing, healthcare, human resources, etc. [35]. Future studies can focus on the effect of technostress in fostering human-AI collaboration in decision-making. For example, a conceptual model constructed by Solberg et al. explored the relationships among trust, perceived risk, and reliance on AI Decision Aids recently [36].

Lastly, a majority of literature focused on the perception of AI adoption (e.g., [4] [32]) and few studies examined the effect of human-AI collaboration through actual co-working with AI. Thus, field research in more industries and functions is needed to investigate the effects of human-AI collaboration in the feeling economy, especially in that have a high level of empathetic skills, such as business and financial operation, and first-line managers [1].

6 Conclusions

Despite increasing interest in human-AI collaboration, there is little research on how working with AI drives technostress and its impact on psychological well-being and organizational behavior. The goal of the current study was to present a holistic review of technostress in relation to human-AI collaboration. In closing, this paper argues that it is important to understand the both risks and potential of human-AI collaboration in today's organizations.

References

1. M.-H. Huang, R. Rust, and V. Maksimovic, "The Feeling Economy: Managing in the Next Generation of Artificial Intelligence (AI)," California Management Review, vol. 61, no. 4, pp. 43–65, Aug. 2019, doi: 10.1177/0008125619863436.
2. J. Amankwah-Amoah and Y. Lu, "Harnessing AI for business development: a review of drivers and challenges in Africa," Production Planning & Control, vol. 0, no. 0, pp. 1–10, Apr. 2022, doi: 10.1080/09537287.2022.2069049.
3. C. Campbell, S. Sands, C. Ferraro, H.-Y. (Jody) Tsao, and A. Mavrommatis, "From data to action: How marketers can leverage AI," Business Horizons, vol. 63, no. 2, pp. 227–243, Mar. 2020, doi: 10.1016/j.bushor.2019.12.002.
4. K. Sowa, A. Przegalinska, and L. Ciechanowski, "Cobots in knowledge work: Human – AI collaboration in managerial professions," Journal of Business Research, vol. 125, pp. 135–142, Mar. 2021, doi: 10.1016/j.jbusres.2020.11.038.
5. M.-H. Huang and R. T. Rust, "Artificial Intelligence in Service," Journal of Service Research, vol. 21, no. 2, pp. 155–172, May 2018, doi: 10.1177/1094670517752459.
6. A. Jaiswal, C. J. Arun, and A. Varma, "Rebooting employees: upskilling for artificial intelligence in multinational corporations," The International Journal of Human Resource Management, vol. 33, no. 6, pp. 1179–1208, Mar. 2022, doi: 10.1080/09585192.2021.1891114.
7. M. H. Jarrahi, "Artificial intelligence and the future of work: Human-AI symbiosis in organizational decision making," Business Horizons, vol. 61, no. 4, pp. 577–586, Jul. 2018, doi: 10.1016/j.bushor.2018.03.007.
8. P. Puranam, "Human–AI collaborative decision-making as an organization design problem," J Org Design, vol. 10, no. 2, pp. 75–80, Jun. 2021, doi: 10.1007/s41469-021-00095-2.
9. N. Malik, S. Tripathi, A. Kar, and S. Gupta, "Impact of Artificial Intelligence on Employees working in Industry 4.0 Led Organizations," International Journal of Manpower, May 2021, doi: 10.1108/IJM-03-2021-0173.
10. C. B. Califf, S. Sarker, and S. Sarker, "The Bright and Dark Sides of Technostress: A Mixed-Methods Study Involving Healthcare IT," p. 49.
11. Y. R. Shrestha, S. M. Ben-Menahem, and G. von Krogh, "Organizational Decision-Making Structures in the Age of Artificial Intelligence," California Management Review, vol. 61, no. 4, pp. 66–83, Aug. 2019, doi: 10.1177/0008125619862257.
12. C. Brod, Technostress : the human cost of the computer revolution. Reading, Mass. : Addison-Wesley, 1984. Accessed: Jun. 22, 2022. [Online]. Available: http://archive.org/details/technostresshuma0000brod
13. R. Ayyagari, V. Grover, and R. Purvis, "Technostress: Technological Antecedents and Implications," MIS Quarterly, vol. 35, pp. 831–858, Dec. 2011, doi: 10.2307/41409963.
14. A. Suh and J. Lee, "Understanding teleworkers' technostress and its influence on job satisfaction," INTR, vol. 27, no. 1, pp. 140–159, Feb. 2017, doi: 10.1108/IntR-06-2015-0181.
15. D. M. Marchiori, E. W. Mainardes, and R. G. Rodrigues, "Do Individual Characteristics Influence

the Types of Technostress Reported by Workers?," International Journal of Human–Computer Interaction, vol. 35, no. 3, pp. 218–230, Feb. 2019, doi: 10.1080/10447318.2018.1449713.

16. M. Tarafdar, Q. Tu, B. S. Ragu-Nathan, and T. S. Ragu-Nathan, "The Impact of Technostress on Role Stress and Productivity," Journal of Management Information Systems, vol. 24, no. 1, pp. 301–328, Jul. 2007, doi: 10.2753/MIS0742-1222240109.
17. M. M. Rahman, T. Ming, T. Baigh, and M. Sarker, "Adoption of artificial intelligence in banking services: an empirical analysis," International Journal of Emerging Markets, vol. ahead-of-print, Dec. 2021, doi: 10.1108/IJOEM-06-2020-0724.
18. P. Meyer, J. Jonas, and A. Roth, "Frontline Employees' Acceptance of and Resistance to Service Robots in Stationary Retail - An Exploratory Interview Study," Journal of Service Management Research, vol. 4, pp. 21–34, Jan. 2020, doi: 10.15358/2511-8676-2020-1-21.
19. A. Arslan, C. Cooper, Z. Khan, I. Golgeci, and I. Ali, "Artificial intelligence and human workers interaction at team level: a conceptual assessment of the challenges and potential HRM strategies," International Journal of Manpower, vol. 43, no. 1, pp. 75–88, Jan. 2021, doi: 10.1108/IJM-01-2021-0052.
20. T.-J. Wu, J.-M. Li, and Y. J. Wu, "Employees' job insecurity perception and unsafe behaviours in human–machine collaboration," Management Decision, vol. ahead-of-print, no. ahead-of-print, Jan. 2022, doi: 10.1108/MD-09-2021-1257.
21. H. Kong, Y. Yuan, Y. Baruch, N. Bu, X. Jiang, and K. Wang, "Influences of artificial intelligence (AI) awareness on career competency and job burnout," IJCHM, vol. 33, no. 2, pp. 717–734, Mar. 2021, doi: 10.1108/IJCHM-07-2020-0789.
22. S. Ransbotham, P. Gerbert, M. Reeves, D. Kiron, and M. Spira, "Artificial Intelligence in Business Gets Real," MIT SMR, Sep. 2018, Accessed: Jun. 23, 2022. [Online]. Available: https://sloanreview.mit.edu/projects/artificial-intelligence-in-business-gets-real/
23. P. Bedué and A. Fritzsche, "Can we trust AI? An empirical investigation of trust requirements and guide to successful AI adoption," JEIM, vol. 35, no. 2, pp. 530–549, Mar. 2022, doi: 10.1108/JEIM-06-2020-0233.
24. G. Cao, Y. Duan, J. S. Edwards, and Y. K. Dwivedi, "Understanding managers' attitudes and behavioral intentions towards using artificial intelligence for organizational decision-making," Technovation, vol. 106, p. 102312, Aug. 2021, doi: 10.1016/j.technovation.2021.102312.
25. E. Kambur and C. Akar, "Human resource developments with the touch of artificial intelligence: a scale development study," IJM, vol. 43, no. 1, pp. 168–205, Apr. 2022, doi: 10.1108/IJM-04-2021-0216.
26. Y. Suseno, C. Chang, M. Hudik, and E. S. Fang, "Beliefs, anxiety and change readiness for artificial intelligence adoption among human resource managers: the moderating role of high-performance work systems," The International Journal of Human Resource Management, vol. 33, no. 6, pp. 1209–1236, Mar. 2022, doi: 10.1080/09585192.2021.1931408.
27. A. P. Henkel, S. Bromuri, D. Iren, and V. Urovi, "Half human, half machine – augmenting service employees with AI for interpersonal emotion regulation," Journal of Service Management, vol. 31, no. 2, pp. 247–265, Jan. 2020, doi: 10.1108/JOSM-05-2019-0160.
28. D. Vorobeva, Y. El Fassi, D. Costa Pinto, D. Hildebrand, M. M. Herter, and A. S. Mattila, "Thinking Skills Don't Protect Service Workers from Replacement by Artificial Intelligence," Journal of Service Research, pp. 1–13, May 2022, doi: 10.1177/10946705221104312.
29. Y. Choi, "A study of employee acceptance of artificial intelligence technology," European Journal of Management and Business Economics, vol. 30, no. 3, pp. 318–330, Jan. 2021, doi: 10.1108/EJMBE-06-2020-0158.
30. E. Glikson and A. W. Woolley, "Human Trust in Artificial Intelligence: Review of Empirical Research," ANNALS, vol. 14, no. 2, pp. 627–660, Jul. 2020, doi: 10.5465/annals.2018.0057.
31. M. Charalampous, C. A. Grant, C. Tramontano, and E. Michailidis, "Systematically reviewing remote e-workers' well-being at work: a multidimensional approach," European Journal of Work and Organizational Psychology, vol. 28, no. 1, pp. 51–73, Jan. 2019, doi: 10.1080/1359432X.2018.1541886.
32. A. Braganza, W. Chen, A. Canhoto, and S. Sap, "Productive employment and decent work: The impact of AI adoption on psychological contracts, job engagement and employee trust," Journal of Business Research, vol. 131, pp. 485–494, Jul. 2021, doi: 10.1016/j.jbusres.2020.08.018.
33. J. Smids, S. Nyholm, and H. Berkers, "Robots in the Workplace: a Threat to—or Opportunity for—Meaningful Work?," Philosophy & Technology, vol. 3, no. 33, pp. 503–522, 2020, doi: https://doi.org/10.1007/s13347-019-00377-4.
34. S. Chandra, A. Shirish, and S. Srivastava, "Does Technostress Inhibit Employee Innovation? Examining the Linear and Curvilinear Influence of Technostress Creators," Communications of the Association for Information Systems, vol. 44, no. 1, Mar. 2019, doi: 10.17705/1CAIS.04419.
35. M.-H. Huang and R. T. Rust, "A strategic framework for artificial intelligence in marketing," J. of the Acad. Mark. Sci., vol. 49, no. 1, pp. 30–50, Jan. 2021, doi: 10.1007/s11747-020-00749-9.
36. E. Solberg, M. Kaarstad, M. H. R. Eitrheim, R. Bisio, K. Reegård, and M. Bloch, "A Conceptual Model of Trust, Perceived Risk, and Reliance on AI Decision Aids," Group & Organization Management, vol. 47, no. 2, pp. 187–222, Apr. 2022, doi: 10.1177/10596011221081238.

Guidelines for Green Design of Televisions

Jianfang Zong[1*], Xin Zhang[1], Yutao Yang[2], Husheng Li[3], Han Yuan[1]

[1]China National Institute of Standardization, No.4 Zhichun Road, Haidian District, Beijing, China
[2]China Electronics Standardization Institute, No.1 Andingmen East Street, Dongcheng District, Beijing, China
[3]Beijing CESI Certification Co, Ltd, No.1 Andingmen East Street, Dongcheng District, Beijing, China
*Corresponding author at: China National Institute of Standardization

Abstract. This paper regulates the purposes and basic principles of green design of televisions, and provides the green design requirements and procedures of televisions. This paper will provide the guiding targets for the television manufacturer, standardize the enterprise behavior of televisions, and lead the television product enterprise to implement the product green design, which is of great significance in promoting the transformation and upgrading of television product, enhancing the green manufacturing level and increasing the green product supply.

1 Introduction

Television is one of the earliest industries in China's home appliance industry to enter the stage of market-oriented development. After decades of development, television products have entered thousands of households. By 2021, the number of colour television sets owned by 100 households in China will reach 121.8. In terms of annual production, the output of colour television sets in China reached 196.262 million in 2020, with a year-on-year increase of 4.7%; In 2021, the output of colour television sets in China reached 175.128 million. Carrying out green design for televisions, meets the needs of consumption upgrading and increases the supply of green products so that consumers can choose and use green products.

Green design of televisions is also known as "design with environmental awareness", "eco-design" or "environmental design", which means the activity of introducing environmental factors to the design and development of products to improve the environmental performance of televisions in their life cycle and optimize the environmental impact of televisions. Researches show that 80% of resources consumption and environmental influence depends on the product design stage, when full consideration will be given to existing technical conditions and raw materials guarantee. Optimizing and solving resources and environmental problems at each stage can realize resource saving to the greatest extent and reduce environmental pollution from the source [1, 2].

2 Purposes and basic principles of green design of televisions

The green design of televisions aims at reducing the environmental influence at the design, raw materials selection, production, packaging, transportation, distribution, utilization, scrapping and subsequent treatment and disposal of products through green design, reducing the pollution from televisions towards the environment, improving the recycling rate of televisions to lessen the adverse impact of televisions on the environment in the whole product life cycle, developing green design methods for modularized, waste-minimized and hazard-free products, and creating ecological, economic and sustainable televisions system.

Basic principles of green design of televisions include: demand orientation, advanced technology, scientific methods, environmentally friendly, economic rationality and compliance with laws.

3 Product green design process and key points

3.1 Outline

The green design of television products usually includes the following processes: a) Product demand analysis; b) Identification of environmental factors; c) Introducing environmental factors into the process of product ecological design; d) Life cycle analysis and evaluation; e) Ecological design information sharing.

3.2 Product Demand Analysis

Designers should carry out product ecological design while meeting product functions, and comprehensively consider factors such as resources, energy and environment, as well as the needs of laws and regulations, technical standards and customer market, so as to meet the goal of product ecological design.

*E-mail address: zongjf@cnis.ac.cn (J.Zong)

Product demand analysis includes but is not limited to the following points: a) Product ecological design objectives; b) Product function design; c) Requirements of laws and regulations; d) Technical standards; e) Customer and market needs.

3.3 Identification of Environmental Factors

Identifying typical environmental factors and corresponding environmental impacts is an important prerequisite for product ecological design. Designers can identify environmental factors according to GB/T 28179[3]. The identification and analysis of these environmental factors can provide a basis for ecological design, identify, evaluate and determine important environmental factors as soon as possible, and consider introducing these factors in the process of product design and development, so as to improve the environmental performance of products. Table 1 gives the identification process of environmental factors of television products.

Table 1. Identification of environmental factors of television products

Environmental factor	Design considerations	Product life cycle phases involved				
		Raw material acquisition	Manufacturing	Transportation and distribution	Use and maintenance	End of life disposal
Hazardous substances	Compliance and reduction of hazardous substances in raw materials (parts, coatings, etc.)	√	√		√	√
	Compliance and reduction of the use of hazardous substances in packaging materials	√		√		√
	Compliance and reduction of the use of hazardous substances in production accessories	√	√			
	Compliance and reduction of the use of hazardous substances in consumables	√			√	√
Material efficiency	Reduce the types of raw materials	√			√	√
	Reduce the consumption of raw materials	√		√		
	Green process of material processing		√			
	Improve the recycling performance of raw materials	√				√
	Reduce the environmental impact of raw material production stage	√				√
	Reduce the consumption of packaging materials	√		√		√
	Improve the recycling performance of packaging materials	√				√
	Reduce consumption of consumables	√			√	√
	Improve the recycling performance of consumables	√			√	√
	Reduce the consumption of production auxiliary materials	√	√			
Battery or power supply	Battery or power supply compliance	√			√	√

Environmental factor	Design considerations	Product life cycle phases involved				
		Raw material acquisition	Manufacturing	Transportation and distribution	Use and maintenance	End of life disposal
	Consider the disassembly of battery or power supply				√	√
	Consider the life of the battery or power supply				√	√
Energy consumption	Improve product energy efficiency				√	
	Reduce energy consumption in product production process		√			
Pollutant emission	Reduce the emission of harmful substances in the use stage of products				√	
	Reduce noise emission during product use				√	
Product longevity	Reasonably extend product longevity (modular design, parts replaceability, software upgrade)				√	
Recycling, utilization and disposal	Consider reuse, recycling and reasonable disposal					√
	Reduce environmental toxicity of final waste					√

3.4 Introducing Environmental Factors into the Process of Product Green Design

3.4.1. Function, Performance and Process Design.

The design and development of products shall meet the requirements of relevant functions in SJ/T 11343 [4], including appearance and structure, function and performance, safety, power adaptability, noise, electromagnetic compatibility, environmental conditions and reliability.

On the basis of meeting the functional requirements of products, ecological design ideas and concepts should be incorporated into the design and development process of the whole product life cycle, including conceptual design and structural design before product finalization.

3.4.2. Improve Material Efficiency.

The choice of materials will affect the environment. The method of improving material efficiency can be considered in the process of product ecological design according to GB/T 34144 [5].

When selecting materials, designers should consider the following factors: reduce the types of materials used; Reduce the amount of materials used, so as to reduce the weight of products; Use materials that have little impact on the environment; Use materials that are easy to recycle.

4.4.3. Restrict the Use of Hazardous Substances.

Reducing or eliminating the use of hazardous substances and products should be one of the priorities in product design. Designers need to know the management list of banned hazardous substances and hazardous products in international, regional and product export target countries in time to ensure that the products sold in the target countries meet the requirements of relevant laws, regulations and standards. Designers should minimize the use of substances that require special separation or disposal at the end of product life.

For substances requiring special treatment or disposal, appropriate information shall be provided to product users and recyclers. The use of restricted substances in products should be identified in accordance with SJ/T 11364 [6], and the inevitable use of hazardous substances and reasons should be recorded in the design process. Appropriate design methods should be adopted to reduce the use of harmful substances in high-risk parts such as television product shell, remote controller and external power cord.

3.4.4. Reduce Chemical Emissions.

Appropriate design methods should be adopted to reduce the possibility of chemical emissions with negative environmental impact during the use of products, such as volatile organic pollutants.

3.4.5. Reduce Noise Emission.

Designers should consider techniques to reduce noise emissions. The method of reducing noise emission can be adopted according to GB/T 34959 [7].

3.4.6. Improve Energy Efficiency.

The design and development of products shall meet the requirements of GB/T 24850 [8] on energy efficiency limit value and energy efficiency grade of flat panel TV. The method in GB/T 34959 should be adopted to improve the energy efficiency of products. In the energy-saving design, the expected use mode of television products should be fully considered, including working mode, energy-saving mode, off mode and no-load mode. Appropriate design methods are adopted to minimize the energy consumption under various modes, such as optimizing the standby mode, which does not provide image or sound, but can be switched to other modes through remote control.

3.4.7. Consumables.

Consider the design and disposal of consumables based on the principle of reduction, modularization and reuse: the function of reducing or saving the use of consumables; Convenient replacement or maintenance of consumables. The manufacturer can provide users with information about the correct use of consumables and, where appropriate, the end of life processing information of consumables.

3.4.8. Battery.

If the battery is used in the product, the design of the battery in the product should be considered from the aspects of battery type, control of harmful substances in the battery, battery durability, easy disassembly and recycling, and so on.

3.4.9. Reasonable Extension of Product Life.

When technically and economically feasible, the service life of the product design shall be reasonably extended and matched with the service life of components, and it is easy to upgrade and repair. The methods to reasonably extend the service life of products include: using common mechanical packaging (such as cover and chassis) or common parts or product types, which are the same parts used by multiple models or the same parts used by the same product for multiple generations, and allowing the reuse of common parts; Use standardized components, which can be easily replaced or repaired; Modular design; Whenever the reuse of modules, components and products is applicable, the reused components (such as for maintenance and spare parts) should be identified.

It is encouraged to optimize the service life of products, and the information of effective options for product upgrading, expansion and maintenance should be available. Examples of design methods for prolonging product life include: modular, compatible and scalable design of television products; If a compatible and extensible interface is set, the operating system can be updated.

3.4.10. Reuse, Recycling and Reasonable Disposal.

The design of products should be convenient for reuse, recycling and reasonable disposal at the end of product life. The requirements for reuse, recycling, recycling, treatment and disposal of waste products shall be implemented in accordance with GB/T 23685 [9]. Designers need to understand the regulations and standards related to the disassembly and recycling of waste electrical and electronic products to ensure the compliance of product sales in the target countries. The end-of-life disposal of products should be designed according to GB/T 31376 [10].

3.4.11. Packaging.

The selection and design of packaging materials will have an impact on the environment. When determining materials and designing packaging, designers should consider the following options: reduce the use of packaging materials through reasonable structural design and material selection, so as to reduce the weight and size of packaging when fully protecting products; Use materials that have little impact on the environment; Use recyclable materials; Use renewable/recyclable materials (considering available recycling technologies).

The minimum requirement for designers is to ensure compliance with international, regional and national regulations related to the following contents: restrictions on hazardous substances and hazardous products; Regeneration capacity, such as reuse or recycling; The appropriate identification (material content) of packaging materials shall be in accordance with the marking requirements of GB/T 18455 [11].

3.5 Life Cycle Analysis, Evaluation and Continuous Improvement

In order to confirm and improve the level of product ecological design, the methods of life cycle analysis and evaluation can be used to establish and implement the procedures of ecological design review and continuous improvement. The process of product life cycle analysis, evaluation and continuous improvement of product ecological design includes the following steps:

a) Product Description: determine the product boundary and goal of the design, including: model, specification, function, energy efficiency grade, life expectancy, etc.

b) Product demand analysis: requirements of interested parties (laws and regulations, standards, customers, etc.) in terms of function, resources, energy, environment, human health, and so on;

c) Product life cycle diagnosis: analyze and diagnose the whole life cycle of the benchmark product, put forward the potential and key points for environmental performance improvement, and put forward the improvement demand;

d) Preliminary evaluation of product ecological design scheme: conduct preliminary evaluation on the improved scheme and confirm the product ecological design scheme;

e) Product life cycle assessment: complete the whole life cycle assessment report of benchmark products and target products according to Appendix A of GB/T 24040 [12], GB/T 24044 [13] and GB/T 34664 [14];

f) Evaluation of product ecological design scheme: analyze and compare the environmental performance of the whole life cycle evaluation report of the target product and the benchmark product, and give the conclusion of the environmental performance of the target product;

g) Scheme selection: select the ecological design scheme of target products and benchmark products;

h) Scheme review: review the best scheme;

i) Scheme output: confirm whether the design scheme of the target product meets the product requirements. If so, carry out design implementation, effect evaluation and information sharing; If the expected product demand is not met, continue to put forward improvement suggestions, and re-evaluate the scheme after adjustment until the expected product demand is met.

4 Conclusions

The green design of televisions provides guidance for enterprise designers in green design when carrying out product design and development, from the basic principles of product green design, product demand analysis, environmental factor identification, design and development process, life cycle analysis and improvement, and information sharing, etc. Identify environmental factors at all stages of the life cycle, such as waste recycling, propose product ecological design requirements in terms of resources, energy and environment, improve the renewable utilization rate of products, reduce adverse environmental impacts in the entire life cycle of products, and develop more ecological, economic and sustainable product systems.

Acknowledgements

This paper is funded by National key research and development plan project "Research on key technology standards of eco-design for consumer electronics and other important products" (2017YFF0207901).

References

1. Jianfang Zong, Jianhua Chen, Comparative Analysis on Domestic and International Mechanism of Product Ecological Evaluation, China Standardization, 2017, (15): 54-58
2. Jianfang Zong, Jianwei Tian, Dongfeng Guo, Liang Chen, A Study on the Green design of Consumer Electronics; E3S Web of Conference 53, 04003 (2018)
3. GB/T 28179 Environmentally concious design for electrical and electronic product - Indentification of environmental aspects
4. SJ/T 11343 General specification for digital television LCD displays
5. GB/T 34144 Guidance on material efficiency considerations in environmentally conscious design of electrical and electronic products
6. SJ/T 11364 Marking for control of pollution caused by electronic information products
7. GB/T 34959 Audio/Video, information and communication technology equipment—Enviromentally conscious design
8. GB/T 24850 Minimum allowable values of energy efficiency and energy efficiency grades for flat panel televisions and set-top boxes
9. GB/T 23685 General technical specifications of recovering for waste electrical and electronic equipment
10. GB/T 31376 Requirements of disassembly and treatment for waste electrical and electronic product—LCD televisions and display devices
11. GB/T 18455 Package recycling marking
12. GB/T 24040 Environmental management - Life cycle assessment - Principles and frameworks
13. GB/T 24044 Environmental management - Life cycle assessment - Requirements and guidelines
14. GB/T 34664 General assessment principles for electrical and electronic eco-design products

 https://doi.org/10.1051/shsconf/202315503024

Guideline for Green Design of Printers and Multi-function Printers

Jianfang Zong[1*], Xin Zhang[1], Yutao Yang[2], Husheng Li[3], Han Yuan[1]

[1]China National Institute of Standardization, No.4 Zhichun Road, Haidian District, Beijing, China
[2]China Electronics Standardization Institute, No.1 Andingmen East Street, Dongcheng District, Beijing, China
[3]Beijing CESI Certification Co, Ltd, No.1 Andingmen East Street, Dongcheng District, Beijing, China
*Corresponding author at: China National Institute of Standardization

Abstract. This paper regulates the purposes and basic principles of green design of printers and multi-function printers, and provides the green design requirements and procedures of printers and multi-function printers. This paper will provide the guiding targets for the printer manufacturer, standardize the enterprise behaviour of printer, and lead the printer product enterprise to implement the product green design, which is of great significance in promoting the transformation and upgrading of printer product, enhancing the green manufacturing level and increasing the green product supply.

1 Introduction

The printer industry is a mature industry, and the market scale is relatively stable. According to CCID Consulting, in 2020, the global printer sales volume was 87.69 million units. Except for 2019, the sales of printers in China have maintained positive growth. According to CCID consultant's prediction, from 2022 to 2025, the global printer sales will remain relatively stable, and the growth rate of printer sales in China will remain above 1%, so there is a large growth space. Carrying out green design for printers and multi-functional all-in-one machines, meets the needs of consumption upgrading, increases the supply of green products, and enables consumers to choose and use green products.

The green design of printers and multi-function printers is also called "environmental conscious design", "green design" or "design for the environment". It intends to optimize the environmental impact of printers and multi-function printers and bring the environmental factors into the product design and development activity, so as to improve the environmental performance of printers and multi-function printers within its life cycle. As shown in the study, 80% resources consumption and environmental impact are subject to the product design phase. At the design phase, it is required to fully consider the existing technical conditions, raw materials guarantee and other factors, and optimize the solution against the resources and environmental problems at each link, which may maximize the resources conservation and reduce the environmental pollution from the source [1, 2].

2 Purposes and basic principles of green design of printers and multi-function printers

The purpose of green design for printers and multi-function printers is to reduce the environmental pollution, and enhance the recyclability rate of printers and multi-function printers, so as to reduce the adverse environmental impact of printers and multi-function printers during its entire life cycle, and develop the more ecologic, economic and sustainable printers and multi-function printers system.

The basic principles for green design of printers and multi-function printers include: the requirements-oriented, advanced technology, scientific methods, environmental-friendly, economic rationality, and abidance by law.

3 Product green design process and key points

3.1 Outline

The green design of printers and all-in-one machines usually includes the following processes: a) Analyze regulations and environmental requirements of stakeholders; b) Identify and evaluate environmental factors and corresponding environmental impacts; c) Design and development; d) Review and continuous improvement.

The green design of printers and all-in-one machines should be based on the PDCA (planning, implementation, inspection and improvement) cycle process, carry out product demand analysis on the basis of full analysis of

*E-mail address: zongjf@cnis.ac.cn(J.Zong)

regulations and environmental requirements of stakeholders, identify environmental factors and corresponding environmental impacts at all stages of the product life cycle, and introduce environmental factors into the product green design process, Review and continuously improve green design through life cycle analysis and evaluation, and promote the whole supply chain and relevant parties to realize green design information sharing.

3.2 Product Demand Analysis

Under the condition of meeting the product function, the product green design should comprehensively consider the factors such as economic and technical feasibility, and the needs of laws and regulations, technical standards and customer market, so as to meet the goal of product green design.

Product demand analysis includes but is not limited to the following points: a) product green design objectives; b) Product function design; c) Requirements of laws and regulations; d) Technical standards; e) Customer and market needs.

3.3 Identification of Environmental Factors

Identifying typical environmental factors and corresponding environmental impacts is an important prerequisite for product green design. Designers can identify environmental factors according to GB/T 28179 [3]. The identification and analysis of these environmental factors can provide a basis for green design, identify, evaluate and determine important environmental factors as soon as possible, and consider introducing these factors in the process of product design and development to improve the environmental performance of products. Table 1 is an example of the environmental factor identification process of printers and all-in-one machines.

Table 1. Example of environmental factor identification process of printer and all-in-one machine products

Environmental factor	Design considerations	Product life cycle phases involved				
		Raw material acquisition	Manufacturing	Transportation and distribution	Use and maintenance	End of life disposal
Hazardous substances	Compliance and reduction of hazardous substances in raw materials (parts, coatings, etc.)	√	√		√	√
	Compliance and reduction of the use of hazardous substances in packaging materials	√		√		√
	Compliance and reduction of the use of hazardous substances in production accessories	√	√			
	Compliance and reduction of the use of hazardous substances in consumables	√			√	√
Material efficiency	Reduce the types of raw materials	√			√	√
	Reduce the consumption of raw materials	√		√		
	Green process of material processing		√			
	Improve the recycling performance of raw materials	√				√
	Reduce the environmental impact of raw material production stage	√				√
	Reduce the consumption of packaging materials	√		√		√
	Improve the recycling performance of packaging materials	√				√
	Reduce consumption of consumables	√			√	√
	Improve the recycling performance of consumables	√			√	√

Environmental factor	Design considerations	Product life cycle phases involved				
		Raw material acquisition	Manufacturing	Transportation and distribution	Use and maintenance	End of life disposal
	Reduce the consumption of production auxiliary materials	√	√			
Battery or power supply	Battery or power supply compliance	√			√	√
	Consider the disassembly of battery or power supply				√	√
	Consider the life of the battery or power supply				√	√
Energy consumption	Improve product energy efficiency				√	
	Reduce energy consumption in product production process		√			
Pollutant emission	Reduce the emission of harmful substances in the use stage of products				√	
	Reduce noise emission during product use				√	
Product longevity	Reasonably extend product longevity (modular design, parts replaceability, software upgrade)				√	
Recycling, utilization and disposal	Consider reuse, recycling and reasonable disposal					√
	Reduce environmental toxicity of final waste					√

3.4 Introducing Environmental Factors into the Process of Product Green Design

3.4.1. Functional Design.

The design and development of products shall meet the requirements of relevant functions in GB/T 9314 [4], GB/T 17540 [5], GB/T 17974 [6] and GB/T 28165 [7], including appearance and structure, function and performance, safety, power adaptability, noise, electromagnetic compatibility, environmental conditions, reliability, etc.

On the basis of meeting the functional requirements of products, green design ideas and concepts should be incorporated into the design and development process of the whole product life cycle, including conceptual design and structural design before product finalization.

3.4.2. Improve Material Efficiency.

The choice of materials will affect the environment. The method of improving material efficiency can be considered in the process of product green design according to GB/T 34144 [8].

When selecting materials, designers should consider the following factors: reduce the types of materials used; Reduce the amount of materials used, so as to reduce the weight of products; Use materials that have little impact on the environment; Use materials that are easy to recycle.

3.4.3. Restrict the Use of Hazardous Substances.

Designers need to reduce or eliminate the use of hazardous substances and products as one of the priorities in product design, and timely understand the international, regional and national management lists of prohibited hazardous substances and products, so as to ensure that the products sold in the target country are legal. Designers should minimize the use of substances that require special separation or disposal at the end of product life.

For substances requiring special treatment or disposal, appropriate information shall be provided to product users and recyclers. When hazardous substances and products are inevitably used (except those that are explicitly restricted by laws and regulations), they should be identified and marked, and the reasons for their inevitable use should be recorded in the design process.

Appropriate design methods should be adopted to reduce the possibility of using harmful substances in high-risk parts such as printer and all-in-one machine product shell, motherboard, paper feeder, print head, ribbon and external power cord.

3.4.4. Reduce Chemical Emissions.

The product is designed to minimize chemical emissions with negative environmental impact, such as volatile organic pollutants, ozone, dust, and so on.

3.4.5. Reduce Noise Emission.

The designer shall consider the technology of reducing noise emission, and the method of reducing noise emission can be adopted in accordance with GB/T 34959 [9].

3.4.6. Improve Energy Efficiency.

The design and development of the product shall meet the requirements of GB/T 21521 [10] on the energy efficiency limit value and energy efficiency grade of the printer. In order to improve the energy efficiency of products, designers should fully investigate and confirm which stage of the product life cycle will consume most of the energy, and take corresponding measures to reduce energy consumption and improve energy efficiency. The method in GB/T 34959 should be adopted to improve the energy efficiency of products.

Examples of energy-saving design methods to improve the energy efficiency of printers and multi-functional all-in-one machines are: design to improve the mechanical transmission efficiency of products and reduce the friction resistance, so as to reduce the energy consumption in the working process.

3.4.7. Consumables.

The design and disposal of consumables (or "consumables") should be guided by reduction, modularization and reuse: the function of reducing or saving the use of consumables; Convenient replacement or maintenance of consumables. The manufacturer can provide users with information about the correct use of consumables and, where appropriate, the end of life processing information of consumables.

Examples of design methods for optimizing consumables include: full consideration of compatibility with consumables (including reusable consumables) in product design. Consumables mainly include ink cartridge of inkjet printer, drum powder box of laser printer and ribbon frame and ribbon of needle printer.

3.4.8. Battery.

If the battery is used in the product, the design of the battery in the product should be considered from the aspects of battery type, control of harmful substances in the battery, battery durability, easy disassembly and recycling, and so on.

3.4.9. Reasonable Extension of Product Life.

When technically and economically feasible, the service life of the product design shall be extended as far as possible and matched with the service life of components, and it is easy to upgrade and repair. However, designers should consider the balance between using the latest efficient technology and prolonging the life of inefficient products.

Designers should consider integrating the following features: using common mechanical packaging (such as cover and chassis) or common components or product types are the same parts used by multiple models or the same parts used by the same product for multiple generations, allowing the reuse of common components; Use standardized components, which can be easily replaced or repaired; Modular design; Whenever the reuse of modules, components and products is applicable, the reused components (such as for maintenance and spare parts) should be identified.

When considering the design method of prolonging the product life, the printer and multifunctional all-in-one machine adopt the design of modularity, compatibility and upgrading, such as the updatable operating system.

3.4.10. End of Life Disposal of Products.

The design of the product shall facilitate the reuse, recovery and correct treatment at the end of the product life. The requirements for reuse, recycling, recycling, treatment and disposal of waste products shall be implemented in accordance with GB/T 23685 [11].

3.4.11. Packaging.

The selection and design of packaging materials will have an impact on the environment. When determining materials and designing packaging, designers should consider the following options: reduce the use of packaging materials through reasonable structural design and material selection, so as to reduce the weight and size of packaging when fully protecting products; Use materials that have little impact on the environment; Use recyclable materials; Use renewable/recyclable materials (considering available recycling technologies).

The minimum requirement for designers is to ensure compliance with international, regional and national regulations related to the following contents: restrictions on hazardous substances and hazardous products; Regeneration capacity, such as reuse or recycling; The appropriate identification (material content) of packaging materials shall comply with the marking requirements of GB/T 18455.

3.5 Product Life Cycle Analysis, Evaluation and Continuous Improvement

In order to confirm and improve the green design level of products, the methods of life cycle analysis and evaluation can be used to establish and implement the procedures of green design review and continuous improvement. The process of product life cycle analysis, evaluation and continuous improvement of product green design includes the following steps:

a) Product Description: determine the product boundary and goal of the design, including: model, specification, function, energy efficiency grade, life expectancy, etc.

b) Product demand analysis: requirements of interested parties (laws and regulations, standards, customers, etc.) in terms of function, resources, energy, environment, human health, and so on;

c) Product life cycle diagnosis: analyze and diagnose the whole life cycle of the benchmark product, put forward the potential and key points for environmental performance improvement, and put forward the improvement demand;

d) Preliminary evaluation of product green design scheme: conduct preliminary evaluation on the improved scheme and confirm the product green design scheme;

e) Product life cycle assessment: complete the whole life cycle assessment report of the benchmark product and the whole life cycle assessment report of the target product;

f) Product green design scheme evaluation: analyze and compare the environmental performance of the whole life cycle evaluation report of the target product and the benchmark product, and give the conclusion of the environmental performance of the target product;

g) Scheme selection: select the green design scheme of target products and benchmark products;

h) Scheme review: review the best scheme;

i) Scheme output: confirm whether the design scheme of the target product meets the product requirements. If so, carry out design implementation, effect evaluation and information sharing; If the expected product demand is not met, continue to put forward improvement suggestions, and re-evaluate the scheme after adjustment until the expected product demand is met.

4 Conclusions

The green design of printers and multi-function machines provides guidance for enterprise designers in green design when carrying out product design and development, from the basic principles of product green design, product demand analysis, environmental factor identification, design and development process, life cycle analysis and improvement, information sharing, etc., and from the selection of raw materials, production and manufacturing, packaging and transportation, use and maintenance Identify environmental factors at all stages of the life cycle, such as waste recycling, propose product ecological design requirements in terms of resources, energy and environment, improve the renewable utilization rate of products, reduce adverse environmental impacts in the entire life cycle of products, and develop more ecological, economic and sustainable product systems.

Acknowledgements

This paper is funded by National key research and development plan project "Research on key technology standards of eco-design for consumer electronics and other important products" (2017YFF0207901).

References

1. Jianfang Zong, Jianhua Chen, Comparative Analysis on Domestic and International Mechanism of Product Ecological Evaluation, China Standardization, 2017, (15): 54-58
2. Jianfang Zong, Jianwei Tian, Dongfeng Guo, Liang Chen, A Study on the Green design of Consumer Electronics; E3S Web of Conference 53, 04003 (2018)
3. GB/T 28179 Environmentally concious design for electrical and electronic product - Indentification of environmental aspects
4. GB/T 9314 General specification of serial impact dot matrix printer
5. GB/T 17540 General specification for desktop laser printers
6. GB/T 17974 General specification for desktop ink-jet printers
7. GB/T 28165 General specification for thermal printer
8. GB/T 34144 Guidance on material efficiency considerations in environmentally conscious design of electrical and electronic products
9. GB/T 34959 Audio/Video, information and communication technology equipment—Enviromentally conscious design
10. GB/T 21521 Minimum allowable values of energy efficiency and energy efficiency grades for copy machines, printers and fax machines
11. GB/T 23685 General technical specifications of recovering for waste electrical and electronic equipment

Guideline for Green Design of Computer Products

Jianfang Zong[1*], Xin Zhang[1], Yutao Yang[2], Husheng Li[3], Han Yuan[1]

[1]China National Institute of Standardization, No.4 Zhichun Road, Haidian District, Beijing, China
[2]China Electronics Standardization Institute, No.1 Andingmen East Street, Dongcheng District, Beijing, China
[3]Beijing CESI Certification Co, Ltd, No.1 Andingmen East Street, Dongcheng District, Beijing, China
*Corresponding author at: China National Institute of Standardization

Abstract. This paper regulates the purposes and basic principles of green design of computer products, and provides the green design requirements and procedures of computer products. This paper will provide the guiding targets for the computer product manufacturer, standardize the enterprise behaviour of computer product, and lead the computer product enterprise to implement the product green design, which is of great significance in promoting the transformation and upgrading of computer product, enhancing the green manufacturing level and increasing the green product supply.

1 Introduction

With the progress of science and technology and the rapid development of various computer technology and network technology, the development of computer products has entered a fast and new era. Computer has been developed from single function and large volume to complex function, small volume, resource networking and so on. At present, China is experiencing an important period of industrial structure adjustment. The industry has changed from resource intensive and labor-intensive to capital intensive and technology intensive. Therefore, there is a huge demand for the development of the computer industry. In 2021, China's total output of computer products has reached 450 million, including about 230 million laptops and 70 million desktops, and others are tablet computers, workstations, industrial control terminals, and so on.

The green design of computer product is also called "environmental conscious design", "green design" or "design for the environment". It intends to optimize the environmental impact of computer product and bring the environmental factors into the product design and development activity, so as to improve the environmental performance of computer product within its life cycle. As shown in the study, 80% resources consumption and environmental impact are subject to the product design phase. At the design phase, it is required to fully consider the existing technical conditions, raw materials guarantee and other factors, and optimize the solution against the resources and environmental problems at each link, which may maximize the resources conservation and reduce the environmental pollution from the source [1, 2].

*E-mail address: zongjf@cnis.ac.cn(J.Zong)

2 Purposes and basic principles of green design of computer products

The purpose of computer product green design is to reduce the environmental pollution, and enhance the recyclability rate of computer product, so as to reduce the adverse environmental impact of computer product during its entire life cycle, and develop the more ecologic, economic and sustainable computer product system.

The basic principles for green design of computer product include: the requirements-oriented, advanced technology, scientific methods, environmental-friendly, economic rationality, and abidance by law.

3 Green design process and key points of computer products

Computer product ecological design usually includes the following processes: a) Product demand analysis; b) Identification of environmental factors; c) Introducing environmental factors into the process of product ecological design; d) Life cycle analysis and evaluation; e) Ecological design information sharing.

3.1 Product Demand Analysis

Designers should carry out product green design while meeting product functions, and comprehensively consider factors such as resources, energy and environment, as well as the needs of laws and regulations, technical standards and customer market, so as to meet the goal of product green design.

Product demand analysis includes but is not limited to the following points: a) Product green design objectives;

b) Product function design; c) Requirements of laws and regulations; d) Technical standards; e) Customer and market needs.

3.2 Identification of Environmental Factors

Identifying typical environmental factors and corresponding environmental impacts is an important prerequisite for computer product green design. Designers can identify environmental factors according to GB/T 28179 [3]. The identification and analysis of these environmental factors can provide a basis for ecological design, identify, evaluate and determine important environmental factors as soon as possible, and consider introducing these factors in the process of product design and development, so as to improve the environmental performance of products. See Table 1 for the identification process of environmental factors of computer products.

Table 1. Identification process of environmental factors of computer products

Environmental factor	Design considerations	Product life cycle phases involved				
		Raw material acquisition	Manufacturing	Transportation and distribution	Use and maintenance	End of life disposal
Hazardous substances	Compliance and reduction of hazardous substances in raw materials (parts, coatings, etc.)	√	√		√	√
	Compliance and reduction of the use of hazardous substances in packaging materials	√		√		√
	Compliance and reduction of the use of hazardous substances in production accessories	√	√			
	Compliance and reduction of the use of hazardous substances in consumables	√			√	√
Material efficiency	Reduce the types of raw materials	√			√	√
	Reduce the consumption of raw materials	√		√		
	Green process of material processing		√			
	Improve the recycling performance of raw materials	√				√
	Reduce the environmental impact of raw material production stage	√				√
	Reduce the consumption of packaging materials	√		√		√
	Improve the recycling performance of packaging materials	√				√
	Reduce consumption of consumables	√			√	√
	Improve the recycling performance of consumables	√			√	√
	Reduce the consumption of production auxiliary materials	√	√			
Battery or power supply	Battery or power supply compliance	√			√	√
	Consider the disassembly of battery or power supply				√	√
	Consider the life of the battery or power supply				√	√
Energy consumption	Improve product energy efficiency				√	

Environmental factor	Design considerations	Product life cycle phases involved				
		Raw material acquisition	Manufacturing	Transportation and distribution	Use and maintenance	End of life disposal
	Reduce energy consumption in product production process		√			
Pollutant emission	Reduce the emission of harmful substances in the use stage of products				√	
	Reduce noise emission during product use				√	
Product longevity	Reasonably extend product longevity (modular design, parts replaceability, software upgrade)				√	
Recycling, utilization and disposal	Consider reuse, recycling and reasonable disposal					√
	Reduce environmental toxicity of final waste					√

3.3 Introducing environmental factors into the process of product ecological design

3.3.1. Product Function, Performance and Process Design.

The design and development of products shall meet the requirements of GB/T 9813.1 [4] and GB/T 9813.2 [5], including the appearance and structure, function and performance, process, safety, power adaptability, noise, electromagnetic compatibility, environmental conditions, reliability, and so on.

On the basis of meeting the requirements of product function and performance, the concept of life cycle management is incorporated into the design and development process of the whole product life cycle, including the conceptual design and structural design before product finalization.

3.3.2. Improve Material Efficiency.

The method of improving material efficiency can be considered in the process of product green design according to GB/T 34144 [6]. When selecting materials, the designer should consider the following factors: a) Reduce the types of materials used; b) Reduce the amount of materials used, so as to reduce the weight of products; c) Use materials that have little impact on the environment; d) Use materials that are easy to recycle.

3.3.3. Restrict the Use of Hazardous Substances.

Reducing or eliminating the use of hazardous substances and products should be one of the priorities in product design. Designers need to know the management list of banned hazardous substances and hazardous products in international, regional and product export target countries in time to ensure that the products sold in the target countries meet the requirements of relevant laws, regulations and standards. Minimize the use of substances that require special separation or disposal at the end of product life. For substances requiring special treatment or disposal, provide appropriate information to product users and recyclers. The use of restricted substances in products should be identified in accordance with SJ/T 11364 [7], and the inevitable use of hazardous substances and reasons should be recorded in the design process. Appropriate design methods should be adopted to reduce the possibility of using harmful substances in parts with high risk, such as microcomputer product shell, keyboard and various keys, mouse shell, touch panel and external power cord.

3.3.4. Reduce Chemical Emissions.

Appropriate design methods should be adopted to reduce the possibility of chemical emissions with negative environmental impact during the use of products, such as volatile organic pollutants.

3.3.5. Reduce Noise Emission.

The designer shall consider the technology to reduce noise emission, so that the product shall meet the requirements on noise in accordance with GB/T 9813.1. The method of reducing noise emission should be adopted in accordance with GB/T 34959 [8].

3.3.6. Improve Energy Efficiency.

The product should be designed and developed to meet the requirements of GB/T 28380 [9] on the energy efficiency limit and energy efficiency grade of microcomputer. The method in GB/T 34959 should be adopted to improve the energy efficiency of products. In the energy-saving design, the expected use mode of microcomputer products should be fully considered, including working mode, energy-saving mode, off mode and no-load mode. Appropriate design

https://doi.org/10.1051/shsconf/202315503025

methods shall be adopted to minimize the energy consumption under various modes, such as passive cooling to avoid the additional energy consumption caused by the use of fans.

3.3.7. Consumables.

Consider the design and disposal of consumables based on the principle of reduction, modularization and reuse: the function of reducing or saving the use of consumables; Convenient replacement or maintenance of consumables. The manufacturer can provide users with information about the correct use of consumables and, where appropriate, the end of life processing information of consumables.

3.3.8. Battery.

The design of batteries in computer products should be considered from the aspects of battery type, control of harmful substances in batteries, battery durability, easy disassembly and recycling, and so on.

3.3.9. Reasonable Extension of Product Life.

When technically and economically feasible, the service life of the product design shall be reasonably extended and matched with the service life of components, and it is easy to upgrade and repair. The methods to reasonably extend the service life of products include: a) Using common mechanical packaging (such as cover and chassis) or common parts or product types, which are the same parts used by multiple models or the same parts used by the same product for multiple generations, and allowing the reuse of common parts; b) Use standardized components, which can be easily replaced or repaired; c) Modular design; d) Whenever the reuse of modules, components and products is applicable, the reused components should be identified.

When considering the design method of prolonging product life, computer products can be designed in terms of modularity, compatibility and upgrading; The CD-ROM drive, hard disk and main memory randomly configured for desktop microcomputer can be upgraded.

3.3.10. Reuse, Recycling and Reasonable Disposal.

The design of products should be convenient for reuse, recycling and reasonable disposal at the end of product life. The requirements for reuse, recycling, recycling, treatment and disposal of waste products shall be implemented in accordance with GB/T 23685 [10].

Designers need to understand the regulations and standards related to the disassembly and recycling of waste electrical and electronic products to ensure the compliance of product sales in the target countries. The end-of-life disposal of products should be designed according to GB/T 31371 [11] and GB/T 31372 [12].

3.3.11. Packaging.

When determining materials and designing packaging, designers should consider the following options: Reduce the use of packaging materials through reasonable structural design and material selection, so as to reduce the weight and size of packaging when fully protecting products; Use materials that have little impact on the environment; Use recyclable materials; Use renewable / recyclable materials.

The minimum requirements for designers shall ensure compliance with relevant domestic and foreign regulations and standards related to the following contents: Restricted use of hazardous substances and hazardous products; Regeneration capacity, such as reuse or recycling; The proper identification of packaging materials should be in accordance with the marking requirements of GB/T 18455 [13].

3.4 Product Life Cycle Analysis, Evaluation and Continuous Improvement

In order to confirm and improve the level of product ecological design, the method of life cycle analysis and evaluation can be used to establish and implement the procedure of ecological design review and continuous improvement. The process of product life cycle analysis, evaluation and continuous improvement of product ecological design includes the following steps:

a) Product Description: determine the product boundary and objectives of the design, including: model, specification, function and performance, product life expectancy, and so on;

b) Product demand analysis: requirements of interested parties in product function, resources, energy, environment, and so on;

c) Product life cycle diagnosis: analyze the whole life cycle of benchmark products, put forward the potential to improve environmental performance, identify key environmental factors and put forward improvement needs;

d) Preliminary evaluation of product ecological design scheme: conduct preliminary evaluation on the improved scheme and confirm the product ecological design scheme;

e) Product life cycle assessment: complete the whole life cycle assessment report of benchmark products and target products according to Appendix A in GB/T 24040 [14], GB/T 24044 [15] and GB/T 34664 [16];

f) Product environmental performance evaluation: analyze and compare the environmental performance of the whole life cycle evaluation report of the target product and the benchmark product, and give the conclusion of the environmental performance of the target product;

g) Scheme selection: select the ecological design scheme of target products and benchmark products;

h) Scheme review: review the best scheme;

i) Scheme output: confirm whether the design scheme of the target product meets the product requirements. If so, carry out design implementation, effect evaluation and information sharing; If the expected product demand is not met, continue to put forward improvement suggestions, and re-evaluate the scheme after adjustment until the expected product demand is met.

3.5 Ecological Design Information Sharing

As part of the ecological design process, organizations in the supply chain should disclose relevant information about products or product types to other organizations related to design and development, so as to enable these organizations to achieve the ecological design objectives.

The contents of information exchange include: products, manufacturing processes and relevant resources used in product operation; Relevant emissions of products; Guidelines for improving environmental performance; Treatment at the end of life; Declaration of compliance with regulations and requirements of interested parties.

4 Conclusions

Green design of computer products is to provide guidance for enterprise designers in green design when carrying out product design and development, from the basic principles of product green design, product demand analysis, environmental factor identification, design and development process, life cycle analysis and improvement, and information sharing, etc. Identify environmental factors at all stages of the life cycle, such as waste recycling, propose green design requirements for products in terms of resources, energy and environment, improve the renewable utilization rate of products, reduce adverse environmental impacts in the entire life cycle of products, and develop more ecological, economic and sustainable product systems.

Acknowledgements

This paper is funded by National key research and development plan project "Research on key technology standards of eco-design for consumer electronics and other important products" (2017YFF0207901).

References

1. Jianfang Zong, Jianhua Chen, Comparative Analysis on Domestic and International Mechanism of Product Ecological Evaluation, China Standardization, 2017, (15): 54-58
2. Jianfang Zong, Jianwei Tian, Dongfeng Guo, Liang Chen, A Study on the Green design of Consumer Electronics; E3S Web of Conference 53, 04003 (2018)
3. GB/T 28179 Environmentally concious design for electrical and electronic product - Indentification of environmental aspects
4. GB/T 9813.1 General specification for computer—Part 1: Desktop microcomputer
5. GB/T 9813.2 General specification for computer—Part 2: Laptop microcomputer
6. GB/T 34144 Guidance on material efficiency considerations in environmentally conscious design of electrical and electronic products
7. SJ/T 11364 Marking for control of pollution caused by electronic information products
8. GB/T 34959 Audio/Video, information and communication technology equipment—Enviromentally conscious design
9. GB/T 28380 Minimum allowable values of energy efficiency and energy grades for microcomputers
10. GB/T 23685 General technical specifications of recovering for waste electrical and electronic equipment
11. GB/T 31371 Requirements of disassembly and treatment for waste electrical and electronic product—Desktop microcomputer
12. GB/T 31372 Requirements of disassembly and treatment for waste electrical and electronic product—Portable microcomputer
13. GB/T 18455 Package recycling marking
14. GB/T 24040 Environmental management - Life cycle assessment - Principles and frameworks
15. GB/T 24044 Environmental management - Life cycle assessment - Requirements and guidelines
16. GB/T 34664 General assessment principles for electrical and electronic eco-design products

Influence and Prospect of Artificial Intelligence on the Development of Cultural Industry

Chao Meng [1, a], Ronaldo Juanatas[1, b], Jasmin Niguidula[1, c]

[1]College of Industrial Education, Technological University of the Philippines, Manila, 0900, Philippines

Abstract: Nowadays, the products carried by artificial intelligence are all over people's life, such as online shopping, intelligent security, remote diagnosis and treatment, life services, driverless, online distance education, and other services. Artificial intelligence not only allows people to enjoy convenient and fast life benefits, but also brings a certain degree of anxiety to the times. In recent years, with the continued popularity of the "artificial intelligence + culture" industry, the integration of cultural industry and artificial intelligence has become an inevitable development trend. The promotion of technology ultimately serves to improve the quality of human life. Artificial intelligence not only brings convenience, but also brings opportunities and challenges to the cultural industry.

1 Introduction

Every scientific and technological leap in the era of mankind has brought about industrial changes and promoted the progress of the times. Today, artificial intelligence has led the world into a new round of industrial change. The social influence of artificial intelligence is increasing day by day. At the same time, it is also accelerating the pace of industrial integration and industrial upgrading. Nowadays, the rapid development of smart home, intelligent transportation and intelligent shopping makes artificial intelligence develop from professional intelligence to general intelligence, and plays a leading role in enhancing the national scientific and technological strength and improving people's quality of life.

Under the influence of 5G technology, the cultural industry will present a new development trend in technology and products, mode and management, change the production mode of cultural products as a whole, integrate intelligence and personality, and give consideration to scene and experience. Changes in the structure and organization of the cultural industry have also led to the increase in the volume of the cultural industry and the expansion of the layout area. Industrial integration is stronger in depth and breadth. The integration and development of emerging industries and traditional industries have become the current trend. The cultural industry is bound to face new challenges and achieve new development in the era of artificial intelligence.

2 Penetration and integration: analysis of the current situation

Facing the development opportunities of digital economy, all countries in the world regard artificial intelligence as the focus of attention and an important symbol to highlight national core competitiveness. The global pattern of artificial intelligence: future trends and China's position comprehensively introduces the layout of governments and economic and financial institutions in the field of artificial intelligence. As the book says, "AI will bring a revolution in scientific research paradigm". In the era of transformation and upgrading with "intelligence and informatization" as the breakthrough, only by keeping up with this round of reform and further promoting industrial development with information intelligence, can a country truly grasp the initiative of development. Industrial integration is a new type of industrial form. It is a change under the condition of technological innovation. It is a process of mutual competition and collaborative evolution between different industrial constituent elements, and finally, a new type of industry is formed. [1] The rise and development of the cultural industry are inseparable from the revolution of science and technology. Industrial integration comes into being in line with the development trend of the times. To reap intelligent dividends and lead the trend of the times, in today's era of vigorous development of artificial intelligence, we must choose a development path in line with our own reality, actively respond to challenges and seize opportunities.

[a] email: edymeng@foxmail.com
[b] email: ronaldo_juanatas@tup.edu.ph
[c] email: Jasmin_niguidula@tup.edu.ph

3 Man-machine coexistence: opportunities and challenges

Artificial intelligence brings new historical development opportunities and is bound to produce new problems. Countries need to explore a road of artificial intelligence in cultural industry in line with national conditions.

3.1 Artificial intelligence brings new opportunities for the development of the cultural industry

In the future development of an intelligent form of cultural industry, man-machine integration will become an important direction. At the same time, with the popularization of 5G technology, cognitive intelligence technology will continue to make breakthroughs, which can produce more cultural formats.

Artificial intelligence promotes the content creation of cultural products. Artificial intelligence was applied earlier in the field of news communication. As early as 2014, Quakebot, a manuscript writing robot of the Los Angeles Times, has been put into use. Since then, internationally renowned news media such as the associated press, the New York Times, and Forbes have also introduced AI writing technology. [2] These artificial intelligence machines applied at the practical level completely avoid the subjective and objective influence commonly seen in artificial writers. They can produce drafts immediately and work all day, greatly improving work efficiency.

In April 2021, Alibaba DAMO Academy released the largest Chinese pre-training language model PLUG (pre-training for language understanding and generation). The model integrates the ability of language understanding and generation. It is outstanding in the field of long text generation such as novel creation, poetry generation, and intelligent question and answer. [3] The technical processing core of this "universal writing artifact" is natural language processing technology, the machine can "understand" human language to the greatest extent, and accelerate the implementation of artificial intelligence in the cultural industry.

Artificial intelligence creates new cultural consumption demand and promotes the intellectualization of cultural product marketing. People's demand for cultural and entertainment consumption is changing towards a more personalized and diversified direction, which gives birth to people's huge demand for high-end cultural products. Experiential and situational consumption mode is a very popular consumption mode since the integration of artificial intelligence and cultural industry. With the upgrading of artificial intelligence technology, the new consumption mode produced by the integration with cultural industry has gradually entered people's vision. For example, the "smart tourism" and other new tourism experiences developed in the tourism industry in recent years have upgraded tourism from "punch card" photo accompanying tourism to "experiential" personalized in-depth tourism, which has brought vitality and power to the cultural tourism industry through experiential consumption.

In the aspect of product marketing, artificial intelligence can carry out large-scale personalized intelligent distributions of content. For example, the intellectualization of advertising, on one hand, targeted advertising through user portrait analysis; on the other hand, it also strengthens the acceptance of consumers. In his book "filter bubbles: what the Internet didn't tell you", Lee·Bariser put forward the concept of "filter bubbles", which directly refers to the problem of algorithm manipulation. In his opinion, the information received by users is often affected by retrieval history, reading behavior, and habits, which will be controlled by algorithms. [4] The growing maturity of algorithms and the increasing computing power of artificial intelligence make the "customized information" exposed to the audience more "palatable".

Artificial intelligence provides personalized services for the cultural industry and promotes the form innovation of cultural products. Artificial intelligence technology has been widely used in film and television industry. Through this technology, creators can create digital actors to complete the difficult actions that real actors cannot complete and the settings of various illusory scenes, such as Alita: Battle angel, Gemini man, etc., these films have not only made considerable box office, but also brought the audience a new film viewing experience.

The new form of cultural products spawned by artificial intelligence technology also provides reference for other cultural industries and creates new development opportunities. At the world Internet Conference held in Zhejiang, China in November 2018, Xinhua news agency released the world's first artificial intelligence anchor. Virtual images created using artificial intelligence technology are almost no different from real people. Just enter the news text and the artificial intelligence synthesis anchor can play. [5] This technology is bound to bring significant changes to the development of various fields of cultural industry.

Holographic projection technology has emerged in recent years. This real and shocking visual experience can easily realize virtual imaging without wearing a bulky VR helmet. Holographic projection technology can not only produce three-dimensional aerial illusion, but also make the illusion interact with the performer to complete the performance together. Among them, the most famous is the Japanese virtual singer " Hatsune Miku". She is the first virtual idol in the world to hold a concert using holographic projection technology. During the performance, she can also change clothes quickly and perform on the same stage with virtual partners. Her charm and economic benefits are no less than real idols.

3.2 Challenges brought by artificial intelligence to the cultural industry

While artificial intelligence injects vitality and kinetic energy into the development of the cultural industry, the possible risks and challenges can not be ignored, which is worthy of our reflection and attention.

Unemployment fear and ethical anxiety. While

artificial intelligence brings progress and convenience to mankind, legal, ethical and social challenges also follow. First, impact the traditional employment structure. In a brief history of the future, Herali once proposed that the wide application of artificial intelligence will certainly lead to large-scale unemployment, and unemployment may dispel the meaning of individual life at a deeper level. In the long run, the development of artificial intelligence will greatly change the employment structure. Secondly, while artificial intelligence reduces the cost of interpersonal communication, the relationship between people is also facing the possibility of alienation, and then the challenges faced by social ethics will rise sharply. Therefore, some scholars and professionals believe that industry rules need to be formulated for this purpose, and the research and development of artificial intelligence must match social ethics. Thirdly, the development of the era of artificial intelligence will enable the intelligent technology class to have the right to speak. This intellectual privilege may bring a new round of social change. While science and technology is the primary productive force, technology has also become the power to rule the society, and the freedom and democracy under the traditional social model will be weakened. [6] Advanced artificial intelligence with deep learning ability may bring technical distortion. Many film works, such as Star Wars, the Matrix, the Terminator, and Artificial Intelligence, present a variety of possibilities for the relationship between artificial intelligence and people in the future, which leads to reflections on artificial intelligence and the essence of life.

Knowledge gap and the formation of information cocoon room. The "knowledge gap hypothesis" put forward by American scholar Tichenor in 1970 holds that "with the increasing information transmitted by mass media to society, people with different socio-economic status have different speeds of acquiring media knowledge, and people with higher socio-economic status will acquire such information faster than people with lower socio-economic status". [7] The development of artificial intelligence technology requires the complex integration and accumulation of different disciplines, and its understanding requires multi-disciplinary background knowledge. In the future development and application of artificial intelligence, the higher the standard and level of personal life, the higher the degree of contact and utilization of artificial intelligence, that is, the formation of "artificial intelligence gap". From a global perspective, the development of artificial intelligence is also mainly concentrated in a few countries and regions, showing an unbalanced state, which will also deepen the intellectual polarization brought by the knowledge gap in the increasingly prosperous future of artificial intelligence. In addition to deepening the expansion of the knowledge gap, intelligent algorithm recommendations in the era of artificial intelligence will also cause the information cocoon effect. Users only follow the information they like recommended by the intelligent system, and an information barrier will be formed over time. At the same time, the information platform will also focus on the characteristics of laziness, curiosity, ugliness and privacy in human nature, and use vulgar and low-quality content to attract network traffic. Such content is strongly recommended by intelligent systems, resulting in the marginalization of truly valuable and meaningful information.

There are hidden dangers in data security and intellectual property rights. In the integrated application of artificial intelligence and cultural industry, the hidden danger of data security increases gradually with the increasing complexity of information in the database. Artificial intelligence needs to rely on a large amount of data for calculation and analysis. The generated intelligent cultural products are recombined by extracting applicable data and elements from the formula. It is difficult to distinguish the part with copyright protection from the part with public data, which is very easy to cause the content infringement of artificial intelligence cultural products, which will greatly reduce the creative enthusiasm and investment of original cultural products. In addition, AI will also lead to other violations, such as some AI face-changing software, which increases the risk of personal privacy and information disclosure by collecting a large amount of face dynamic information.

In fact, the development of artificial intelligence is still in a relatively primary stage, and there are still many outstanding problems. Thomas Sargent, the winner of the Nobel Prize in economics, once pointed out that "artificial intelligence is actually statistics, which only uses a gorgeous language. Compared with the brain, artificial intelligence still has many shortcomings, such as' algorithm black box ', large demand for data, poor noise resistance and high energy consumption, which is far from being fully applied." [8] Although artificial intelligence is commonly used, there is still a technical "ceiling" at this stage. Although the risk impact and ethical anxiety it brings are not fully revealed, it still provides a cautionary guide.

4 Regulation and supervision: thoughts and suggestions on the development

4.1 Technological progress and talent training

The key to the development of cultural industry in the era of artificial intelligence is to cultivate high-quality innovative talents. Compared with artificial, in some basic types of work with complex operation, artificial intelligence is more efficient and advantageous, and it is easier to replace auxiliary posts in the cultural industry. Nowadays, in the field of cultural industry, the demand for labor force in the market is a compound talent who knows both technology and cultural and creative output under the "one specialty and multiple abilities". The cultivation of practical ability and thinking ability in the stage of basic education is an important stage for the cultural industry to cultivate high-quality compound talents. We should deepen the combination of industry and education and transform the power of knowledge into economic power. This requires education to be ahead of schedule, have a forward-looking awareness of the cultural industry market, and reserve and cultivate innovative talents suitable for the market and technological development in advance.

4.2 Standardizing the data and strengthening supervision

The world produces a huge amount of data and information every day. Behind these data is the development trend of all walks of life and even the future development direction. At the legislative level, the protection mechanism of personal data information needs to be further improved. The development of artificial intelligence is based on data. In the cultural industry, the development of artificial intelligence should standardize the data in order to effectively improve the performance of the integration of the artificial intelligence industry. At the same time, it is necessary to clarify the definition of intellectual property rights of artificial intelligence, improve the relevant legal system, combat the excessive acquisition and illegal disclosure of public-private data, and make the rational application of artificial intelligence technology within the scope specified by law.

5 Conclusion

Looking around the world, artificial intelligence technology has penetrated into all walks of life. The arrival of 5G era, as a communication infrastructure, has also opened a new application mode of "interconnection of all things". With the support and assistance of technology, the transformation of artificial intelligence to the cultural industry will enter a new stage. However, over reliance on technology has always been a double-edged sword. The convenience brought by artificial intelligence provides creativity and nutrients for people's life and production, but potential hidden dangers and risks also surge in this rapidly developing era of science and technology. At present, the technology of artificial intelligence still has obvious limitations, and the integration and reconstruction with the cultural industry is also a long and arduous road. We should rationally analyze the needs, reasonably formulate goals, and take practical development measures to make artificial intelligence move forward in innovation and revision and benefit mankind.

Reference

1. Liu J., Liu Z. (2019) Research review on the development of cultural industry in the era of artificial intelligence. Journal of Sichuan cadre correspondence college, No. 6.
2. Na F. Z. (2016) automation news, printing today, issue 5.
3. Zhao C. (2021) Ali DAMO Academy releases "universal writing artifact", AI technology moves towards "modeling era". China Electronics News, April.
4. Eli P. (2011) The Filter Bubble: What the Internet Is Hiding from You. Penguin.
5. Li J.P. (2019) Problems and risk prevention of AI's deep involvement in the cultural industry. Journal of Shenzhen University (HUMANITIES AND SOCIAL SCIENCES EDITION), Issue 9.
6. Zhou Z. (2019) Aesthetic challenge and reflection of artificial intelligence art. Shandong Social Sciences, issue 10.
7. Hu S.G., Chen C.F. (2019) "concepts and norms: media ethics dilemma and its guidance in the era of artificial intelligence". China publishing, issue 2.
8. Wu H.q. (2019) An objective view of artificial intelligence, Beijing Daily, October.

AUTHOR INDEX

AUTHOR INDEX